T0221167

Serpentine

Serpentine

The Evolution and Ecology of a Model System

Edited by

Susan Harrison
Nishanta Rajakaruna

UNIVERSITY OF CALIFORNIA PRESS

Berkeley   Los Angeles   London

University of California Press, one of the most distinguished university presses in the United States, enriches lives around the world by advancing scholarship in the humanities, social sciences, and natural sciences. Its activities are supported by the UC Press Foundation and by philanthropic contributions from individuals and institutions. For more information, visit www.ucpress.edu.

University of California Press
Berkeley and Los Angeles, California

University of California Press, Ltd.
London, England

Library of Congress Cataloging-in-Publication Data

Serpentine : the evolution and ecology of a model system / edited by Susan Harrison and Nishanta Rajakaruna.
    p.   cm.
Includes bibliographical references and index.
ISBN 978-0-520-26835-7 (cloth)
1. Serpentine plants.   2. Plants—Adaptation.   3. Plants—Evolution.
4. plant-soil relationships.   5. Soils—Serpentine content.
I. Harrison, Susan (Susan Patricia), 1961–   II. Rajakaruna, Nishanta, 1969–
QK938.S45.S47 2010
581.4—dc22   2010019373

19   18   17   16   15   14   13   12   11
10   9   8   7   6   5   4   3   2   1

Cover illustration: Stream bank revegetation trial at New Idria serpentine barrens, San Benito County, CA. Photo by Nishanta Rajakaruna.

CONTENTS

CONTRIBUTORS

BRAIN L. ANACKER
University of California, Davis
blanacker@ucdavis.edu

BROOKE S. BAYTHAVONG
University of California, Davis
bbaythavong@gmail.com

ROBERT S. BOYD
Auburn University
boydrob@auburn.edu

DAWN CARDACE
University of Rhode Island
cardace@uri.edu

VICTOR P. CLAASSEN
University of California, Davis
vpclaassen@ucdavis.edu

ELLEN I. DAMSCHEN
University of Wisconsin, Madison
damschen@wisc.edu

KENDI F. DAVIES
University of Colorado, Boulder
Kendi.Davies@Colorado.edu

SARAH C. ELMENDORF
University of British Columbia
scelmend@interchange.ubc.ca

BARBARA M. GOING
University of California, Davis
bmgoing@ucdavis.edu

SUSAN HARRISON
University of California, Davis
spharrison@ucdavis.edu

TORI M. HOEHLER
NASA Ames Research Center
Tori.M.Hoehler@nasa.gov

DAVID HOOPER
Western Washington University
hooper@biol.wwu.edu

PATRICK R. HUBER
University of California, Davis
prhuber@ucdavis.edu

KATHLEEN M. KAY
University of California, Santa Cruz
kay@biology.ucsc.edu

CHRIS R. MALLEK
University of California, Davis
crmallek@ucdavis.edu

KARA A. MOORE
University of California, Davis
kmoore@ucdavis.edu

ELDRIDGE M. MOORES
University of California, Davis
emmoores@ucdavis.edu

RYAN E. O'DELL
Bureau of Land Management
rodell@blm.gov

NISHANTA RAJAKARUNA
College of the Atlantic
nrajakaruna@coa.edu

KEVIN J. RICE
University of California, Davis
kjrice@ucdavis.edu

HUGH D. SAFFORD
Pacific Southwest Research Station, USDA Forest Service
hughsafford@fs.fed.us

DOUGLAS W. SCHEMSKE
Michigan State University
schem@msu.edu

MAUREEN L. STANTON
University of California, Davis
mlstanton@ucdavis.edu

SHARON Y. STRAUSS
University of California, Davis
systrauss@ucdavis.edu

DIANE M. THOMSON
Claremont Colleges
dthomson@jsd.claremont.edu

JAMES H. THORNE
University of California, Davis
jhthorne@ucdavis.edu

ROBBIN THORP
University of California, Davis
rwthorp@ucdavis.edu

ERIC VON WETTBERG
Florida International University and Fairchild Tropical Botanical Garden
eric.vonwettberg@gmail.com

KIMIORA L. WARD
University of California, Davis
kimiora_w@yahoo.com

LORNA R. WATT
Michigan State University
wattlron@msu.edu

AMY T. WOLF
University of Wisconsin, Green Bay
wolfa@uwgb.edu

JESSICA W. WRIGHT
Pacific Southwest Research Station, USDA Forest Service
jessicawwright@fs.fed.us

PREFACE

We are fortunate to belong to a thriving international community of scientists interested in serpentine ecosystems. In the past several decades, the efforts of this community have laid a solid foundation for understanding serpentine rocks, soils, floras, and faunas (see reviews in Proctor and Woodell, 1975; Kruckeberg, 1984, 2005; Brooks, 1987; Baker et al., 1992; Roberts and Proctor, 1992; Jaffré et al., 1997; Balkwill, 2001; Boyd et al., 2004; Alexander et al., 2006; Chiariucci and Baker, 2007; Rajakaruna and Boyd, 2009; Rajakaruna et al., 2009).

We are extremely grateful to our contributors for the expertise and dedication that made this book possible. We also thank our reviewers, who include David Ackerly, Paul Aigner, Bruce Baldwin, Robert Boyd, Emilio Bruna, Jean Burns, Sharon Collinge, Stella Copeland, Curtis Daehler, Katrina Duglosch, Elizabeth Elle, Anu Eskelinen, Paul Fine, Lila Fishman, Sophie Karrenberg, Scott Loarie, D'Arcy Meyer-Dombard, Risa Sargent, Matthew Schrenk, Mark Stromberg, and many of the chapter authors.

Finally, we especially thank three of the leaders in interdisciplinary studies of serpentine—soil scientist Earl Alexander, geologist Robert Coleman, and botanist Arthur Kruckeberg—for the incomparable generosity and enthusiasm with which they have shared their knowledge.

LITERATURE CITED

Alexander, E. A., Coleman R. G., Keeler-Wolf, T., and Harrison, S. (2006) *Serpentine Geoecology of Western North America*. Oxford University Press, Oxford.

Baker, A. J. M., Proctor, J., and Reeves, R. D. (1992) *The Vegetation of Ultramafic (Serpentine) Soils*. Intercept, Andover, U.K.

Balkwill, K. (2001) Proceedings: Third international conference on serpentine ecology. *South African Journal of Science,* 97 (special issue).

Boyd, R. S., Baker, A. J. M., and Proctor, J. (2004) *Ultramafic Rocks: Their Soils, Vegetation, and Fauna*. Science Reviews, St. Albans, U.K.

Brooks, R. R. (1987) *Serpentine and Its Vegetation: A Multidisciplinary Approach*. Dioscorides Press, Portland, OR.

Chiarucci, A., and Baker, A. J. M. (2007) Proceedings of the fifth international conference on serpentine ecology. *Plant and Soil,* 293 (special issue).

Jaffré, T., Reeves, R. D., and Becquer, T. (1997) The ecology of ultramafic and metalliferous areas. Proceedings of the second international conference on serpentine ecology. ORSTOM Noumea, *Documents Scientifi ques et Techniques III* (special issue).

Kruckeberg, A. R. (1984) *California Serpentines: Flora, Vegetation, Geology, Soils and Management Problems*. University of California Press, Berkeley.

Kruckeberg, A. (2005) *Geology and Plant Life*. University of Washington Press, Seattle.

Proctor, J., and Woodell, S. R. J. (1975) The ecology of serpentine soils. *Advances in Ecological Research,* 9, 255–366.

Rajakaruna, N., and Boyd, R. (2009) Soil and biota of serpentine: A world view. Proceedings of the Sixth International Conference on Serpentine Ecology. *Northeastern Naturalist,* 16 (special issue 5). Eaglehill Press, Steuben, ME.

Rajakaruna, N., Harris, T. B., and Alexander, E. B. (2009). Serpentine geoecology of eastern North America: A review. *Rhodora,* 111, 21–108.

Roberts, B. A., and Proctor, J. (1992) *The Ecology of Areas with Serpentinized Rocks: A World View*. Kluwer, Dordrecht.

INTRODUCTION

Terrestrial life, perched on the Earth's continental crust, has evolved on soils formed from relatively low-density rocks such as granite that are rich in silica, calcium, potassium, and phosphorous. The chemistry of these soils is usually amenable to plant growth almost by definition. Deeper in the Earth, forming its mantle and most of its oceanic crust, are darker and denser ultramafic (high-iron and -magnesium) rocks and minerals. Near the surface they may become serpentinized—altered in contact with water. These submarine rocks are seldom seen on land but occasionally become stranded on the edges of continents during the process of subduction (the disappearance of one crustal plate beneath another). The resulting terrestrial islands of ultramafic rock, or serpentine outcrops, are truly "unearthly" in their appearance. (Serpentine is technically a mineral, but the same word is often used for all ultramafic rocks, the soils that form from them, and the unique ecosystems that form on them.) Serpentine soils are deficient in plant-essential nutrients and often also in organic matter, cation exchange capacity, and water availability, whereas they are enriched in magnesium and sometimes in nickel, chromium, and cobalt. This unusual chemistry gives rise to rocky, sparsely vegetated landscapes that form striking boundaries with the lusher vegetation on neighboring soils. In some parts of the world, serpentine has given rise to spectacular levels of plant endemism.

Serpentine ecosystems have long fascinated scientists because of the unique flora they support, the adaptive challenges they pose to plants, and the difficulties they create for agriculture and more recently for ecological restoration. In the wake of the modern synthesis in the 1940s, classic work by Arthur Kruckeberg, G. Ledyard Stebbins, and others treated serpentine flora as a model system for understanding mechanisms of adaptation, ecotypic differentiation, and the linkage between natural selection and speciation. The advent of molecular techniques has brought renewed energy to the study of serpentine plant evolution, and recent studies have confirmed the remarkable power of serpentine to shed light on fundamental questions. Moreover, this power has not been confined to evolutionary biology. Interpreting the geologic origins of serpentine played a critical role in the 1960s plate tectonics revolution, and geochemists and microbiologists today use serpentinizing systems to investigate microbial life in the "deep biosphere" of Earth and the potential for life on other planets. Ecological research has used serpentine to investigate the effects of two of its hallmark characteristics—very low plant productivity and spatially complex landscape structure—on the structure and function of natural communities, as well as their conservation and management.

Our novel goal in this book is to ask what serpentine studies have revealed about broader theoretical questions in geology, evolution, ecology, and other fields. Our contributors were encouraged to take a concept-centered approach, explaining the key unresolved questions in their disciplines before describing how they and others used serpentine studies to address them; they were also encouraged to evaluate strengths and weaknesses of serpentine as a model system. In some cases, comparisons of the same process or pattern in serpentine and nonserpentine ecosystems helped to reveal the effects of low productivity, patchy distribution, or other general properties of serpentine. We believe that the contents of this book, which we briefly review and synthesize in our final chapter, bear out the value of serpentine as a model system in multiple disciplines. Serpentine environments are not just a fascinating and unique phenomenon in themselves but offer premier opportunities to study the origins, development, diversity, and function of life on a heterogeneous planet.

PART ONE

Serpentine as a Model in Earth History and Evolution

Serpentinites and Other Ultramafic Rocks

Why They Are Important for Earth's History and Possibly for Its Future

Eldridge M. Moores, *University of California, Davis*

Geology is a historical science, one of the "storytelling sciences," not simply a laboratory science. As such, geologists try to not only understand basic and timeless principles related to the rocks being studied but also give an account of what has happened in the past, and when possible, use this past history to forecast future events (Primack and Abrams, 2006: 17).

Serpentine, strictly speaking, is a mineral. Rocks formed mostly of serpentine are called serpentinites. Serpentine forms chiefly by the alteration (hydration) of the minerals olivine and pyroxene, found mostly in rocks called peridotites, a type of ultramafic rock. The term *ultramafic* indicates that the rocks are more than 90% olivine and pyroxene; most ultramafic rocks were derived from the Earth's mantle, the layer below the topmost layer or crust (Figure 1.1). Thus, both peridotite and serpentinite are ultramafic.

The olivine- or pyroxene-rich rocks from which serpentinites come are common in the Earth's mantle. Exposures of serpentinite at the Earth's surface indicate special tectonic action to move the rocks from 10–50 km deep to the surface. Serpentinite is relatively widespread in oceanic crust, which comprises about 70% of the Earth's surface. Oceanic crust in the oceans is not more than about 185 million years old. Most exposures of serpentinite at the Earth's continental surface come from ophiolites—exposures of oceanic crust and mantle formed at oceanic spreading centers. This process requires placement of oceanic crust and mantle on the

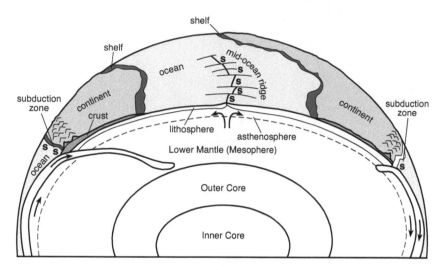

FIGURE 1.1. Diagrammatic cross section of the Earth showing the inner and outer core, the mantle with three main layers (the mesosphere, asthenosphere, and lithosphere), and the crust, which is the top part of the lithosphere. The spreading centers and subduction zones of plate tectonics are also indicated. New lithosphere is produced by upwelling along oceanic spreading centers. Subduction zones recycle lithospheric plates back into the interior. Some plates descend to the lower mantle. Areas denoted by s indicate areas of possible formation of serpentinite. Modified after Twiss and Moores (2007).

continental crust or in exposure of subduction accretionary complexes above sea level. Mélanges, so-called stratiform mafic-ultramafic complexes, and subcontinental mantle represent subordinate sources of serpentinites.

The general tale of peridotites and their derivative serpentinites involves a long detective story of geologists trying to understand the origin of these rocks. This investigation started in the nineteenth century, continued through the twentieth century and up to the present time. The study of ophiolites led in part to the plate tectonic revolution that transformed our understanding of how the Earth evolved and continues to do so.

In this chapter, I concentrate on the general nature and geologic history of serpentine and its antecedent related rocks and how regional differences in serpentinites relate to the specific history of a particular region. I begin with the basic structure of the Earth and the nature of peridotites and ultramafic rocks. I give the history of the ophiolite concept and how it influenced plate tectonics, an account of our understanding of ophiolites today, ophiolites through time, and other occurrences of serpentinite. The history section partly includes my personal story, as I have worked on ophiolites since the mid-1960s.

GENERAL EARTH STRUCTURE

Earth is composed of a series of layers, determined principally by different chemical and mineral compositions (see Figure 1.1). From the surface to the center, these layers include the crust, the mantle, and the core. The two outer layers, the crust and mantle, are composed mostly of silicate minerals, that is, minerals composed of Si, O, and other elements; carbonate rocks (containing carbon, in addition to oxygen and other elements); and lesser amounts of rocks composed dominantly of oxide minerals, sulfates, phosphates, or related rocks.

The recognition of these layers comes chiefly from the study of the passage through the Earth of seismic waves that are generated in earthquakes. The velocity of seismic waves—that is, how fast seismic energy passes through a rock—varies with respect to the composition of the material, whether it is solid or liquid, its density, and its stiffness. Seismic waves are faster in rocks with olivine and pyroxene than in rocks containing the minerals quartz and feldspar. Molten rock or magma passes seismic energy more slowly than solid rocks; indeed, some types of seismic waves do not pass through liquids at all. Study of these seismic effects, coupled with measurement of the attraction of the Earth's gravity below a point on the surface, as well as slight perturbations in the Earth's magnetic field near the surface, have contributed to the layered model of the Earth.

The boundaries of the layered model are reasonably sharp on a global scale; on a more local scale, they become fuzzy and complex. In places there is considerable mixing of rocks from two layers. This mixing adds to the complexity of analysis of the boundary.

The crust of the Earth consists of two main parts—continental crust and oceanic crust. Continental crust consists of diverse sedimentary, metamorphic, and igneous rocks ranging in age from about 4 billion years old (4 Ga) to recent. Continents average approximately 35 km thick, but range from approximately 15–30 km thick along their margins and in rifted regions (such as the U.S. Basin and Range Province or the East African Rift) to about 70–100 km thickness under high mountain regions, such as the Himalayas or the Andes. The average composition of a continent is approximately that of a granitic rock, with Na approximately equal to K. Minerals in such rocks chiefly include quartz (SiO_2), feldspar including potassium feldspar ($KAlSi_3O_8$) and plagioclase ($(Ca,Na)Si_2O_6$), subordinate mica (biotite or muscovite), and minor iron- and magnesium-bearing minerals, such as amphibole.

As mentioned, oceanic crust underlies some 70% of the area of the Earth's surface. It is thinner than continental crust, about 5–7 km thick on average, and is considerably different in composition from continental crust. A typical oceanic crust comprises a sequence of fine- and coarse-grained rocks of basaltic composition (about 50% SiO_2, 10–20% Al_2O_3, a few percent CaO, and small amounts of

K_2O and Na_2O). The average composition and seismic velocities of oceanic crust are that of a basalt. Chief minerals in basaltic rocks include plagioclase, pyroxene, olivine, and amphibole (another Mg-Fe-bearing mineral).

Both continental and oceanic crust overlie the Earth's mantle. The mantle comprises the main volume of the Earth. At shallow levels it is thought to be mostly composed of olivine, with subordinate pyroxene and spinel, an oxide mineral. The crust together with the uppermost part of the mantle is the lithosphere, a dense, strong layer that forms at mid-oceanic ridges or other spreading centers and thickens away from them to an average thickness of 100 km. Beneath continents, the lithosphere may be as much as 250 km thick.

Beneath the lithosphere is a weak zone in the mantle, the asthenosphere, where the rock is closer to its melting temperature than it is in the overlying lithosphere. It contains the zone on which the plates slide during plate motion. The asthenosphere may attain thicknesses of up to 150 km under the oceans, but it is thinner or possibly absent under continents (e.g., Fjeldskaar, 1994).

Beneath the asthenosphere, the olivine and pyroxenes in mantle rocks change to denser crystalline forms and form a dense layer, the mesosphere, which extends to the core-mantle boundary. The inferred "hot abyssal layer" in the lowermost mantle is a region thought to contain relatively primordial mantle.

Three principal types of plate boundaries exist. Divergent margins, chiefly mid-oceanic ridges, occur where plates move apart and new oceanic lithosphere develops; convergent or subduction margins are where one plate descends beneath another into the Earth's interior (*subduction* is the term given to the process by which one plate slides beneath another. It is the English translation of the German word *Verschluchung*, meaning "underthrusting," a term that was used in the early twentieth century to explain the formation of the Alps; Sengor, 1977); conservative or transform margins are where two plates slide past one another without creation or destruction of lithosphere. Figure 1.1 shows these margins schematically, and Figure 1.2 shows the current distribution of continents, oceans, island arcs, and plate boundaries. The plate boundaries are the locations where most of the Earth's earthquakes occur.

An island arc is a curved chain of islands, usually with active volcanoes, that lies above a subduction zone within the oceans. Examples include the Japanese islands, the Philippines, the Aleutians, the Lesser Antilles, and the Marianas. Subduction zones beneath a continent also cause curved chains of volcanoes, such as those on the Kamchatka Peninsula, Mexico and Central America, and the Cascades of the Pacific Northwest, but because they lie on continents, these chains are called continental arcs. Island arcs as defined are distinct from more linear island chains, such as the Hawaiian Islands, that form above a "hot spot" or point-like source of magma that produces a succession of volcanic islands as the plate moves over it. Island arc volcanoes characteristically are cone-shaped, such as Mt. Shasta,

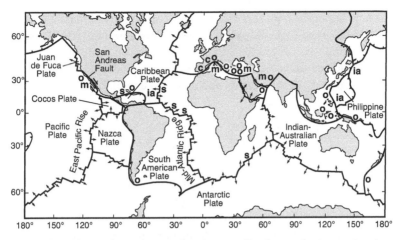

FIGURE 1.2. World map showing main distribution of land ocean basins, major plate boundaries, and tectonic plates on the surface of the Earth. Selected average relative plate velocities are illustrated. General areas with serpentinite on the ocean floor indicated by s. These areas include transform faults, fracture zones, faulted spreading centers, especially in the central Atlantic and along the southwest Indian Ocean Ridge and the continental margin west of the Iberian Peninsula. Modified after Twiss and Moores (2007). Also shown are major areas of ophiolite complexes (o), mélanges (m), select island arcs (ia), and subcontinental peridotites (c). See text for discussion.

and have more Si-rich magmas (andesites) than those of Hawaii-like hot spot islands, which have more rounded or "shield-like" volcanoes and are chiefly of basaltic composition (see later discussion).

The depth of subduction varies from place to place. As illustrated in Figure 1.1, some down-going (subducting) plates descend into the mantle all the way to the core-mantle boundary. Others seem to get stuck at the top of the mesosphere. In other places, down-going plates descend only 100–300 km in depth.

Most of the time, the plates move fairly smoothly, with more or less constant angular velocity with respect to each other. In some places, however, collisions occur that interrupt this smooth motion. For example, a continent on a down-going plate eventually may collide with a subduction zone and subduct a short distance, until the buoyancy of the continental crust arrests the subduction. Modern examples of subduction zone–continent collisions include the northern margin of Australia, which is colliding with the Indonesian subduction zone near Timor, and East China, which is colliding with the east-dipping West Luzon subduction zone near Taiwan.

In other cases, two continents collide with each other. The Alpine-Himalayan mountain belt displays the best examples of such a situation, with continental collisions taking place at present along the Taurus-Zagros Mountains of Turkey, Iraq, and Iran, where Arabia is colliding with Eurasia; and the Himalayas, where India is colliding with central Asia. Previous collisions include the Alps, the Appalachians, and the Urals. Collisions interrupt the smooth action of plates: they change plate motions or the location and nature of boundaries.

PERIDOTITES, SERPENTINITES, AND ASSOCIATED ROCKS

Peridotites are composed of silicon-oxygen–containing minerals called silicates. A peridotite consists principally of olivine, with lesser amounts of one or two pyroxenes and minor oxide and sulfide minerals of chromium, aluminum, and nickel. Olivine is a silicate mineral containing chiefly Mg and Fe, as well as Si and O. Pyroxene comprises chiefly two separate silicate minerals, one containing chiefly Mg and Fe, and the other with significant amounts of Ca as well. Neither olivine nor pyroxene contains large amounts of Al, K, or Na. Olivine, pyroxene, and feldspar are particular silicate minerals with a specific structure and composition.

Serpentinites are called "ultramafic" or "mafic" because they are relatively high in magnesium and iron, as well as silicon, and lack large amounts of aluminum, calcium, sodium, and potassium. As mentioned before, "ultramafic" is reserved for rocks composed of at least 90% olivine and pyroxene or their alteration products, including serpentine; "mafic" refers to rocks that contain olivine and/or pyroxene, with approximately equal amounts of plagioclase feldspar. Ultramafic rocks variously are called peridotite, for a mixture of olivine and pyroxene, and dunite, for a rock composed mostly of olivine. A peridotite, in turn, is a harzburgite if the principal pyroxene is Ca-poor, a lherzolite if there is roughly equal amounts of Ca-poor pyroxene (enstatite) and Ca-rich pyroxene (diopside), and a wehrlite if the pyroxene is mostly diopside. An olivine pyroxenite contains at least 50% pyroxene and a pyroxenite if the pyroxene content is over 90%.

Serpentinites can form in any environment where water and peridotites come in contact with each other at temperatures lower than 500°C. Thus they can form along active plate margins, during emplacement of mantle rocks into the Earth's crust, as will be discussed, or even after emplacement as a result of reaction of peridotite with hot ground water.

Mafic rocks include many diverse types, particularly extrusive basalts, the most common volcanic rock on Earth. Additional rocks include shallow intrusive diabase, a medium fine–grained rock of basaltic composition, and intrusive gabbro, a coarse-grained rock of basaltic composition composed of variable amounts of olivine, pyroxene, and plagioclase.

Ultramafic rocks have high density (3.0–3.3 g/cc), high strength, and high seismic velocities. Hydration of olivine or pyroxene to serpentine in a rock produces a change in the structure of the minerals, weakens the rocks, and lowers its density from 3.3 g/cc for fresh peridotite to 2.4–2.9 g/cc for a serpentinite, depending on the amount of water added. Because serpentine is a "sheet silicate" mineral, similar to mica, zones of planar weakness develop in formerly strong rocks. Under high confining pressure, the rocks maintain considerable strength. At conditions of low confining pressure (near the Earth's surface), serpentinites lose their strength, are easily faulted, and turn into the weak, slippery, sheared bodies that are common in some regions. With low density and planar weakness, serpentinite is easily mobilized in Earth movements and thus becomes detached from its original location. Therefore, its movement may complicate our understanding of the nature and origin of any particular serpentine.

The velocity of seismic waves in the mantle increases sharply from crustal values at the crust-mantle boundary. This sharp increase is called the Moho or M discontinuity after Croatian seismologist Andrija Mohorovičić, who discovered it in 1909.

The complex structural and mineralogical changes involved in serpentinization are beyond the scope of this short chapter. They are well covered in Alexander et al. (2007), and O'Hanley (1996). Bear in mind, however, that the available experimental information indicates that serpentine can form at any temperature from room temperature up to about 500°C. In some places, for example, the northern California Coast Ranges and the Mariana forearc (the region just east of the Mariana Islands; Mottl et al., 2008), serpentine is thought to be forming at shallow levels within the crust. This ongoing reaction produces highly alkaline waters that come out on the surface as springs that on land give rise to deposition of calcium carbonate and special plant communities (e.g., O'Hanley, 1996: 196–97; Alexander et al., 2007). Interestingly, these reactions seem to be a model for some of the rocks found on the Martian surface (Mottl et al., 2008).

HISTORICAL DEVELOPMENT OF THE OPHIOLITE CONCEPT

Ophiolite sequences are on-land exposures of oceanic crust and mantle. The word comes from the classical Greek words *ophis*, "snake," and *lithos*, "rock." The name arose in studies of European exposures in the early nineteenth century and refers to the mottled-green snake-like appearance of many serpentinites found in ophiolites.

Ophiolites were first defined by Brongniart (1813) as representing the "common serpentines," the rocks that had been used as dark green-black dimension stone in many Renaissance and pre-Renaissance constructions. In his first writings, Brongniart made no distinction between igneous and metamorphic rocks, but in his later classifications (e.g. Brongniart, 1827), the igneous-metamorphic distinction had become clear, and he grouped the ophiolites with igneous rocks. This classification led to confusion for over a century.

By the end of the nineteenth century, the presence of "greenstones (including ophiolites)," "roches vertes," "pietri verdi," and "grünsteins" was well known throughout the world (Suess, 1909). Suess argued that these rocks commonly displayed igneous contacts, they were present as discordant or concordant sheet-like igneous masses in highly folded terranes, and they never cropped out in large batholith-sized masses, in contrast to granitic rocks. (Batholiths, from the Greek words *bathy*, "deep," and *lithos*, "rock," are defined as regions of intrusive igneous rocks larger than 50 square miles in area. A smaller area is a "pluton," after Pluto, the Roman god of the Underworld.)

Steinmann (1905, 1927) added considerably to the understanding of ophiolites. Beginning with his work in the late nineteenth century, Steinmann pointed to the ubiquitous association of serpentinite, "diabase," including altered volcanic rocks (so-called spilites, keratophyres), hypabyssal and plutonic mafic rocks, and radiolarian chert. He described this association, which became known as Steinmann's trinity, in the western Alps and Italy, where the sequence abundantly crops out but generally lacks gabbro and nowhere includes dike complexes (Bernoulli, 2001). Interestingly, Steinmann (1905) also recognized the association in the Golden Gate region, San Francisco, and on Mt. Diablo, California, based on a field trip led by Andrew C. Lawson of University of California, Berkeley, in 1892. Steinmann's other important contribution was drawing attention to the deep-water environment of the immediately overlying radiolarian sediments. He also observed that shallow water sediments overlay the radiolarites and argued for massive uplift in Cretaceous time and for the ophiolite emplacement during folding. After Steinmann's (1927) synthesis, nearly all European (but not Anglophone) workers recognized the importance of the association, but its true significance was not generally recognized until the mid-1960s (Hess, 1965). Steinmann also argued that ophiolites represented the remnants of a 500–700-km-wide Tethyan ocean that formerly existed between Africa and Europe (Oreskes, 1999).

Contemporaneously with Steinmann's work, two other lines of thought developed as to the origin and tectonic significance of peridotites and/or serpentines. American petrologist N. L. Bowen conducted experiments as early as 1914 on the crystallization of olivine from a basaltic (mafic) melt. Bowen (1927) argued that peridotites formed chiefly by stratiform accumulation of early formed olivine in the bottom of a magma chamber as the magma cooled.

A contrasting view was that of Australian geologist W. N. Benson. Drawing on his own fieldwork in eastern Australia and a worldwide survey of peridotite/serpentinite occurrences, Benson (1926) argued that field relations indicated that peridotites were of magmatic origin and different from peridotites formed by

fractional crystallization processes. Benson coined the term "Alpine-type" peridotites because of the widespread presence of such peridotite/serpentinite bodies in
Alpine-type orogenic belts. He argued that peridotites were intruded during the
initiation of deformation of an orogen in its marginal parts, as opposed to the
central core region where granitic intrusions predominated.

Thus by 1927, three conflicting lines of opinion existed. First was the continental European view, which emphasized the relationship of peridotites to mafic pillow lavas and radiolarian cherts. The second, an "Anglophone" view, espoused
principally by workers in the United States, the United Kingdom, Australia, and
New Zealand, argued that the "Alpine" peridotites were unrelated to the other
rocks of Steinmann's trinity, were igneous, and were intruded into the margins of
the core regions of "Alpine" mountain systems. Third was Bowen's view that the
temperatures of formation of an olivine-rich magma were too high to explain the
field relations of general lack of metamorphism and that peridotites were formed
by fractional crystallization of basaltic magma. With the benefit of hindsight, it is
clear that all three lines of thought were partly correct. Part of the controversy was
cultural and was aided and abetted by the tendency for authors to cite mainly literature in the language most familiar to them. This conflict raged on for four decades and was resolved only in the 1960s.

The global tectonic model or paradigm that existed through much of the nineteenth and the early twentieth centuries was the geosynclinal model. Mountain
belts, also called "orogenic belts (after the Greek words *oros*, "mountain," and *genesis*, "birth"), are linear regions of folded, faulted, metamorphic, and igneous intrusive rocks (including both peridotites or ophiolites) and granitic rocks that are
present in the Earth's continents. A geosyncline was thought to be a long, deep
trough in the Earth's continental crust that developed and filled with shallow marine sedimentary rocks at the margins and deep-marine sedimentary and volcanic
rocks at its center. Continents and ocean basins were fixed on the Earth, and the
oceans were as old as the continents.

At some point, for an unknown reason, the depression of the geosyncline re-
versed, and the geosynclinal rocks rose out of the interior, became folded and
faulted, metamorphosed, and were intruded by abundant granitic and ultramafic
(or ophiolitic) rocks. This was the tectonic theory that I learned as a student in the
late 1950s and early 1960s; as graduate students, we wondered where the modern
geosynclines were.

Ophiolites were long considered to be igneous rocks. Hess (1939, 1955) tried to
relate the emplacement of ophiolites to the beginnings of deformation of geosynclinal belts, arguing that ophiolites were intruded during the first deformation of a
geosynclinal pile of sediments. Based primarily on the pattern of ultramafic rocks
(not all of ophiolitic origin) in the Appalachians, Hess argued that there were two

belts of rocks and they represented the two flanks of the geosyncline. Hess, how-ever, was speaking mostly of peridotites, and he considered the peridotites to be magmatic.

After World War II, much investigation of the deep ocean crust ensued. The oceanic crust turned out to be thinner than continental crust, a fact not known until the late 1940s. The centers of many oceans were shallower than the flanks, and these mid-oceanic ridges had more heat coming from the Earth's interior than the flanks did. Many island arcs, such as the Philippines, the Marianas, the Aleu-tians, and the Antilles, had deep ocean troughs or trenches on their oceanward sides. These island arcs had planar zones of seismic activity that were inclined be-neath the islands, descending deep within the mantle from shallow levels near the trench.

The 1950s brought new developments. Magnetic evidence accumulated sug-gesting that the continents were not fixed in position, as heretofore assumed, but had moved with respect to the Earth's magnetic pole. This so-called polar wander-ing was a serious problem for the idea of fixed continents, an important corollary of the geosynclinal model. In addition, evidence accumulated that the ocean ba-sins were much younger than the continents.

Thus by 1959, it became clear that there were mounting problems with the geo-synclinal paradigm. In the early 1960s, an innovative hypothesis by Hess emerged in preprint form (ultimately published as Hess, 1962) that swept away the old model: mobile continents moved about by a process ultimately called "sea floor spreading" (Hess, 1962; Vine and Matthews, 1963) that involved movement of the mantle as well as the crust. New ocean crust formed at mid-ocean ridges. Crust was somehow recycled back into the Earth's interior, but Hess did not address that issue in detail.

Independent of these developments, measurements of the Earth's magnetic field for the past several million years showed that it changed its polarity about every million years or so. Marine magnetic measurements over the oceans re-vealed the presence of strip-like magnetic anomalies (changes in the strength of the Earth's magnetic field) that were symmetrical about mid-oceanic ridges. In clear support for Hess's new hypothesis, Vine and Matthews (1963) argued that the symmetrical magnetic anomalies formed as new oceanic crust was produced at spreading centers and moved outward, recording the Earth's magnetic field at the time of cooling of the new oceanic crust below a specific temperature.

A new global seismic network, installed in the late 1950s to monitor under-ground nuclear explosions, resulted in a map of the Earth showing that earth-quakes were closely arranged in zones. The seismic information showed that the top of the mantle was dense and strong, thin at mid-oceanic ridges, and thickening away from the ridges as the oceanic crust aged. This thick, dense, strong layer was shown to be inclined beneath the island arcs. A new class of faults, transform faults, was recognized that separated two plates between segments of mid-oceanic ridges (Wilson, 1965). By 1968, all these new data were published, and the new global tectonics or plate tectonics was born.

Meanwhile, work on ophiolites in the 1960s contributed significantly to the plate tectonic revolution. Field studies of the Vourinos complex in northern Greece (Moores et al., 1966; Moores, 1969a) showed that there was a close connection between the peridotites at the base, the gabbros in the middle, and the extrusive rocks and shallow intrusive basaltic rocks at the top. Most of the peridotites were not igneous but metamorphic (so-called tectonites), having been deformed at a high temperature.

While writing up the Vourinos work, I discovered reports of the Troodos complex in Cyprus. Troodos maps revealed a remarkable set of ophiolitic rocks, with tectonite peridotite at the base, overlain by magmatic peridotites and gabbros, a remarkable set of mafic dikes, tabular intrusions of diabase standing more or less vertically and overlain by extrusive submarine lavas and deep sea sediments. Fred Vine and I (Moores, 1969b; Moores and Vine, 1969, 1971) studied these rocks in terms of their possible formation by sea floor spreading. We essentially established that we were looking at a fragment of ocean crust and mantle formed by sea floor spreading.

The results of this work quickly became famous. The Troodos complex's excellent exposures give a clear view of how sea floor spreading might take place and what kinds of rocks can be formed in the process.

A conference was held at Asilomar, California, in December 1969 to evaluate the effect of the new discoveries of plate tectonics on geology (Dickinson, 1970). This conference was a watershed event. It soon became clear that the features of the newly discovered plate tectonics applied to geologic history. At the end of the conference, its organizer, W. R. Dickinson of Stanford University, reinterpreted geosynclines in terms of modern oceanic environments (Dickinson, 1971). Ophiolites, especially those in Cyprus but also in Papua New Guinea and Newfoundland, became the standard model for the formation of oceanic crust, and the issue of ophiolite emplacement was much discussed. The nature of the three principal types of plate boundaries became clear (Figure 1.3).

After hearing Dickinson's talk on geosynclines, in a flash of insight I conceived of a model to account for the tectonic development of the western margin of the United States for the past 500 million years (Moores, 1970). It was one of the most exciting moments of my professional life! The model proposed that there has been a subduction zone beneath the western United States for much of this time. Periodically, island arcs have migrated toward the continental margin and collided with it. These collisions have emplaced many of the ophiolites in the western United States and caused the development of many of the older (pre–60 Ma) folds and faults present there. Some aspects of the original model have proven incorrect.

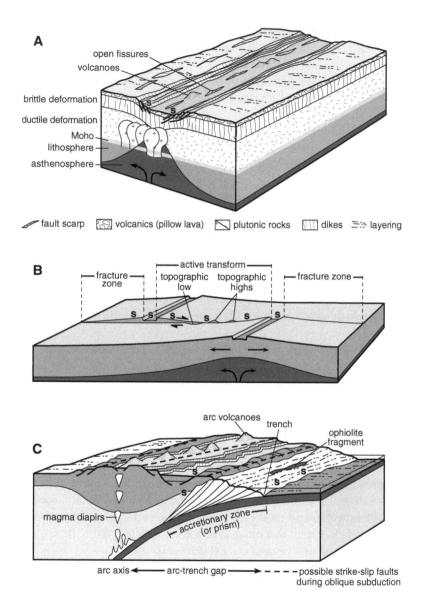

A

open fissures
volcanoes
brittle deformation
ductile deformation
Moho
lithosphere
asthenosphere

⟋ fault scarp   🔲 volcanics (pillow lava)   ◺ plutonic rocks   ⦚ dikes   ⩵ layering

B

— fracture zone —   — active transform —   — fracture zone —
topographic low   topographic highs

C

arc volcanoes   trench   ophiolite fragment

magma diapirs   accretionary zone (or prism)

arc axis ← — arc-trench gap — → – – – possible strike-slip faults during oblique subduction

FIGURE 1.3. Schematic block diagrams showing major features of three main types of plate boundaries. A: Divergent margin at mid-ocean ridge. B: Transform fault margin between two ridge segments. C: Subduction zone boundary within an ocean. s = areas of possible serpentinite exposure. Modified after Twiss and Moores (2007).

Yet collision of the western United States with island arcs coming toward it from the ocean has been accepted as part of the tectonic development of the western United States. One of these major collisions (about 160–175 Ma) emplaced many of the ophiolites in the Klamath Mountains, the Sierra Nevadas, and in Oregon and Washington.

OPHIOLITES AND OCEANIC CRUST SINCE THE REVOLUTION

Since the plate tectonic revolution, studies of ophiolites extended the new paradigm to global occurrences and the entire geologic record. Thousands of studies have been conducted of ophiolite complexes throughout the world in various tectonic regions, and our understanding of them has improved enormously. Particularly important have been international conferences and field trips investigating ophiolite complexes in various areas, summary monographs devoted to ophiolites, and increasingly detailed comparisons of ophiolites with oceanic crust, particularly as documented through the Integrated Ocean Drilling Project (IODP).

An important international Penrose (named after a 1920s benefactor of the Geological Society of America) ophiolite conference in 1972 consisted of a 1600-mile (2500 km) road trip of newly recognized ophiolite complexes in the western United States, specifically Oregon and northern California. Twelve informal seminars during the trip culminated in the so-called Penrose definition of ophiolites— "a distinctive assemblage of mafic to ultramafic rocks," consisting of an ultramafic complex, a gabbroic complex, a mafic sheeted dike complex, and a mafic volcanic complex, commonly pillowed. So-called associated rocks types include an overlying sedimentary section of chert, minor shale and limestone, and/or volcaniclastic sediments (see Figure 1.4A). The report called for more careful mapping of the various members within ophiolites and more petrologic studies (Anonymous, 1972: 25). With the benefit of hindsight, one can note that conspicuously absent from the discussion of this field trip is any mention of the need for consideration of the ophiolite in its regional context.

A second international ophiolite conference the following year (May 31–June 14, 1973) in the Soviet Union focused on Hercynian (Paleozoic) ophiolitic complexes in the Alai Range and the Kyzyl Kum Desert (in Uzbekistan), as well as Mesozoic ophiolitic complexes of the Lesser Caucasus (in Nagorno-Karabakh). In addition, the conference provided a detailed exchange of views between Soviet and Western geologists and introduced many Soviet geologists to plate tectonic concepts (Coleman, 1973).

Several workers have written books devoted to ophiolites (e.g., Coleman, 1977; Nicolas, 1989). My own contributions have included considerations of the tectonic significance of ophiolite emplacement (Moores, 1970, 1982), the reinterpretation

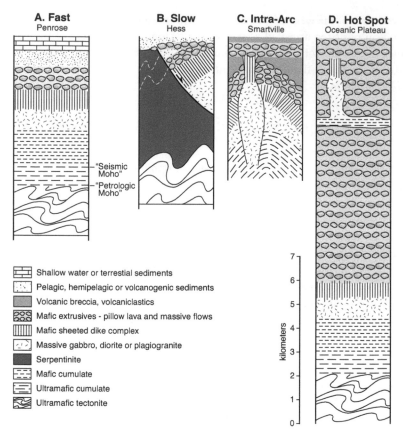

FIGURE 1.4. Schematic columnar sections of representative oceanic crust and corresponding ophiolite types. Serpentinites may form in ultramafic tectonites, ultramafic cumulates, as well as in scattered spots within pillow lavas and flows. A: Complete idealized (or Penrose ophiolite sequence, outlined in a conference in 1972; Anonymous, 1972) on ophiolites. B: Faulted, incomplete sequence (or Hess-type crust, after H. H. Hess [1962] who proposed such an oceanic crust) from a magma-starved spreading center where tectonic processes dominate. C: Complex composite section of an oceanic island arc developed within or on oceanic crust (Smartville type from the Smartville complex, northwest Sierra Nevadas, California). D: Possible oceanic crust in a "hot spot" or oceanic plateau section. Serpentinites will form as alterations of ultramafic cumulates or ultramafic tectonites. Note large region of serpentinite in B. After Moores (2002).

of all ultramafic rocks in the light of plate tectonics (Moores, 1973), and an early attempt to relate the structure of oceanic crust and ophiolites to spreading rate (Moores and Jackson, 1974). Studies of the Vourinos complex with students and colleagues led to the recognition of cyclic accumulations of olivine, pyroxene, and plagioclase in magma bodies, further detailing the nature of magmatic processes within an ophiolite (Jackson et al., 1975; Harkins et al., 1980; Rassios et al., 1983). Detailed studies of the Troodos dike complex in Cyprus led to discovery and elaboration of curved (listric; from the Greek word *listron*, "shovel") normal fault systems (e.g., Varga and Moores, 1985); similar features were discovered in the Josephine complex in California–Oregon (Harper, 1982). Thus some ophiolites and some oceanic spreading centers have faults similar to those in rifted regions such as the Basin and Range province of Nevada and Utah.

Studies of ophiolitic rocks in the western North American Cordillera have led to the use of ophiolites in a reinterpretation of the structural evolution of that margin (e.g., Moores, 1970; Dilek et al., 1988). A global review of ophiolites and their significance led to separation of them into Tethyan and Cordilleran types based on the presence or absence of a continental substrate, an island arc edifice, or other geologic criteria (Moores, 1982). Exploration of the nature of Precambrian, especially pre–1000 million years old (pre–1 Ga) oceanic crust has led to the hypothesis that earlier oceanic crust was thicker and thinned abruptly about 1000 million years ago (Moores, 1973, 1986, 1993, 2002).

Several more international conferences have contributed to our understanding of ophiolites, including conferences on the Troodos ophiolite in 1979 and 1987 (Panayiotou, 1980; Malpas et al., 1990), the Oman ophiolite in 1990 (Peters et al., 1991), Circum-Pacific ophiolites (Ishiwatari et al., 1994), and a second ophiolite Penrose conference comparing ophiolites and Ocean Drilling Project (ODP) results (Dilek et al., 2000). The latter conference was particularly valuable because it brought together workers concentrating on the ODP and those more focused on land-based ophiolite studies. Through such comparisons, new insights develop.

A major postrevolution discussion began with Miyashiro's (1973) focus on the island arc–like chemistry of the Troodos complex. Most subsequent petrological and geochemical discussions have focused on the geochemical evidence in ophiolites for a mantle source already depleted of its MORB components (MORB is an acronym for the chemical composition of average "mid-ocean ridge basalts"; e.g., Robinson and Malpas, 1990; Bloomer et al., 1995). These environments are especially present in spreading centers in so-called forearc or back-arc regions, regions between the active volcanic arc axis and the trench (or subduction zone) or in the basin behind the arc, respectively. Their position on the overriding plate in a subduction zone has led to yet another subdivision: so-called suprasubduction zone (SSZ) ophiolites.

OCEANIC CRUST STUDIES

Studies of the oceans have proceeded apace with studies of ophiolites. It has become clear that a complete ophiolitic sequence is present in some regions, especially where magma is abundant and spreading is fast, such as at the East Pacific

Rise. Oceanic island chains in the middle of oceanic plates are thought to be the product of hot spot volcanism, arising from deep within the mantle. In other mid-ocean ridge regions, however, the oceanic crust sequence is incomplete, and serpentinized-ultramafic rock is present at the sea floor surface.

The map in Figure 1.2 shows the present land, ocean, island arcs, and plate boundaries. Specific locations where serpentine has been dredged include the inner wall of the Mariana Trench, the Southwest Indian Ridge and associated transform faults, the central Atlantic transform faults, and on the edge of the continental shelf offshore from Spain and Portugal. There are many similar sites where serpentinite has been recovered or is suspected as well. In a few places, fresh mantle is exposed at the ocean bottom.

PLATE BOUNDARIES AND SERPENTINITE

Figure 1.3 shows the three main types of plate boundary—divergent margin (Figure 1.3A), convergent margin (Figure 1.3B), and transform margin (Figure 1.3C). In divergent margins, rocks capable of forming serpentinite are those of the lithosphere and any olivine-rich rocks in the crust. Faults and cracks penetrating into the lithosphere, which is chiefly peridotite, can produce serpentinite. A few scattered olivine-rich volcanic rocks or intrusions can also become serpentinized. Convergent margins can have serpentinite in the accretionary zone or prism, formed by incorporation of serpentinite from the down-going plate, as well as in any oceanic crust that might be present in the overriding plate. Transform faults within oceanic crust ubiquitously show serpentinite (some transform fault zones are over 100 km wide); and the oceanic crust offset by the transform fault can contain serpentinite, as well.

CURRENT STATUS OF THE OPHIOLITE QUESTION

Three points stand out: the environment of formation of ophiolites and what they can tell us about oceanic crust formation, the mechanism of emplacement of ophiolites and its significance for interpretation of the tectonic development of orogenic systems, and the change in ophiolites and thus oceanic spreading processes through time.

Environment of Formation

Ophiolites represent ocean crust and mantle formed at oceanic spreading centers. These centers occur either at mid-oceanic ridges in pull-apart intra-arc oceanic basins such as those within the Philippines or at active back-arc basins, which are oceanic basins behind (i.e., on the other side of) an island arc from a subduction zone. For example, the Philippine Sea lies west of the Mariana Islands, and the Mariana Trench and subduction zone lie east of the Marianas. Finally some island arcs, especially the Marianas and possibly the Tonga-Kermadec Islands, exhibit active extensional zones in the centers of forearcs (the region between an active island arc and its subduction zone) during the initial development of island arcs. The nature of the environment of formation of ophiolites cannot be obtained by geochemistry alone; rather, a comprehensive set of data, including geologic relations, associated deposits, internal structure, and geochemistry, is necessary to evaluate the significance of ophiolite complexes. Some ophiolites from magma-rich spreading centers display a complete sequence (Penrose-type ophiolites; Figure 1.4A), which may imply a fast-spreading environment (Dilek et al., 1998). Other complexes formed in a magma-starved environment display incomplete sequences, as seen on modern slow-spreading ridges and in the Alps and Apennines where Steinmann did his work (Bernoulli, 2001). The Hess-type complexes (Figure 1.4B) imply a significant portion of the oceanic crust was serpentinized peridotite, as advocated by Hess (1962) and Vine and Hess (1970); see also Moores (2002) and Dilek et al. (1998). The double discovery of faulted structures in ophiolites and oceanic crust strengthens the link between these different environments.

Some ophiolite complexes represent the remnants of island arcs. In such cases, intrusive rocks at the base are succeeded by thick lavas and in places a sequence of later intrusive rocks. Such a complex has been called a Smartville-type, after exposures in the northwest Sierra Nevadas, California (Figure 1.4C).

Finally, many hot spots that erupt on oceanic plates produce an oceanic crustal sequence overlain by thick sequences of pillow lavas (see Figure 1.4D). In places these sequences also have ultramafic rocks in high-level small magma chambers. This type of ophiolite is especially well developed in the Solomon Islands, southwest Pacific, and in Colombia (Figure 1.4D). The Caribbean plate is thought to be composed of a thick oceanic crustal sequence, as in Figure 1.4D.

The environment of formation of ophiolites will continue to be controversial. Moores et al. (2000) attempted to resolve the difficulties between universal application of this model and lack of geologic evidence for any island arc in many Tethyan ophiolites by suggesting that the magma source compositions were "historically contingent," that is, a product of prior history and not necessarily reflective, a priori, of modern environments. Metcalf and Shervais (2001) criticized the historical contingency concept on geochemical grounds, but they did not consider many data from the mid-oceanic ridges and the Tethyan region. The issues are not whether some ophiolites formed in the overlying plate of a subduction zone (so-called suprasubduction zone) but whether all ophiolites are so formed; the inadequacy of geochemistry to determine the tectonic environment in the absence of geologic evidence; and the increasing evidence for long-lived mantle heterogeneity. Recent discovery of silicic lavas at active mid-oceanic ridges (e.g., Stoffers et al.,

2001) at the very least should encourage use of multiple working hypotheses and careful fieldwork before any assertion of a tectonic interpretation of a particular ophiolite.

Mechanism of Emplacement of Ophiolites and Its Tectonic Significance

The tectonic significance of ophiolite emplacement was signaled early, principally by Hess's (1939) and Stille's (1939) observations that ophiolites (or ultramafics) were intruded in the initial stages of orogeny. However, tectonics of ophiolite emplacement was not considered in the initial ophiolite Penrose conference (Anonymous, 1972), even though I had previously published a paper that dealt with ophiolite emplacement (Moores, 1970). The tectonic significance of ophiolites has received relatively short shrift in most discussions of ophiolite complexes. The latter have concentrated on petrology (i.e., rock types) and geochemistry.

Viewed in a plate tectonic context, the issue becomes how to get little-deformed oceanic crust and unserpentinized mantle emplaced over continental platforms (in the case of Tethyan-type ophiolites) or island arc crust (in the case of Cordilleran-type ophiolites). Many researchers (e.g., Temple and Zimmerman, 1969; Moores, 1970, 1973; see also Coleman, 1971; Dewey, 1976) have argued that such emplacement, given the presence of the topographic difference between continental and oceanic crust, must be by collision of a continental margin with a subduction zone dipping *away* from the continental margin (see Figure 1.5). The continent on the down-going plate subducts a short distance, until the buoyancy of the continental crust arrests the subduction (Figure 1.5A), the leading edge of the overriding plate is thrust up over the continent and then is broken off and preserved as an ophiolite complex (Figure 1.5B). Ophiolites thus formed are generally little affected by the emplacement. They are characteristically overlain by shallow water or continental deposits or even an erosion surface. Examples of such ophiolites include the Troodos complex (Cyprus), the Bay of Islands complex (Newfoundland), and the Papuan ophiolite (Papua New Guinea). In a few places, such as the Solomon Islands, the top of very thick oceanic crust has been scraped off and thrust landward as the plate goes down.

In this collisional interpretation of ophiolite complexes, the basal sole thrusts of ophiolites represent the remnants of former subduction zones; their displacements are indeterminate but may be very large. Sedimentary fold-thrust, so common along the margins of orogenic belts, are secondary to the main ophiolite thrusts. Rare is the synthesis of any orogenic belt that has adequately taken this fact into account.

Some ophiolite complexes or incomplete complexes are present within accretionary prisms of ancient or modern subduction zones. In such cases, faulting of the down-going oceanic plate forms fragments that become incorporated within

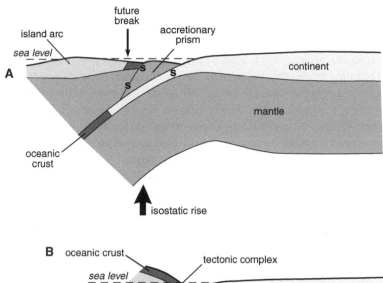

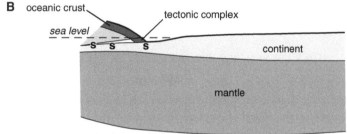

FIGURE 1.5. Emplacement of ophiolites. A: Schematic diagram depicting collision of continental margin on down-going plate with subduction zone. B: Subduction eventually stops because of buoyancy of continental crust, after which a part of the overriding plate is preserved on continental margin as an ophiolite complex. Areas of possible serpentinite denoted by s. After Moores (1982).

the accretionary complex of the subduction zone (see Figure 1.5B). These complexes characteristically display some effect of the process, and they are not as complete or well preserved as ophiolites emplaced by collision, as outlined.

Ophiolites, Serpentines, and Oceanic Crust through Time

The oldest oceanic crust preserved in the oceans is about 185 Ma, and is present in the western Pacific just east of the Mariana Trench. Thus for earlier times in Earth's history, one is dependent completely on land-based ophiolites for an idea of what the oceanic crust might have looked like. Ophiolites in orogenic belts as old as about 1000 Ma appear to be similar in aspect to Mesozoic-Cenozoic ophiolites. In these belts, serpentine can be expected to be present wherever ultramafic rocks are exposed, either as parts of ophiolites or as part of accretionary complexes.

Many authors have suggested (e.g., Sleep and Windley, 1982; Hoffman and Ranalli, 1988) that on theoretical grounds oceanic crust should be thicker in early Earth history than at present. I suggested that oceanic crust thinned abruptly about 1000 Ma (Moores, 1973, 1986, 1993, 2002). Recent reports of an ophiolite of about 1020 Ma in the East Sayan belt, Siberia (Khain et al., 2002). may alter the timing of this possible abrupt thinning. Whatever the nature of the pattern, there is increasing recognition that thicker oceanic crustal sequences are present in mid-lower Proterozoic and Archean (1000–3500 Ma) terrains. Thus interpretation of Proterozoic-Archean orogens can find guidance from the consideration of much better known ophiolites in Phanerozoic orogens.

In these older regions of mid-lower Proterozoic (1000–2500 Ma) and Archean (2500–3800 Ma) age, ophiolites may contain thicker magmatic sequences, and tectonite peridotites may not be present. However, many magmatic sequences of this age are ultramafic, and many extrusive volcanic rocks are also olivine-rich. Thus, these sequences will develop serpentinite, whenever the appropriate rocks are exposed.

ULTRAMAFIC EXPOSURES OTHER THAN OPHIOLITES

Mélanges

Mélanges are chaotic mixtures of diverse rock types of different origin. Originally recognized on the island of Anglesey in the early twentieth century, they were the subject of several studies prior to the plate tectonic revolution, particularly in the early 1950s in Turkey and in the 1960s in the California Coast Ranges, the Italian Apennines, and Iran. They are widespread in the Earth's orogenic belts, and they are also present in several settings in the modern oceans.

In the modern oceans, mélanges form along some wide transform fault zones, as well as in the accretionary complexes of some subduction zones. Oceanic islands, such as the Hawaiian Islands, also have large landslide complexes surrounding them, which contain debris of the islands themselves. Some forearc regions, particularly the Marianas, display the presence of serpentinite volcanoes, wherein the ultramafic rocks of the overriding plate have become hydrated and rise along a crack or other conduit to the ocean bottom. From there, the finely ground-up serpentinite flows slowly downhill as a landslide or earth-flow. Figure 1.2 generally shows areas of the Earth where mélanges are widespread (marked on the map with an m).

Many mélanges contain tectonic blocks of ultramafic rocks that become serpentinized. Others have a matrix made up of serpentinite. The origin of the latter is problematic, but they may form as parts of transform faults, of subduction zones, as landslide deposits off forearc exposures of ultramafic rock, or from serpentinite volcanoes.

Stratiform Mafic-Ultramafic Complexes

In some places, so-called stratiform mafic-ultramafic complexes are present. These are large thick (tens to hundreds of kilometers in extent) semicircular intrusive-extrusive complexes that crystallized from a magma of a composition that allowed for the formation and settling out of crystals of olivine and pyroxene. Notable examples include the Stillwater complex in Montana, the Muskox complex in Canada, the Bushveld complex in South Africa, the Skaergaard complex in east Greenland, several smaller intrusions in the northwest Highlands of Scotland, and the Dufek intrusion in Antarctica. Many layers of these complexes, especially their lower levels, are ultramafic in composition. These ultramafic rocks also will form serpentinites when hydrated. Some of these complexes—notably the Bushveld complex—have layers that are high in elements such as Pt and Ni. In fact, the Bushveld complex contains a large amount of the world's reserves of platinum.

The most recent large stratiform complexes (Skaergaard complex, Scottish complexes) are on the continental extension of hot spot traces. The tectonic significance of older stratiform complexes is more problematic. Some researchers (e.g., Windley, 1995) have argued that there is a concentration of such complexes in the Proterozoic, around 2 Ga, and that they indicate a special time in the evolution of the Earth.

Subcontinental Mantle

In a few places in the world, subcontinental mantle is present at the Earth's surface. Such places include the continental shelf west of the Iberian Peninsula, in the Pyrenees, the western Alps, the Betic Cordillera (southern Spain), and the Riff Mountains (Morocco). Here, one finds generally lherzolite present beneath highly metamorphosed lower crustal rocks. Generally, such peridotites contain a small amount of Al and Ca, in contrast to oceanic varieties. Serpentinite has formed from these rocks as well.

SUMMARY AND CONCLUSIONS

On-land ultramafic rocks that produce serpentinites are of great tectonic interest and importance. Most represent fragments of oceanic crust and mantle preserved on land. Studies of these sequences provide insight into the processes of sea floor spreading and oceanic lithosphere formation. Because the oldest oceanic crust is less than 200 Ma, only ophiolites older than this provide information about the nature of the oceanic crust in earlier times of Earth history.

The emplacement of oceanic crust and mantle (ophiolites) on land requires special conditions, generally the collision of a continental margin with a subduction zone, that is, the aborted subduction of a continent. Such events are important parts of the orogenic process. They record the interaction of oceanic subduction zones and continental masses in geologic history. Other ophiolites may form as fragments of the down-going plate in subduction zones that become incorporated into the accretionary complex itself.

Mélanges may contain fragments of serpentinite, or the matrix of mélanges may be serpentinite. Ultramafic portions of stratiform complexes and exposures of subcontinental mantle may also form serpentinite. In a few places, variably serpentinized subcontinental mantle has been exposed at the surface.

Thus, serpentinites have complex geologic settings and histories. The history of a particular ultramafic rock determines its structure and composition, and thus the nature of the resulting serpentinite and in turn the nature and composition of the soils formed from them. As the soils influence the ecology of biological communities, any study of the latter needs to take into account the nature of the rocks on which they develop.

Final Note: Carbon Dioxide Sequestration and Serpentinites

The growing amount of carbon dioxide and the resultant global changes in temperature are sources of increasing concern. Nearly all industrialized and developing countries use fossil fuels as a source of energy, and they will probably continue to do so. If the global community is going to avoid climatic disaster, it is imperative that we take steps to reduce atmospheric CO_2. Many researchers (e.g., Goff and Lackner, 2002) have pointed out that serpentinites are an ideal locality for sequestering carbon dioxide. Chemical reactions produce magnesite ($MgCO_3$) by reacting especially serpentine and olivine with CO_2. It is relatively easy to produce this reaction in the laboratory, but scaling it up to industrial size involves large production of magnesium hydroxide, grinding large amounts of serpentine and olivine to a fine size, and large amounts of water. The potential for environmental damage is clear. Serpentinite exposures will increasingly be viewed as possible repositories of waste CO_2. Any area where this process is instituted will necessarily lose its plant cover.

Acknowledgments

Many colleagues have enriched my understanding of ophiolites, notably R. G. Coleman, H. W. Day, Y. Dilek, H. H. Hess, C. A. Hopson, S. J. Hurst, J. Malpas, J. C. Maxwell, A. Moiseyev, A. Rassios, R. J. Varga, J. Wakabayashi, and C. Xenophontos. Janice Fong drafted the figures.

LITERATURE CITED

Alexander, E. B., Coleman, R. G., Keeler-Wolf, T., and Harrison, S. (2007) *Serpentine Geoecology of Western North America.* Oxford University Press, New York.

Anonymous. (1972) Penrose field conference on ophiolites. *Geotimes*, 17, 24–25.

Benson, W. N. (1926) The tectonic conditions accompanying the intrusion of basic and ultrabasic igneous rocks. *Memoirs of the National Academy of Sciences*, 19, 1

Bernoulli, D. (2001) Where did Gustav Steinmann see the trinity? *Geological Society of America Abstracts with Programs*, 33, 6, A172.

Bloomer, S. H., Taylor, B., MacLeod, C. J., Stern, R. J., Fryer, P., Hawkins, J. W., and Johnson, L. (1995) Early arc volcanism and the ophiolite problem: A perspective from drilling in the western Pacific. *American Geophysical Union Monograph*, 88, 1–30.

Brongniart, A. (1813) Essai de classification minéralogique des roches mélanges. *Journal des Mines*, 34, 190–99.

Brongniart, A. (1827) *Classification et Caractère Minéralogique des Roches Homogène et Heterogènes*. F. G. Levrault, Paris.

Bowen, N. L. (1927) The origin of ultrabasic and related rocks. *American Journal of Science*, 14, 89–108.

Coleman, R. G. (1971) Plate tectonic emplacement of upper mantle peridotites along continental edges. *Journal of Geophysical Research*, 76, 1212–22.

Coleman, R. G. (1973) Ophiolites in the Earth's crust: A symposium, field excursions and cultural exchange in the USSR. *Geology*, 1, 51–54.

Coleman, R. G. (1977) *Ophiolites*. Springer-Verlag, New York.

Dewey, J. F. (1976) Ophiolite obduction. *Tectonophysics*, 31, 93–120.

Dickinson, W. R. (1970) The new global tectonics, 2nd Penrose Conference. *Geotimes*, 154, 18–22.

Dickinson, W. R. (1971) Plate tectonic models of geosynclines. *Earth and Planetary Science Letters*, 10, 165–74.

Dilek, Y., Moores, E. M., Elthon, D., and Nicolas, A., eds. (2000) Ophiolites and oceanic crust: New insights from field studies and the Ocean Drilling Program. Geological Society of America Special Paper, 349.

Dilek, Y., Moores, E. M, and Erskine, M. C. (1988) Ophiolitic thrust nappes in western Nevada: Implication for the Cordilleran orogen. *Journal of the Geological Society, London*, 145, 969–975.

Dilek, Y., Moores, E. M., and Furnes, H. (1998) Structure of modern oceanic crust and ophiolites and implications for faulting and magmatism at oceanic spreading centers. In *Faulting and Magmatism at Mid-Ocean Ridges* (eds. R. Buck, J. Karson, P. Delaney, and Y. Lagabrielle). American Geophysical Union Monograph, 106, 219–66.

Fjeldskaar, W. (1994) Viscosity and thickness of the asthenosphere detected from the Fennoscandian uplift. *Earth and Planetary Science Letters*, 126, 399–410; DOI: 10.1016/0012-821X(95)90120-1.

Goff, F., and Lackner, K. S. (2002) Carbon dioxide sequestering using ultramafic rocks. *Environmental Geosciences*, 5, 89–102.

Harkins, M. E., Green, H. W., and Moores, E. M. (1980) Multiple intrusive events documented from the Vourinos ophiolite complex, northern Greece. *American Journal of Science*, 280-A, 284–95.

Harper, G. D. (1982) Evidence for large-scale rotations at spreading centers from the Josephine ophiolite. *Tectonophysics*, 82, 25–44.

Hess, H. H. (1939) Island arcs, gravity anomalies, and serpentinite intrusions, a contribution to the ophiolite problem. *International Geological Congress Moscow 1937, Report 17*, 263–83.

Hess, H. H. (1955) Serpentines, orogeny and epeirogeny. Geological Society of America Special Paper, 62, 391–408.

Hess, H. H. (1962) History of ocean basins. In *Petrologic Studies: A Volume to Honor A. F. Buddington* (eds. A. E. J. Engel, H. L. James, B. F. Leonard), pp. 599–620. Geological Society of America, New York.

Hess, H. H. (1965) Mid-oceanic ridges and tectonics of the sea floor. *Submarine Geology and Geophysics. Proceedings of the Seventeenth Symposium of the Colston Research Society* (eds. W. F. Whittard and R. Bradshaw), pp. 317–34. Butterworths, London.

Hoffman, P. F., and Ranalli, G. (1988) Archean oceanic flake tectonics. *Geophysical Research Letters*, 15, 1077–80.

Ishiwatari, A., Malpas, J., and Ishizuka, H. (1994) *Circum-Pacific Ophiolites: Proceedings of the 29th International Congress, Part D*. VSP, Utrecht.

Jackson, E. D., Green, H. W. II, and Moores, E. M. (1975) The Vourinos ophiolite, Greece: Cyclic units of lineated cumulates overlying harzburgite tectonite. *Geological Society of America Bulletin*, 86, 390–98.

Khain, E. V., Bibikova, E. V., Kröner, A., Zhuravlev, D. Z., Sklyarov, E. V., Fedotova, A. A., and Kravchenko-Berezhnoy, I. R. (2002) The most ancient ophiolite of the central Asian fold belt: U-Pb and Pb-Pb zircon ages for the Dunzhugur complex, eastern Sayan, Siberia, and geodynamic implications. *Earth and Planetary Sciences Letters*, 1399, 311–25.

Malpas, J., Moores, E. M., Panayiotou, A., and Xenophontos, C., eds. (1990) *Ophiolites, Oceanic Crustal Analogues*. Proceedings of the Troodos Symposium, Geological Survey Department, Nicosia.

Metcalf, R. V., and Shervais, J. (2001) Supra-subduction zone (SSZ) ophiolites: Is there really an "ophiolite conundrum?" *Geological Society of America Abstracts with Programs*, 33, 6, A-173.

Miyashiro, A. (1973) The Troodos complex was probably formed in an island arc. *Earth and Planetary Science Letters*, 19, 218–24.

Moores, E. M. (1969a) Petrology and structure of the Vourinos ophiolite complex, northern Greece. Geological Society of America Special Paper, 118.

Moores, E. M. (1969b) The Troodos, Cyprus, and Vourinos, Greece, ultramafic complexes, and an evaluation of ophiolites as ocean floor slices. *Program, Discussion Meeting of the Royal Society of London*. Royal Society, London, pp. 12–14.

Moores, E. M. (1970) Ultramafics and orogeny, with models for the U.S. Cordillera and the Tethys. *Nature*, 228, 837–42.

Moores, E. M. (1973) Geotectonic significance of ultramafic rocks. *Earth Science Reviews*, 9, 241–258.

Moores, E. M. (1982) Origin and emplacement of ophiolites. *Reviews of Geophysics and Space Physics*, 20, 735–60.

Moores, E. M. (1986) The Proterozoic ophiolite problem, continental emergence, and the Venus connection. *Science*, 234, 65–68.

Moores, E. M. (1993) Neoproterozoic oceanic crustal thinning, emergence of continents, and origin of the Phanerozoic ecosystem: A model. *Geology*, 21, 5–8.

Moores, E. M. (2002) Pre-1 Ga (pre-Rodinian) ophiolites: Their tectonic and environmental implications. *Geological Society of America Bulletin*, 114, 80–95.

Moores, E. M., and Jackson, E. D. (1974) Ophiolites and oceanic crust. *Nature*, 250, 136–38.

Moores, E. M., and Vine, F. J. (1969) Troodos Massif, Cyprus, as ocean floor: Preliminary petrologic and structural evidence. *Eos (Transactions, American Geophysical Union)*, 50, 333.

Moores, E. M., and Vine, F. J. (1971) The Troodos massif, Cyprus, and other ophiolites as oceanic crust: Evaluation and implications. *Philosophical Transactions of the Royal Society London*, 268A, 443–66.

Moores, E. M., Hess, H. H., and Maxwell, J. C. (1966) Petrology of and ultramafic (ophiolite) sequence. *Geological Society of America, Abstracts with Programs* 143.

Moores, E. M., Kellogg, L. H., and Dilek, Y. (2000) Tethyan ophiolites, mantle convection, and tectonic "historical contingency": A resolution of the "Ophiolite Conundrum." In *Ophiolites and Oceanic Crust: New Insights from Field Studies and the Ocean Drilling Program* (eds. Y. Dilek, E. M. Moores, D. Elthon, and A. Nicolas), pp. 3–12. Geological Society of America Special Paper, 349.

Mottl, M. J., Glazer, B. T., Kaiser, R. I., and Meech, K. J. (2008) Water and astrobiology. *Chemie der Erde*, 67, 253–82.

Nicolas, A. (1989) *Structures of Ophiolites and Dynamics of Oceanic Lithosphere*. Kluwer, Boston.

O'Hanley, D. S. (1996) *Serpentinites Records of Tectonic and Petrological History*. Oxford University Press, New York.

Oreskes, N. (1999) *The Rejection of Continental Drift: Theory and Method in American Earth Science*. Oxford University Press, New York.

Panayiotou, A., ed. (1980) *Ophiolites: Proceedings International Ophiolite Symposium Cyprus 1979*. Geological Survey Department, Nicosia.

Peters, T., Nicolas, A., and Coleman, R. G., eds. (1991) *Ophiolite Genesis and Evolution of the Oceanic Lithosphere*. Proceedings of the Ophiolite Conference, Muscat, Oman, January 7–18, 1990. Springer, Dordrecht.

Primack, J. R., and Abrams, N. E. (2006) *The View from the Center of the Universe: Discovering Our Extraordinary Place in the Cosmos*. Riverhead Books, New York.

Rassios, A., Moores, E. M., and Green H. W. II. (1983) Magmatic structure and stratigraphy of the Vourinos ophiolite cumulate zone, northren Greece. *Ofioliti*, 8, 377–410.

Robinson, P. T., and Malpas, J. (1990) The Troodos ophiolite of Cyprus: New perspectives on its origin and emplacement. In *Ophiolites, Oceanic Crustal Analogues* (eds. J. Malpas, E. M. Moores, A. Panayiotou, and C. Xenophontos), pp. 13–26. Proceedings of the Troodos 1987 Symposium. Nicosia, Cyprus. Geological Survey Department, Nicosia.

Sengor, A. M. C. (1977) New historical data on crustal subduction. *Journal of Geology*, 85, 631–34.

Sleep, N. H., and Windley, B. F. (1982) Archean plate tectonics: Constraints and inferences. *Journal of Geology*, 90, 363–80.

Steinmann, G. (1905) Geologische Beobachtungen in den Alpen, II. Die schart'sche Überfaltungstheorie und die geologischen Bedeutung der Tiefseeabsätze und der ophiolitische Massengesteine. *Bericht Naturforschung Gesellschaft Freiburg 1*, 16, 44–67.

Steinmann, G. (1927) Der ophiolitischen Zonen in der mediterranean Kettengebirgen. Proceedings, 14th International Geological Congress, Madrid, 2, 638–67.

Stille, H. (1939) Bemerkungen betreffend die "sardische" Faltung und den Ausdruck "ophiolitisch." *Zeitschrift der Deutschen Geologischen Gesellschaft*, 91, 771–73.

Stoffers, P., Worthington, T., Hekinian, R., Peterson, S., Hannington, M., Türkay, M., and the SO 157 Shipboard Scientific Party. (2001) Silicic volcanism and hydrothermal activity documented at Pacific-Antarctic Ridge. *Eos (Transactions, American Geophysical Union)*, 82, 38, 301–4.

Suess, E. (1909) *The Face of the Earth*, 4. Clarendon Press, Oxford.

Temple, P.G., and Zimmerman, J. (1969) Tectonic significance of Alpine ophiolites in Greece and Turkey. *Geological Society of America Abstracts with Programs*, 7, 221–22.

Twiss, R. J., and Moores, E. M. (2007) *Structural Geology*, 2nd ed. Freeman, New York.

Varga, R. J., and Moores, E. M. (1985) Spreading structure of the Troodos ophiolite, Cyprus. *Geology*, 13, 846–50.

Vine, F. J., and Hess, H. H. (1970) Sea-floor spreading. *The Sea*, 4, 2, 587–622.

Vine, F. J., and Matthews, D. H. (1963) Magnetic anomalies over oceanic ridges. *Nature*, 199, 947–49.

Wilson, J. T. (1965) A new class of faults and their bearing on continental drift. *Nature*, 203, 343–47.

Windley, B. F. (1995) *The Evolving Continents*, 3rd ed. Wiley, New York.

2
———

Microbes in Extreme Environments

Implications for Life on the Early Earth and Other Planets

Dawn Cardace, *University of Rhode Island*
Tori M. Hoehler, *NASA Ames Research Center*

Delineating the boundaries of the biosphere is an area of exciting research, integrating work in biology, geology, and chemistry. Of particular interest is how microorganisms adapt to survive under stress (e.g., extreme acidity or alkalinity, temperature fluctuations, changing activity of water) and persist as communities, perhaps over geologically meaningful time scales. Considering how terrestrial environments support and/or challenge microbial life in Earth's most extreme settings also fuels scientific investigation as we recognize habitable zones (where conditions appear appropriate for life as we know it) in our solar system and around stars other than our own Sun. In this chapter, we discuss how serpentinizing systems serve as habitat for extremophile microbes (those inhabiting the high pH, Ca^{2+}-rich waters circulating in serpentine bodies) and may provide novel ground for scientific investigation into extremophile evolution.

GEOLOGICAL PERSPECTIVE ON SERPENTINE MINERALS

The solid Earth consists of three layers—core, mantle, and crust—and the inherent instability of mantle rocks positioned at the Earth's surface brings about serpentinization. Recall that the core holds the densest Earth materials, with an extremely iron-rich solid inner core (likely an iron-nickel alloy, ~13 g/cm³) and liquid outer core (likely molten iron with ~10% light elements such as sulfur, oxygen, or carbon, ~10–12 g/cm³). The mantle (3.3–5.7 g/cm³) makes up most of the planet, with olivine and its high-pressure structural counterparts, γ-spinels and perovskites, covering the bulk of its mineralogy. The planetary crust holds less dense mineral assemblages (continental crust, 2.7–2.8 g/cm³, and oceanic crust, ~3 g/cm³), sorted by planetary differentiation, volcanic activity, continent building, and weathering processes over geologic time (all density data from Lodders and Fegley, 1998).

When rocks from the higher temperature and pressure planetary interior are brought to the surface by, for example, tectonic processes, they are no longer at equilibrium with their surroundings. The reaction of higher temperature and pressure ultramafic mantle rocks (i.e., rocks rich in magnesium and iron) with water near the planet's surface generates a suite of products featuring serpentine group minerals. Such ultramafic rocks are typically emplaced as very large slabs, forced upward as tectonic plates collide and converge, sometimes resting on top of another like shingles on a roof. An intact stack of ocean floor with mantle rock, igneous complexes, and sediments, once emplaced on land, is called an ophiolite. As these mantle-derived ultramafics alter, in oceanic or continental settings, the serpentinizing subsurface represents a large-volume reaction zone in the planetary interior: a region in which disequilibrium spurs geochemical and mineralogical transformations at habitable temperatures (defined here as ~≤121°C, as indicated by microbial growth experiments; Kashefi and Lovley, 2003). When microbial processes catalyze these transformations or take advantage of their products, energy can be harvested from the chemical system to fuel metabolic reactions and provide the means of survival for subsurface microbial communities.

GEOSPHERE–BIOSPHERE INTERACTIONS IN THE PLANETARY SUBSURFACE

Survey of Life's Distribution and Survival Strategies below Ground

Subsurface life may account for a daunting proportion of all the life on Earth. In particular, the absolute extent of the deep subsurface marine biosphere is unknown, but is estimated at a tenth to a third of Earth's biomass (Whitman et al., 1998; D'Hondt et al., 2002). Historical assumptions of a barren subseafloor have been challenged by studies showing diverse microbial communities thriving in this extreme setting (e.g., Lovley and Chapelle, 1995). Microbes residing in continental soils and pore spaces and fractures of rocks (i.e., endolithic life) have been underestimated for decades, and only presently are we beginning to understand their distribution and environmental interactions. Though the specifics of microbial metabolisms vary, all subsurface life must derive energy from the chemical environment (i.e., via chemosynthesis) rather than the Sun, as in photosynthesis. Subsurface life exists in conditions more reducing than those at Earth's surface, requiring organisms have the capacity to metabolize in the absence of oxygen and at challenging pH, pressures, and temperatures.

Lessons from Submarine Settings

Decades of collaborative scientific research by the Deep Sea Drilling Program and the Ocean Drilling Program (now the Integrated Ocean Drilling Program, IODP, oceandrilling.org) have characterized the submarine subsurface in terms of sediment and rock composition and properties. Deep sea sediments range from thin veneers of sediment on young oceanic crust to hundreds of meters of sediment near convergent plate boundaries at subduction zones. Where sediment-hosted microbes are buried and isolated from the ocean reservoir, they depend on interstitial fluids for nutrients and the maintenance of a habitable environment. Interstitial fluids themselves react with sediment grains in situ and mix with laterally advected fluids, changing over time, perhaps benefiting or stressing microorganisms.

Only recently—with the landmark ODP Leg 201, "Controls of Microbial Communities in Deeply Buried Sediments, Eastern Equatorial Pacific and Peru Margin, Sites 1225–1231," in 2005—have dedicated studies of microbial community structure, makeup, and environmental controls been possible. Jørgensen et al. (2006) report life to 420 m below seafloor, even in ~35 Ma sediment, finding that microbial population sizes respond to chemical zonation in the sediments and characteristics of the hydrosphere at the time of sediment deposition. In addition, Jørgensen et al. (2006) note higher than expected cell densities (up to 10^{10} cells/ml) occur in deeply buried sediments with predominantly uncultured microorganisms. The microbiological characterization of deeply buried sediments continues with increasingly sophisticated gene- and lipid-based surveys (e.g., Parkes et al., 2000; D'Hondt et al., 2004; Biddle et al., 2006, 2008; Cardace et al., 2006; Dang et al., 2009; Nunoura et al., 2009; Roussel et al., 2009).

The deep marine rocky subsurface holds its own challenges to endolithic life. As microbes deplete pore spaces of their nutrient load, there may be a figurative minimum in the habitability of serpentinizing systems. However, fracturing of the host rock, induced by tectonic activity or local volume increases due to mineral transformations, can allow passage of fresh fluids through the rock body, and can potentially rejuvenate the microbial community with new aqueous raw material and cells. Note that the transformation of olivines to serpentine mineral phases is accompanied by a volume increase of 25–45% (O'Hanley, 1992), thus as serpentinization proceeds, it creates new fracture networks, forming new microhabitats. In general, fluid–sediment–rock interfaces (e.g., fractures) show richer biomass with respect to unaltered bedrock in field studies (Mason et al., 2009; Rathsack et al.,

2009). Furthermore, biomass concentrations on rock surfaces may be several orders of magnitude greater than ambient natural waters, with pronounced alteration, perhaps driven by organic acids produced by microbes (Santelli et al., 2008, 2009). Extrapolating from these studies, serpentinizing rocks, with intricate fracture networks and ongoing water–rock reactions, may have biological concentrations along fractures and/or faults where serpentinization is focused.

Additionally, where temperatures allow life, where geochemical energy is available from water–rock reactions, and where there is available carbon, the habitable space of subseafloor sediments and rocks is constantly limited by indomitable mineral precipitation processes that fill pore spaces. Cements of calcite, silica, or infillings of salts and phyllosilicates (including diverse clay minerals and serpentine) are common in modern and ancient materials, at once testifying to the changing chemical environment experienced by microbes and also decreasing the local habitable volume within sediment and rock formations.

Windows into the deeper interior are also provided by mud volcanoes, oceanic spreading centers, and deep sea vents. Serpentinite mud volcanism at convergent plate boundaries is driven by the density loss of serpentine minerals with respect to their ultramafic parent minerals; the relative buoyancy of serpentine mineral phases causes them to rise through overlaying sediments or fracture zones to the seafloor, often carrying clasts of ultramafic rocks entrained from depth. These clasts frequently reflect not just parent rock physicochemical characteristics but also transport microbial life (Takai et al., 2005b) or mineral phases diagnostic of changing geochemical conditions to the seafloor (as in Alt and Shanks, 2006; Wheat et al., 2008), and are thus of particular importance.

At spreading centers, deeply incised faults in the cooling magma of the new ocean crust can provide glimpses of the fresh ultramafic rocks in situ; hot new crust is flushed with hydrothermal fluids, presumably too hot to be habitable. However, the cooling over time of hydrothermal circulation (in temperature and vigor) brings a new and more clement habitat for microbes. Where faulting associated with tectonic activity and/or fractures from the cooling crust occur, exposed interior samples may be targeted for stable isotope systematics indicative of biosphere activity (Alt et al., 2007) or geomicrobiological study.

Deep sea vents in the ocean floor reveal the changing chemistry (and by implication habitability) of basement waters. Hot, acidic conditions associated with oceanic crust near spreading centers host communities adapted to that environment. Wherever fluids escape to the ocean, hot vents develop biofilm-coated chimney structures composed largely of sulfur-rich minerals (e.g., Huber et al., 2003; Edwards et al., 2005; Nakagawa and Takai, 2008; Takai et al., 2009; Schrenk et al., 2010) with instantaneous precipitation of minerals resulting from the mixing of vent fluids with seawater. Fine grains of minerals entrained in flowing streams of vent fluids invite the nicknames "black" or "white smokers," depending on the mineralogical makeup of the grains—typically black sulfides or white silica/oxide/sulfate/clay assemblages.

Lower temperature, alkaline conditions, such as those formed by submarine serpentinization, offer different challenges to resident microbes. The Lost City Hydrothermal Field, discussed shortly, is a site near the Mid-Atlantic Ridge where a large block of mantle rock bounded by faults has seawater reacting within it, generating a distinct serpentine mineralogy and attendant aqueous characteristics (e.g., high pH, high dissolved hydrogen and methane; Kelley et al., 2005, 2008). Cold seeps at convergent margin aprons likely represent a similar setting in terms of chemical ecology, in that fluids often tap into serpentinizing regions below the seafloor and are alkaline and methane-rich. Seeps may be small in spatial scope and somewhat isolated in time and geography, fed by deeply sourced fluids interacting with basement rocks, shallow fluids altered by sediment diagenesis, or ocean water inputs, and they may have minor to very pronounced microbial mats, faunal communities, and/or mineralization near seep points at the seafloor (e.g., Lewis and Marshall, 1996; Valentine et al., 2005).

Lessons from Subaerial Settings

To date, life has been found in the continental subsurface wherever it is sought at habitable temperatures. Hot springs plumb habitats that represent some of the most acidic and high-temperature conditions known to occur naturally. Fault exposures (Lindsay et al., 2005) and rift valley settings (e.g., Osanjo et al., 2009) can provide fortuitous windows to the deep biosphere, as can outcrops of uncommon geologic formations including banded iron formations (e.g., Brown et al., 1995) and black shales (e.g., Piper and Calvert, 2009). Caves showcase life in the cool, humid subsurface (e.g., Hose et al., 2000), whereas continental drilling projects (Takeuchi et al., 2009) and deep mines (Wanger et al., 2006) commonly explore extremely high-temperature environments. As for serpentinization, points of access to weathering oceanic crust emplaced on land, as encountered in ophiolites, reveal the impact of serpentinization beyond the modern ocean. The diversity of the continental subsurface biosphere is reflected in observations in these types of settings.

Geomicrobiological studies of actively serpentinizing continental sites are few, limited to case studies discussed later. More common are biogeochemical investigations of serpentinizing groundwaters and gas inventories for the deep Earth. Barnes et al. (1967, 1972, 1978) presented field data for "ultrabasic" Ca^{2+}-OH^- type waters from California, New Caledonia, the former Yugoslavia, and Oman, characterized by calcium levels up to 120 ppm, low temperatures of ~10–~31°C, and pH values in the range of 9.2–12.07, concluding that they originated in actively serpentinizing subsurface environments. This body of work has laid a strong foundation for ongoing research in ophiolite-sourced waters, although gas data are not reported. However, deeply sourced fluids from Precambrian Shield rocks of Canada and Finland and the Witwatersrand Basin in South Africa all have dissolved hydrogen in the mM (millimolar) range, and free gases from fractures in continental shields have abundant abiogenic methane and 9–58% hydrogen by volume (Lollar et al., 2007). This hydrogen reservoir is a result of the subsurface serpentinization of ultramafic rocks and/or hydrolysis and other water–rock reactions, with fault- and fracture-controlled transfer to the point of sampling. Future studies will continue to broaden scientific understanding of the limits to life in serpentinite ecosystems in terms of chemistry (of rocks, aqueous solutions, and gas), temperature, and the utility of hydrogen for microbial metabolisms.

Surface Serpentine = Indicator of Subsurface Ultramafic Rock Weathering

The production of serpentine minerals requires the transformation of constituent minerals of parent ultramafic rocks, olivine or pyroxene. See Moody (1976) for a thorough review. The iron in olivine (a solid solution of minerals ranging from pure fayalite to pure forsterite) reacts as follows:

$$Fe_2SiO_{4(s)} + 5Mg_2SiO_{4(s)} + 9H_2O_{(l)} \rightarrow 3Mg_3Si_2O_5(OH)_{4(s)}$$
$$+ Mg(OH)_{2(s)} + 2Fe(OH)_{2(s)}. \qquad (2.1)$$
Fayalite + forsterite + water → serpentine
+ brucite + iron hydroxide.

Hydrogen, a powerful reducing agent, is generated when Fe^{2+} in $Fe(OH)_2$ from (2.1) is oxidized to magnetite, coupled to the reduction of water, shown in (2.2):

$$3Fe(OH)_{2(s)} \rightarrow Fe_3O_{4(s)} + 2H_2O_{(l)} + H_{2(g)}. \qquad (2.2)$$
Iron hydroxide → magnetite + water + hydrogen.

More generally, the iron transformation can be expressed as:

$$2(FeO)_{rock} + H_2O_{(l)} \rightarrow (Fe_3O_4)_{rock} + H_{2(g)}. \qquad (2.3)$$

The production of magnetite results in a serpentine-dominated assemblage that has very small magnetite grains, often aligned along the preexisting olivine mineral grain boundary, and can be strongly magnetic. Overall, serpentinization yields a body of rock that inherits chemical characteristics of its ultramafic parent and reflects the chemistry of throughgoing groundwaters.

CHEMISTRY OF SERPENTINIZING FLUIDS

General Chemical Characteristics of Serpentinization

Serpentinization produces reducing solutions that are enriched in Ca^{2+} and OH^- and are postulated also to be carbon-poor. Elevated Ca^{2+} results from crystal-chemical constraints: the mineral products of serpentinization do not accommodate calcium well in their structures—the calcium literally has nowhere to go and remains in the aqueous phase until precipitation as carbonate or sulfate minerals, for example, may be spurred by aqueous chemistry. OH^- abundances result from the ample hydroxide phases produced alongside serpentine minerals and serve to drive up pH dramatically; recall that the pH of household ammonia at room temperature is ~11, similar to serpentinizing groundwaters. Contradictory data for metal contents (especially nickel, chromium, and cobalt) in groundwaters leave open the possibility that elevated concentrations could prove a biogeochemical challenge. Analytical and experimental data show there is an inventory of dissolved hydrogen, methane, and other hydrocarbons/organic compounds produced abiotically in the deep Earth (Lollar et al., 2007; Fiebig et al., 2009; McCollom and Bach, 2009), and growing interest in complex organics will certainly bring better parameterization of these compounds in the future.

Biogeochemically Necessary Components

Fundamentally, all life requires carbon and energy to build biomass, with requisite levels of other elements necessary for growth. Carbon is largely available at the Earth's surface as CO_2 in the atmosphere and carbonic acid, bicarbonate, and diverse organic compounds in aqueous solutions. In the ultramafic subsurface, where water is largely inherited from the rock's history at the seabed, there may be little communication with surface Earth fluids (such as precipitation), and thus carbon inputs from surface vegetation are few if any. There is generally little carbon associated with the rock itself (0.01 wt% C in ultramafic rocks; Lodders and Fegley, 1998), barring significant carbonate mineralization. To date, dissolved carbon loads have been observed to be minimal: Barnes et al. (1967, 1972, 1978) documented CO_3^{2-} typically near zero, with a few outstanding groundwaters reaching 5450 ppm (likely due to interactions with adjacent sedimentary rocks); Tiago et al. (2004) report 6.6 ppm CO_3^{2-} in Portuguese high pH, "ophiolite-type" waters; and Cardace and Hoehler (in preparation) find serpentinizing waters from California, New Zealand, and Newfoundland to be typically near or below detection for CO_3^{2-}. Attention is turning to more complex carbon compounds as candidate carbon sources, particularly formate ions (McCollom and Seewald, 2001), and it may be that available carbon resides in organic acids and other molecules in this subsurface environment. Whatever the carbon source, energy for metabolic activity must be derived from disequilibrium, enabling electron shuttling between atoms of different oxidation states, as will be discussed.

Redox Couples that Can Drive Metabolisms: The Elements S, P, O, N, C, and H

The utility of these elements to organisms resides in their abundance, electron affinity, and diversity of possible valence states. Consider sulfur: natural occurrences include S^{2-} (sulfide, -2 oxidation state, in the rotten egg smell of so many hot spring and volcanic emissions), elemental S (o oxidation state), SO_2 (sulfur dioxide, $+4$ oxidation state), and SO_4^{2-} (sulfate, $+6$ oxidation state, the predominant form in oxygen-rich surface waters and mineral precipitates like gypsum). There is great potential for oxidation-reduction activity targeting sulfur atoms with such a wide range of oxidation states in nature. Similarly, nitrogen ranges from a -3 oxidation state in NH_3 (ammonia) to a $+5$ oxidation state in NO_3^- (nitrate), carbon from a -4 oxidation state in CH_4 (methane) to $+4$ in CO_2 (carbon dioxide), and phosphorus in the oxidation states of $+5$, $+3$, and -3 in diverse natural compounds. Furthermore, phosphorus is part of the biologically critical molecules DNA, RNA, and ATP, as well as necessary to the construction of membrane phospholipids (Westheimer, 1987) and is particularly scarce in deep biosphere habitats, likely due to its predominance in the fully oxidized state throughout the rock cycle (Pasek, 2008). Oxygen and hydrogen vary from the o valence state at diatomic compounds to -2 and $+1$, respectively, and make up water, the universal solvent and nearly omnipresent reactant for biogeochemical reactions.

ENERGY SOURCES FOR MICROBIAL METABOLISMS

How Is Chemical (Not Light) Energy Furnished?

In the absence of sunlight and associated energy inputs, photosynthetic life on the surface of the Earth would stall. In the subsurface, organisms take advantage of available energy sources, independent of sunlight. Oxidation-reduction reactions are the key to all chemosynthetic energy mining. Where there is redox disequilibrium, there is potential for life to harvest energy and carry on with carbon fixing and the work of cellular repair, maintenance, and growth; modeling of water–rock reactions in ultramafic rocks has shown that such disequilibrium is feasible and ample chemical energy production quite likely (McCollom, 2007).

Likely Metabolisms Spurred by Serpentinization

Based on observed geochemical data for the serpentinizing subsurface, we predict that hydrogen oxidation coupled to methanogenesis, sulfate reduction, iron reduction (and possibly that of other metals), and nitrate reduction are feasible. We have calculated the Gibbs energy (i.e., the energy available to do useful chemical work) associated with each proposed metabolic reaction given the observed activities of necessary reactants in this setting. Where the Gibbs energy $<$o, there is energy to be harvested, and over the range of hydrogen abundances expected in nature, these metabolic reactions are reasonable (Cardace and Hoehler, 2009). It remains to be seen whether organisms that can carry out these reactions inhabit the serpentinizing subsurface and do this metabolic work. Bear in mind that serpentine-associated waters have also been shown to be rich in nickel, chromium, and other metals, which may reach levels detrimental to microorganisms. The metal tolerance of subsurface extremophiles is far from fully understood and may well impact the full exploitation of the metabolic landscape as we envision it at present.

METABOLIC DIVERSITY, EXPECTED AND OBSERVED

Preview of the Microbial Metabolic Landscape

Where serpentinization is ongoing, the abiotic production of hydrogen can fuel hydrogen oxidation–based metabolisms. Additionally, the abiotic production of methane and other hydrocarbons has been documented for ultramafic environments (such as Precambrian Shield rocks, as in Lollar et al., 2007) and is also expected to occur in submarine basement rocks. Thus, hydrogen oxidation coupled to methane production is an exciting putative strategy for energy and carbon cycling in serpentinites. Any available carbon sources, such as organic acids (e.g., acetate, formate, propionate) could also anchor carbon cycling schema in this environment.

Case Study: Lost City Hydrothermal Field, Near the Mid-Atlantic Ridge

A recent discovery in interdisciplinary oceanography is the Lost City Hydrothermal Field, about 15 km from the Mid-Atlantic Ridge (Kelley et al., 2005). At this site, faulting in ultramafic basement rocks (uppermost mantle peridotite) has exposed serpentinite, or rocks composed dominantly of serpentine minerals, to the ocean reservoir. There is active hydrothermal circulation pumping ocean water through the serpentinite body, reacting along its path with presumably less altered or unaltered parent rock. As they escape the serpentinite block, very alkaline (pH ~11) hydrothermal fluids with high calcium, dissolved hydrogen, and dissolved methane levels and intermediate temperatures (~40–90°C) react with ocean water, producing elaborate chimneys and pinnacles of carbonate minerals up to 60 m high. The mixing of such fluids creates chemical disequilibrium and opens a range of redox possibilities for life to exploit. At Lost City, bacteria capable of oxidizing sulfur and methane and reducing sulfate coexist with Archaeal methanogens and methane-oxidizers in vent fluids and chimney biofilms (Brazelton et al., 2006).

Case Study: Submarine Serpentinite Mud Volcano, Marianas Forearc

Leg 195 of the Ocean Drilling Program drilled Site 1200 into an active serpentine mud volcano, the South Chamarro Seamount, associated with the forearc of the Marianas convergent margin. Mud volcanism is driven by the buoyancy of less dense serpentine phases formed by the reaction of fault-controlled fluids with subseafloor ultramafics in shallow subduction zones. Shipboard scientists discovered an active serpentinization-associated microbial community at Site 1200 up to 30 m below the seafloor with interstitial waters up to pH 12.5 (Fryer and Salisbury, 2006). Here, Mottl et al. (2003) reported evidence for methane-oxidizing, sulfate-reducing, Archaea-dominated subsurface microbiological community. Samples from this drilling expedition also enabled characterization of a novel sediment-hosted microbe, *Marinobacter alkaliphilus*, with optimum growth temperature of 30–35°C, optimum pH between 8.5 and 9.0, and optimum NaCl concentration of 2.5–3.5 wt% by volume (Takai et al., 2005a).

Case Study: Coast Range Ophiolite, Lake County, California

Ongoing field studies at the McLaughlin Reserve (Lower Lake, California) indicate that a large block of the mantle unit of the Coast Range Ophiolite is actively serpentinizing. Groundwaters, accessed via heritage wells operated by the Homestake Mine, historically have alkaline pH readings from 9 to 10.5 and elevated calcium loads. Regional rock cores archived at the site show a diversity of serpentine-associated minerals, clays, and carbonate vein fillings and also have intervals of relatively unaltered ultramafics (Cardace and Hoehler, in preparation). Cardace and Hoehler (2009) document dissolved hydrogen and methane approaching experimental predictions for these groundwaters (see also Sleep et al., 2004), and find intact microorganisms on slides incubated in groundwater wells as well as on filters used to process many liters of groundwater. Functional genes detected in DNA extracted from these samples include those in Eubacteria (Eub338 and Eub518) and those necessary for methanogenesis (ME1 and ME2), sulfate reduction (dsrp2060 and dsr4r), and iron reduction (GEO494f/GEO825r; Cardace and Hoehler, in preparation).

Case Study: Cabeço de Vide, an Ophiolite-Type
Ultramafic Setting, South Portugal

In the Cabeço de Vide, a 130-m-deep artesian well drilled into ophiolite-type rocks tapped a groundwater system with a pH of 11.4 (Tiago et al., 2004), apparently at the contact of ultramafic and mafic rocks with dolomites and limestones. Waters were drawn out with hydraulic pump, transported to the laboratory at 4°C, and assayed under aerobic conditions for microbial abundance, diversity, and pH tolerance. Microorganisms similar to, for example, *Dietzia natrolimnae, Frigoribacterium/Clavibacter* lineages, and *Microbacterium kitamiense* were documented, and experiments with a variety of growth media produced colonies of diverse color, morphology, Gram stain result, G+C content, and pH tolerance. In fact, only two strains were able to grow at pH 11, suggesting that the in situ groundwater pH itself presented a stress factor to the microbial communities. Ongoing work will better characterize the geochemistry of the site (I. Tiago, personal communication, 2009) and isolate novel microorganisms and further characterize their growth requirements and ecological interactions (as in Tiago et al., 2006).

Case Study: The Cedars Ultramafics,
Sonoma County, California

Dewatering ultramafic rock in Sonoma County, California, at The Cedars peridotite locality generates seasonal travertine deposits as deeply sourced fluids react
with the atmosphere at seepage points (K. Nealson, personal communication,
2007). At this site, workers report dilute solution chemistry with pH values near
11.5, with dominant cations Ca^{2+} and Na^+ at $\leq$ 2 mM, and find cultures of seep
microbes possible at pH near 10, with the lowest species diversity in the most dilute, highest pH samples (Johnson et al., 2004). Additionally, rapid transformation
of minerals artificially immersed in the seeps are noteworthy, likely due to redox
chemistry mediated by microbes, and culturing studies of seep organisms continue
(O. Johnson, unpublished data, 2008). Gas chemistry and isotope systematics may
indicate ongoing microbial cycling of methane in this setting (Morrill et al., 2008).

EVOLUTIONARY ASPECTS OF EXTREMOPHILE
COMMUNITY DIVERSITY AND SUCCESS

Impacts of the Serpentinizing Subsurface on
Microbial Communities

Extremely high pH groundwaters with limited carbon and redox-available chemical constituents equates to a harsh setting for deep subsurface life. Successful microbial communities must cope with fluctuating salinity, chemistry, and temperature, possibly complicated by episodic serpentinization. There may occur fracturing
of bedrock $\rightarrow$ access of fluids to fracture zones $\rightarrow$ serpentinization reactions $\rightarrow$
infilling/healing of fractures by product minerals $\rightarrow$ new fracturing of bedrock
(promoted by volume expansion of serpentines/local tectonics/changes in hydro-
logic regimen), and so forth. The serpentinization process involves exothermic
metamorphic phase transformations, resulting in localized temperature increases
(perhaps to 260°C in some subsurface settings; Allen and Seyfried, 2004); temperature oscillations related to the evolving mineralogy must fundamentally constrain
the habitability of this setting. The production of serpentines and attendant hydrogen production thus may literally proceed in stages and also slow considerably
when the parent bedrock is very thoroughly serpentinized. Adaptations to this
changing environment must include cell membrane fluidity (Yumoto et al., 2000).
Regulation of proton or cation pumps that enable nutrients' access to the cytoplasm
and maintain internal cell pH is also necessary. Cell surface adhesion strategies
may play an additional role in microscale habitability, because cell "stickiness" may
respond to cation abundances in aqueous solutions (Dass et al., 2009), and serpentinizing groundwaters have variable calcium, magnesium, and sodium loads.

Gene transfer between microbial populations could well bring novel function
to the subsurface environmental genome. A population exposed to changing pH

and variable levels of carbon-bearing compounds and redox couples may benefit from genetic material that brings new or changing metabolic capacity. For context, an analysis of the environmental genomics associated with a salt-loving microorganism has found abundant genes sequenced from a saltern environment that are not present in the genome of the dominant isolate (Legault et al., 2006). Indeed, Brazelton and Baross (2009) report elevated abundances and diversity of genes in "extragenomic molecules" from Lost City Hydrothermal Field samples, perhaps a snapshot of vigorous lateral gene transfer in a serpentinizing system.

Isolation of populations exists on various scales in serpentinizing systems. On a cell-to-cell scale, pore spaces in rocks infill and are lost over time with indefatigable cementation processes. On a crystal-to-crystal scale (grain diameters may vary from some microns to 1 cm), mineral substrates transform with aqueous alteration, creating new spatial distributions of constituent atoms with different surface chemistries. Within a single block of ultramafic rock, there may be more olivine-rich zones, more iron-rich zones, fracture localization, and serpentinized regions. On a regional landscape scale, imbricate slabs of ultramafic rock may be closely transposed or several kilometers apart. Around the globe, ultramafics associated with uplifted oceanic crust at convergent plate boundaries exist in a discontinuous belt of rock at continental edges and suture zones. This brings us to the question of micro- and tectonic plate–scale biogeography for life dependent on serpentinizing systems.

THOUGHT EXPERIMENT: TECTONICS AND EXTREMOPHILE BIOGEOGRAPHY

The biogeography of deep sea vent biota has come under discussion as having a strong tectonic control (Tunnicliffe et al., 1998; Van Dover, 2002). Given that seafloor spreading at ridges and back-arc spreading build all new oceanic crust and that a feasible migration path is *along* a ridge, the notion that vent biota biogeography depends on this aspect of plate tectonics is supported. Because lithospheric plate motions carry new oceanic crust in definite lateral directions, it is of interest here to consider the fate of endolithic communities carried *away* from the ridge. Does transporting the crust (and any vent-sourced communities) as necessitated by plate tectonics deliver some remnant of the original community to the convergent plate boundary across the ocean basin? If so, does tectonics also exert control on the biogeography of subsurface extremophiles in serpentinizing systems in ophiolites and/or other ultramafic rock units emplaced on land? Certainly strong differences exist between modern submarine serpentinites and ophiolite-associated serpentinites: aqueous characteristics like sulfate, salinity, and total dissolved inorganic carbon must vary, as will thermal gradients, depending largely on the age of oceanic crust, which cools over time. Compilation of mature subsurface biogeochemical and microbiological studies determine whether biogeographic relatedness exists in the subsurface of continental serpentinizing terrains.

SUMMARY AND IMPLICATIONS FOR ASTROBIOLOGY

Chemolithoautotrophy and Ultramafic-Hosted Extremophiles

Serpentinizing systems consist of ultramafic rocks, aqueous phases, ambient gases, and deep biosphere microbes. In submarine settings, such as the Lost City Hydrothermal Field near the Mid-Atlantic Ridge, active serpentinization of a fault-bounded block of peridotite is evident in the geochemistry and mineralization at fluid escape points. The disequilibrium mixing of serpentinizing fluids with seawater creates a strong thermodynamic drive for certain chemolithotrophic metabolic reactions, including those involved in methane cycling, which constitute an opportunity for extremophiles. On land, parallel investigations in the geomicrobiology of ultramafics in the Coast Range (California), Bay of Islands (Newfoundland, Canada), and Dun Mountain (South Island, New Zealand) ophiolites, The Cedars peridotite (California), and Cabeço de Vide (Portugal) have begun to characterize the continental serpentinizing subsurface. Primary challenges to such communities are the low carbon inventory, limited redox couples documented in the groundwater chemistry, and possibly the episodic changes in serpentinizing system mineralogy and structure (pore scale to landscape scale) over time.

Serpentine-Associated Settings Serving as Model Systems
for Extremophile Evolution

Serpentine settings have great potential as model systems for research in evolutionary biology in general and have specific utility in studying microbial evolution. Natural variability in groundwater chemistry (including temperature, pH, gas contents, cation dominance, and metal loads) provides ready material for study in terms of how alkaline-loving microbes adapt to environmental change, with relevance across interdisciplinary lines (e.g., environmental microbiology and groundwater remediation). The primary limitation of the study of evolution in the serpentinizing subsurface is presently the lack of access: very few areas on the planet provide access even to the surface expressions of the subsurface serpentinizing reaction zone. One continental drill site is currently under consideration in northern California, otherwise, access is limited to a few boreholes in Portugal and California, all drilled for industrial purposes and highly impacted. Until access to active sites of continental serpentinization is possible, with potential for time-series monitoring of system chemistry and microbial ecology, laboratory studies will have to suffice by tackling issues facing microbial interactions with ultramafic minerals and serpentine phases.

Serpentinizing Systems as Habitat: On Earth and Beyond

Understanding biosphere–geosphere interactions in serpentinizing systems on Earth is an area of exciting interdisciplinary research with enticing astrobiological implications. As pioneering studies document the geochemistry, mineralogy, and geomicrobiology of this setting and consider the importance of the serpentine subsurface to microbial ecology and evolution, the as-yet-unconstrained volume and productivity of this component of the deep biosphere can be estimated. Additionally, ultramafic terrains and alteration assemblages on other planetary surfaces continue to be described (e.g., olivine mineral, serpentine, carbonate associations with attendant plumes of methane gas, as in the Nili Fossae region on Mars; Ehlmann et al., 2009; Mumma et al., 2009) and may indicate a habitable subsurface in the past and/or present. Critical questions about how serpentinizing systems uniquely support the deep biosphere include the following.

- What microbes are present in the serpentinizing subsurface, what metabolic machinery do they possess, and what metabolisms are they carrying out?
- Is hydrogen oxidation in fact driving metabolic reactions, and if so, is that hydrogen derived from water–rock reactions such as serpentinization or the hydrolysis of water (as suggested for deeply buried sediments on Earth, as in Blair et al., 2007), or in what proportions are both processes represented?
- What budgets of hydrocarbons are available to deep biosphere microbes profiting from serpentinization reactions, and how stable are they over time?
- How do the extreme conditions (in terms of chemistry and carbon limitation) experienced by the deep biosphere in serpentinizing ultramafic rocks impact microbial adaptations in the subsurface?
- How are modern microbial systems encountered in the serpentine subsurface directly analogous to those hypothesized in ancient rocks? How are they analogous to those hypothesized in other planetary bodies with weathering ultramafics?
- What impacts does a serpentine-associated deep biosphere have on planetary scale cycles of hydrogen, carbon, oxygen, nitrogen, and perhaps sulfur and some transition metals?

As the microbial metabolic opportunities afforded by serpentinization are better understood in terrestrial subsurface settings, Earth system cycles can be tuned to accommodate this facet of the deep biosphere. Additionally, other solid-bodied planets can be evaluated with serpentinization in mind: where there are ultramafic rocks, water, and permeability, serpentinization can proceed, to some depth, signaled by the prevalence of serpentine phases at the planetary surface. As such, surface serpentine minerals flag a subsurface reaction zone that probably hosts an active deep biosphere on Earth (an area of current research) and could also do so on other planetary bodies. New findings of surficial geology from, for example, NASA's

Mars Science Laboratory (to be deployed in 2011 on the Martian surface) will bring new relevance to these landmark studies of the serpentine subsurface on Earth.

Acknowledgments

Aspects of this work were supported by the NASA Astrobiology Institute and Exobiology Program, and D. Cardace's fellowship with the NASA Postdoctoral Program, through Oak Ridge Associated Universities.

LITERATURE CITED

Allen, D. E., and Seyfried, W. E. (2004) Serpentinization and heat generation: Constraints from Lost City and rainbow hydrothermal systems. *Geochimica et Cosmochimica Acta,* 68, 1347–54.

Alt, J. C., and Shanks, W. C. (2006) Stable isotope compositions of serpentinite seamounts in the Mariana forearc: Serpentinization processes, fluid sources and sulfur metasomatism. *Earth and Planetary Science Letters,* 242, 272–85.

Alt, J. C., Shanks, W. C., Bach, W., Paulick, H., Garrido, C. J., and Beaudoin, G. (2007) Hydrothermal alteration and microbial sulfate reduction in peridotite and gabbro exposed by detachment faulting at the Mid-Atlantic Ridge, 15 degrees 20' N (ODP Leg 209): A sulfur and oxygen isotope study. *Geochemistry Geophysics Geosystems,* 8, Q08002.

Barnes, I., Lamarche, V. C., and Himmelberg, G. (1967) Geochemical evidence of present-day serpentinization. *Science,* 156, 830–32.

Barnes, I., O'Neil, J. R., and Trescases, J. J. (1978) Present day serpentinization in New Caledonia, Oman and Yugoslavia. *Geochimica et Cosmochimica Acta,* 42, 144–45.

Barnes, I., Sheppard, R. A., Gude, A. J., Rapp, J. B., and O'Neil, J. R. (1972) Metamorphic assemblages and direction of flow of metamorphic fluids in 4 instances of serpentinization. *Contributions to Mineralogy and Petrology,* 35, 263–76.

Biddle, J. F., Fitz-Gibbon, S., Schuster, S. C., Brenchley, J. E., and House, C. H. (2008) Metagenomic signatures of the Peru Margin subseafloor biosphere show a genetically distinct environment. *Proceedings of the National Academy of Sciences of the United States of America,* 105, 10583–88.

Biddle, J. F., Lipp, J. S., Lever, M. A., Lloyd, K. G., Sorensen, K. B., Anderson, R., Fredricks, H. F., Elvert, M., Kelly, T. J., Schrag, D. P., Sogin, M. L., Brenchley, J. E., Teske, A., House, C. H., and Hinrichs, K. U. (2006) Heterotrophic Archaea dominate sedimentary subsurface ecosystems off Peru. *Proceedings of the National Academy of Sciences of the United States of America,* 103, 3846–51.

Blair, C. C., D'Hondt, S., Spivack, A. J., and Kingsley, R. H. (2007) Radiolytic hydrogen and microbial respiration in subsurface sediments. *Astrobiology,* 7, 951–70.

Brazelton, W. J., and Baross, J. A. (2009) Abundant transposases encoded by the metagenome of a hydrothermal chimney biofilm. *ISME Journal,* 3, 1420–24.

Brazelton, W. J., Schrenk, M. O., Kelley, D. S., and Baross, J. A. (2006) Methane- and sulfur-metabolizing microbial communities dominate the Lost City hydrothermal field ecosystem. *Applied and Environmental Microbiology,* 72, 6257–70.

Brown, D. A., Gross, G. A., and Sawicki, J. A. (1995) A review of the microbial geochemistry of banded iron-formations. *Canadian Mineralogist*, 33, 1321–33.

Cardace, D., and Hoehler, T. M. (2009) Serpentinizing fluids craft microbial habitat. *Northeastern Naturalist* 16(sp5), 272–84.

Cardace, D., Morris, J. D., Peacock, A. D., and White, D. C. (2006) Habitability of subseafloor sediments at the Costa Rica convergent margin. *Proceedings of the Ocean Drilling Program, Scientific Results Volume 205* (eds. J. D. Morris, H. W. Villinger, and A. Klaus), pp. 1–26. Texas A&M University, College Station.

D'Hondt, S., Jørgensen, B. B., Miller, D. J., Batzke, A., Blake, R., Cragg, B. A., Cypionka, H., Dickens, G. R., Ferdelman, T., Hinrichs, K. U., Holm, N. G., Mitterer, R., Spivack, A., Wang, G. Z., Bekins, B., Engelen, B., Ford, K., Gettemy, G., Rutherford, S.D., Sass, H., Skilbeck, C. G., Aiello, I. W., Guerin, G., House, C. H., Inagaki, F., Meister, P., Naehr, T., Niitsuma, S., Parkes, R. J., Schippers, A., Smith, D. C., Teske, A., Wiegel, J., Padilla, C. N., and Acosta, J. L. S. (2004) Distributions of microbial activities in deep subseafloor sediments. *Science*, 306, 2216–21.

D'Hondt, S., Rutherford, S., and Spivack, A. J. (2002) Metabolic activity of subsurface life in deep-sea sediments. *Science*, 295, 2067–70.

Dang, H. Y., Li, J., Chen, M. N., Li, T. G., Zeng, Z. G., and Yin, X. B. (2009) Fine-scale vertical distribution of bacteria in the East Pacific deep-sea sediments determined via 16S rRNA gene T-RFLP and clone library analyses. *World Journal of Microbiology and Biotechnology*, 25, 179–88.

Dass, C. L., Walsh, M. F., Seo, S., Shiratsuchi, H., Craig, D. H., and Basson, M. D. (2009) Irrigant divalent cation concentrations influence bacterial adhesion. *Journal of Surgical Research*, 156, 57–63.

Ehlmann, B. L., Mustard, J. F., Swayze, G. A., Clark, R. N., Bishop, J. L., Poulet, F., DesMarais, D. J., Roach, L. H., Milliken, R. E., Wray, J. J., Barnoin-Jha, O., and Murchie, S. L. (2009) Identification of hydrated silicate minerals on Mars using MRO-CRISM: Geologic context near Nili Fossae and implications for aqueous alteration. *Journal of Geophysical Research*, 114, E00D08, doi:10.1029/2009JE003339.

Fiebig, J., Woodland, A. B., D'Alessandro, W., and Puttmann, W. (2009) Excess methane in continental hydrothermal emissions is abiogenic. *Geology*, 37, 495–98.

Edwards, K. J., Bach, W., and McCollom, T. M. (2005) Geomicrobiology in oceanography: Microbe-mineral interactions at and below the seafloor. *Trends in Microbiology*, 13, 449–56.

Fryer, P. B., and Salisbury, M. H. (2006) Leg 195 synthesis: Site 1200—Serpentinite seamounts of the Izu-Bonin/Mariana convergent plate margin (ODP Leg 125 and 195 drilling results). *Proceedings of the Ocean Drilling Program, Scientific Results Volume 195* (eds. M. Shinohara, M.H. Salisbury, and C. Richter), pp. 1–30. Ocean Drilling Program, Texas A&M University, College Station.

Hose, L. D., Palmer, A. N., Palmer, M. V., Northup, D. E., Boston, P. J, and DuChene, H. R. (2000) Microbiology and geochemistry in a hydrogen-sulphide-rich karst environment. *Chemical Geology*, 69, 399–423.

Huber, J. A., Butterfield, D. A., and Baross, J. A. (2003) Bacterial diversity in a subseafloor habitat following a deep-sea volcanic eruption. *FEMS Microbiology Ecology*, 43, 393–409.

Johnson, O. J., Rye, R., Namsaraev, Z. B., Han, S. , Lanoil, B. D., and Nealson, K. H. (2004) Microbial ecology of ultrabasic springs on the actively serpentinizing cedars peridotite, Sonoma County, California. *Astrobiology,* 4(2), 318–19.

Jørgensen, B. B., D'Hondt, S. L., and Miller, D. J. (2006) Leg 201 synthesis: Controls on microbial communities in deeply buried sediments. *Proceedings of the Ocean Drilling Program, Scientific Results Volume 201* (eds. B. B. Jørgensen, S. D. D'Hondt, and D. J. Miller), pp. 1–45. Texas A&M University, College Station.

Kashefi, K., and Lovley, D. R. (2003) Extending the upper temperature limit for life. *Science,* 301, 934–34.

Kelley, D. S., Brazelton, W. J., and Baross, J. A. (2008) Lost City: Serpentinization and life. *Geochimica et Cosmochimica Acta,* 72, A460.

Kelley, D. S., Karson, J. A., Fruh-Green, G. L., Yoerger, D. R., Shank, T. M., Butterfield, D. A., Hayes, J. M., Schrenk, M. O., Olson, E. J., Proskurowski, G., Jakuba, M., Bradley, A., Larson, B., Ludwig, K., Glickson, D., Buckman, K., Bradley, A. S., Brazelton, W. J., Roe, K., Elend, M. J., Delacour, A., Bernasconi, S. M., Lilley, M. D., Baross, J. A., Summons, R. T., and Sylva, S. P. (2005) A serpentinite-hosted ecosystem: The Lost City hydrothermal field. *Science,* 307, 1428–34.

Legault, B. A., Lopez-Lopez, A., Alba-Casado, J. C., Doolittle, W. F., Bolhuis, H., Rodriguez-Valera, F., and Papke, R. T. (2006) Environmental genomics of "Haloquadratum walsbyi" in a saltern crystallizer indicates a large pool of accessory genes in an otherwise coherent species. *BMC Genomics,* 7, 171–83.

Lewis, K. B., and Marshall, B. A. (1996) Seep faunas and other indicators of methane-rich dewatering on New Zealand convergent margins. *New Zealand Journal of Geology and Geophysics,* 39, 181–200.

Lindsay, J. F., Brasier, M. D., McLoughlin, N., Green, O. R., Fogel, M., Steele, A., and Mertzman, S. A. (2005) The problem of deep carbon—an Archean paradox. *Precambrian Research,* 143, 1–22.

Lodders, K., and Fegley, B. (1998) *The Planetary Scientist's Companion.* Oxford University Press, New York.

Lollar, B. S., Voglesonger, K., Lin, L. H., Lacrampe-Couloume, G., Telling, J., Abrajano, T. A., Onstott, T. C., and Pratt, L. M. (2007) Hydrogeologic controls on episodic H_2 release from Precambrian fractured rocks—energy for deep subsurface life on Earth and Mars. *Astrobiology,* 7, 971–86.

Lovley, D. R., and Chapelle, F. H. (1995) Deep subsurface microbial processes. *Reviews of Geophysics,* 33, 365–81.

Mason, O. U., Di Meo-Savoie, C. A., Van Nostrand, J. D., Zhou, J. Z., Fisk, M. R., and Giovannoni, S. J. (2009) Prokaryotic diversity, distribution, and insights into their role in biogeochemical cycling in marine basalts. *ISME Journal,* 3, 231–42.

McCollom, T. M. (2007) Geochemical constraints on sources of metabolic energy for chemolithoautotrophy in ultramafic-hosted deep-sea hydrothermal systems. *Astrobiology,* 7, 933–50.

McCollom, T. M., and Bach, W. G. (2009) Thermodynamic constraints on hydrogen generation during serpentinization of ultramafic rocks. *Geochimica et Cosmochimica Acta,* 73, 856–75.

McCollom, T. M., and Seewald, J. S. (2001) A reassessment of the potential for reduction of dissolved CO_2 to hydrocarbons during serpentinization of olivine. *Geochimica et Cosmochimica Acta*, 65, 3769–78.

Moody, J. B. (1976) Serpentinization: A review. *Lithos*, 9, 125–38.

Morrill, P. L., Johnson, O. J., Cotton, J., Eigenbrode, J. L., Nealson, K. H., Lollar, B. S., and Fogel, M. L. (2008) Isotopic evidence of microbial methane in ultrabasic reducing waters at a continental site of active serpentinization in N. California. *Geochimica et Cosmochimica Acta*, 72, A652.

Mottl, M. J., Komor, S. C., Fryer, P., and Moyer, C. L. (2003) Deep-slab fluids fuel extremophilic Archaea on a Mariana forearc serpentinite mud volcano: Ocean Drilling Program Leg 195. *Geochemistry, Geophysics, Geosystems*, 4, 9009–22.

Mumma, M. J., Villanueva, G. L., Novak, R. E., Hewagama, T., Bonev, B. P., DiSanti, M. A., Mandell, A. M., and Smith, M. D. (2009) Strong release of methane on Mars in northern summer 2003. *Science*, 323, 1041–45.

Nakagawa, S., and Takai, K. (2008) Deep-sea vent chemoautotrophs: Diversity, biochemistry and ecological significance. *FEMS Microbiology Ecology*, 65, 1–14.

Nunoura, T., Soffientino, B., Blazejak, A., Kakuta, J., Oida, H., Schippers, A., and Takai, K. (2009) Subseafloor microbial communities associated with rapid turbidite deposition in the Gulf of Mexico continental slope (IODP Expedition 308). *FEMS Microbiology Ecology*, 69, 410–24.

O'Hanley, D. S. (1992) Solution to the volume problem in serpentinization. *Geology*, 20, 705–8.

Osanjo, G. O., Muthike, E. W., Tsuma, L., Okoth, M. W., Bulimo, W. D., Lunsdorf, H., Abraham, W. R., Dion, M., Timmis, K. N., Golyshin, P. N., and Mulaa, F. J. (2009) A salt lake extremophile, *Paracoccus bogoriensis* sp nov., efficiently produces xanthophyll carotenoids. *African Journal of Microbiology Research*, 3, 426–33.

Parkes, R. J., Cragg, B. A., and Wellsbury, P. (2000) Recent studies on bacterial populations and processes in subseafloor sediments: A review. *Hydrogeology Journal*, 8, 11–28.

Pasek, M.A. (2008) Rethinking early Earth phosphorus geochemistry. *Proceedings of the National Academy of Sciences of the United States of America*, 105, 853–58.

Piper, D. Z., and Calvert, S. E. (2009) A marine biogeochemical perspective on black shale deposition. *Earth Science Reviews*, 95, 63–96.

Rathsack, K., Stackebrandt, E., Reitner, J., and Schumann, G. (2009) Microorganisms isolated from deep sea low-temperature influenced oceanic crust basalts and sediment samples collected along the Mid-Atlantic Ridge. *Geomicrobiology Journal*, 26, 264–74.

Roussel, E. G., Sauvadet, A. L., Chaduteau, C., Fouquet, Y., Charlou, J. L., Prieur, D., and Bonavita, M. A. C. (2009) Archaeal communities associated with shallow to deep subseafloor sediments of the New Caledonia Basin. *Environmental Microbiology*, 11, 2446–62.

Santelli, C. M., Edgcomb, V. P., Bach, W., and Edwards, K. J. (2009) The diversity and abundance of bacteria inhabiting seafloor lavas positively correlate with rock alteration. *Environmental Microbiology*, 11, 86–98.

Santelli, C. M., Orcutt, B. N., Banning, E., Bach, W., Moyer, C. L., Sogin, M. L., Staudigel, H., and Edwards, K. J. (2008) Abundance and diversity of microbial life in ocean crust. *Nature*, 453, 653–57.

Schrenk, M.O., Huber, J.A., and Edwards, K.J. (2010) Microbial provinces in the sub-seafloor. *Annual Reviews of Marine Science*, 2, 279–304.

Sleep, N.H., Meibom, A., Fridriksson, T., Coleman, R.G., and Bird, D.K. (2004) H_2-rich fluids from serpentinization: Geochemical and biotic implications. *Proceedings of the National Academy of Sciences of the United States of America*, 101, 12818–23.

Takai, K., Moyer, C.L., Miyazaki, M., Nogi, Y., Hirayama, H., Nealson, K.H., and Horikoshi, K. (2005a) *Marinobacter alkaliphilus* sp. nov., a novel alkaliphilic bacterium isolated from subseafloor alkaline serpentine mud from Ocean Drilling Program Site 1200 at South Chamorro Seamount, Mariana Forearc. *Extremophiles*, 9, 17–27.

Takai, K., Hirayama, H., Nakagawa, T., Suzuki, Y., Nealson, K.H., and Horikoshi, K. (2005b) *Lebetimonas acidiphila* gen. nov., sp nov., a novel thermophilic, acidophilic, hydrogen-oxidizing chemolithoautotroph within the *Epsilonproteobacteria*, isolated from a deep-sea hydrothermal fumarole in the Mariana Arc. *International Journal of Systematic and Evolutionary Microbiology*, 55, 183–89.

Takai, K., Miyazaki, M., Hirayama, H., Nakagawa, S., Querellou, J., and Godfroy, A. (2009) Isolation and physiological characterization of two novel, piezophilic, thermophilic chemolithoautotrophs from a deep-sea hydrothermal vent chimney. *Environmental Microbiology*, 11, 1983–97.

Takeuchi, M., Komai, T., Hanada, S., Tamaki, H., Tanabe, S., Miyachi, Y., Uchiyama, M., Nakazawa, T., Kimura, K., and Kamagata, Y. (2009) Bacterial and Archaeal 16S rRNA genes in Late Pleistocene to Holocene muddy sediments from the Kanto Plain of Japan. *Geomicrobiology Journal*, 26, 104–18.

Tiago, I., Chung, A.P., and Verissimo, A. (2004) Bacterial diversity in a nonsaline alkaline environment: Heterotrophic aerobic populations. *Applied and Environmental Microbiology*, 70, 7378–87.

Tiago, I., Mendes, V., Pires, C., Morais, P.V., and Verissimo, A. (2006) *Chimaereicella alkaliphila* gen. nov., sp nov., a Gram-negative alkaliphilic bacterium isolated from a nonsaline alkaline groundwater. *Systematic and Applied Microbiology*, 29, 100–108.

Tunnicliffe, V., McArthur, A.G., and McHugh, D. (1998) A biogeographical perspective of the deep-sea hydrothermal vent fauna. *Advances in Marine Biology*, 34, 353–442.

Valentine, D.L., Kastner, M., Wardlaw, G.D., Wang, X.C., Purdy, A., and Bartlett, D.H. (2005) Biogeochemical investigations of marine methane seeps, Hydrate Ridge, Oregon. *Journal of Geophysical Research-Biogeosciences*, 110, G02005.

Van Dover, C.L. (2002) Community structure of mussel beds at deep-sea hydrothermal vents. *Marine Ecology—Progress Series*, 230, 137–58.

Wanger, G., Southam, G., and Onstott, T.C. (2006) Structural and chemical characterization of a natural fracture surface from 2.8 kilometers below land surface: Biofilms in the deep subsurface. *Geomicrobiology Journal*, 23, 443–52.

Westheimer, F.H. (1987) Why nature chose phosphates. *Science*, 235, 1173–78.

Wheat, C.G., Fryer, P., Fisher, A.T., Hulme, S., Jannasch, H., Mottl, M.J., and Becker, K. (2008) Borehole observations of fluid flow from South Chamorro Seamount, an active serpentinite mud volcano in the Mariana forearc. *Earth and Planetary Science Letters*, 267, 401–9.

Whitman, W. B., Coleman, D. C., and Wiebe, W. J. (1998) Prokaryotes: The unseen majority. *Proceedings of the National Academy of Sciences of the United States of America*, 95, 6578–83.

Yumoto, I., Yamazaki, K., Hishinuma, M., Nodasaka, Y., Inoue, N., and Kawasaki, K. (2000) Identification of facultatively alkaliphilic *Bacillus* sp strain YN-2000 and its fatty acid composition and cell-surface aspects depending on culture pH. *Extremophiles*, 4, 285–90.

3
_____

Phylogenetic Patterns of Endemism and Diversity

Brian L. Anacker, *University of California, Davis*

Every species in nature uses a subset of the habitats available to it, but those with narrow ranges and unique adaptations have traditionally captured the attention of naturalists and ecologists (Futuyma and Moreno, 1988; Stevens, 1989; Brown, 1995; Berenbaum, 1996; Losos et al., 1998; Gaston and Blackburn, 2000; Schluter, 2002; Fine et al., 2005; Grant and Grant, 2007). In plants, endemism to a particular soil type is an especially common and important form of habitat specialization. As Kruckeberg pointed out, "Endemism is the hallmark of specialized edaphic habitats" (Kruckeberg, 2002). Edaphic endemism contributes to species diversity by promoting spatial turnover in community composition (beta diversity), especially in areas of high edaphic and topographic heterogeneity (Whittaker, 1960; Harrison and Inouye, 2002; Legendre et al., 2005; Qian and Ricklefs, 2007). In addition to limiting species ranges and increasing beta diversity, habitat specialization affects the origin and extinction of species (Kruckeberg, 1986; Futuyma and Moreno, 1988; Rajakaruna, 2004; Fine et al., 2005).

The steep ecological gradients in serpentine soils promote a remarkable level of habitat specialization in plants worldwide (Figure 3.1) (Alexander et al., 2006; Rajakaruna et al., 2009). Serpentine species can be placed into three categories or geographic states based on their degree of restriction: endemic, tolerator, and nontolerator (Harrison and Inouye, 2002; Safford et al., 2005). Serpentine endemics

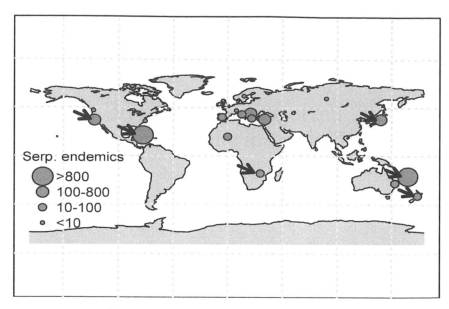

FIGURE 3.1. Map of the worldwide distribution of serpentine endemic diversity in 18 of the world's serpentine-containing regions. Filled circles are sized relative to the approximate number of serpentine endemics found in the region. Arrows indicate the six regions for which endemic species richness per family was tabulated and compared in the text (left to right: California, Cuba, Great Dyke of Zimbabwe, Japan, New Caledonia, and New Zealand).

(or "bodenstadt"; Kruckeberg, 2002) are species that are geographically restricted to the soil type. Serpentine tolerators (or "bodenvag") are species found on- and off-serpentine. Serpentine nontolerators are species that either cannot maintain populations on serpentine or have not been exposed to serpentine (i.e., no range overlap). Endemics have classically been placed into two subcategories, paleo- and neoendemics, reflecting their age and the timing of serpentine adaptation. Paleoendemics arise from the extinction of nonserpentine populations of older pre-adapted lineages (biotype depletion) (Stebbins and Major, 1965; Raven and Axelrod, 1978; Fiedler, 1992), whereas neoendemics originate from adaptive evolution in younger lineages (Kruckeberg, 1986; Rajakaruna, 2004).

Most studies to date have examined serpentine adaptation in plants from a population perspective, identifying locally adapted serpentine ecotypes that may be the first steps toward speciation (Hughes et al., 2001; Berglund et al., 2004; Sambatti and Rice, 2006; Wright et al., 2006). These studies highlight the striking ability of serpentine outcrops and edaphic heterogeneity in general to drive population divergence in the face of high levels of gene flow. Although this work is essential to gaining a mechanistic understanding of serpentine adaptation, questions from a deep time perspective regarding adaptation and specialization remain largely unaddressed, including the following. How often does serpentine adaptation lead to speciation? Is speciation more likely in preadapted lineages? Does restriction to serpentine soils promote adaptive radiations or lead to evolutionary dead-ends? Does the age and area of serpentine outcrops influence the endemic diversity in an area? How does this history of lineage diversification and preadaptation influence regional species pool and phylogenetic, functional, and species diversity of local communities? These and related questions can be addressed using phylogenies, which contain information on the relatedness of species and relative timing of branching events (Harvey and Pagel, 1991).

Serpentine is an ideal system for using comparative phylogenetic analysis to gain insight into the evolutionary consequences of plant adaptation and specialization (Brady et al., 2005). The strong selection and island-like nature of these soils drive adaptive evolution and increase ecological and spatial isolation, setting the stage for several modes of speciation (Kruckeberg, 1991, 2002; Ackerly, 2003; Rajakaruna, 2004). Following speciation, island-like adaptive radiations or evolutionary dead-ends may occur. Serpentine adaptations and endemism have apparently evolved numerous times in many lineages, providing evolutionary replication for comparative analysis. From a community perspective, serpentine enables strong inferences to be made about how the biogeographic and evolutionary histories of lineages in a regional species pool influence the assembly of local communities on adjacent, contrasting soil types.

In the few lineages where serpentine tolerance has been mapped onto phylogenies (mostly in California), we have learned that its evolutionary origin is highly labile and is associated with elevated rates of evolutionary diversification in some cases and evolutionary dead-ends in others (Spencer and Porter, 1997; Patterson and Givnish, 2002; Baldwin, 2005; Nguyen et al., 2008). We know much less about the evolutionary history of serpentine adaptation and specialization outside of California (but see De Kok, 2002). Finally, there are no investigations to date on how phylogenetic community structure changes across the boundaries of serpentine and nonserpentine soils.

In this chapter, I assess how often and in which plant lineages serpentine endemism arose and the corresponding levels of extant taxonomic and phylogenetic diversity. Then I review recent work on phylogenetic patterns associated with origin and consequences of serpentine endemism in California. Finally, I discuss serpentine community phylogenetics, describing two recent analyses for Californian serpentine plant communities. I conclude by evaluating the appropriateness of serpentine as a model system for answering questions about adaptation, lineage diversification, and coexistence.

GLOBAL TAXONOMIC AND PHYLOGENETIC
PATTERNS OF SERPENTINE ENDEMISM

How often and in which plant lineages did serpentine endemism arise? What were the consequences for diversification? To address these questions, I examine the taxonomic and phylogenetic position of endemism in each of six serpentine floras (indicated with arrows in Figure 3.1). These preliminary analyses may be helpful in developing hypotheses about adaptation and endemism that can be tested using fully resolved phylogenies, detailed species distribution maps, functional trait data, and historical biogeography.

The global comparative view of serpentine endemism is at the family level and higher, due to the lack of species-level lists for endemics outside of California (Table 3.1). The main challenges in generating these lists are insufficient botanical observation and collection, geographic variation in the degree of species restriction, taxonomic disagreement (e.g., endemic species versus variety), and simply insufficient reporting. However, even if the lists existed, my initial survey suggested that there is a nearly total lack of species-level phylogenies for most serpentine clades outside of North America. Furthermore, it would be intractable to deal with species-level patterns when comparing entire floras across the globe. It should be noted that the family-level estimates for New Caledonia should be considered maximum estimates because the data source did not clearly state whether the numbers provided per family were for species restricted to serpentine or for species associated with ultramafic rocks (Jaffré et al., 1987).

Examination of the six regional floras, which includes 2315 species, shows that endemism occurs in 105 unique families and 41 orders, including angiosperms, conifers, and ferns (Table 3.1). These 105 families represent 23% of the ~450 known plant families. Given that there are at least 10 other important serpentine areas in the world (Figure 3.1), this is a minimum estimate. This tabulation excludes the thousands of species that tolerate serpentine but are not excluded from it. Serpentine adaptation and endemism have clearly evolved independently multiple times in plants. In some cases an entire family (12 and 4 cases, respectively) or even order of plants has just one serpentine endemic, whereas in others a complex of closely related species are all endemics (Fiedler, 1992). On average, there were more than 10 serpentine endemics per family, but regional variance was considerable. The most endemic-rich family documented was Myrtaceae, with 174 endemics (some of which may actually be tolerators) in New Caledonia and 117 endemics in Cuba.

Serpentine regions of the world have different levels of serpentine endemic diversity, with some having fewer than 10 and others having more than 900 endemics (Figure 3.1). Interestingly, the serpentine flora of endemic-rich Sri Lanka has no known serpentine endemics, based on preliminary work (Rajakaruna and Bohn,

TABLE 3.1 Endemic Diversity Patterns in Six Serpentine Floras, Separated into "Endemic Rich" (>200) and "Endemic Poor" (<200) Categories

	Serpentine							Total		
Area	No. Endemic Taxa	No. Families	Mean Tax on Richness per Family	No. Orders	Area (km²)	Min. Age	Max. Age	No. Endemic Taxa	Area (km²)	Refs.
Endemic-rich										
California	215	39	5.5	24	6000	1	50	2387	423,970	4, 10
Cuba	854	24	35.6	14	7500	1	30	3153	110,922	1, 2
New Caledonia	1150[a]	64[b]	18.0[c]	31	5500	33	55	2323	19,060	3, 5–7
Endemic-poor										
Great Dyke	14	8	1.8	6	3000	144	144	232	386,670	2, 8, 11
Japan	50	19	2.6	13	5256	50	50	1866	377,873	2, 8
New Zealand	32	18	1.8	13	309	65	250+	1865	268,671	9
Summary	2920[d]	105[e]	12.4[f]	41[e]	4594[f]	—	—	2054[f]	264,528[f]	—

REFERENCES: 1. Borhidi, 1996. 2. Brooks, 1987. 3. Dawson, 1963. 4. Harrison et al., 2004. 5. Jaffré, 1980. 6. Jaffré et al: 1987. 7. Jaffré, 1992. 8. Kruckeberg, 2002. 9. Lee, 1992. 10. Safford et al., 2005. 11. Wild and Bradshaw, 1977.

[a] This is an approximation based on the cited references; the exact number is not known.

[b] The number of families, in this case, is the number with serpentine endemics or tolerators. Thus, it should be considered a *maximum* estimate of the number of families with endemics.

[c] The "mean richness" value, in this case, should be interpreted as a *minimum*.

[d] Sum of six rows above.

[e] Number of unique families or orders, not the sum of six rows above.

[f] Mean of six rows above.

2002). The areas with endemics can be classified into two groups with significantly different levels of endemic species richness: "endemic-rich" and "endemic-poor" areas (means: 941.3 and 32, respectively; two-sample Wilcoxon test $p = 0.04$) (Table 3.1). The serpentine floras of New Caledonia and Cuba make major contributions to these islands' total plant diversities, likely due to the tropical climates and the sizes and ages of serpentine exposures (Table 3.1). In the temperate zone, California has the greatest number of endemics, probably reflecting the quantity of serpentine, the spatial variation in climate and topography, and the lack of glaciation (Brooks, 1987; Alexander et al., 2006). The place where endemism evolved in the taxonomic hierarchy differs between these three regions (Table 3.1, Figure 3.2). Although California ranks third in total endemic diversity, it ranks second in the number of families with endemics (having more families with endemics than Cuba) and last in endemics per family among the endemic-rich areas (Table 3.1, Figure 3.2). Numerous families in California have just one or two endemics

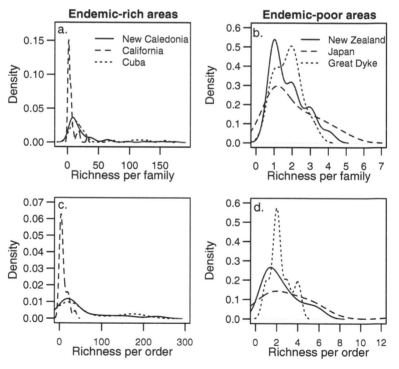

FIGURE 3.2. Density plots of the distribution of serpentine endemic richness per family (a–b) and richness per order (c–d) in each of six serpentine floras. Floras are separated into endemic-rich (a, c) and endemic-poor (b, d) areas for graphing purposes.

(8 with one, 10 with two). The same is true at the next taxonomic level (species per order), which is not surprising given the number of families and number of orders are highly correlated (Pearson's correlation = 0.98). The endemic-poor areas typically have around two species per family.

To assess the phylogenetic diversity of endemism within and among the six floras, I created a supertree for the 110 families based on the Angiosperm Phylogeny Group classification, using the program Phylomatic, with branch lengths estimated using Bladj (Webb and Donoghue, 2005). For each flora, two indices of phylogenetic diversity were then calculated, mean pairwise distance (MPD) and Faith's phylogenetic diversity (PD) using the Picante package in R (Kembel et al., 2008) (Figures 3.3–3.5). In addition, for each of the serpentine floras, expected MPD values (MPD_{null}) were calculated as the average MPD of 1000 random draws from the pool of families in the supertree, where each draw includes the same number of families as observed in the flora. In this phylogenetic context, the contrast of California and Cuba remains: the endemics of California are more phylogenetically diverse after controlling for their species diversity than those in Cuba, as measured by both MPD and PD (Figure 3.3). An additional explanation for the low phylogenetic diversity in Cuba, Japan, and Zimbabwe is that these regions have no serpentine-endemic conifers, despite *Juniperus*, *Pinus*, and Podocarpaceae often being present in the nonendemic flora. This contrasts with the conifer-rich serpentine endemic flora of New Caledonia ($n = 39$ species) and the conifer-containing endemic flora of California ($n = 2$ species) (Page, 1999). Conifers contribute disproportionately to phylogenetic diversity because of their distant relatedness to other plants. However, New Caledonia and California are more diverse than the other three areas even without conifers.

Clearly, some of the families with concentrations of endemics simply reflect high diversity of that family in the region. For example, Asteraceae has more than 30 endemics in California, but that is a small proportion of the over 1200 species in the state (Figure 3.3). Likewise, in Cuba, the 27 endemics in Orchidaceae are a small fraction of the more than 600 in the West Indies. Thus, endemics often reflect patterns of dominance in the surrounding flora, suggesting that they are likely closely related to nonendemics in their region. However, some lineages rank very high as a proportion of the species in the region, including Buxaceae in Cuba (26 of 43 species) and Linaceae in California (9 of 21 species). These families may either be foci of neoendemism or have preadaptations favoring serpentine adaptation of endemism, which could be tested with phylogenetic information once it is available (Springer, 2006). Two of the most important families of endemics in Cuba are closely related: Melastomataceae and Myrtaceae of the order Myrtales (Figure 3.3). Melastomataceae is a neotropical family, observed only in Cuba in this sample, whereas Myrtaceae has a much broader tropical

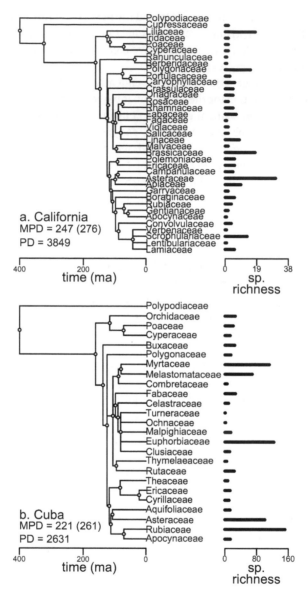

FIGURE 3.3. Phylogenetic patterns of serpentine endemics in (a) California and (b) Cuba. White dots indicate internal nodes. Scale bar is time before present in millions of years (ma). Bars indicate the number of serpentine endemics per family. The two measures of phylogenetic diversity are mean pairwise distance (MPD), which is the average branch length between family pairs, and Faith's phylogenetic diversity (PD), which is the total branch length for each flora. Parenthetical values are MPD$_{null}$ (see text). Polypodiaceae added for scaling purposes only.

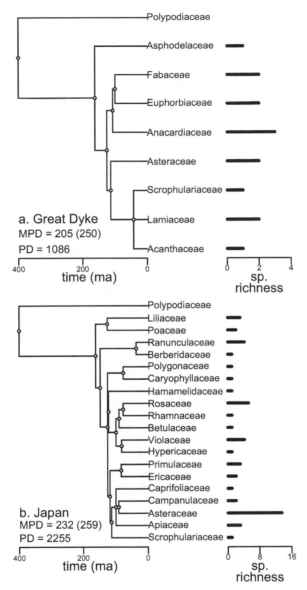

FIGURE 3.4. Phylogenetic patterns of serpentine endemics in the (a) Great Dyke and (b) Japan. Scale bar is time before present in millions of years (ma). Bars indicate the number of serpentine endemics per family. The two measures of phylogenetic diversity are mean pairwise distance (MPD), which is the average branch length between family pairs, and Faith's phylogenetic diversity (PD), which is the total branch length for each flora. Parenthetical values are MPD$_{null}$ (see text). Polypodiaceae added for scaling purposes only.

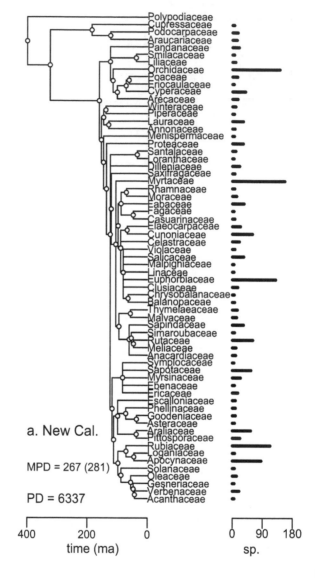

Polypodiaceae
Cupressaceae
Podocarpaceae
Araucariaceae
Pandanaceae
Smilacaceae
Liliaceae
Orchidaceae
Poaceae
Eriocaulaceae
Cyperaceae
Arecaceae
Winteraceae
Piperaceae
Lauraceae
Annonaceae
Menispermaceae
Proteaceae
Santalaceae
Loranthaceae
Dilleniaceae
Saxifragaceae
Myrtaceae
Rhamnaceae
Moraceae
Fabaceae
Fagaceae
Casuarinaceae
Elaeocarpaceae
Cunoniaceae
Celastraceae
Violaceae
Salicaceae
Malpighiaceae
Linaceae
Euphorbiaceae
Clusiaceae
Chrysobalanaceae
Balanopaceae
Thymelaeaceae
Malvaceae
Sapindaceae
Simaroubaceae
Rutaceae
Meliaceae
Anacardiaceae
Symplocaceae
Sapotaceae
Myrsinaceae
Ebenaceae
Ericaceae
Escalloniaceae
Phellinaceae
Goodeniaceae
Asteraceae
Araliaceae
Pittosporaceae
Rubiaceae
Loganiaceae
Apocynaceae
Solanaceae
Oleaceae
Gesneriaceae
Verbenaceae
Acanthaceae

a. New Cal.

MPD = 267 (281)

PD = 6337

400   200   0   0   90 180
time (ma)   sp.

(*continued*)

FIGURE 3.5. Phylogenetic patterns of serpentine endemics in (a) New Caledonia and (b) New Zealand. Scale bar is time before present in millions of years (ma). Bars indicate the number of serpentine endemics per family. The two measures of phylogenetic diversity are mean pairwise distance (MPD), which is the average branch length between family pairs, and Faith's phylogenetic diversity (PD), which is the total branch length for each flora. Parenthetical values are MPD$_{null}$ (see text). Polypodiaceae added for scaling purposes only.

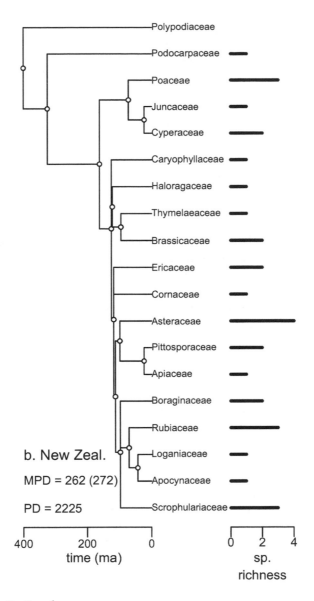

FIGURE 3.5. *Continued.*

range and contributes a large number of endemics in New Caledonia as well (Stevens, 2001 onward). Assuming endemism arose repeatedly in these families, they may have characteristics promoting serpentine adaptation and specialization. A final interesting pattern is that in the endemic flora of New Zealand, 10 families are from the asterid clade (bottom 10 in Figure 3.5b) and just two from the rosid clade (Thymelaeaceae and Brassicaceae), whereas in the other five floras, there is roughly equal representation of these two clades. Rosids and asterids are the two principal clades of eudicots, diverging over 100 Ma (Stevens, 2001 onward). The great diversity within both the asterid and rosid clades makes it difficult to infer how their ecological differences would influence the evolution of endemism without extensive study.

Overlap of Endemic Diversity between California and Cuba

Endemic diversity in different lineages is likely a function of biogeographic and evolutionary history and regional environmental conditions; visualizing the overlap of two serpentine floras on a phylogeny helps develop hypotheses about the evolution of endemism with this complexity in mind. For California and Cuba, I was able to obtain lists for the nonendemic as well as endemic floras (CalJep, Viers et al., 2006; Flora of the West Indies, Acevedo-Rodríguez and Strong, 2007), allowing consideration of the overlap in the two region's endemic floras in the context of general floristic patterns. Cuba and California have serpentine endemics in eight shared families (Figure 3.6). All 27 families that have endemics only in California have species in or near Cuba; in contrast, 9 Cuban endemic-containing families are not found in California. Families with serpentine endemics that belong to the rosid clade are almost completely unshared (with the exception of Fabaceae), but overlap is common in asterids. In other words, endemism arose in similar families of asterids but different families of rosids in these two floras.

In rosids, the low overlap between California and Cuba is driven by different patterns of dominance in two major subclades of rosids: the N-fixing subclade and the group containing Celastrales, Oxalidales, and Malpighiales (COM subclade) (Wang et al., 2009). The California endemics are in four families (15 species) from the N-fixing subclade and three families (13 species) from the non–N-fixing COM subclade, whereas the Cuban endemics come from just one family (28 species) of the N-fixing subclade and five families (161 species) from the COM subclade. The high dominance of Cuban serpentine endemics from the COM subclade is due largely to Euphorbiaceae (129 endemics), a family with no serpentine endemics in California despite the presence of 15 genera and 62 species. Although most Euphorbiaceae are tropical shrubs and trees, California species are typically small herbs or desert shrubs. Clearly, the conditions that affected the evolution of serpentine adaptation and growth forms of Euphorbiaceae in Cuba were widely

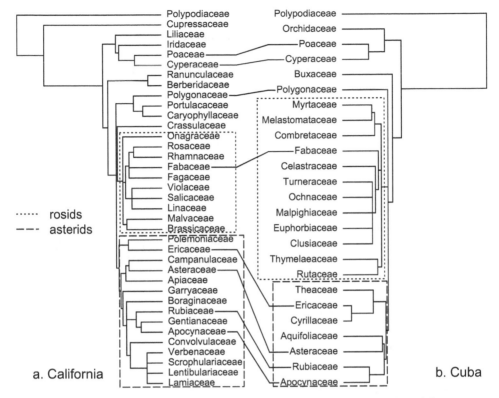

FIGURE 3.6. Phylogenetic overlap of families with serpentine endemics in (a) California and (b) Cuba. Solid connecting lines indicate families that have serpentine endemics in both floras. Rosids and asterids, two major clades of eudicots, are indicated. Polypodiaceae added for scaling purposes only.

different than what was found in California. These results combine to paint a complicated picture of serpentine endemism in rosids: Cuban rosid families are largely confined to the tropics, while California rosid families are cosmopolitan but promote the evolution of endemics only in California.

In asterids, endemics are found in four of the same families in Cuba and California (compared with one for rosids). Two of the four families are very large (Asteraceae and Rubiaceae) and therefore have a greater likelihood of having family members, and therefore endemics, in each area. Alternatively, the overlap may reflect niche conservatism (Pearman et al., 2008) for traits that similarly favor the evolution of serpentine endemics (Brady et al., 2005) in environments as different as Cuba and California.

Endemism and Serpentine Area

The percentage of serpentine endemics in regional flora is strongly related to the percentage of land area covered by serpentine ($r^2 = 0.83$, $p < 0.01$) (Figure 3.7). The increased exposure may provide more chances for species to adapt to the soil and eventually become restricted to it.

Endemism and Time for Speciation

The amount of time serpentine soils have been available for colonization varies greatly among (Table 3.1) and within regions (Alexander et al., 2006). The mean age of exposure of serpentine in endemic-rich areas is significantly lower than in endemic-poor areas (means: 11.7 and 86.3 for minimum ages, respectively; two-sample Wilcoxon test $p = 0.038$) (Raven and Axelrod, 1978; Lee, 1992; Borhidi, 1996; Harrison et al., 2004). Furthermore, endemic age, range size, and geographic distance to closely related nonendemics has been suggested to reflect the origin of endemics (i.e., neoendemics versus paleoendemics) (Harrison and Inouye, 2002; Kruckeberg, 2002). These patterns imply that the origin and diversity of endemics in an area is partly a function of time for speciation.

Phylogenetics offers a way to investigate the influence of the age of exposure of serpentine on the origin and diversity of habitat specialists more directly. Such an analysis minimally requires species-level phylogenies, careful estimation of di-

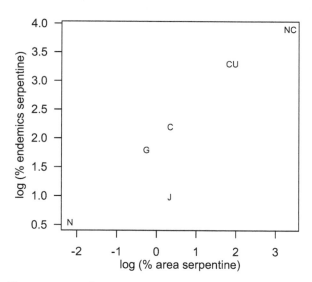

FIGURE 3.7. The percentage of serpentine endemics as a function of the percentage of area serpentine in six floras: N = New Zealand; J = Japan; G = Great Dyke, Zimbabwe; C = California; CU = Cuba; and NC = New Caledonia.

vergence times, and the age of exposure of serpentine outcrops. It is possible to estimate the age of species from a phylogeny by applying a molecular clock combined with rate smoothing and fossil calibration (Klicka and Zink, 1997; Sanderson, 2002). Species' ages can then be compared to the time of exposure of serpentine in an area. Such an analysis is not yet possible, however.

THE ORIGINS AND CONSEQUENCES OF ENDEMISM IN CALIFORNIA FLORA

To examine the origin and consequences of serpentine endemism in the California flora, we conducted a detailed phylogenetic study for 23 endemic-containing genera: *Allium, Aquilegia, Arctostaphylos, Balsamorhiza, Calochortus, Calycadenia, Ceanothus, Cirsium, Collinsia, Cupressus* (= *Hesperocyparis*), *Ericameria, Erythronium, Iris, Layia, Lessingia, Mimulus, Navarretia, Orthocarpus, Perideridia, Sanicula, Sidalcea, Trichostema,* and *Trifolium* (Anacker et al., in press). For each genus, we created phylogenies based on molecular sequence data and categorized each species (total = 798) as endemic, tolerator, or nontolerator. We tested for biased evolutionary transition rates among the three states and for shifts in diversification associated with endemism.

Transitions between Nontolerator, Tolerator, and Endemic

We found that transitions to or from serpentine tolerance were significantly more common than transitions to or from endemism (mean 7.2 versus 3.1, respectively; paired Student's t-test: df = 21, $p < 0.003$). This implies that serpentine tolerance may be more easily attained and lost (i.e., labile) over evolutionary time than habitat specialization. We also found that the majority of the endemic-containing plant genera examined (21 of 23) also included tolerators. However, transitions to serpentine endemism were equally likely to occur from nontolerator as from tolerator ancestors, even after controlling for the frequencies of tolerance and intolerance in each genus. Notably, adaptation to serpentine is only implicated as a cause of speciation when the transition is from nontolerator to endemic, as in the case of *Layia discoidea* and its ancestor *L. glandulosa* (see Chapter 4). In other cases, tolerance to serpentine may be a preadaptation for endemism, as has been examined for *Calochortus* (Fiedler, 1992; Patterson and Givnish, 2002) and several genera in New Caledonia (De Kok, 2002).

Lineage Diversification

The relationship of a particular trait such as serpentine endemism to the rate of evolutionary diversification (speciation minus extinction) can be examined by a sister clade comparison, in which species numbers are compared between two clades that are one another's closest relatives and differ with respect to the trait of interest (Vamosi and Vamosi, 2005). For example, if clade A consists of six serpentine endemics and clade B consists of four nonendemics, the difference in diversification associated with endemism is $\log(6) - \log(4) = 0.41$, with the positive value indicating that endemism is associated with increased diversification. A phylogeny is necessary to determine which clades to compare.

For sister clade comparisons, we found that the transition to serpentine endemism resulted in either no change (15 of 20) or decreased diversification (5 of 20) relative to the sister clade (Wilcoxon signed ranks, df $= 19$, $p = 0.0313$). This is consistent with the low number of endemics per genus (mean $= 2.4$) and family (mean $= 5.5$) (Table 3.1) and with studies on diversification in *Navarretia* and other genera (Spencer and Porter, 1997). These and other results led us to conclude that serpentine endemism generally is an evolutionary dead-end in California flora. The leading exception was *Allium*, in which diversification occurred following the transition to endemism. There are two additional genera in California flora in which the shift to serpentine endemism may have promoted diversification but which we did not use for lack of phylogenetic information (a subclade of *Streptanthus*, Mayer and Soltis, 1994, and *Hesperolinon*, Kruckeberg, 2002; Nguyen et al., 2008).

Many aspects of serpentine soils lead to the expectation that diversification might be limited by high levels of extinction, rather than limited speciation. Although edaphic stress and insularity may promote lineage divergence and isolation, they also contribute to several interrelated factors associated with extinction: small population size, spatial isolation, narrow geographic distribution, low genetic diversity, and high levels of inbreeding (Lande, 1993; Mills and Smouse, 1994; Stockwell et al., 2003). We also found that endemics were younger, on average, than nonendemics across 23 lineages, suggesting short lineage persistence times and/or recent origins (Wilcoxon signed ranks, df $= 21$, $p = 0.0066$). Further work is being done to estimate the speciation and extinction rates separately for these lineages using recent analytical advances in phylogenetics (Maddison et al., 2007).

These results may or may not generalize to other edaphic endemic hot spots, such as Cuba and New Caledonia, where the numbers of endemics per family are much higher but where there still appear to be very few congeneric endemics and the endemic-only genera and family are species-poor (possibly reflecting taxonomic inflation; Kruckeberg, 2002). Sister clade comparisons are not possible for these floras because they lack both species-level lists of endemics and phylogenies. However, it will be valuable to extend this type of analysis to other stressful, insular habitat types with high degrees of specialization, such as vernal pools.

PHYLOGENETIC DIVERSITY AND COMMUNITY STRUCTURE ACROSS ECOLOGICAL STRESS GRADIENTS

Evolutionary and biogeographic history determine what species are available in a region (Ricklefs, 1987; Ricklefs and Schluter, 1993), and in turn, species traits influence how species from the regional pool assemble into local communities (Darwin, 1859; Brown and Wilson, 1956; Hardin, 1960; Webb et al., 2002). Because close relatives often share similar traits, the phylogenetic composition of local communities is not a random draw from the regional pool; rather, coexisting species may be either more closely or distantly related than expected by chance. With the availability of phylogenetic trees for large groups of taxa, indices of mean relatedness for species within communities are frequently calculated and interpreted with regard to community assembly processes (Prinzing et al., 2001). Serpentine soils offer promising opportunities for investigating the role of environmental heterogeneity on phylogenetic community assembly, as illustrated by two examples.

Phylogenetic Diversity in California Serpentine Communities

In a climatic gradient across California, represented by 50×20 m plots at 109 serpentine sites (Harrison et al., 2006), plant communities in arid regions contained species that were more closely related to one another than by chance, that is, they showed low phylogenetic diversity or phylogenetic clustering. In cooler and wetter regions, communities contained species that were more distantly related to one another than by chance, that is, they showed high phylogenetic diversity or phylogenetic overdispersion. For example, wet sites (those with >150 cm annual precipitation) had significantly lower net relatedness values than dry sites (those with <50 cm annual precipitation) (mean -0.48 and 0.34, respectively; two-sample Wilcoxon test $p < 0.01$), indicating a shift from phylogenetic overdispersion to clustering across California's precipitation gradient. This pattern was also found at the regional scale, where serpentine plant communities drew from more phylogenetically diverse regional pools of species in wetter than in drier regions. The relationship of diversity to climate was much stronger for phylogenetic diversity ($p < 0.001$, $r^2 = 0.36$) than for species richness ($p = 0.53$, $r^2 = 0$), suggesting that phylogenetic diversity may be a better surrogate than species number for information on functional diversity (Cadotte et al., 2008), which is an important aspect of coexistence theories (Hutchinson, 1957, 1961; Loehle, 2000). Communities in wetter climates showed higher diversity in several functional traits, including plant height, seed mass, and life form.

Postfire Regeneration Strategy and Serpentine Endemism

Chaparral shrubs differ in their regeneration niche, falling into one of three categories: obligate postfire seeders, sprouters, or facultative seeders (Keeley and Zedler, 1978). Fire regimes of serpentine and neighboring sandstone chaparral differ considerably, with serpentine having less frequent and severe fires due to its low biomass and productivity (Safford and Harrison, 2004). Using data from Safford and Harrison (2004), we found that postfire seeders dominate serpentine communities, which is expected under infrequent fire and high water stress, whereas postfire sprouters dominate sandstone communities and were more closely related than expected. We also found that serpentine (but not sandstone) communities were more phylogenetically diverse (less closely related) than expected based in the combined species pool for the two soils (mean 0.98 versus −0.42, respectively; paired Student's t-test: df $= 42$, $p < 0.001$; Anacker, Rajakaruna, Ackerly, Harrison, Keeley, unpublished data). In part, this is because the dominant character state on serpentine, postfire seeding, is found in four distantly related clades, and the dominant character state on sandstone, sprouting, is dispersed within a few clades of closely related species. Also, postfire seeders that were closely related co-occurred infrequently on serpentine while closely related sprouters co-occurred frequently on sandstone. Thus, the shared ancestry of species in the regional pool strongly influenced the structure of communities from contrasting soil types.

CONCLUSIONS

Serpentine is an excellent model system for studying adaptation and specialization, given its degree of abiotic stress and insularity and high degree of habitat specialization. However, given that serpentine represents a syndrome of challenges, studies will not likely have simple conclusions, such as identifying the gene for serpentine adaptation or the trait for serpentine tolerance. Likewise, given the variability in regional evolutionary histories and climate, we are unlikely to uncover the pathway or the consequence of specialization to serpentine. Numerous ecological and evolutionary strategies are likely associated with edaphic specialization in this system. Rather than detracting from the utility of serpentine as a model system, these challenges present opportunities for ongoing research. The serpentine system will also be a useful case study for the "move versus evolve" debate in regard to climate change. Given that serpentine can foster adaptive evolution, and the spatial isolation limits plant migration, we may detect on serpentine the early signs of climate change–induced evolution. Phylogeny will be an important tool in pursing such research, both in identifying key principles on how habitat specialization influences diversification and adaptation and predicting how lineages will respond to perturbation.

LITERATURE CITED

Acevedo-Rodríguez, P., and Strong, M. T. (2007) Catalogue of the seed plants of the West Indies Web site. http://botany.si.edu/antilles/WestIndies/catalog.htm. Accessed September 2009.

Ackerly, D. D. (2003) Community assembly, niche conservatism, and adaptive evolution in changing environments. *International Journal of Plant Sciences,* 164, 165–84.

Alexander, E. B., Coleman, R. G., Keeler-Wolf, T., and Harrison, S. P. (2006) *Serpentine Geoecology of Western North America: Geology, Soils, and Vegetation.* Oxford University Press, New York.

Anacker, B. L., Whittall J. W., Goldberg, E. B., and Harrison, S. P. (in press) Origins and consequences of serpentine endemism in the California flora. *Evolution.*

Baldwin, B. G. (2005) Origin of the serpentine-endemic herb *Layia discoidea* from the widespread *L. glandulosa* (Compositae). *Evolution,* 59, 2473–79.

Berenbaum, M. R. (1996) Introduction to the symposium: On the evolution of specialization. *American Naturalist,* 148, 78–83.

Berglund, A. B. N., Dahlgren, S., and Westerbergh, A. (2004) Evidence for parallel evolution and site-specific selection of serpentine tolerance in *Cerastium alpinum* during the colonization of Scandinavia. *New Phytologist,* 161, 199–209.

Borhidi, A. (1996) *Phytogeography and Vegetation Ecology of Cuba.* Akademiai Kiado, Budapest.

Brady, K. U., Kruckeberg, A. R., and Bradshaw, H. D. Jr. (2005) Evolutionary ecology of plant adaptation to serpentine soils. *Annual Review of Ecology Evolution and Systematics,* 36, 243–66.

Brooks, R. R. (1987) *Serpentine and Its Vegetation: A Multidisciplinary Approach.* Dioscorides Press, Portland, OR.

Brown, J. H. (1995) *Macroecology.* University of Chicago Press, Chicago.

Brown, W. L. Jr., and Wilson, E. O. (1956) Character displacement. *Systematic Biology,* 5, 49.

Cadotte, M. W., Cardinale, B. J., and Oakley, T. H. (2008) Evolutionary history and the effect of biodiversity on plant productivity. *Proceedings of the National Academy of Sciences,* 105, 17012.

Darwin, C. (1859) *On the Origin of Species by Means of Natural Selection, or the Preservation of Favoured Races in the Struggle for Life.* Appleton, New York.

Dawson, J. W. (1963) New Caledonia and New Zealand—A botanical comparison. *Tuatara,* 11, 178.

De Kok, R. (2002) Are plant adaptations to growing on serpentine soil rare or common? A few case studies from New Caledonia. *Adansonia,* 24, 229–38.

Fiedler, P. L. (1992) Cladistic test of the adaptational hypothesis for serpentine endemism. In *The Vegetation of Ultramafic (Serpentine) Soils: Proceedings of the First International Conference on Serpentine Ecology,* pp. 421–34. Intercept, Andover, Hampshire, UK.

Fine, P. V. A., Daly, D. C., Villa Munoz, G., Mesones, I., and Cameron, K. M. (2005) The contribution of edaphic heterogeneity to the evolution and diversity of Burseraceae trees in the western Amazon. *Evolution,* 59, 1464–78.

Futuyma, D. J., and Moreno, G. (1988) The evolution of ecological specialization. *Annual Reviews in Ecology and Systematics,* 19, 207–33.

Gaston, K. J., and Blackburn, T. M. (2000) *Pattern and Process in Macroecology*. Blackwell, Oxford.

Grant, P. R., and Grant, B. R. (2007) *How and Why Species Multiply: The Radiation of Darwin's Finches*. Princeton University Press, Princeton, NJ.

Hardin, G. (1960) The competitive exclusion principle. *Science*, 131, 1292–97.

Harrison, S. P., and Inouye, B. D. (2002) High ß diversity in the flora of Californian serpentine "islands." *Biodiversity and Conservation*, 11, 1869–76.

Harrison, S. P., Safford, H. D., Grace, J. B., Viers, J. H., and Davies, K. F. (2006) Regional and local species richness in an insular environment: Serpentine plants in California. *Ecological Monographs*, 76, 41–56.

Harrison, S. P., Safford, H. D., and Wakabayashi, J. (2004) Does the age of exposure of serpentine explain variation in endemic plant diversity in California? *International Geology Review*, 46, 235–42.

Harvey, P. H., and Pagel, M. D. (1991) *The comparative method in evolutionary biology*. Oxford University Press, Oxford.

Hughes, R., Bachmann, K., Smirnoff, N., and Macnair, M. R. (2001) The role of drought tolerance in serpentine tolerance in the *Mimulus guttatus* Fischer ex DC. complex. *South African Journal of Science*, 97, 581–86.

Hutchinson, G. E. (1957) Concluding remarks. *Cold Spring Harbor Symposium on Quantitative Biology*, 22, 415–27.

Hutchinson, G. E. (1961) The paradox of the plankton. *American Naturalist*, 95, 137–45.

Jaffré, T. (1980) Etude écologique du peuplement végétal des sols dérivés de roches ultrabasiques en Nouvelle Calédonie. *Travaux et Documents de l'ORSTOM 124*. ORSTOM, Paris.

Jaffré, T. (1992) Floristic and ecological diversity of the vegetation on ultramafic rocks in New Caledonia. In *The Vegetation of Ultramafic (Serpentine) Soils: Proceedings of the First International Conference on Serpentine Ecology*, pp. 101–7. Intercept, Andover, Hampshire, UK.

Jaffré, T., Morat, P., Veillon, J. M., and MacKee, H. S. (1987) Changements dans la végétation de la Nouvelle-Calédonie au cours du Tertiaire: La végétation et la flore des roches ultrabasiques. *Adansonia*, 4, 365–91.

Keeley, J. E., and Zedler, P. H. (1978) Reproduction of chaparral shrubs after fire: A comparison of sprouting and seeding strategies. *American Midland Naturalist*, 142–61.

Kembel, S. W., Ackerly, D. D., Blomberg, S. P., Cowan, P., Helmus, M. R. and Webb, C. O. (2008) Picante: Tools for integrating phylogenies and ecology. Available online at http://picante.r-forge.r-project.org.

Klicka, J., and Zink, R. M. (1997) The importance of recent ice ages in speciation: A failed paradigm. *Science*, 277, 1666–69.

Kruckeberg, A. R. (1986) An essay: The stimulus of unusual geologies for plant speciation. *Systematic Botany*, 11, 455–63.

Kruckeberg, A. R. (1991) An essay: Geoedaphics and island biogeography for vascular plants. *Aliso*, 13, 225–38.

Kruckeberg, A. R. (2002) *Geology and Plant Life: The Effects of Landforms and Rock Types on Plants*. University of Washington Press, Seattle.

Lande, R. (1993) Risks of population extinction from demographic and environmental stochasticity and random catastrophes. *American Naturalist*, 142, 911.

Lee, W.G. (1992) The serpentinized areas of New Zealand, their structure and ecology. In *The Ecology of Areas with Serpentinized Rocks: A World View* (eds. B.A. Roberts, J. Proctor). Kluwer Academic Publishers, Dordrecht.

Legendre, P., Borcard, D., and Peres-Neto, P.R. (2005) Analyzing beta diversity: Partitioning the spatial variation of community composition data. *Ecological Monographs, 75,* 435–50.

Loehle, C. (2000) Strategy space and the disturbance spectrum: A life-history model for tree species coexistence. *American Naturalist, 156,* 14–33.

Losos, J.B., Jackman, T.R., Larson, A., de Queiroz, K., and Rodriguez-Schettino, L. (1998) Contingency and determinism in replicated adaptive radiations of island lizards. *Science, 279,* 2115–18.

Maddison, W.P., Midford, P.E., and Otto, S.P. (2007) Estimating a binary character's effect on speciation and extinction. *Systematic Biology, 56,* 701–10.

Mayer, M.S., and Soltis, P.S. (1994) The evolution of serpentine endemics: A chloroplast DNA phylogeny of the *Streptanthus glandulosus* complex (Cruciferae). *Systematic Botany, 19,* 557–74.

Mills, L.S., and Smouse, P.E. (1994) Demographic consequences of inbreeding in remnant populations. *American Naturalist, 144,* 412–31.

Nguyen, N.H., Driscoll, H.E., and Specht, C.D. (2008) A molecular phylogeny of the wild onions (*Allium*; Alliaceae) with a focus on the western North American center of diversity. *Molecular Phylogenetics and Evolution, 47,* 1157–72.

Page, C.N. (1999) The conifer flora of New Caledonia—stasis, evolution and survival in an ancient group. *International Conifer Conference, 149–155,* ISHS.

Patterson, T.B., and Givnish, T.J. (2002) Phylogeny, concerted convergence, and phylogenetic niche conservatism in the core Liliales: Insights from rbcL and dhF sequence data. *Evolution, 56,* 233–52.

Pearman, P., Guisan, A., Broennimann, O., and Randin, C. (2008) Niche dynamics in space and time. *Trends in Ecology and Evolution, 23,* 149–58.

Prinzing, A., Durka, W., Klotz, S., and Brandl, R. (2001) The niche of higher plants: Evidence for phylogenetic conservatism. *Proceedings of the Royal Society of London, Series B: Biological Sciences, 268,* 2383–89.

Qian, H., and Ricklefs, R.E. (2007) A latitudinal gradient in large-scale beta diversity for vascular plants in North America. *Ecology Letters, 10,* 737–44.

Rajakaruna, N. (2004) The edaphic factor in the origin of plant species. *International Geology Review, 46,* 471–78.

Rajakaruna, N., and Bohn, B.A. (2002) Serpentine and its vegetation: A preliminary study from Sri Lanka. *Journal of Applied Botany, 76,* 20–28.

Rajakaruna, N., Harris, T.B., and Alexander, E.B. (2009) Serpentine geoecology of eastern North America: A review. *Rhodora, 111,* 21–108.

Raven, P.H., and Axelrod, D.I. (1978) *Origin and Relationships of the California Flora.* University of California Press, Berkeley.

Ricklefs, R.E. (1987) Community diversity: Relative roles of local and regional processes. *Science, 235,* 167–71.

Ricklefs, R.E., and Schluter, D. (1993) *Species Diversity in Ecological Communities: Historical and Geographical Perspectives.* University of Chicago Press, Chicago.

Safford, H. D., and Harrison, S. (2004) Fire effects on plant diversity in serpentine vs. sandstone chaparral. *Ecology,* 85, 539–48.

Safford, H. D., Viers, J. H., and Harrison, S. P. (2005) Serpentine endemism in the California flora: A database of serpentine affinity. *Madroño,* 52, 222–57.

Sambatti, J. B. M., and Rice, K. J. (2006) Local adaptation, patterns of selection, and gene flow in the Californian serpentine sunflower (*Helianthus exilis*). *Evolution,* 60, 696–710.

Sanderson, M. J. (2002) Estimating absolute rates of molecular evolution and divergence times: A penalized likelihood approach. *Molecular Biology and Evolution,* 19, 101–9.

Schluter, D. (2002) *The Ecology of Adaptive Radiation.* Oxford Series in Ecology and Evolution. Oxford University Press, Oxford.

Spencer, S. C., and Porter, J. M. (1997) Evolutionary diversification and adaptation to novel environments in *Navarretia* (Polemoniaceae). *Systematic Botany,* 22, 649–68.

Springer, Y. P. (2006) *Epidemiology, Resistance Structure, and the Effects of Soil Calcium on a Serpentine Plant-Pathogen Interaction.* PhD dissertation, University of California, Santa Cruz.

Stebbins, G. L., and Major, J. (1965) Endemism and speciation in the California flora. *Ecological Monographs,* 35, 1–35.

Stevens, G. C. (1989) The latitudinal gradient in geographical range: How so many species coexist in the tropics. *American Naturalist,* 133, 240–56.

Stevens, P. F. (2001 onward) Angiosperm phylogeny Website. Version 9, June 2008. http://www.mobot.org/MOBOT/research/APweb.

Stockwell, C. A., Hendry, A. P., and Kinnison, M. T. (2003) Contemporary evolution meets conservation biology. *Trends in Ecology and Evolution,* 18, 94–101.

Vamosi, S. M., and Vamosi, J. C. (2005) Endless tests: Guidelines for analysing non-nested sister-group comparisons. *Evolutionary Ecology Research,* 7, 567–79.

Viers, J. H., Thorne, J. H., and Quinn, J. F. (2006) CalJep: A spatial distribution database of Calflora and Jepson plant species. *San Francisco Estuary and Watershed Science,* 4, 1–18.

Wang, H., Moore, M. J., Soltis, P. S., Bell, C. D., Brockington, S. F., Alexandre, R., Davis, C. C., Latvis, M., Manchester, S. R., and Soltis, D. E. (2009) Rosid radiation and the rapid rise of angiosperm-dominated forests. *Proceedings of the National Academy of Sciences,* 106, 3853–58.

Webb, C. O., Ackerly, D. D., McPeek, M. A., and Donoghue, M. J. (2002) Phylogenies and community ecology. *Annual Review of Ecology and Systematics,* 33, 475–505.

Webb, C. O., and Donoghue, M. J. (2005) Phylomatic: Tree assembly for applied phylogenetics. *Molecular Ecology Notes,* 5, 181–83.

Whittaker, R. H. (1960) Vegetation of the Siskiyou mountains, Oregon and California. *Ecological Monographs,* 30, 279–338.

Wild, H., and Bradshaw A. D. (1977) The evolutionary effects of metalliferous and other anomalous soils in south central Africa. *Evolution,* 31, 282–93.

Wright, J. W., Stanton, M. L., and Scherson, R. (2006) Local adaptation to serpentine and non-serpentine soils in *Collinsia sparsiflora. Evolutionary Ecology Research,* 8, 1–21.

4
_____

Plant Speciation

Kathleen M. Kay, *University of California, Santa Cruz*
Kimiora L. Ward, *University of California, Davis*
Lorna R. Watt, *Michigan State University*
Douglas W. Schemske, *Michigan State University*

The world's serpentine regions are known for their striking levels of endemism and the distinctive flora they possess relative to surrounding areas. Although much work has been done to catalog the plant diversity on serpentine, including taxonomic, morphological, and physiological diversity, relatively little has been done to understand the evolutionary origins of serpentine species. Yet serpentine species present an excellent system to study the general processes of plant speciation. We critically evaluate the theory and evidence for the mechanisms of plant speciation on serpentine. We highlight the contributions that studies of serpentine plants have made to the general understanding of speciation processes, suggest directions for future research, and call for efforts to conserve serpentine habitats and the unique opportunities they provide for ecological and evolutionary studies.

Serpentine habitats possess a seemingly insurmountable set of obstacles to successful colonization by plants. Among the many characteristics of serpentine that might limit adaptation are high concentrations of toxic metals, low calcium:magnesium ratios, thin soils prone to rapid desiccation, high gene flow with adjacent nonserpentine habitats, and spatial isolation from source populations. Nevertheless, there is ample evidence that serpentine adaptation has evolved repeatedly and independently in a wide diversity of plants (Kruckeberg and Rabinowitz, 1985; Rajakaruna, 2004; Anacker, 2010). Furthermore, in proportion to their land area, serpentine habitats harbor more endemic species per area than surrounding habitats (Kruckeberg, 1984).

PATTERNS OF ENDEMISM SUPPORT A ROLE FOR SERPENTINE IN SPECIATION

Serpentine plants show a gradient of tolerance and restriction to serpentine, from widely tolerant to narrowly endemic. In early studies of the evolutionary ecology of serpentine plants, Kruckeberg (1951, 1954) characterized plants found on serpentine as bodenvag, indicator, or endemic species. Bodenvag species appeared indifferent to the soil but often showed differences in tolerance to serpentine at a population scale. Indicators were typically found on serpentine but also occurred occasionally off serpentine, and endemics were restricted wholly to serpentine soil. He reinterpreted Stebbins's (1942) ideas about the origins of endemism through the prism of serpentine soils. Stebbins proposed two routes to endemism. Depleted species, or paleoendemics, were once widespread and genetically diverse but had lost many or most of their biotypes, resulting in endemism on a narrow set of environmental conditions. In contrast, insular species, or neoendemics, developed on a habitat island from a small group of initial founders. Kruckeberg (1986) thought that both routes could be represented among serpentine species, with bodenvag and indicator species possibly showing the process of biotype depletion and narrow endemics representing insular species.

Raven and Axelrod (1978), in a comprehensive treatment of the evolutionary and fossil history of the California flora, largely supported this view. They believed that Californian serpentine exposures were very young, less than 10 Ma (but see Harrison et al., 2004), and that most bodenvag and indicator species predated the exposures. Thus, indicator species that are largely but not wholly restricted to serpentine, such as *Cupressus sargentii* (now classified as *Hesperocyparis sargentii*) and *Quercus durata*, were interpreted as paleoendemics. Although serpentine was not likely involved in the origin of these species, the patchy and isolated distribution of serpentine habitat islands could contribute to further divergence within these species. Strict endemic species, on the other hand, could have diversified substantially since the exposures of California serpentine and were suggested to primarily represent neoendemic or insular taxa.

Raven and Axelrod (1978) further noted patterns in the types of California plants that fall into these categories. Putative neoendemics are comprised chiefly of herbaceous lineages, whereas many paleoendemics are woody. They hypothesized that mesic-adapted woody taxa were gradually outcompeted in nonserpentine areas as the climate became warmer and drier during the summer, whereas herbaceous lineages with Madro-tertiary affinities gave rise to many serpentine neoendemics. Notable exceptions to these categories exist, such as the wide-ranging herb *Streptanthus glandulosus*, which shows a genetic structure consistent with paleoendemism followed by allopatric divergence among regions (Kruckeberg, 1957; Mayer et al., 1994; Mayer and Soltis, 1994, 1999).

The patterns of endemism found on serpentine are consistent with recent, insular origins of many serpentine endemics. Again, the most comprehensive data on endemism are available for the California Floristic Province. California serpentine outcrops have low species richness compared to nonserpentine areas of similar size, but the turnover in species among serpentine regions, or beta diversity, is extremely high (Harrison and Inouye, 2002). Whereas there are many serpentine endemic species, composing approximately 10% of the plants unique to California (Kruckeberg, 1984), many of these are narrowly distributed. Although serpentine species are rare in absolute terms, owing to the small land area comprised of serpentine habitats, their diversity is far greater than might be expected given the apparent harshness of the habitat and the small land area involved. From a phylogenetic perspective, tolerance to serpentine is often gained and lost, but endemism is less common (Anacker, 2010). The number of endemic serpentine species in a region increases with the proportion of the region exhibiting serpentine soil and the time since exposure (Harrison et al., 2004; Anacker, 2010). These patterns suggest that adaptation to serpentine provides a stimulus for speciation and is not simply a refuge for relictual taxa. Notably, however, the radiation of serpentine clades is uncommon, except where nonserpentine substrates are absent (Spencer and Porter, 1997; Pepper and Norwood, 2001; Patterson and Givnish, 2004; Heads, 2008).

WHAT DOES IT TAKE TO LIVE ON SERPENTINE?

Serpentine soils are characterized by a low Ca:Mg ratio, deficiency in essential macronutrients, elevated concentrations of heavy metals, and low water-holding capacity. They are typically rocky, shallow soils vulnerable to erosion. These inhospitable soils harbor sparse vegetation, further contributing to nutrient limitation and erosion. Serpentine soils vary from bare, rocky outcrops to deeper, more fertile grasslands even within local areas (Alexander et al., 2007). This variation likely creates an evolutionary mosaic that is oversimplified by a dichotomous view of serpentine versus nonserpentine (Brady et al., 2005; Alexander et al., 2007; Springer, 2007; Kazakou et al., 2008).

Adaptations to serpentine soils have been extensively described and recently reviewed (Brady et al., 2005; Kazakou et al., 2008). Evolution of physiological tolerance to serpentine soils appears to carry a corresponding morphology (Kruckeberg, 1954). Plants having the "serpentine syndrome" are typically adapted to dry soils and are of smaller stature than their nonserpentine relatives. Many have strongly developed root systems, presumably to facilitate uptake of water and nutrients (Brooks, 1987). Other species down-regulate lateral root growth in high Mg soils, allocating more resources to deep-growing roots important in dry conditions.

The low Ca:Mg ratio presents a physiological challenge not only because of Ca's importance in plant growth and signal transduction but also because high levels of Mg are antagonistic to plant uptake of Ca (Marschner, 2002; Brady et al., 2005). The challenge of low Ca:Mg ratio in serpentine soils has elicited a wide range of adaptive responses based on either ion exclusion at the root/soil interface, selective translocation of Ca from root to shoot, sequestration of Mg in the vacuole, or internal mechanisms of tolerance. Selective uptake or transport of Ca, and sometimes lowered uptake of Mg, have been demonstrated in several serpentine lineages (Walker et al., 1955; O'Dell et al., 2006; Asemaneh et al., 2007). Several species adapted to serpentine environments have higher external and internal Mg requirements than their nonserpentine relatives (Brady et al., 2005). Serpentine forms of *Lasthenia californica* and *L. gracilis* show physiological tolerance to internal ionic stresses, rather than exclusion of Mg ions (Rajakaruna et al., 2003b). Although these many studies have shown a wide range of gross adaptations to the low Ca:Mg ratio in serpentine soil, we still have much to learn about the precise physiological mechanisms of the toxic action of Mg and plant resistance (Asemaneh et al., 2007). Furthermore, the variety of responses shown suggests that the physiological basis for tolerating low Ca:Mg may involve more than one mechanism in a given species (Brady et al., 2005).

Nutrient limitation is a major stressor in serpentine environments, and the mechanisms of adaptation range widely. The deficiency of macronutrients results from low amounts of organic material and the lack of P and K in parent materials. The primary macronutrient deficit appears to vary globally (reviewed in Kazakou et al., 2008)—California serpentine is typically deficient in N, whereas K is the primary deficiency in Europe.

Serpentine soil often exhibits elevated levels of heavy metals, such as iron, nickel, zinc, cadmium, cobalt, chromium, and manganese, which present another set of challenges for plants. The presence of heavy metals in soils can affect plants through direct toxicity, resulting in stunting and chlorosis, antagonism with other nutrients (which can lead to iron deficiency), and inhibition of root penetration and growth (Antonovics et al., 1971). Soil pH mediates heavy metal levels and can both ameliorate and increase their effect on plants (Wang et al., 2006). Adaptive mechanisms include exclusion of metals (either by restricting them to the roots or through the absence of any uptake mechanism), compartmentalization of metals in various organs, or toxicity tolerance. Some plants concentrate heavy metals in their tissues at levels higher than in the soil. These hyperaccumulators (especially of Ni) have received considerable attention, due in part to their economic potential for the bioextraction of valuable metals. Hyperaccumulation as an adaptation to serpentine soils is uncommon. For example, the majority of Ni accumulators (~400 species) are serpentine endemics, yet these species represent a mere 2% of all serpentine species worldwide (Proctor, 1999; Kazakou et al., 2008; Reeves and Adiguzel, 2008).

Water deficiency has been suggested as another stressor on serpentine soils (Proctor and Woodell, 1975; MacNair and Gardner, 1998; Proctor, 1999; Gardner and MacNair, 2000; Sambatti and Rice, 2007), but this aspect of serpentine tolerance has received relatively little attention (Brady et al., 2005). Water is sometimes (but not always) less available in serpentine sites than in surrounding habitats (Alexander et al., 2007).

Biotic factors may also contribute to serpentine adaptation. For example, plants on serpentine experience reduced competition from invasive species (Kruckeberg, 1984; Harrison, 1999; Gram et al., 2004; Going et al., 2009). Nickel hyperaccumulation may confer a defense against herbivory (Martens and Boyd, 1994), and symbioses with serpentine-tolerant ectomycorrhizal communities may facilitate adaptation to edaphic stressors on serpentine (Schechter and Bruns, 2008; Urban et al., 2008; Gonçalves et al., 2009; Moser et al., 2009).

The pathogen refuge hypothesis (Kruckeberg, 1992) suggests that plants may escape pathogen pressure on serpentine, either through reduced horizontal transmission rates in sparse serpentine plant communities (Thrall et al., 2007) or through lowered symptom-associated damage. This has been confirmed for extreme serpentine specialists in the genus *Hesperolinon* (Springer, 2009), perhaps through an ability to selectively uptake Ca, which is required to initiate an effective immune response. Less specialized species, therefore, may be more susceptible to disease on serpentine soil (Springer, 2007, 2009). Rather than receiving refuge from pathogens, plants on serpentine may face increased biotic as well as abiotic stressors.

HOW IS SERPENTINE INVOLVED IN SPECIATION?

Serpentine is an excellent system for examining some of the most fundamental questions about speciation. First, how is adaptation involved in speciation? Most studies of speciation indicate a central role of natural selection, but our understanding of the strength of selection and its mode of action is incomplete (Coyne and Orr, 2004; Sobel et al., 2009). Second, how geographically isolated do populations need to be in order to become new species? The idea that speciation can proceed between geographically proximate populations is one of the most controversial in evolutionary biology.

To examine the role of serpentine in the formation of new species, it is first necessary to define what we mean by species. Like most plant taxa, most serpentine species are initially recognized by their consistent morphological differences from related species (Cronquist, 1978). With the rise of biosystematics, in-depth crossing studies were made of many plant groups to determine the interfertility of species. These studies at times provided surprising results, with the discovery of strong crossing barriers within some taxonomic species but weak barriers between others. Biosystematic studies of groups including serpentine species were no exception (e.g., Lewis and Lewis, 1955; Kruckeberg, 1957; Ornduff, 1966).

Using the biological species concept, speciation can be considered the process whereby populations evolve genetically based barriers to gene exchange (Dobzhansky, 1940; Mayr, 1942). A major aim of speciation studies has been to identify and explain the traits or genes that cause reproductive isolation. Isolating mechanisms can include ecological factors, such as genetically based differences in habitat affinity or flowering time, in addition to lower success of hybrid crosses and lower fertility of hybrids. Speciation may be most likely to occur when there is divergent selection for traits that confer reproductive isolation as a by-product, such as flowering time differences. Unfortunately, most studies consider only a subset of all possible isolating barriers, and this can severely bias the understanding of speciation mechanisms (Ramsey et al., 2003; Lowry et al., 2008). Moreover, the traits that cause reproductive isolation and their fitness effects are rarely estimated. Identifying and explaining the evolution of isolating barriers is crucial to explaining the origin of species.

Serpentine soils can contribute to speciation in two primary ways. First, adaptation to serpentine soils can contribute indirectly to pre- or postzygotic reproductive barriers that genetically isolate serpentine populations from nonserpentine relatives. Second, the patchy distribution of serpentine can contribute to the geographic isolation of populations. We examine each of these in turn.

We surveyed the literature for empirical studies that have documented mechanisms of reproductive isolation between closely related serpentine and nonserpentine species or between serpentine and nonserpentine populations of the same taxonomic species (Table 4.1). We searched the ISI Web of Science database with the criteria Topic = (serpentine AND ultramafic) and Timespan = All Years through October 2009, and looked for pertinent references cited within these publications. Studies were included if they documented a form of pre- or postzygotic isolation between serpentine and nonserpentine populations or species.

Habitat Isolation

Adaptive trade-offs are a common theme in evolutionary biology—adaptation to one environment is expected to reduce performance in other environments. When accompanied by trade-offs, adaptive differentiation on a local spatial scale can be an effective barrier to gene flow because migrants between habitats have reduced fitness and assortative mating between similarly adapted individuals is increased (Dobzhansky, 1937; Mayr, 1947; Coyne and Orr, 2004). If migrants between serpentine and nonserpentine soils are selected against, most matings will occur between similarly adapted individuals. In this way, adaptation to serpentine can

TABLE 4.1 Forms of Reproductive Isolation Documented between Serpentine (S) and Nonserpentine (NS) Sister Species or Populations

Taxa	Habitat Isolation	Temporal Isolation	Pollinator Isolation	Postmating Isolation	References
Mimulus nudatus (S) and *M. guttatus* (NS)	X	X	X	X	Macnair and Gardner, 1998; Gardner and Macnair, 2000; Hughes et al., 2001
Mimulus pardalis (S) and *M. guttatus* (NS)	X	X	—	—	Macnair and Gardner 1998; Hughes et al., 2001
Mimulus pardalis (S) and *M. marmoratus* (NS)	X	—	—	X	Hughes et al., 2001
Gilia capitata ssp. *capitata* populations	X	—	—	—	Kruckeberg, 1951
Cleome heratensis (S) and *C. foliolosa* (NS)	X	—	—	—	Asemaneh et al., 2007
Calochortus umbellatus (S) and *C. uniflorus* (NS)	—	—	—	X	Ness et al., 1990
Layia glandulosa ssp. *lutea* (NS) and *L. discoidea* (S)	—	—	—	X	Baldwin, 2005
Streptanthus glandulosus complex	X	—	—	X	Kruckeberg, 1951, 1957
Achillea borealis populations	X	—	—	—	Kruckeberg, 1951
Achillea millefolium populations	X	—	—	—	O'Dell and Claassen, 2006
Collinsia sparsiflora populations	X	X	—	—	Wright et al., 2006; Wright and Stanton 2007
Alyssum inflatum populations	X	—	—	—	Ghasemi and Ghaderian, 2009
Cerastium alpinum populations	X	—	—	—	Berglund et al., 2004
Gilia capitata ssp. *capitata* populations	X	—	—	—	Kruckeberg, 1951
Helianthus bolanderi ssp. *exilis* populations	X	X	—	—	Sambatti and Rice, 2007
Mimulus guttatus populations	X	—	—	—	Macnair and Gardner, 1998; Gardner and Macnair, 2000; Hughes et al., 2001
Thlaspi goesingense populations	X	—	—	—	Reeves and Baker, 1984

NOTE: Dashes indicate no known data. Based on an ISI Web search; see text for search criteria.

contribute significantly to prezygotic isolation between the serpentine lineage and a nonserpentine progenitor and protect less abundant serpentine forms from the swamping effects of gene flow. Furthermore, hybrids between differently adapted populations may be relatively unfit, not because of intrinsic genetic incompatibilities but because they are poorly adapted to available habitats (Dobzhansky, 1937). Such habitat isolation, or reproductive isolation based on selection against migrants between habitats and/or hybrid unfitness in available habitats, is particularly relevant for serpentine adaptation, where gene flow between adjacent populations on and off serpentine could otherwise homogenize populations.

We know relatively little about the specific costs of serpentine tolerance. The cost, and thus the magnitude of trade-offs, should vary according to the specific adaptations and peculiarities of the physiology of different plant lineages, as well as with the biotic and abiotic conditions in the surrounding environment (Elmendorf and Moore, 2007). On serpentine, a small, drought-adapted stature and deep roots are advantageous traits but may reduce the growth rate and competitive ability off serpentine. Indeed, Kruckeberg (1954) showed that serpentine endemics are competitively excluded by nonserpentine plants on "normal" soil. More recent ecological studies have also shown that serpentine plants are poor competitors on higher nutrient soils (Rice, 1989; Huenneke et al., 1990; Jurjavcic et al., 2002). A pleiotropic trade-off between early reproduction, which should confer a fitness advantage in drought conditions, and growth has been confirmed in the serpentine-tolerant *Microseris douglasii* (Gailing et al., 2004). There is some suggestion that plants adapted to serpentine have intrinsically lowered growth rates even when grown on more fertile soil (Sambatti and Rice, 2007; Brady unpublished data; Schemske unpublished data), which is expected from plants adapted to stressful environments (Grime, 1977). Adaptation to serpentine may result in an increased demographic susceptibility to herbivory in *Collinsia sparsiflora* (Lau et al., 2008). In contrast, no evidence has been found for a cost to metal tolerance or tolerance of low Ca:Mg ratios in serpentine plants (reviewed in Brady et al., 2005), but this has rarely been directly addressed.

Adaptation to different edaphic habitats likely constitutes an important form of reproductive isolation between serpentine plants and their nonserpentine relatives. Table 4.1 includes examples in which reciprocal soil treatments or ion addition experiments were performed, some fitness component was measured, and fitness trade-offs were quantified between soil types. The apparent strength of the trade-off in performance on serpentine and nonserpentine habitats varied among studies, but some habitat isolation was found in nearly every study (Table 4.1).

Phenological Isolation

Shifts in flowering time are often associated with adaptation to serpentine soils, most likely as result of selection for earlier reproduction in drought conditions or as a phenotypically plastic response to earlier drying of the soil (Hughes et al., 2001; Brady et al., 2005; Wright et al., 2006). The serpentine endemics *Mimulus pardalis* and *M. nudatus* flower earlier than their progenitor *M. guttatus* (MacNair and Gardner, 1998), as do serpentine populations of *Helianthus bolanderi* ssp. *exilis* (Sambatti and Rice, 2007) and *Collinsia sparsiflora* (Wright et al., 2006; Wright and Stanton, 2007). These shifts in flowering time are sometimes enough to reproductively isolate differentially adapted populations (Rajakaruna and Whitton, 2004). Dry conditions and early flowering may also be correlated with a reduction in flower size, and flower size differences confer mechanical floral isolation between the serpentine endemic *M. nudatus* and its progenitor *M. guttatus* (Gardner and MacNair, 2000). Although in some cases, earlier flowering may be a purely plastic response to the serpentine habitat, it may allow further divergence to proceed in habitat affinity or other isolating factors and can therefore be important in isolating nearby populations (Levin, 2009).

Postzygotic Isolation

If adaptation to serpentine involves catastrophic selection leading to genomic reorganization in a small founder population (Lewis, 1962), then the process of serpentine adaptation could also confer postzygotic reproductive isolation. This was proposed to occur during the formation of the rare serpentine endemic *Clarkia franciscana*, which is isolated from its nearest relatives by intrinsic postzygotic barriers (Lewis and Raven, 1958). However, further genetic work supported an older origin and a more widespread former distribution of *C. franciscana* (Gottlieb and Edwards, 1992). It is unclear how often catastrophic selection contributes to reproductive isolation in serpentine systems. Postzygotic isolation could also evolve over time between geographically isolated serpentine and nonserpentine relatives or different isolated populations of serpentine plants through the accumulation of intrinsic genetic incompatibilities (Kruckeberg, 1957) or from differential adaptation that renders hybrids unfit in the available niches. Few studies have addressed the strength of postmating or postzygotic isolation between serpentine and nonserpentine adapted plants (Table 4.1).

Spatial Isolation among Serpentine Outcrops

The fragmented and patchy distribution of serpentine outcrops also contributes to the genetic isolation and divergence of serpentine species. Paleoendemics that have become restricted to serpentine will lose the intervening nonserpentine populations and experience reduced gene flow. For example, populations of *Streptanthus glandulosus* show substantial genetic isolation and often partial postzygotic isolation among distant serpentine outcrops (Kruckeberg, 1957; Mayer and Soltis, 1994, 1999; Mayer et al., 1994). Neoendemics that arise from the same progenitor species may similarly be isolated from each other and proceed along independent evolutionary trajectories. Again, we find an example in *Streptanthus*, in which several local endemics have been derived from within *S. glandulosus* and show low interfertility with each other (Kruckeberg, 1957).

THE BIOGEOGRAPHY OF SPECIATION
ON SERPENTINE

Traditional models of speciation are classified geographically, and it is a central goal in evolutionary biology to understand their relative importance. We outline these models and showcase empirical examples from serpentine species.

Allopatric Speciation

The well-accepted allopatric model of speciation is that populations become geographically isolated, and through selection or genetic drift, gradually acquire genetic differences that pleiotropically confer reproductive isolation (Muller, 1942; Mayr, 1947, 1963). If incipient species come into secondary sympatry, under this model they will coexist only if the reproductive isolation acquired in allopatry is completely effective in preventing gene flow. The isolating mechanisms can act prezygotically in preventing the species from successfully mating, for example, (Funk, 1998; Rundle et al., 2000) or postzygotically by reducing the fitness of their hybrids, either through intrinsic genetic incompatibilities that cause inviability or infertility (Dobzhansky, 1937) or through poor competitive or mating success in nature (Hatfield and Schluter, 1999). Because of extensive empirical support and a lack of theoretical objections, it is widely acknowledged that allopatric speciation is the most common mechanism of speciation (Coyne and Orr, 2004).

Among serpentine species, a couple of well-studied systems support allopatry as the root cause of divergence. First, there is evidence for incipient allopatric divergence in the widespread species *S. glandulosus*, as both genetic distance and hybrid infertility increase with geographic distance, and geographically isolated populations show morphological divergence as well (Kruckeberg, 1957; Mayer and Soltis, 1994, 1999; Mayer et al., 1994). In addition, phylogenetic work on *Lasthenia* revealed the presence of two cryptic species on serpentine outcrops in the California Coast Ranges, each comprising populations of different edaphic races (Chan et al., 2002; Rajakaruna et al., 2003a). The phylogenetic structuring of these two species across a latitudinal range suggests that although they are now partially sympatric, allopatry was likely during early stages of divergence.

Speciation with Gene Flow: The Theory of Sympatric and
Parapatric Divergence

In contrast to allopatry, sympatric and parapatric speciation involves divergence in the face of gene flow between geographically overlapping or adjacent populations. Theoretically, however, divergence in the face of gene flow is problematic. As Felsenstein (1981) pointed out, recombination between genes for mating preferences and other population-specific traits should lead to a breakdown in species divergence. Imagine a simple situation with a two-allele locus for mating preference and another two-allele locus for a population-specific trait like habitat affinity. With gene flow and recombination, some individuals adapted to the habitat of population A will prefer to mate with population B, and vice versa, reducing the likelihood of divergence.

The most controversial mode of divergence with gene flow is sympatric speciation, in which selection initiates assortative mating without any geographic isolation. Even with strong disruptive selection, sympatric speciation is theoretically implausible except via polyploidization, since linkage disequilibrium must build up de novo (Mayr, 1947; Coyne and Orr, 2004). It is also difficult to exclude past allopatry and range shifts as explanations of presently sympatric species (Mayr, 1947; Templeton, 1981; Coyne and Price, 2000). The few well-documented empirical examples suggest that it may occur under unusual ecological or genetic situations (e.g., Bush, 1969; Ramsey and Schemske, 2002; Seehausen et al., 2008). Because serpentine occurs in discrete patches, sympatric speciation has not been proposed for serpentine plants; however, serpentine endemics often occur within plausible dispersal distance of nonserpentine relatives, and parapatric speciation is often posited.

It is possible that strong selection might overcome the homogenizing effects of gene flow between parapatric populations that are not isolated by geographic barriers. The pioneering work of Clausen, Keck, and Hiesey (1958) demonstrated that local adaptation is common in plant populations and species distributed across altitudinal and climatic gradients. Parapatric populations can be distributed along gradual or steep environmental gradients (as in the boundary between serpentine and nonserpentine soils) or in distinct habitat patches distributed across the landscape in a stepping-stone–like pattern (as are serpentine outcrops throughout many tectonic contact zones). Serpentine systems are ideally suited to test the theory of parapatric speciation because populations adapted to serpentine soils are typically distributed as a mosaic of distinct patches surrounded by adjacent populations growing on nonserpentine soils. Indeed, Coyne and Orr (2004) proposed that edaphic plant specialists provide an excellent opportunity to assess the likelihood of parapatric speciation.

The conditions for parapatric speciation are very restrictive, requiring some combination of strong selection and assortative mating (Coyne and Orr, 2004). Population genetic models of speciation investigate the conditions under which barriers to gene flow can arise between populations that are initially exchanging genes (Endler, 1977; Kirkpatrick and Ravigne, 2002; Gavrilets, 2003; Bolnick and Fitzpatrick, 2007). Two main types of models can be described, distinguished by the spatial distribution of populations. In clinal models, the underlying environmental factors are assumed to vary along a geographic gradient, whereas in stepping-stone models, the habitats are discrete patches distributed in a mosaic across the landscape. In both models, it is assumed that there is geographic variation in selection on a local spatial scale and that gene flow between populations experiencing different selective regimes is the major obstacle to both adaptive differentiation and speciation. The major difference is that the magnitude of gene flow between locally adapting populations is greater for clinal than stepping-stone models, the latter being nearly equivalent to allopatric divergence but without a geographic barrier (sometimes referred to as peripatric divergence).

In clinal models, the steepness of the environmental gradient and the magnitude of selection and gene flow are key factors that determine the opportunity for adaptive divergence and speciation. Steep gradients may restrict adaptive gene substitutions to mutations of large phenotypic effect because the spatial extent of intermediate habitat is limited. Although it is theoretically possible that strong local selection can eliminate all foreign genes that migrate from neighboring populations, achieving complete reproductive isolation between parapatric populations probably requires the evolution of additional isolating barriers. For example, Caisse and Antonovics (1978) modeled assortative mating in plants as a flowering time difference genetically linked to local adaptation. Complete isolation only evolved between populations at the poles of the cline. Although this model had restrictive conditions, the parameter values may be realized in some serpentine habitats.

Divergence and speciation of serpentine plants via a pure stepping-stone model is somewhat analogous to the evolution of species that colonize island archipelagos in marine systems. In both cases, there is a low probability of migration between favorable "islands" of habitat. Successful colonization may provide an opportunity for rapid divergence due to both low gene flow between ancestral and derived populations and novel environmental conditions in the new habitat. Environmental heterogeneity and the geographic area, both of which are probably limiting factors in the case of serpentine adaptation, will determine the opportunity for further diversification.

Figure 4.1 illustrates some of the many possible routes to plant speciation on serpentine. In stage I, plants from the ancestral nonserpentine population colonize the adjacent serpentine habitat and either evolve into a new species in situ (B_s; clinal parapatric speciation) or become locally adapted to serpentine soil without speciation (A'_s; clinal parapatric divergence). In stage II, seeds from the serpentine habitat of the ancestral range colonize nearby serpentine. If the founding population is the new species (B_s), it may persist without further divergence. If the founding population had evolved serpentine tolerance in the ancestral habitat (A'_s), it may speciate in the new site ($A'_s \rightarrow B_s$; stepping-stone parapatric speciation). In

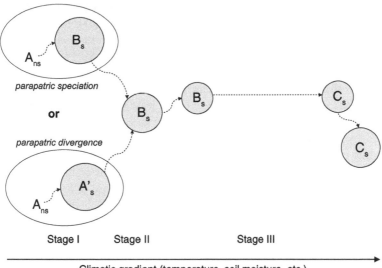

FIGURE 4.1. Hypothetical scenarios for parapatric divergence and speciation on serpentine soils. Shaded circles denote distinct serpentine habitats, and the ellipse represents the geographic distribution of an ancestral species adapted to nonserpentine soils (A_{ns}). Dashed lines indicate migration/colonization. The geographic region depicted spans a climatic gradient. In stage I, plants from the ancestral nonserpentine population colonize the adjacent serpentine habitat and either evolve into a new species (B_s; clinal parapatric speciation) or become locally adapted to serpentine soil (A'_s; clinal parapatric divergence). In stage II, seeds from the serpentine habitat of the ancestral range colonize a serpentine habitat within a similar climatic region, and the founding population either persists without further divergence if migration is from the new species ($B_s \rightarrow B_s$), or speciates in the case of migration from the population adapted to serpentine ($A'_s \rightarrow B_s$; stepping-stone speciation). In stage III, seeds from the new serpentine-adapted species colonize a serpentine habitat in a different climatic region and evolve into a new serpentine-adapted species ($B_s \rightarrow C_s$; stepping-stone speciation).

stage III, seeds from the new serpentine-adapted species (B_s) colonize a distinct serpentine habitat in a different climatic region and evolve into a new serpentine-adapted species ($B_s \rightarrow C_s$; stepping-stone parapatric speciation).

What is the relative importance of clinal and stepping-stone mechanisms of plant speciation on serpentine? If clinal speciation predominates, then we expect repeated and independent colonization of serpentine habitats followed by adaptation and the evolution of isolating barriers in situ. If stepping-stone speciation predominates, we expect that the initial adaptation to serpentine habitats is followed by subsequent diversification as serpentine-adapted species colonize other sites. The opportunities for speciation on serpentine via a pure stepping-stone model may be somewhat limited because it requires both dispersal to a new island and serpentine adaptation simultaneously. Rather, stepping-stone speciation may be more likely after a population has acquired some serpentine adaptation in parapatry. Phylogenetic data may provide the best opportunity for evaluating the relative importance of clinal and stepping-stone models (Anacker, 2010).

A major challenge to demonstrating a parapatric origin for species in nature is excluding the possibility that speciation (or at least the initial evolution of traits involved in assortative mating) occurred during an earlier period of allopatry. For example, young species that evolve in allopatry and whose ranges subsequently come into contact may appear identical to parapatric species in being young, ecologically divergent, and reproductively isolated along their contact zone. Genetic data may be able to distinguish these situations because substantial allopatry should lead to divergence at both neutral and selected loci, whereas neutral loci should be fairly homogenized during parapatric divergence. Extensive hybridization and introgression on secondary contact, however, would obscure these differences and make a coalescent-based approach that incorporates variation in the timing of gene tree divergence among loci more appropriate (Hey, 2006; Becquet and Przeworski, 2009).

A Case Study of the Early Stages of Parapatric Divergence: Leptosiphon parviflorus

Leptosiphon parviflorus (Polemoniaceae) provides a clear example of differentiation among parapatric populations growing on and off adjacent serpentine soils. *L. parviflorus* is a small, spring-flowering annual herb abundant in open and wooded habitats in the California Floristic Province (Hickman, 1993). Populations exhibit a single gene polymorphism for flower color, which ranges from white to deep pink, with pink dominant to white (Schemske, unpublished data). At Jasper Ridge Biological Preserve in San Mateo County, serpentine and nonserpentine (sandstone) soils are found in close proximity (<10 m), and *L. parviflorus* grows on both soil types. Populations on serpentine are typically pink-flowered and flower earlier than the adjacent white-flowered populations on nonserpentine soil. In addition, *L. parviflorus* exhibits striking differentiation in tolerance to serpentine soils. In reciprocal transplant experiments conducted in the greenhouse using field-collected soil, plants performed best on their native soil type. "Sandstone" plants grown on serpentine soil have very low survival, and the few survivors are small and produce few flowers. In comparison, "serpentine" plants grown on sandstone soil have high survival, equivalent to that on serpentine soil, but reach a smaller size than "sandstone" plants (Schemske, unpublished data).

We used the *L. parviflorus* system at Jasper Ridge to investigate the dynamics of parapatric divergence across a local contact zone. Although the transplant experiments demonstrate that populations are locally adapted to their native soil, we wanted to know whether the sharp boundary between the pink- and white-flowered plants was maintained by strong selection or simply a lack of migration. Moreover, we wished to understand how the strength of selection varied on different putatively adaptive plant traits. To accomplish these objectives, we estimated differentiation in putatively neutral molecular markers (F_{ST}) across the contact zone and compared this differentiation to estimates of quantitative genetic differentiation (Q_{ST}) for a variety of morphological and life history traits and components of fitness on serpentine. High F_{ST} for neutral markers indicates low levels of gene flow, whereas low F_{ST} indicates extensive migration and gene flow. If estimates of Q_{ST} are higher than the estimates of F_{ST}, then this is strong evidence that quantitative trait differences are maintained by selection in the face of gene flow. The magnitude of the difference between Q_{ST} and F_{ST} can further be used as evidence for the strength of selection on different traits (McKay and Latta, 2002).

Details of this work can be found in Ward (2000) and Figure 4.2. The parental generation was collected in the field as seed and raised in the greenhouse before crossing to eliminate maternal effects. One hundred individuals were sampled from each of two populations: one on serpentine soil and one on sandstone soil, 109 m apart. Morphological and life history traits were assayed in progeny on standard greenhouse soil mix, and several measures of fitness on serpentine were assayed in progeny sown directly on field-collected serpentine soil in the greenhouse. An estimate of molecular genetic differentiation (F_{ST}) was derived from analysis of 101 amplified fragment length polymorphism (AFLP) markers in the same progeny (Ward and Schemske, unpublished data).

The results show that differentiation in several plant traits is maintained by selection in the face of substantial migration and gene flow. First, we found that differentiation in the putatively neutral AFLP molecular markers was very low (F_{ST} = 0.120), which is evidence for high gene flow. Second, we found that measures of population differentiation in flower color, flowering phenology, plant size, and fitness on serpentine were significantly higher than the neutral expectation provided by F_{ST} (Figure 4.2), suggesting that these phenotypic traits are subject to strong disruptive selection on the two soil types. Thus, these populations maintain marked adaptive differentiation over a microspatial scale despite considerable gene flow.

L. parviflorus is an interesting model system for the study of serpentine adaptation and speciation. Populations at Jasper Ridge meet the conditions required for parapatric speciation in that strong divergent selection leads to adaptive genetic differentiation despite considerable gene flow. These populations experience strong selection on traits associated with use of the serpentine habitat, selection against migrants from sandstone to serpentine soils, and at least an association of phenotypes affecting fitness (serpentine tolerance) and mate choice (as yielded by differing flowering times). Hence, the system presents an ideal opportunity to examine

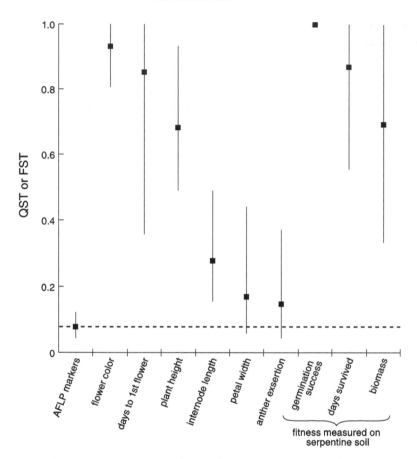

FIGURE 4.2. Estimates of *Leptosiphon parviflorus* population differentiation for putatively neutral AFLP markers (F_{ST}) and quantitative traits (Q_{ST}) measured on progeny grown in the greenhouse on regular nursery mix or field-collected serpentine soil. The dashed line tracks the value of F_{ST} for easy comparison. Controlled crosses for Q_{ST} measures followed the North Carolina II breeding design. Variance components were calculated using restricted maximum likelihood with SAS Proc MIXED METHOD=REML, with the model trait = pop dam(pop) sire(pop) dam*sire(pop). Additive genetic variance was calculated as V_a = 4*sire(pop), and Q_{ST} was calculated from variance components as Q_{ST} = pop/(2* V_a + pop). F_{ST} was calculated from the software program Arlequin (Schneider et al., 2000). Values are the mean of 1000 bootstrap replicates, with 95% bootstrap confidence intervals.

the adaptation that parapatric populations undergo in what must be the first steps of speciation. Additional studies are needed to determine whether the adaptive differentiation among populations on and off serpentine at Jasper Ridge may ultimately result in parapatric speciation or whether additional external factors are required to further isolate the populations.

A Case Study of Putative Completion of Parapatric Speciation:
Layia discoidea

Layia discoidea, a rare annual herb known only from a small area of serpentine in the inner South Coast Ranges of California, presents perhaps the most well-documented example of an insular neoendemic arising from within the geographic range of its progenitor. Although morphologically distinctive enough to be initially assigned to a new genus, crossing studies and phylogenetic analysis of *L. discoidea* show that it was derived from within the widespread *L. glandulosa* and is most closely related to populations of *L. glandulosa* growing on sandy nonserpentine soils a few kilometers away (Clausen et al., 1947; Gottlieb et al., 1985; Ford and Gottlieb, 1989, 1990; Baldwin, 2005). Moreover, molecular clock dating and the young age of the isolated South Coast Range serpentine exposures support a Quaternary origin of *L. discoidea* (Baldwin, 2005). It will be important to better understand the mechanisms of reproductive isolation between *L. discoidea* and nearby *L. glandulosa*. The two species are highly interfertile in artificial crosses but are presumably strongly isolated by ecological differences because there is no evidence of hybridization and gene flow (Baldwin, 2005). For these species, a prior period of large-scale allopatry is highly unlikely, but it is unclear how much spatial isolation was involved in their initial divergence. Whether they diverged, like *Leptosiphon*, across a contact zone that is no longer present or followed a stepping-stone or peripatric model may never be deciphered. Studies of the strength of different isolating mechanisms, like habitat and floral isolation, would better indicate the likelihood of clinal parapatric divergence.

Assigning speciation events an allopatric or parapatric label may impose an artificial dichotomy on what is essentially a continuum. The amount of spatial separation necessary to disrupt gene flow for divergence to proceed depends on the strength of disruptive selection and whether any selected traits confer reproductive isolation. With limited spatial isolation and strong disruptive selection on traits that pleiotropically confer reproductive isolation, speciation may occur rapidly. Serpentine plants have played and will continue to play a prominent role in biologists' attempts to understand the interplay between gene flow and selection in speciation.

FUTURE DIRECTIONS

Although the plants growing on serpentine have fascinated botanists for many decades, the work of connecting the patterns of endemism and adaptation to speciation processes has just begun. Much of what we understand comes from the well-studied flora of California; we know relatively little about the origin of serpentine taxa in other geographic regions, even though serpentine is a widespread phenomenon. Regions vary in the age and extent of serpentine exposures and in the amount of endemism. For example, the relatively small island of New Caledonia has 1755 species of serpentine plants, whereas the relatively large island of Japan has only 50 endemics (reviewed in Anacker, 2010). California is on the endemic-rich side of the spectrum, and it is unclear how much of what we know from California applies across the broad range of serpentine regions. It will be important to conduct studies in other regions to understand the aspects of serpentine that promote and constrain neoendemism.

We also need to understand better the physiological and genetic basis of adaptation to serpentine to elucidate the trade-offs involved and the effects on reproductive isolation. How many and what types of mutations are involved, and what pleiotropic effects do they have on other plant traits? What are the exact physiological mechanisms that allow serpentine tolerance, and how do these affect the ability to live off serpentine? How do changes in these traits affect mating patterns? Prior studies of the trade-offs of serpentine tolerance are few, are limited taxonomically and geographically, and often involve transplants or ion addition treatments that do not address the effects of biotic interactions, such as competitors, herbivores, and pathogens, which are likely to be very important in the wild.

Some questions may best be answered in a comparative context. Does the strength of trade-offs to living on versus off of serpentine differ between tolerator and endemic species? This could be addressed in a comparative transplant study with multiple pairs of serpentine and nonserpentine sister taxa, with the expectation that endemics would show a greater cost to living on serpentine than tolerators. We could also ask whether the magnitude of trade-offs vary between types of serpentine habitats, that is, serpentine grasslands versus serpentine barrens or across a moisture gradient. Comparative studies could also better circumscribe the physiological or ecological traits that are coincident with serpentine tolerance or endemism, such as intrinsic growth rates, flowering time, or root depth, and whether these traits are typically plastic responses to the soil environment or genetically based.

Serpentine endemics may provide the best opportunity to address parapatric speciation models with real data. The many evolutionarily independent examples of serpentine endemism should make it possible to find all the different stages of the process of parapatric divergence and speciation. It may be most productive to further develop some model serpentine systems that are experimentally tractable and have extensive genetic tools. *Mimulus, Layia, Helianthus, Thlaspi, Lasthenia*, and *Leptosiphon* are all good candidates. All have genetic tools in development (or close relatives with genetic tools), their natural history is well documented, they are experimentally tractable, and they represent a range of stages of divergence and degrees of endemism. In these systems, it should be possible to determine the traits targeted by selection on serpentine soil, the effects of those traits on reproductive isolation, and the genetic and/or physiological basis of those traits.

The work on serpentine plant origins to date confirms the hypothesis that serpentine is a driver of speciation and that neoendemics make up an important portion of serpentine plant diversity. It will be important to protect as many serpentine patches as possible to conserve the evolutionary potential of serpentine plants, as each patch may be unique. We also know that it will be important to conserve these areas as natural laboratories for understanding general processes of adaptation and speciation.

Acknowledgments

We are grateful to Doug Ewing for expert greenhouse care of *Leptosiphon*, Dena Grossenbacher for greenhouse assistance, Barb Frewen for lab assistance, and Nona Chiarello for facilitating fieldwork at Jasper Ridge Biological Preserve. We also thank the editors, Maureen Stanton, Bruce Baldwin, and Jennifer Yost for helpful comments on the manuscript.

LITERATURE CITED

Alexander, E. B., Colemena, R. G., Keeler-Wolf, T., and Harrison, S. (2007) *Serpentine Geoecology of Western North America*. Oxford University Press, New York.

Anacker, B. L. (2010) Phylogenetic patterns of serpentine plant endemism and diversity. In *Serpentine as a Model in Evolution and Ecology* (eds. S. Harrison and N. Rajakaruna). University of California Press, Berkeley.

Antonovics, J., Bradshaw, A. D., and Turner, R. G. (1971) Heavy metal tolerance in plants. *Advances in Ecological Research, 7*, 1–85.

Asemaneh, T., Ghaderian, S. M., and Baker, A. J. M. (2007) Responses to Mg/Ca balance in an Iranian serpentine endemic plant, *Cleome heratensis* (Capparaceae) and a related non-serpentine species, *C. foliolosa*. *Plant and Soil, 293*, 49–59.

Baldwin, B. G. (2005) Origin of the serpentine-endemic herb *Layia discoidea* from the widespread *L. glandulosa* (Compositae). *Evolution, 59*, 2473–79.

Becquet, C., and Przeworski, M. (2009) Learning about modes of speciation by computational approaches. *Evolution, 63*, 2547–62.

Berglund, A. B. N., Dahlgren, S., and Westerbergh, A. (2004) Evidence for parallel evolution and site-specific selection of serpentine tolerance in *Cerastium alpinum* during the colonization of Scandinavia. *New Phytologist, 161*, 199–209.

Bolnick, D. I., and Fitzpatrick, B. M. (2007) Sympatric speciation: models and empirical evidence. *Annual Review of Ecology, Evolution, and Systematics, 39*, 459–87.

Brady, K. U., Kruckeberg, A. R., and Bradshaw, H. D. (2005) Evolutionary ecology of plant adaptation to serpentine soils. *Annual Review of Ecology, Evolution, and Systematics, 36*, 243–66.

Brooks, R. R. (1987) *Serpentine and Its Vegetation: A Multidisciplinary Approach*. Dioscorides, Portland, OR.

Bush, G. L. (1969) Sympatric host race formation and speciation in frugivorous flies of genus *Rhagoletis* (Diptera, Tephritidae). *Evolution, 23*, 237–51.

Caisse, M., and Antonovics, J. (1978) Evolution in closely adjacent plant populations 9. Evolution of reproductive isolation in clinal populations. *Heredity*, 40, 371–84.

Chan, R., Baldwin, B.G., and Ornduff, R. (2002) Cryptic goldfields: A molecular phylogenetic reinvestigation of Lasthenia californica sensu lato and close relatives (Compositae: Heliantheae sensu lato). *American Journal of Botany*, 89, 1103–12.

Clausen, J., Keck, D.D., and Hiesey, W.M. (1947) Heredity of geographically and ecologically isolated races. *American Naturalist*, 81, 114–33.

Clausen, J., Keck, D.D., and Hiesey, W.M. (1958) *Experimental Studies on the Nature of Species. IV. Genetic Structure of Ecological Races.* Carnegie Institute, Washington, DC.

Coyne, J., and Orr, H.A. (2004) *Speciation.* Sinauer, Sunderland, MA.

Coyne, J.A., and Price, T.D. (2000) Little evidence for sympatric speciation in island birds. *Evolution*, 54, 2166–71.

Cronquist, A. (1978) Once again, what is a species? In *Biosystematics in Agriculture* (ed. J.A. Romberger), pp. 3–20. Allenheld, Osmun, Montclair, NJ.

Dobzhansky, T. (1937) *Genetics and the Origin of Species.* Columbia University Press, New York.

Dobzhansky, T. (1940) Speciation as a stage in evolutionary divergence. *American Naturalist*, 74, 312–21.

Elmendorf, S.C., and Moore, K.A. (2007) Plant competition varies with community composition in an edaphically complex landscape. *Ecology*, 88, 2640–50.

Endler, J.A. (1977) *Geographic Variation, Speciation and Clines.* Princeton University Press, Princeton, NJ.

Felsenstein, J. (1981) Skepticism towards Santa Rosalia, or why are there so few kinds of animals? *Evolution*, 35, 124–38.

Ford, V.S., and Gottlieb, L.D. (1989) Morphological evolution in Layia (Compositae): Character recombination in hybrids between *L. discoidea* and *L. glandulosa*. *Systematic Botany*, 14, 284–96.

Ford, V.S., and Gottlieb, L.D. (1990) Genetic studies of floral evolution in *Layia*. *Heredity*, 64, 29–44.

Funk, D.J. (1998) Isolating a role for natural selection in speciation: Host adaptation and sexual isolation in *Neochlamisus bebbianae* leaf beetles. *Evolution*, 52, 1744–59.

Gailing, O., Macnair, M.R., and Bachmann, K. (2004) QTL mapping for a trade-off between leaf and bud production in a recombinant inbred population of *Microseris douglasii* and *M-bigelovii* (Asteraceae, Lactuceae): A potential preadaptation for the colonization of serpentine soils. *Plant Biology*, 6, 440–46.

Gardner, M., and MacNair, M. (2000) Factors affecting the co-existence of the serpentine endemic *Mimulus nudatus* Curran and its presumed progenitor, *Mimulus guttatus* Fischer ex DC. *Biological Journal of the Linnean Society*, 69, 443–59.

Gavrilets, S. (2003) Perspective: Models of speciation: What have we learned in 40 years? *Evolution*, 57, 2197–215.

Ghasemi, R., and Ghaderian, S.M. (2009) Responses of two populations of an Iranian nickel-hyperaccumulating serpentine plant, *Alyssum inflatum* Nyar., to substrate Ca/Mg quotient and nickel. *Environmental and Experimental Botany*, 67, 260–68.

Going, B.M., Hillerislambers, J., and Levine, J.M. (2009) Abiotic and biotic resistance to grass invasion in serpentine annual plant communities. *Oecologia*, 159, 839–47.

Gonçalves, S. C., Martins-Loução, M. A., and Freitas, H. (2009) Evidence of adaptive toler-ance to nickel in isolates of *Cenococcum geophilum* from serpentine soils. *Mycorrhiza*, 19, 221–30.

Gottlieb, L. D., and Edwards, S. (1992) An electrophoretic test of the genetic independence of a newly discovered population of *Clarkia franciscana*. *Madroño*, 39, 1–7.

Gottlieb, L. D., Warwick, S. I., and Ford, V. S. (1985) Morphological and electrophoretic di-vergence between *Layia discoidea* and *L. glandulosa*. *Systematic Botany*, 10, 484–95.

Gram, W. K., Borer, E. T., Cottingham, K. L., Seabloom, E. W., Boucher, V. L., Goldwasser, L., Micheli, F., Kendall, B. E., and Burton, R. S. (2004) Distribution of plants in a California serpentine grassland: Are rocky hummocks spatial refuges for native species? *Plant Ecology*, 172, 159–71.

Grime, J. P. (1977) Evidence for the existence of three primary strategies in plants and its relevance to ecological and evolutionary theory. *American Naturalist*, 11, 1169–94.

Harrison, S. (1999) Native and alien species diversity at the local and regional scales in a grazed California grassland. *Oecologia*, 121, 99–106.

Harrison, S., and Inouye, B. D. (2002) High beta diversity in the flora of Californian serpen-tine "islands." *Biodiversity and Conservation*, 11, 1869–76.

Harrison, S., Safford, H., and Wakabayashi, J. (2004) Does the age of exposure of serpentine explain variation in endemic plant diversity in California? *International Geology Review*, 46, 235–42.

Hatfield, T., and Schluter, D. (1999) Ecological speciation in sticklebacks: Environment-dependent hybrid fitness. *Evolution*, 53, 866–73.

Heads, M. (2008) Panbiogeography of New Caledonia, south-west Pacific: Basal angio-sperms on basement terranes, ultramafic endemics inherited from volcanic island arcs and old taxa endemic to young islands. *Journal of Biogeography*, 35, 2153–75.

Hey, J. (2006) Recent advances in assessing gene flow between diverging populations and species. *Current Opinion in Genetics and Development*, 16, 592–96.

Hickman, J. C. (1993) *The Jepson Manual*. University of California Press, Berkeley.

Huenneke, L. F., Hamburg, S. P., Koide, R., Mooney, H. A., and Vitousek, P. M. (1990) Effects of soil resources on plant invasion and community structure in Californian serpentine grassland. *Ecology*, 71, 478–91.

Hughes, R., Bachmann, K., Smirnoff, N., and Macnair, M. R. (2001) The role of drought tolerance in serpentine tolerance in the *Mimulus guttatus* Fischer ex DC. complex. *South African Journal of Science*, 97, 581–86.

Jurjavcic, N. L., Harrison, S., and Wolf, A. T. (2002) Abiotic stress, competition, and the distribution of the native annual grass *Vulpia microstachys* in a mosaic environment. *Oecologia*, 130, 555–62.

Kazakou, E., Dimitrakopoulos, P. G., Baker, A. J. M., Reeves, R. D., and Troumbis, A. Y. (2008) Hypotheses, mechanisms and trade-offs of tolerance and adaptation to serpen-tine soils: From species to ecosystem level. *Biological Reviews*, 83, 495–508.

Kirkpatrick, M., and Ravigne, V. (2002) Speciation by natural and sexual selection: Models and experiments. *American Naturalist*, 159, S22–S35.

Kruckeberg, A. R. (1951) Intraspecific variability in the response of certain native plant spe-cies to serpentine soil. *American Journal of Botany*, 33, 408–19.

Kruckeberg, A. R. (1954) The ecology of serpentine soils. 3. Plant species in relation to serpentine soils. *Ecology,* 35, 267–74.

Kruckeberg, A. R. (1957) Variation in fertility of hybrids between isolated populations of the serpentine species, *Streptanthus glandulosus* Hook. *Evolution,* 11, 185.

Kruckeberg, A. R. (1984) *California Serpentines: Flora, Vegetation, Geology, Soils and Management Problems.* University of California Press, Berkeley.

Kruckeberg, A. R. (1986) An essay: The stimulus of unusual geologies for plant speciation. *Systematic Botany,* 11, 455–63.

Kruckeberg, A. R. (1992) Plant life of western North America ultramafics. In *The Ecology of Areas with Serpentinized Rocks* (eds. B. A. Roberts and J. Proctor), pp. 31–73. Kluwer, Dordrecht.

Kruckeberg, A. R., and Rabinowitz, D. (1985) Biological aspects of endemism in higher plants. *Annual Review of Ecology and Systematics,* 16, 447–79.

Lau, J. A., McCall, A. C., Davies, K. F., McKay, J. K., and Wright, J. W. (2008) Herbivores and edaphic factors constrain the realized niche of a native plant. *Ecology,* 89, 754–62.

Levin, D. A. (2009) Flowering-time plasticity facilitates niche shifts in adjacent populations. *New Phytologist,* 183, 661–66.

Lewis, H. (1962) Catastrophic selection as a factor in speciation. *Evolution,* 16, 257.

Lewis, H., and Lewis, M. E. (1955) *The Genus Clarkia.* University of California Press, Berkeley.

Lewis, H., and Raven, P. H. (1958) Rapid evolution in *Clarkia. Evolution,* 12, 319–36.

Lowry, D. B., Modliszewski, J. L., Wright, K. M., Wu, C. A., and Willis, J. H. (2008) The strength and genetic basis of reproductive isolating barriers in flowering plants. *Philosophical Transactions of the Royal Society B-Biological Sciences,* 363, 3009–21.

MacNair, M. R., and Gardner, M. P. (1998) The evolution of edaphic endemics. In *Endless Forms* (eds. D. J. Howard and S. H. Berlocher), pp. 157–71. Oxford University Press, Oxford.

Marschner, H. (2002) *Mineral Nutrition of Higher Plants.* Academic Press, San Diego.

Martens, S. N., and Boyd, R. S. (1994) The ecological significance of nickel hyperaccumulation: A plant chemical defense. *Oecologia,* 98, 379–84.

Mayer, M. S., and Soltis, P. S. (1994) The evolution of serpentine endemics—a chloroplast DNA phylogeny of the *Streptanthus-Glandulosus* complex (Cruciferae). *Systematic Botany,* 19, 557–74.

Mayer, M. S., and Soltis, P. S. (1999) Intraspecific phylogeny analysis using ITS sequences: Insights from studies of the *Streptanthus glandulosus* complex (cruciferae). *Systematic Botany,* 24, 47–61.

Mayer, M. S., Soltis, P. S. and Soltis, D. E. (1994) The evolution of the *Streptanthus-Glandulosus* complex (Cruciferae)—genetic-divergence and gene flow in serpentine endemics. *American Journal of Botany,* 81, 1288–1299.

Mayr, E. (1942) *Systematics and the Origin of Species.* Columbia University Press, New York.

Mayr, E. (1947) Ecological factors in speciation. *Evolution,* 1, 263–288.

Mayr, E. (1963) *Animal Species and Evolution.* Harvard University Press, Cambridge, MA.

McKay, J. K., and Latta, R. G. (2002) Adaptive population divergence: markers, QTL and traits. *Trends in Ecology and Evolution,* 17, 285–291.

Moser, A. M., Frank, J. L., D'Allura, J. A., and Southworth, D. (2009) Ectomycorrhizal communities of *Quercus garryana* are similar on serpentine and nonserpentine soils. *Plant and Soil*, 315, 185–94.

Muller, H. J. (1942) Isolating mechanisms, evolution, and temperature. *Biological Symposia*, 6, 71–125.

Ness, B. D., Soltis, D. E., and Soltis, P. S. (1990) An examination of polyploidy and putative introgression in *Calochortus* subsection *Nudi* (Liliaceae). *American Journal of Botany*, 77, 1519–31.

O'Dell, R. E., and Claassen, V. P. (2006) Serpentine and nonserpentine *Achillea millefolium* accessions differ in serpentine substrate tolerance and response to organic and inorganic amendments. *Plant and Soil*, 279, 253–69.

O'Dell, R. E., James, J. J., and Richards, J. H. (2006) Congeneric serpentine and nonserpentine shrubs differ more in leaf Ca : Mg than in tolerance of low N, low P, or heavy metals. *Plant and Soil*, 280, 49–64.

Ornduff, R. (1966) *A Biosystematic Survey of the Goldfield Genus* Lasthenia (*Compositae: Helenieae*). University of California Press, Berkeley.

Patterson, T. B., and Givnish, T. J. (2004) Geographic cohesion, chromosomal evolution, parallel adaptive radiations, and consequent floral adaptations in *Calochortus* (Calochortaceae): Evidence from a cpDNA phylogeny. *New Phytologist*, 161, 253–64.

Pepper, A. E., and Norwood, L. E. (2001) Evolution of *Caulanthus amplexicaulis* var. *barbarae* (Brassicaceae), a rare serpentine endemic plant: A molecular phylogenetic perspective. *American Journal of Botany*, 88, 1479–89.

Proctor, J. (1999) Toxins, nutrient shortages and droughts: The serpentine challenge. *Trends in Ecology and Evolution*, 14, 334–35.

Proctor, J., and Woodell, S. R. J. (1975) The ecology of serpentine soils. *Advances in Ecological Research*, 9, 255–365.

Rajakaruna, N. (2004) The edaphic factor in the origin of plant species. *International Geology Review*, 46, 471–78.

Rajakaruna, N., Baldwin, B. G., Chan, R., Desrochers, A. M., Bohm, B. A., and Whitton, J. (2003a) Edaphic races and phylogenetic taxa in the *Lasthenia californica* complex (Asteraceae : Heliantheae): An hypothesis of parallel evolution. *Molecular Ecology*, 12, 1675–79.

Rajakaruna, N., Siddiqi, M. Y., Whitton, J., Bohm, B. A., and Glass, A. D. M. (2003b) Differential responses to Na$^+$/K$^+$ and Ca^{2+}/Mg^{2+} in two edaphic races of the *Lasthenia californica* (Asteraceae) complex: A case for parallel evolution of physiological traits. *New Phytologist*, 157, 93–103.

Rajakaruna, N., and Whitton, J. (2004) Trends in the evolution of edaphic specialists with an example of parallel evolution in the *Lasthenia californica* complex. In *Plant Adaptation: Molecular Biology and Ecology* (eds. Q. C. B. Cronk, I. E. P. Taylor, R. Ree, and J. Whitton), pp. 103–10. NRC Research Press, Ottowa.

Ramsey, J., Bradshaw, H. D., and Schemske, D. W. (2003) Components of reproductive isolation between the monkeyflowers *Mimulus lewisii* and *M. cardinalis* (Phrymaceae). *Evolution*, 57, 1520–34.

Ramsey, J., and Schemske, D. W. (2002) Neopolyploidy in flowering plants. *Annual Review of Ecology and Systematics*, 33, 589–639.

Raven, P. H., and Axelrod, D. I. (1978) *Origin and Relationships of the California Flora.* University of California Press, Berkeley.

Reeves, R. D., and Adiguzel, N. (2008) The nickel hyperaccumulating plants of the serpentines of Turkey and adjacent areas: A review with new data. *Turkish Journal of Biology,* 32, 143–53.

Reeves, R. D., and Baker, A. J. M. (1984) Studies on metal uptake by plants from serpentine and non-serpentine populations of *Thlaspi goesingense* Halacsy (Cruciferae). *New Phytologist,* 98, 191–204.

Rice, K. (1989) Competitive interactions in California annual grasslands. In *Grassland Structure and Function: California Annual Grassland* (eds. L. F. Huenneke and H. A. Mooney), pp. 59–72. Kluwer, Dordrecht.

Rundle, H. D., Nagel, L., Boughman, J. W., and Schluter, D. (2000) Natural selection and parallel speciation in sympatric sticklebacks. *Science,* 287, 306–8.

Sambatti, J. B. M., and Rice, K. J. (2007) Functional ecology of ecotypic differentiation in the Californian serpentine sunflower (*Helianthus exilis*). *New Phytologist,* 175, 107–19.

Schechter, S. P., and Bruns, T. D. (2008) Serpentine and non-serpentine ecotypes of *Collinsia sparsiflora* associate with distinct arbuscular mycorrhizal fungal assemblages. *Molecular Ecology,* 17, 3198–210.

Schneider, S., Roessli, D., and Excoffier, L. (2000) Arlequin: A software for population genetics data analysis, v. 2.001. Genetics and Biometry Lab, Department of Anthropology, University of Geneva.

Seehausen, O., Terai, Y., Magalhaes, I. S., Carleton, K. L., Mrosso, H. D. J., Miyagi, R., van der Sluijs, I., Schneider, M. V., Maan, M. E., Tachida, H., Imai, H., and Okada, N. (2008) Speciation through sensory drive in cichlid fish. *Nature,* 455, 620–23.

Sobel, J. M., Chen, G. F., Watt, L. R., and Schemske, D. W. (2009) The biology of speciation. *Evolution,* 64, 295–315.

Spencer, S. C., and Porter, J. M. (1997) Evolutionary diversification and adaptation to novel environments in *Navarretia* (Polemoniaceae). *Systematic Botany,* 22, 649–68.

Springer, Y. P. (2007) Clinal resistance structure and pathogen local adaptation in a serpentine flax-flax rust interaction. *Evolution,* 61, 1812–22.

Springer, Y. P. (2009) Edaphic quality and plant-pathogen interactions: Effects of soil calcium on fungal infection of a serpentine flax. *Ecology,* 90, 1852–62.

Stebbins, G. L. (1942) The genetic approach to problems of rare and endemic species. *Madroño,* 6, 241–72.

Templeton, A. R. (1981) Mechanisms of speciation: A population genetic approach. *Annual Review of Ecology and Systematics,* 12, 23–48.

Thrall, P. H., Hochberg, M. E., Burdon, J. J., and Beyer, J. D. (2007) Coevolution of symbiotic mutualists and parasites in a community context. *Trends in Ecology and Evolution,* 22, 120–26.

Urban, A., Puschenreiter, M., Strauss, J., and Gorfer, M. (2008) Diversity and structure of ectomycorrhizal and co-associated fungal communities in a serpentine soil. *Mycorrhiza,* 18, 339–54.

Walker, R. B., Walker, H., and Ashworth, P. R. (1955) Calcium-magnesium nutrition with special reference to serpentine soils. *Plant Physiology,* 30, 214–22.

Wang, A. S., Angle, J. S., Chaney, R. L., Delorme, T. A., and Reeves, R. D. (2006) Soil pH effects on uptake of Cd and Zn by *Thlaspi caerulescens*. *Plant and Soil*, 281, 325–37.

Ward, K. L. (2000) *Quantitative Genetic Differentiation among Adjacent Populations of* Linanthus parviflorus *(Polemoniaceae): Natural Selection or Genetic Drift?* MS thesis, University of Washington, Seattle.

Wright, J. W., and Stanton, M. L. (2007) *Collinsia sparsiflora* in serpentine and nonserpentine habitats: Using F2 hybrids to detect the potential role of selection in ecotypic differentiation. *New Phytologist*, 173, 354–66.

Wright, J. W., Stanton, M. L., and Scherson, R. (2006) Local adaptation to serpentine and non-serpentine soils in *Collinsia sparsiflora*. *Evolutionary Ecology Research*, 8, 1–21.

5

Intraspecific Variation, Adaptation, and Evolution

Ryan E. O'Dell, *Bureau of Land Management*
Nishanta Rajakaruna, *College of the Atlantic*

Intraspecific variation refers to the genotypic (genetic) and resulting phenotypic (morphological and physiological) variation found within a species. Variation within a species is crucial for adaptation and evolution by natural selection. Over time, selection can result in genetically distinct populations of a species that are adapted to specific environmental conditions. Such populations are referred to as ecotypes. The study of distinct climatic, elevational, latitudinal, geographic, and edaphic ecotypes has provided much insight into the process of evolution by natural selection (Briggs and Walters, 1997; Levin, 2000; Silvertown and Charlesworth, 2001).

Edaphic islands, characterized by distinct soil characteristics, provide unique settings to study the factors and mechanisms promoting the generation and maintenance of intraspecific variation (Kruckeberg, 1986; Rajakaruna, 2004). Soil characteristics are strongly influenced by topography and underlying geology. Topography affects soil depth, water availability, and solar exposure, whereas geology influences soil development rate (mineral weathering), soil physical properties including texture and water-holding capacity, and soil chemical properties including plant essential nutrients, as well as toxic elements (Mason, 1946a; Kruckeberg, 1986; Alexander et al., 2007; Rajakaruna and Boyd, 2008). Extreme soil physical or chemical properties are typical of edaphically stressful habitats. Plants often respond to such edaphic stresses by ecotypic differentiation and subsequent speciation. Various methods have been used to detect ecotypic differentiation. These include reciprocal or unilateral transplant studies conducted in field, glasshouse, or laboratory settings, as well as genetic analysis. Although reciprocal transplant studies (see Chapter 7 in this volume) compare the response of two or more populations to their own, as well as other populations' environments, unilateral transplant studies only examine the response of two or more populations to one of the population's environments (typically serpentine and nonserpentine populations to serpentine soil). Due to the difficulty in establishing reciprocal or unilateral field transplant studies and the large time investment required to conduct them, transplant studies are more commonly carried out in controlled glasshouse or laboratory settings. Habitat-correlated differences in morphology, growth rates, physiology, or allele frequencies between populations documented by these studies are regarded as evidence for ecotypic differentiation.

This chapter focuses on how distinct soil types have resulted in ecotypic differentiation and adaptation. We begin with evidence for ecotypic differentiation in chemically stressful edaphic environments, including saline soils and metalliferous mine tailings, and then proceed to highlight ecotypic differentiation in response to serpentine soils. The discussion includes a summary of the adverse physical and chemical characteristics of serpentine soils, key plant morphological and physiological mechanisms involved in serpentine soil tolerance, how ecotypic differentiation leads to the origin of new species (also see Chapter 4 in this volume), and an extensive review of plant intraspecific variation found within serpentine ecosystems worldwide. We conclude by summarizing major trends in plant adaptation to serpentine soils as demonstrated by examples of intraspecific variation.

SALINE AND SELENIUM-ENRICHED HABITATS

Soils derived from marine sedimentary rocks in arid or semiarid environments are often laden with sodium chloride (NaCl) and selenium (Se), rendering them stressful or toxic to plant establishment and growth (Läuchli and Lüttge, 2002). Some species exhibit ecotypic differentiation, with saline-tolerant ecotypes displaying distinct physiological adaptations, such as leaf vacuolar NaCl accumulation, to cope with high soil NaCl concentrations (Rozema et al., 1978; Cheeseman, 1988; Staal et al., 1991). Other plant species contain Se-tolerant ecotypes that have adapted to tolerate high concentrations of bioavailable Se in soil (Enberg and Wu, 1995; Wu et al., 1997). In addition to tolerance, some species of *Astragalus* (Fabaceae) have evolved selective root uptake and shoot translocation mechanisms to hyperaccumulate Se (Sors et al., 2005). Se hyperaccumulation has been demonstrated to deter herbivory by insects and rodents (Hanson et al., 2004; Freeman et al., 2007, 2009; Quinn et al., 2008).

METALLIFEROUS MINE TAILINGS

Intraspecific variation has led to ecotypic differentiation of plant species growing on metalliferous mine tailings. Metalliferous mine tailings typically have low water-holding capacity, nutrients, and pH and contain high concentrations of bioavailable heavy metals, including aluminum (Al), arsenic (As), cadmium (Cd), cobalt (Co), chromium (Cr), copper (Cu), iron (Fe), manganese (Mn), nickel (Ni), lead (Pb), and/or zinc (Zn), depending on the host rock and ore type. These conditions make mine tailings a hostile environment for plant establishment and growth. Due to their widely dispersed, discontinuous distribution and stressful edaphic conditions that drastically differ from the surrounding environment, mines may in effect be regarded as isolated island environments. Ecotypic differentiation along mine tailings has been demonstrated in numerous plant species across diverse families.

Some species exhibit metal tolerance at low levels within populations not growing on mine tailings, indicating preadaptation (Macnair and Gardner, 1998). Mine ecotypes of some species also exhibit early flowering compared with adjacent normal soil ecotypes, thereby reducing gene flow between the populations (McNeilly and Antonovics, 1968; Macnair and Gardner, 1998; Antonovics, 2006). Physiological adaptations identified for metal-tolerant ecotypes include metal exclusion or sequestration at the root or shoot level (Smith and Bradshaw, 1979; Macnair, 1983; Baker, 1987; Hall, 2002). Ecotypes of some species exhibit heavy metal hyperaccumulation (He et al., 2002; Yang et al., 2006), and some metal-hyperaccumulating species show ecotypic differentiation with respect to metal uptake rates (Zhao et al., 2002; Richau and Schat, 2009). Metal-tolerant ecotypes of some species presumably originated independently multiple times via parallel evolution and convergence (Nordal et al., 1999).

Extensive studies have been conducted on *Mimulus guttatus* (Phrymaceae; formerly Scrophulariaceae) found on 150-year-old Cu mines in California (Macnair and Gardner, 1998). These studies have demonstrated ecotypic differentiation with respect to Cu, including the evolution of a new Cu-tolerant species, *M. cupriphilus*, from the Cu-tolerant ecotypes of *M. guttatus* (Macnair, 1989a, 1989b; Macnair and Cumbes, 1989; Macnair et al., 1989). The studies of the *M. guttatus* complex demonstrate the capacity for rapid evolution of both ecotypes and species within a relatively short time span when a species is subjected to the extreme directional selective pressures of an edaphically stressful environment.

SERPENTINE SOIL PROPERTIES AND PLANT TOLERANCE MECHANISMS

Serpentine ecosystems bear many similarities to mine tailings and other edaphically unusual soil types. Serpentine soils are often shallow and rocky; frequently deficient in nitrogen (N), phosphorus (P), potassium (K), and calcium (Ca); contain toxic concentrations of bioavailable magnesium (Mg) and heavy metals including Ni, Cr, and Co; and exhibit low water-holding capacity (Brooks, 1987; Roberts and Proctor, 1992; Alexander et al., 2007). Exceptionally low Ca:Mg molar ratios (typically $\ll 1.0$) are a key characteristic of serpentine soils. The properties of serpentine soils are inherited from Fe-Mg-rich (ferromagnesian silicates) parent materials, which may also contain Ni-, Cr-, and Co-bearing accessory minerals (Burt et al., 2001).

Plant adaptations to serpentine soils are varied and include those that convey tolerance to drought stress, macronutrient deficiency (N, P, K, S, and Ca), macronutrient toxicity (Mg), and heavy metal toxicity (Ni, Cr, Co). Typical drought stress adaptations include reduced leaf size, sclerophylly, succulence, and high root:shoot biomass ratio (Kruckeberg, 1984; Tibbetts and Smith, 1992; Brady et al., 2005). Elevated concentrations of cations associated with osmotic adjustment, including K^+, Na^+, and less frequently, Ni^{2+} as found in some hyperaccumulators, are also associated with drought tolerance (Boyd and Martens, 1992; Bhatia et al., 2005; Kazakou et al., 2008). Inherently slow growth rates and high root:shoot biomass ratios are common strategies to cope with nutrient deficiency (Brady et al., 2005; Kazakou et al., 2008).

Enhanced physiological regulation of nutrient uptake, particularly with respect to Ca and Mg, is also a common serpentine tolerance feature (Brady et al., 2005; Kazakou et al., 2008). The exceedingly low Ca:Mg ratios typical of serpentine soils are antagonistic to plant Ca uptake, requiring specialized physiological mechanisms to maintain adequate internal concentrations of Ca. In dicotyledonous angiosperms, a high proportion of the plant's total Ca is located in the cell wall, bound to pectins at the middle lamella (Demarty et al., 1984; Marschner, 2002). Pectins play an important role in plant cell wall stability and cell-to-cell adhesion (Jarvis, 1984; Jarvis et al., 2003). Typical symptoms of Ca deficiency are cessation of cell wall extension, disintegration of cell walls, and tissue collapse (necrosis) due to pectin breakdown (Demarty et al., 1984; Jarvis, 1984; Marschner, 2002). These symptoms are most apparent in actively growing tissues, such as root tips, new shoots, and leaves, because Ca is a phloem-immobile element. Physiological tolerance mechanisms to cope with the low Ca:Mg ratio of serpentine soils, including selective Ca uptake and translocation, Mg exclusion, and Mg sequestration, have been identified in diverse serpentine plant species and ecotypes globally (Kruckeberg, 1950; Brady et al., 2005; Kazakou et al., 2008).

Heavy metal tolerance mechanisms identified in serpentine-tolerant plant species include metal exclusion at the root level, sequestration to various plant organs, and toxicity tolerance (Baker, 1987; Shaw, 1990). Some plant species and ecotypes hyperaccumulate heavy metals to exceptionally high concentrations ($\geq$ 1000 µg/g; Reeves, 1992; Brooks, 1998). Heavy metal hyperaccumulation has been demonstrated to reduce insect herbivory in several plant species and eco-types (Brooks, 1998; Boyd, 2007) and also serve in osmotic regulation and drought tolerance (Boyd and Martens, 1992; Bhatia et al., 2005; Kazakou et al., 2008).

EDAPHIC ECOTYPE DIFFERENTIATION AND MODES OF SPECIATION

Acquisition of environmental stress tolerance and subsequent ecotypic differentiation has been proposed as the first steps toward speciation (Figure 5.1; Mason, 1946a, 1946b; Lewis, 1962; Kruckeberg, 1986; Macnair and Gardner, 1998; Rajakaruna, 2004; see Chapter 4 in this volume). In step 1, individuals of the species acquire edaphic tolerance (preadaptive trait) through random genetic mutation that confers serpentine tolerance within a nonserpentine population. In step 2, disruptive selection, catastrophic selection, or gradual divergence separates the species into distinct serpentine-tolerant and serpentine-intolerant ecotypes. In step 3, further genetic divergence in structural and functional traits occurs within the serpentine-tolerant ecotype. At step 4, isolation between serpentine-tolerant

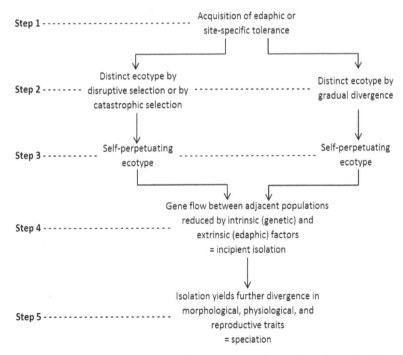

FIGURE 5.1. Hypothesized speciation pathway. Redrawn from Kruckeberg (1986).

and serpentine-intolerant ecotypes of the species becomes genetically fixed. Finally, at step 5, further divergence of the serpentine-tolerant ecotype leads to a distinct species. The recurrent process of independent ecotype differentiation for the same adaptations to the same environmental factors within the same species is referred to as parallel evolution (Levin, 2001).

Four speciation modes are generally recognized, including parapatric, allopatric, peripatric, and sympatric (Figure 5.2; Briggs and Walters, 1997). Parapatric, peripatric, and sympatric speciation are the most logical modes of ecotypic differentiation in the case of plant species that are not strict serpentine endemics and have populations that may exist on both serpentine as well as adjacent or nearby nonserpentine soils. Allopatric speciation is the most logical mode of differentiation in the case of strict serpentine endemic species that are entirely restricted to serpentine (Kruckeberg, 1984).

Similar to mines and other edaphically unusual substrates, serpentine soils have a discontinuous, widespread distribution across the landscape and may be regarded as island environments with geographic separation acting as a barrier to gene flow (Figure 5.3). As isolated environments, they are subject to the selective processes that occur with respect to island biogeography (Kruckeberg, 1991). Extension of the concepts of island biogeography to serpentine habitats may be limited because unlike oceanic islands, mainland edaphic islands are surrounded by

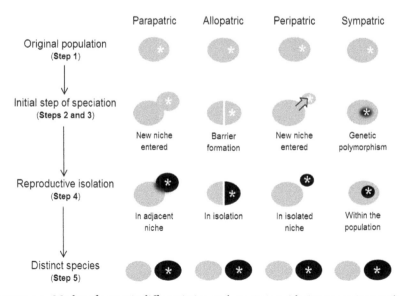

FIGURE 5.2. Modes of ecotypic differentiation and speciation. *designates unique trait acquired by ecotype and subsequently the resulting distinct species.

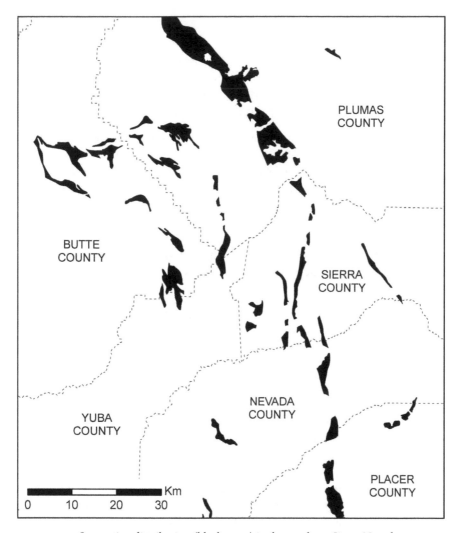

FIGURE 5.3. Serpentine distribution (black areas) in the northern Sierra Nevada mountain range, California. Note the discontinuous, widespread, island-like distribution. This type of discontinuous distribution is typical of most serpentine areas around the world (Brooks, 1987).

habitats occupied by potential recruits. On mainland edaphic islands, recruit pre-adaptation to edaphic stress restricts successful colonization. In contrast, recruit dispersal distance is the primary factor that limits successful colonization on oceanic islands. As a result, extension of island biogeography concepts to edaphic islands is most suitable for examining the influence of spatial isolation on population genetic attributes, such as reduced gene flow, inbreeding, and genetic drift (Mayer et al., 1994).

INTRASPECIFIC VARIATION: PLANT TAXA USED IN STUDIES OF PLANT ADAPTATION AND EVOLUTION TO SERPENTINE SOILS

Formal studies of intraspecific variation involving serpentine soils have largely been limited to a small number of genera within a limited number of families. Most plant taxa used are found in the following families: Asteraceae, Brassicaceae, Caryophyllaceae, Cupressaceae, Hydrophyllaceae, Lamiaceae, Phrymaceae, Pinaceae, Plantaginaceae, Poaceae, Polemoniaceae, and Rosaceae. Serpentine tolerance adaptations vary by species, with some ecotypes having several adaptive traits. A summary of adaptations by species is provided in Table 5.1. Adaptation patterns have emerged within several families. To highlight those patterns, each major plant taxon used for intraspecific variation studies is discussed in turn by plant family.

ASTERACEAE: DROUGHT TOLERANCE, CA AND MG REGULATION, PARALLEL AND CONVERGENT EVOLUTION

Plant species with widespread distribution on both serpentine and nonserpentine soil types have provided ideal models to study evolutionary concepts related to ecotypic differentiation, as well as the underlying adaptations in ecotypes that convey tolerance to serpentine soils. For example, *Achillea millefolium*, an obligately outcrossing, insect-pollinated species, has a widespread Northern Hemisphere distribution and occurs across a wide range of climates and soil types, including serpentine (Löfgren, 2002). Reciprocal and unilateral transplant studies of *A. millefolium* serpentine and nonserpentine soil populations in California and Washington revealed that the species has distinct serpentine-tolerant ecotypes (Figure 5.4; Kruckeberg, 1950, 1967; McMillan, 1956; Higgins and Mack, 1987; Cooke, 1994). Identified adaptations include reduced growth rate, nutrient deficiency tolerance, and tolerance of low Ca:Mg ratios and elevated Ni. Curiously, some nonserpentine ecotypes were fully or partially serpentine-tolerant, suggesting preadaptation (Kruckeberg, 1950).

Unilateral transplant glasshouse studies (O'Dell and Claassen, 2008) conducted with a large number of serpentine and nonserpentine ecotypes from North America (eastern and western United States and Canada) and Europe (United Kingdom and Austria) revealed the same trend as found by Kruckeberg (1950) on a much broader spatial scale. When grown on serpentine, serpentine ecotype seedlings always exhibited healthy radicle growth, whereas the radicles of nonserpentine ecotype seedlings typically became necrotic (O'Dell and Claassen, 2006, 2008). A comparison of nutrient uptake between serpentine and nonserpentine ecotypes revealed that nearly all serpentine-tolerant ecotypes displayed selective Ca uptake at the root level and translocation to the shoot, resulting in a significantly higher shoot Ca:Mg ratio than found in nonserpentine ecotypes, which lack this mechanism (O'Dell and Claassen, 2006, 2008). This tolerance mechanism was nearly ubiquitous for serpentine ecotypes of both North America and Europe, suggesting that it has originated multiple times in distant serpentine ecotypes via parallel evolution and convergence (O'Dell and Claassen, 2008).

Parallel evolution has also been found for races of the *Lasthenia californica* complex. *L. californica* is an obligately outcrossing, insect-pollinated species. It has a widespread distribution in California grasslands, occurring on both serpentine and nonserpentine soils. Two distinct races (races A and C) of this complex have been identified (Desrochers and Bohm, 1995; Rajakaruna and Bohm, 1999). Race A occupies ionically extreme environments (coastal bluffs, alkaline flats, and serpentine), whereas race C occurs mostly on more ionically moderate and drier sites (pastures and oak woodland). Although race A is tolerant of the nonserpentine soils that support race C, race C is generally not found on extreme serpentine soils that support race A.

Race A has significantly greater Ca uptake and Na and Mg tolerance compared to race C, as measured by root growth, survival, and nutrient uptake (Rajakaruna et al., 2003b). Elevated Na uptake and root-to-shoot translocation in race A is believed to confer greater tolerance of soil ionic stress owing to greater capacity for osmotic adjustment. Elevated Ca uptake and root-to-shoot translocation in race A allows these plants to maintain a favorable shoot tissue Ca:Mg ratio in serpentine soils. The races appear to have evolved barriers to reproduction as well, especially via flowering time differences in their parapatric occurrences in northern California (Rajakaruna and Bohm, 1999). Phylogenetic analysis has revealed that the races occur in parallel in two geographically based clades within the complex, suggesting the parallel evolution of one or both races in response to contrasting soil conditions under which these races are found (Rajakaruna et al., 2003a).

PHRYMACEAE AND PLANTAGINACEAE: DROUGHT TOLERANCE, FLOWER MODIFICATION, EARLY FLOWERING, CA AND MG REGULATION, NI TOLERANCE

Genus *Mimulus* (Phrymaceae; formerly Scrophulariaceae) has emerged as an important model in plant evolution studies (Wu et al., 2008). *Mimulus guttatus* is insect-pollinated with high variation of pollination mode in individual plants, ranging from fully outcrossing to fully selfing (Gardner and Macnair, 2000). The

TABLE 5.1 Summary of Studies on Intraspecific Variation and Adaptations

Species	Family	Comparison Populations	Comparison Serpentine Tolerance	Differentiation Supported by Phyogenetics	Serpentine Tolerance Adaptations			
					Drought Tolerance	Reduced Growth Rate	Reduced Leaf Size	Earlier Flowering
Achillea millefolium	Asteraceae	S and NS	S > NS			X		
Lasthenia califonica	Asteraceae	S and NS	S > NS	X[b]	X			X
Helianthus exilis[a]	Asteraceae	S only	...		X		X	X
Erigeron thunbergii	Asteraceae	S and NS	No data	X[b]				
Erigeron compositus	Asteraceae	S and NS	S > NS					
Senecio paupercaulus	Asteraceae	S and NS	S > NS					
Senecio coronatus	Asteraceae	S and NS	S > NS					
Berkheya zeyeri subsp. rehmannii	Asteraceae	S and NS	No data	X[b]				
Fragaria virginiana	Rosaceae	S and NS	S > NS					
Potentilla glandulosa	Rosaceae	S and NS	S > NS					
Spirea douglasii	Rosaceae	S and NS	S > NS					
Mimulus guttatus	Phrymaceae	S and NS	S > NS		X	X	X	X
Mimulus nudatus[a]	Phrymaceae	S only	...				X	X
Collinsia sparsifolia	Plantaginaceae	S and NS	S > NS					
Plantago erecta	Plantaginaceae	S and NS	S > NS					
Salvia columbariae	Lamiaceae	S and NS	S > NS					
Prunella vulgaris	Lamiaceae	S and NS	S > NS					
Teucrium chamaedrys	Lamiaceae	S and NS	No data			X		
Silene dioica	Caryophyllaceae	S andNS	S = NS	X[b]				
Silene vulgaris	Caryophyllaceae	S and NS	S > NS			X	X	
Silene armeria	Caryophyllaceae	S and NS	S > NS					
Silene paradoxa	Caryophyllaceae	S and NS	S > NS					
Silene italica	Caryophyllaceae	S and NS	S > NS					
Cerastium alpinum	Caryophyllaceae	S and NS	S > NS	X[b]				
Arenaria katoana[a]	Caryophyllaceae	S only	...	X				
Gilia capitata	Polemoniaceae	S and NS	S > NS					X
Navarretia squarrosa	Polemoniaceae	S and NS	S > NS					
Navarretia mellita	Polemoniaceae	S and NS	S > NS					
Navarretia rosulata[a]	Polemoniaceae	S only	...	X[b]				
Navarretia jepsonii[a]	Polemoniaceae	S only	...	X[b]				
Navarretia jaredii[a]	Polemoniaceae	S only	...	X[b]				
Leptosiphon androsaceus	Polemoniaceae	S and NS	S > NS					X
Leptosiphon parviflorus	Polemoniaceae	S and NS	S > NS	X				X
Leptosiphon bicolor	Polemoniaceae	S and NS	No data					X
Phacelia californica	Hydrophyllaceae	S and NS	S > NS					
Phacelia dubia	Hydrophyllaceae	NS only	var. georgiana[c]					
Agropyron spicatum	Poaceae	S and NS	S > NS					
Festuca rubra	Poaceae	S and NS	S = NS					
Agrostis stolonifera	Poaceae	S and NS	S = NS					
Agrostis canina	Poaceae	S and NS	S = NS					
Agrostis hallii	Poaceae	S and NS	S = NS					

Serpentine Tolerance Adaptations

| Flower Modification | | Ca Regulation | | Mg Regulation | | | Ni Regulation | | | | |
Shape	Color	Selective Uptake	Translocation to Shoot	Tolerance	Root Level Exclusion	Higher Shoot Ca:Mg Molar	Tolerance	Exclusion	Root Level Sequestration	Translocation to Shoot (accumulation)	References
		X[b]	X[b]	X		X[b]	X				1–5
		X[b]	X[b]	X[b]		X[b]					6–9
											10, 11
											12
											13
											13
										X	14–16
										X	17
											13
											13
											13
		X	X			X	X				18–22
											23
X	X										24, 25
											1, 26
					X	X					1
											13
											27
											28–30
							X				31, 32
											33
							X		X		34
							X		X		35
				X			X				36–38
											39
					X	X					1
											40
											40
											41
											41
											41
	X	X	X			X					42–44
	X										45
											44
		X	X			X					1
				X[c]			X[c]				46
		X	X	X	X	X					47
											48, 49
											48, 49
											48, 49
											40

(continued)

TABLE 5.1 *(continued)*

Species	Family	Comparison Populations	Comparison Serpentine Tolerance	Differentiation Supported by Phyogenetics	Serpentine Tolerance Adaptations			
					Drought Tolerance	Reduced Growth Rate	Reduced Leaf Size	Earlier Flowering
Avena fatua	Poaceae	S and NS	S > NS		X			
Bromus madritensis	Poaceae	S and NS	S = NS					
Bromus tectorum	Poaceae	S and NS	S = NS					
Bromus hordeaceus	Poaceae	S and NS	S > NS		X	X		
Bromus diandrus	Poaceae	S and NS	S = NS					
Bromus laevipes	Poaceae	S and NS	S = NS					
Elymus multisetus	Poaceae	S and NS	S = NS					
Vulpia microsatchys	Poaceae	S and NS	S > NS		X			
Aegilops triuncialis	Poaceae	S and NS	S = NS					
Pinus ponderosa	Pinaceae	S and NS	S > NS					
Pinus contorta	Pinaceae	S and NS	S > NS					
Pinus sabiniana	Pinaceae	S and NS	S = NS					
Pinus attenuata	Pinaceae	S and NS	S = NS					
Pinus virginiana	Pinaceae	S and NS	S = NS					
Pinus jeffreyii	Pinaceae	S and NS	No data	X				
Pinus balfouriana subsp. *balfouriana*	Pinaceae	S and NS	No data	X				
Picea glehnii	Pinaceae	NS only	NS					
Picea jezoensis	Pinaceae	NS only	NS					
Hesperocyparis macnabiana	Cupressaceae	S and NS	S = NS					
Juniperus communis	Cupressaceae	S and NS	S = NS					
Arabidopsis thaliana	Brassicaceae	NS only	CAX mutant[c]					
Arabidopsis lyrata	Brassicaceae	S and NS	S > NS	X[b]		X		
Erysimum capitatum	Brassicaceae	S and NS	S > NS					
Streptanthus polygaloides[a]	Brassicaceae	S only	...	X			X	
Streptanthus glandulosus	Brassicaceae	S and NS	S > NS	X[b]				
Caulanthus amplexicaulis	Brassicaceae	S and NS	S > NS	X				
Noccaea fendlerii subsp. *glauca*	Brassicaceae	S and NS	S = NS[d]					
Noccaea fendlerii subsp. *siskiyouense*[a]	Brassicaceae	S only	...					
Noccaea fendlerii subsp. *californica*[a]	Brassicaceae	S only	...					
Noccaea goesingense	Brassicaceae	S and NS	S = NS[d]					
Noccaea cochleariforme	Brassicaceae	S and NS	S = NS[d]					
Noccaea caerulescens	Brassicaceae	S and NS	No data					

Serpentine Tolerance Adaptations

Flower Modifcation		Ca Regulation		Mg Regulation		Higher Shoot Ca:Mg Molar	Ni Regulation				
Shape	Color	Selective Uplake	Translocation to Shoot	Tolerance	Root Level Exclusion		Tolerance	Exclusion	Rool Level Sequestration	Translocation to Shoot (accumulation)	References
											50
											1
											13
											1, 40, 51
											1
											1
											1
											52
											53,54
		X	X			X					55, 56
											13
											57
											40
											58
											59
											60
											61
											61
											40
											13, 62, 63
					X[c]	X[c]					64
		X[b]	X	X	X[b]						65–68
											1
X	X									X (variation)	69–71
X	X										1, 72–75
X	X										76
										X[d] (both S and NS)	77–79
										X	77
										X	77
										X[d] (both S and NS)	80
										X[d] (both S and NS)	81, 82
										X (S > NS)	83, 84

(continued)

TABLE 5.1 *(continued)*

Species	Family	Comparison Populations	Comparison Serpentine Tolerance	Differentiation Supported by Phyogenetics	Drought Tolerance	Reduced Growth Rate	Reduced Leaf Size	Earlier Flowering
						Serpentine Tolerance Adaptations		
Noccaea pindica[a]	Brassicaceae	S only	...					
Cochlearia pyrenaica	Brassicaceae	S and NS	No data					
Alyssum murale	Brassicaceae	Ecotypes	S > NS					
Alyssum bertolonii[a]	Brassicaceae	Ecotypes	...					
Alyssum inflatum	Brassicaceae	Ecotypes	S > NS					

REFERENCES: 1: Kruckeberg, 1950; 2: Higgins and Mack, 1987; 3: Cooke, 1994; 4: O'Dell and Claassen, 2006; 5: O'Dell and Claassen, 2008; 6: Desrochers and Bohm, 1995; 7: Rajakaruna and Bohm, 1999; 8: Rajakaruna et al., 2003a; 9: Rajakaruna et al., 2003b; 10: Sambatti and Rice, 2006; 11: Sambatti and Rice, 2007; 12: Kawase et al., 2007; 13: Kruckeberg, 1967; 14: Morrey et al., 1992; 15: Mesjasz-Przybyłowicz et al., 2007; 16: Boyd et al., 2008; 17: Williamson et al., 1997; 18: Tilstone and Macnair, 1997; 19: Hughes et al., 2001; 20: Murren et al., 2006; 21: Hendrick, 2008; 22: Selby et al., 2008; 23: Macnair, 1992; 24: Wright et al., 2006; 25: Wright and Stanton, 2007; 26: Espeland and Rice, 2007; 27: Pavlova, 2009; 28: Westerbergh and Saura, 1992; 29: Westerbergh and Saura, 1994; 30: Westerbergh, 1994; 31: Bratteler et al., 2006a; 32: Bratteler et al., 2006b; 33: Lombini et al., 2003; 34: Gonnelli et al., 2001; 35: Gabbrielli et al., 1990; 36: Nyberg Berglund and Westerbergh, 2001a; 37: Nyberg Berglund and Westerbergh, 2001b; 38: Nyberg Berglund et al., 2003; 39: Kawase et al., 2009; 40: McMillan, 1956; 41: Spencer and Porter, 1997; 42: Proctor and Woodell, 1975; 43: Woodell et al., 1975; 44: Schmitt, 1983; 45: Chapter 4 in this volume; 46: Taylor and Levy, 2002; 47: Main, 1974; 48: Proctor, 1971; 49: Johnston and Proctor, 1981; 50: Harrison et al., 2001; 51: Freitas and Mooney, 1996; 52: Jurjavcic et al., 2002; 53: Thomson, 2007; 54: Lyons et al., in press; 55: Jenkinson, 1974; 56: Wright, 2007; 57: Griffin, 1962; 58: Miller and Cumming, 2000; 59: Fur nier and Adams, 1986; 60: Oline et al., 2000; 61: Kayama et al., 2005; 62: Ashworth et al., 2001; 63: Adams and Nyugen, 2007; 64: Bradshaw, 2005; 65: Roche et al., 2004; 66: Von Wettberg, 2008; 67: Turner et al., 2008; 68: Turner et al., 2010; 69: Reeves et al., 1981; 70: Rodman et al., 1981; 71: Boyd et al., 2009; 72: Kruckeberg, 1957; 73: Mayer et al., 1994; 74: Mayer and Soltis, 1994; 75: Mayer and Soltis, 1999; 76: Pepper and Norwood, 2001; 77: Reeves et al., 1983; 78: Boyd and Martens, 1998; 79: Peer et al., 2006; 80: Reeves and Baker, 1984; 81: Mizuno et al., 2003; 82: Mizuno et al., 2005; 83: Assunção et al., 2003; 84: Assunção et al., 2008; 85: Taylor and Macnair, 2006; 86: Nagy and Proctor, 1997; 87: Asemaneh et al., 2006; 88: Galardi et al., 2007; 89: Ghasemi and Gahderian, 2009.

[a] Strict serpentine endemic species.

[b] Evidence of parallel and convergent evolution within species.

[c] Example of preadaptation.

[d] Constitutive trait.

| Serpentine Tolerance Adaptations | | | | | | | | | | | |
| Flower Modifcation | | Ca Regulation | | Mg Regulation | | | Ni Regulation | | | | |
Shape	Color	Selective Uplake	Translocation to Shoot	Tolerance	Root Level Exclusion	Higher Shoot Ca: Mg Molar	Tolerance	Exclusion	Rool Level Sequestration	Translocation to Shoot (accumulation)	References
										X (S > other S)	85
							X				86
							X				87
										X (S > other S)	88
				X							89

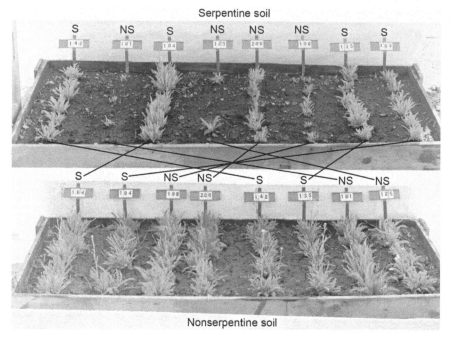

FIGURE 5.4. Response of serpentine (S) and nonserpentine (NS) ecotypes of *Achillea millefolium* (Asteraceae) grown on serpentine (top) and nonserpentine (bottom) soil (photo credit: Kruckeberg, 1950). Note the poor growth of the NS ecotypes on the serpentine soil and normal growth of all ecotypes on the nonserpentine soil. Also note, however, that sporadic individuals of the NS ecotypes appear to have tolerance to the serpentine soil. This is particularly evident for ecotype 206 (fifth row from left) where most if not all individuals of the nonserpentine ecotype appear to be serpentine-tolerant.

species has a widespread North American distribution and is found in a variety of climates and on diverse soil types, including serpentine.

Reciprocal and unilateral transplant studies of serpentine and nonserpentine populations of *M. guttatus* in California have revealed that the species has undergone ecotypic divergence with respect to serpentine tolerance. *M. guttatus* serpentine ecotypes have an earlier flowering time, reduced leaf size and shoot biomass, and higher Ca uptake and root-to-shoot translocation when compared to nonserpentine ecotypes (Hughes et al., 2001; Murren et al., 2006; Hendrick, 2008; Selby et al., 2008). Additionally, serpentine ecotypes have greater Ni tolerance than nonserpentine ecotypes (Tilstone and Macnair, 1997).

M. guttatus is believed to be the progenitor of *M. nudatus*, a strict serpentine endemic in California (Macnair, 1992; Safford et al., 2005). Although *M. nudatus* does not appear to differ significantly with respect to Ca and Mg regulation as compared to nonserpentine *M. guttatus* ecotypes, the species does have drought tolerance mechanisms, including significantly earlier flowering and reduced leaf size (Macnair and Gardner, 1998; Gardner and Macnair, 2000).

Reciprocal field transplant studies of *Collinsia sparsifolia* and *Plantago erecta* (Plantaginaceae; formerly Scrophulariacae) showed that both species display ecotypic differentiation, with serpentine ecotypes displaying greater tolerance of serpentine soils than nonserpentine ecotypes (Kruckeberg, 1950; Wright et al., 2006; Espeland and Rice, 2007). *C. sparsifolia* ecotypes display differentiation of floral traits, including larger flowers with more intense pigmentation in the serpentine ecotype (Wright and Stanton, 2007; Chapter 7 in this volume).

CARYOPHYLLACEAE: CA AND MG REGULATION, NI TOLERANCE, PARALLEL AND CONVERGENT EVOLUTION

Ecotypic differentiation has been studied in several species of *Silene* in Europe. *S. dioica* is an insect-pollinated, dioecious, obligately outcrossing species (Westerbergh and Saura, 1994). In Sweden, serpentine and nonserpentine populations of *S. dioica* do not significantly differ in either serpentine or Ni tolerance in a glasshouse setting (Westerbergh, 1994). Both serpentine and nonserpentine populations were equally tolerant of serpentine soils, and thus serpentine tolerance is regarded as a constitutive trait in the species. Phylogenetic analysis of several distant serpentine and nonserpentine populations revealed that although both populations in northern Sweden were genetically similar, these populations in southern Sweden exhibited genetic differentiation (Westerbergh and Saura, 1992). Although the serpentine and nonserpentine populations do not display phenotypic differentiation with respect to serpentine tolerance, genetic differentiation between them suggests that differences between the environments have restricted gene flow between the populations (Westerbergh and Saura, 1994). *S. dioica* primarily grows on fragmented habitat types, including serpentine and meadows. It is hypothesized that differences in the dominant vegetation cover between the serpentine (grassland) and nonserpentine (forest) environments have restricted the movement of pollinator guilds of *Silene* between them (Westerbergh and Saura, 1994).

Quantitative trait loci analysis of serpentine and nonserpentine populations of *S. vulgaris* in Switzerland showed that the populations display ecotypic differentiation with respect to Ni tolerance and morphological traits, including leaf size and shoot length (Bratteler et al., 2006a, 2006b). *S. armeria* exhibits ecotypic differentiation with the serpentine ecotype displaying greater tolerance of low substrate Ca:Mg ratios than the nonserpentine ecotype (Lombini et al., 2003). Likewise, *S. paradoxa* exhibits ecotypic differentiation, with the serpentine ecotype displaying greater tolerance of Ni (as measured by root growth) than the nonserpentine ecotype (Gonnelli et al., 2001). The Ni tolerance mechanism identified in the serpentine ecotype was root sequestration. *S. italica* also appears to use root sequestration as a Ni tolerance mechanism (Gabbrielli et al., 1990).

Unlike *S. dioica*, *Cerastium alpinum* from Sweden, Norway, and Finland does display ecotypic differentiation with serpentine ecotypes having greater Ni and Mg tolerance (Nyberg Berglund et al., 2003). Phylogenetic analysis has revealed that serpentine ecotypes are genetically more similar to nonserpentine ecotypes within the same region than other distant serpentine ecotypes, suggesting that like *S. dioica*, the species has repeatedly evolved serpentine-tolerant ecotypes via parallel evolution (Nyberg Berglund, 2005; Nyberg Berglund and Westerbergh, 2001a, 2001b; Nyberg Berglund et al., 2003).

POLEMONIACEAE: EARLY FLOWERING, CA AND MG REGULATION, PARALLEL AND CONVERGENT EVOLUTION

Gilia capitata has a widespread distribution on numerous soil types in western North America (Kruckeberg, 1950). Reciprocal transplant glasshouse studies of *G. capitata* from California have revealed the presence of distinct serpentine and nonserpentine ecotypes (Kruckeberg, 1950). Shoot tissue analysis of serpentine and nonserpentine ecotypes grown on serpentine soil revealed that the serpentine ecotype had significantly lower Mg concentrations than the nonserpentine ecotype (Kruckeberg, 1950). This suggests that the primary tolerance mechanism used by the serpentine ecotype to maintain a balanced shoot Ca:Mg ratio is restricted root-to-shoot translocation of Mg. Under natural field conditions, *G. capitata* growing on serpentine soil flowers at least two weeks earlier than plants growing on nonserpentine soil in the same region, alluding to another possible serpentine tolerance trait (Kruckeberg, 1950).

Leptosiphon androsaceus, *L. parviflorus*, and *L. bicolor* have a limited western U.S. distribution, with populations occurring on both serpentine and nonserpentine soil (Woodell et al., 1975; Schmitt, 1983; Chapter 4 in this volume). Serpentine populations of *L. androsaceus* typically have purple flowers, whereas nonserpentine populations typically bear white flowers (Proctor and Woodell, 1975). Reciprocal field transplant studies of serpentine and nonserpentine populations in southern California showed that the species has undergone ecotypic differentiation (Woodell et al., 1975). Serpentine ecotypes grew significantly larger on serpentine soil than did nonserpentine ecotypes. The tolerance mechanism identified in the serpentine ecotype was the ability to maintain an adequate shoot Ca:Mg ratio via selective root-level Ca uptake and translocation to the shoot.

Like *L. androsaceus*, *L. parviflorus* has serpentine and nonserpentine populations that differ in flower color (Chapter 4 in this volume). Serpentine populations in northern California typically bear pink flowers, whereas adjacent nonserpentine populations bear white flowers. Reciprocal transplant studies involving serpentine and nonserpentine populations in the glasshouse demonstrated ecotypic differentiation with the nonserpentine populations being almost completely intolerant of serpentine soil. Genetic analysis of these populations revealed that flower color, flowering phenology, plant size, and serpentine tolerance were strongly influenced by soil type and that the trait differences may be reinforced by differing flowering time and restricted gene flow between the two populations.

Studies of pollen dispersal between serpentine and nonserpentine ecotypes of *L. bicolor* in northern California have showed that the serpentine ecotype flowers two weeks earlier than the nonserpentine ecotype (Schmitt, 1983). *L. bicolor* is self-compatible, facultatively autogamous, and primarily pollinated by beeflies (Bombylliidae). Flowering time difference between the two populations had important consequences for the availability of pollinators and outcrossing between the populations. The early flowering serpentine population received significantly fewer visits from beeflies (44%) compared with the later flowering nonserpentine population (95%) since the serpentine population's flowering period was largely outside of the beeflies' activity period. As a result, the proportion of flowers available to outcross with the other population was only 7% for the serpentine population and 88% for the nonserpentine population. The study suggests that because serpentine soils tend to dry faster than nonserpentine soils, early flowering of the serpentine ecotype may have been selected for, at the expense of outcrossing opportunity, in a response to mortality from drought (Schmitt, 1983). Like *L. bicolor*, *L. androsaceus* serpentine ecotypes have also been noted to flower earlier than nonserpentine ecotypes (Schmitt, 1983).

HYDROPHYLLACEAE: CA AND MG REGULATION, NI TOLERANCE

Genus *Phacelia* occurs throughout North and South America and consists of herbaceous annual or perennial species. Reciprocal glasshouse transplant studies of serpentine and nonserpentine populations of *P. californica* from California revealed ecotypic differentiation (Kruckeberg, 1950). Serpentine ecotypes maintained a significantly higher shoot Ca:Mg concentration than nonserpentine ecotypes, primarily by selective Ca uptake and translocation to the shoot.

P. dubia is found throughout the eastern United States and includes three allopatric edaphic endemic varieties: var. *georgiana*, var. *interior*, and var. *dubia*. *P. dubia* var. *georgiana* grows on shallow granite soils, var. *interior* grows on shallow limestone soils, and var. *dubia* grows on relatively fertile woodland soils (Taylor and Levy, 2002). All three varieties were tested for serpentine soil tolerance as well as low Ca:Mg ratio and elevated Ni in culture solution (Taylor and Levy, 2002). Of them, only var. *georgiana* displayed some tolerance of serpentine soil, low Ca:Mg ratios, and elevated Ni. Since var. *georgiana* is not known to naturally grow on serpentine soils, elevated tolerance of serpentine soil factors in the variety is regarded as an example of preadaptation to serpentine soils.

POACEAE: CA AND MG REGULATION, TYPE II CELL WALL, DROUGHT TOLERANCE

Poaceae is a global family of herbaceous, wind-pollinated, monocotyledon angiosperms. Studies of serpentine tolerance and ecotypic differentiation in Poaceae have largely focused on species found in North America and Europe. Serpentine and nonserpentine populations of *Agropyron spicatum* from Washington grown in serpentine soil and solution culture simulating the chemical conditions of serpentine soil have been shown to exhibit ecotypic differentiation (Kruckeberg, 1967; Main, 1974). The serpentine ecotype was able to produce significantly more shoot biomass than the nonserpentine ecotype at lower solution Ca:Mg ratios. Shoot tissue analysis revealed that the serpentine ecotype was able to maintain a significantly higher shoot Ca:Mg ratio due to selective Ca uptake at the root and translocation to the shoot, as well as Mg exclusion at the root or shoot level.

Studies of serpentine and nonserpentine populations of *Festuca rubra*, *Agrostis stolonifera*, and *A. canina* from the United Kingdom revealed little difference between the populations with respect to serpentine soil tolerance or Ca, Mg, and Ni uptake in culture solution (Proctor, 1971; Johnston and Proctor, 1981). Like *A. stolonifera* and *A. canina*, serpentine and nonserpentine populations of *A. hallii* from California were also found not to differ in serpentine tolerance (McMillan, 1956). Likewise, growth comparisons of serpentine and nonserpentine populations of *Bromus hordeaceus*, *B. madritensis*, *B. tectorum*, *B. diandrus*, *B. laevipes*, and *Elymus multisetus* (Figure 5.5) from California on serpentine soil did not reveal any ecotypic differentiation within those species (Kruckeberg, 1950, 1967). Some reciprocal transplant studies involving *Avena fatua* (Harrison et al., 2001), *Bromus hordeaceus* (Freitas and Mooney, 1996; Harrison et al., 2001), *Vulpia microstachys* (Jurjavcic et al., 2002), and *Aegilops triuncialis* (Thomson, 2007; Lyons et al., in review) from California revealed some ecotypic differentiation, but the evidence implicated soil texture and water availability (drought tolerance) rather than soil chemistry. Drought stress tolerance of the *B. hordeaceus* serpentine ecotype was attributed to its ability to maintain a slow growth rate and high root:shoot biomass ratio (Freitas and Mooney, 1996).

Species in order Poales have type II cell walls (Jarvis et al., 1988; Marschner, 2002; Vogel, 2008). Studies have demonstrated that species of order Poales have a lower Ca requirement than dicot angiosperms (type I cell walls), primarily due to lower cell wall pectin content (Jarvis et al., 1988; Marschner, 2002; Broadley et al.,

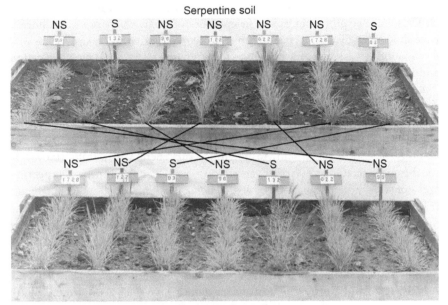

FIGURE 5.5. Response of serpentine (S) and nonserpentine (NS) ecotypes of *Elymus mul-tisetus* (Poaceae; monocotyledon) to being grown on serpentine (top) and nonserpentine (bottom) soil (photo credit: Kruckeberg, 1950). Note equal growth of NS and S ecotypes on serpentine soil. Species of Poaceae have type II cell walls, which contain much less Ca than the type I cell walls possessed by dicotyledonous plants. The lower Ca requirement of monocotyledonous plants like grasses confers tolerance of the low Ca:Mg ratio of serpentine soils. As a result, many or most grass species typically display tolerance of serpentine soil even if they do not naturally grow on serpentine.

2003; Vogel, 2008). Compared to the higher Ca requirement of dicot angiosperms, the lower Ca requirement of species in order Poales could represent a physiological advantage when growing on Ca-deficient serpentine soils. Wind pollination with its high outcrossing potential (Freidman and Barrett, 2009) combined with the low Ca requirement may explain why species of Poales infrequently exhibit differentiation into distinct serpentine and nonserpentine ecotypes (general indifference to serpentine) and why there are relatively few serpentine endemic species within the order.

PINACAE AND CUPRESSACEAE: CA AND MG REGULATION, LOW PECTIN TYPE I CELL WALL

Like Poaceae, ecotypic differentiation of species in the wind-pollinated gymnosperm genera *Pinus* (Pinaceae), *Picea* (Pinaceae), *Hesperocyparis* (formerly

Cupressus; Cupressaceae), and *Juniperus* (Cupressaceae) appears to be weak to nonexistent. Reciprocal transplant field studies of *Pinus ponderosa* in California have shown that distinct serpentine-tolerant ecotypes exist, however, ecotypic differentiation is weak and only became evident after 20 years of growing serpentine and nonserpentine ecotypes on serpentine soil (Jenkinson, 1974; Wright, 2007). The serpentine ecotype was shown to maintain significantly higher shoot Ca concentrations than the nonserpentine ecotype by selective Ca uptake by roots and translocation to the shoot (Jenkinson, 1974).

Similar delayed expression of serpentine and nonserpentine ecotypes has been demonstrated for *P. contorta* in Washington (Kruckeberg, 1967). In shorter reciprocal transplant studies (<1 year) comparing serpentine tolerance of serpentine and nonserpentine populations of *P. sabiniana* (Griffin, 1962) and *P. attenuata* (McMillan, 1956) in California and *P. virginiana* in Maryland (Miller and Cumming, 2000), no ecotypic differentiation was detected. Similarly, nonserpentine populations of *Picea glehnii* and *P. jezoensis* were found to be equally tolerant of serpentine and nonserpentine soils in Hokkaido, Japan (Kayama et al., 2005). Although serpentine and nonserpentine populations of *P. jeffreyi* (Furnier and Adams, 1986) and *P. balfouriana* ssp. *balfouriana* (Oline et al., 2000) from California exhibit genetic differentiation, no serpentine soil tolerance studies have been conducted with those species to verify ecotypic differentiation.

Like Pinaceae, ecotypic differentiation of species in Cupressaceae is weak to nonexistent. Little difference in serpentine tolerance was found for serpentine and nonserpentine populations of *Hesperocyparis macnabiana* (formerly *Cupressus macnabiana*) from California (McMillan, 1956). *Juniperus communis* has a widespread, Northern Hemisphere distribution. Taxonomy of the species is confounded with disagreement of specific recognition of variety and subspecies ranks (Ashworth et al., 2001; Adams and Pandey, 2003; Adams et al., 2003). At least one recognized variety, var. *jackii*, is believed to be exclusive to serpentine soils in the northwestern United States (Kruckeberg, 1984; Adams and Nyugen, 2007). The variety exhibits a unique growth form of the species consisting of elongated, sparsely branched, lateral branches (Kruckeberg, 1984). Although some phylogenetic analyses support taxonomic recognition of var. *jackii* (Adams and Nguyen, 2007), others do not (Ashworth et al., 2001). Additionally, comparison of serpentine tolerance of serpentine and nonserpentine populations has revealed that both are equally serpentine-tolerant (Kruckeberg, 1967). When var. *jackii* was grown on nonserpentine soil, its growth form reverted to that more typical of nonserpentine var. *alpina*, suggesting that the unique growth form of var. *jackii* is environmentally induced rather than genetically fixed (Ashworth et al., 2001).

Similar to Poacae and other species in order Poales, the general indifference of species of Pinaceae and Cupressaceae to serpentine soils may be related to cell wall pectin content. Gymnosperms possess type I cell walls like dicotyledon angiosperms, however, studies of Pinaceae and Cupressaceae cell wall composition have shown that they have a substantially lower pectin content (15–35%) than dicotyledon (40–60%) angiosperms (Simson and Timell, 1978a, 1987b; Thomas et al., 1987; Edashige et al., 1995; Piro and Dalessandro, 1998; Hafrén et al., 2000). The lower cell wall pectin content of species of Pinaceae and Cupressaceae may confer a lower Ca requirement. Low Ca requirement has been noted for *Pinus radiata* growing on serpentine in New Zealand (Mead and Gadgil, 1978; Lee et al., 1991).

BRASSICACEAE: FLOWER MODIFICATION, CA AND MG REGULATION, NI TOLERANCE AND HYPERACCUMULATION, PARALLEL AND CONVERGENT EVOLUTION

Arabidopsis thaliana is one of the most widely used models in the study of plant genetics and development. In a study to identify the underlying physiological tolerance mechanisms involved in serpentine tolerance, mutants of *A. thaliana* were screened for tolerance to low solution Ca:Mg ratios (Bradshaw, 2005). Intolerant mutants perished at the cotyledon stage. Survivors were grown to provide tissue for genetic analysis. Genetic analysis of surviving mutants revealed that tolerance of the mutants to low solution Ca:Mg ratios was due to the loss of Ca^{2+}–H^+ antiporter CAX1 function. CAX1 is a high capacity Ca-proton antiporter that helps maintain cytoplasmic homeostasis by pumping excess Ca^{2+} from the cytoplasm into the vacuole (Cheng et al., 2003). Mutation of CAX1 effectively acts as a Mg^{2+} exclusion mechanism, allowing the plant to maintain adequate cytoplasmic Ca^{2+} concentrations when subjected to low substrate Ca:Mg ratios.

Arabidopis lyrata has been found to have natural serpentine-tolerant ecotypes (Turner et al., 2008; Von Wettberg, 2008). Recently identified candidate genes for serpentine tolerance in *A. lyrata* serpentine populations from the eastern United States and United Kingdom include K^+ transporters (KUP9 and KUP10), cation Ca^{2+} exchanger CCX1 (formerly CAX7; Shigaki et al., 2006), CorA-like Mg^{2+} transporters (MRS2-2, MRS2-6, and MRS2-7), voltage-gated Ca^{2+} channel 1S, and metal tolerance proteins (MTPc3, MTPc1) (Turner et al., 2008, 2010). These candidate genes suggest that Ca and Mg regulation and heavy metal detoxification are important tolerance mechanisms in the serpentine ecotypes. Loci sequencing of *A. lyrata* from the United Kingdom has revealed evidence of parallel and convergent evolution of serpentine tolerance within ecotypes of the species. Comparison studies of the growth rates of serpentine and nonserpentine ecotypes of *A. lyrata* on potting soil showed that the serpentine ecotype had significantly slower growth rate than the nonserpentine ecotype, revealing another serpentine tolerance trait in the species (Roche et al., 2004).

Brassicaceae has a large number of serpentine endemic species, particularly in the genus *Streptanthus*. Eighteen out of a total of 44 species, subspecies, and varieties in the genus *Streptanthus* in California are strict serpentine endemics (Safford et al., 2005; Al-Shehbaz and Mayer, 2008).

Streptanthus polygaloides is a strict serpentine endemic and the only known species in the genus that hyperaccumulates Ni (Reeves et al., 1981; Boyd et al., 2009). Ni hyperaccumulation in the species has been demonstrated to deter insect herbivory (Jhee et al., 2005). The range of *S. polygaloides* is limited to the Sierra Nevada mountains in California (Figure 5.6). There is substantial ecotypic variation, especially for flower color and leaf morphology, within *Streptanthus* species (Table 5.2). *S. polygaloides* has at least four flower sepal color morphs, including yellow (Y), purple (P), yellow maturing to purple (Y-P), and yellow undulate (U) throughout its range (Boyd et al., 2009). Whereas Y and P morphs have broad but elevationally divided (Y at lower elevations; P at higher elevations) distributions in the northern Sierra Nevada mountains, the Y-P and U populations are geographically restricted, especially the U population, which is almost 100 airline kilometers away from any other color morph populations.

Plants of all four color morphs of *S. polygaloides* were grown in a glasshouse on soil treated with Ni to examine elemental uptake and morphological differences. Significant differences in elemental uptake and leaf morphology were found. The P morph accumulated significantly higher shoot Ni and Mg concentrations than the other three morphs. The geographically isolated U morph had significantly higher K and Zn than the other morphs. Significant differences in plant height and leaf morphology were also found between all four flower color morphs. Morphological and physiological differences between the color morphs suggest that the species is diverging under genetic isolation (Boyd et al., 2009). Chemotaxonomic analysis of two widely separated populations of *S. polygaloides* within the Y and Y-P morph population range strongly suggests that at least two taxonomically distinct variants are present within the species (Rodman et al., 1981).

Further evidence of genetic isolation is provided by phylogenetic studies of the *Streptanthus glandulosus* complex (Mayer et al., 1994; Mayer and Soltis, 1994, 1999; Mayer and Beseda, 2010). *S. glandulosus* is found throughout the Coast Range mountains in California (Figure 5.6). The species complex consists of ten subspecies with great variation in flower color, leaf morphology, and affinity for serpentine soils (Table 5.2). Phyogenetic studies have resulted in the identification of four major clades, each corresponding to a geographic area: Northwest, East, Bay Area, and South (Mayer and Soltis, 1994, 1999). Subspecies are distributed within the clades as follows: Northwest—*raichei, hoffmanii,* and *sonomensis*; East—*arkii*; Bay Area—*secundus, pulchellus,* and *niger*; South—*albidus* and *glandulosus* (Mayer and Beseda, 2010). Subspecies *josephinensis* is a distant disjunct, occurring in southwestern Oregon, USA. Serpentine soil tolerance tests of serpentine and

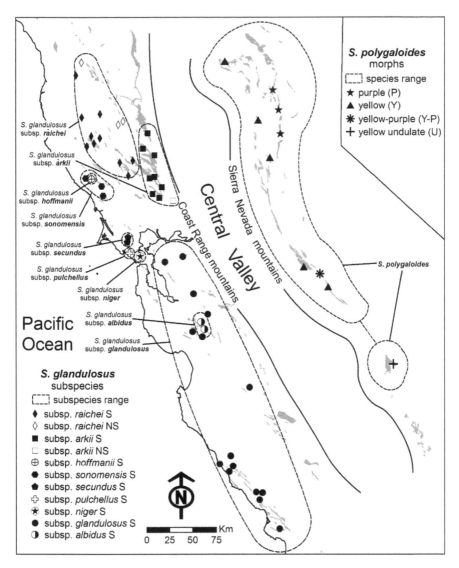

FIGURE 5.6. Distribution of *Streptanthus polygaloides* flower color morphs and *S. glandulosus* subspecies in California. Gray areas = serpentine, S = serpentine population, NS = nonserpentine population. All populations of *S. polygaloides* (strict serpentine endemic) are from serpentine. *S. glandulosus* ssp. *josephinensis*, a serpentine endemic disjunct located in southwestern Oregon, is not shown. It occurs approximately 200 airline km north of the northernmost known populations (approximately delineated by the top edge of this figure) of *S. glandulosus* in California. Figure drawn from Kruckeberg (1957), Mayer and Soltis (1994), and Boyd et al. (2009).

TABLE 5.2 Features of Morphs, Subspecies, and Varieties of *Streptanthus polygaloides*, *S. glandulosus* complex, and *Caulanthus amplexicaulis*

Streptanthus polygaloides (Boyd et al., 2009)

Species	Morph	Flower Color	Leaf Dimensions	Serpentine Endemicity*
S. polygaloides	P	purple	shorter, narrower	Strict
S. polygaloides	Y	yellow	shorter, narrower	Strict
S. polygaloides	Y-P	yellow maturing to purple	longer, wider	Strict
S. polygaloides	U	yellow, undulate	longer, narrower	Strict

Streptanthus glandulosus complex (Kruckeberg, 1957; Mayer and Beseda, 2010)

Species	Subspecies	Flower Color	Leaf Pubescence	Serpentine Endemicity*
S. glandulosus	*raichei*	rose, magenta, purple	hairy - hispid	Weak
S. glandulosus	*hoffmanii*	rose, magenta	pubescent - hispid	Broad
S. glandulosus	*sonomensis*	pale yellow, creamy white	pubescent - hispid	Strong
S. glandulosus	*arkii*	dark maroon, purplish black	hairy - hispid	Weak
S. glandulosus	*secundus*	yellow, creamy white	pubescent - hispid	Strong
S. glandulosus	*pulchellus*	red, reddish purple	pubescent - hispid	Broad
S. glandulosus	*niger*	purplish black	glabrous	Strict
S. glandulosus	*albidus*	cream, greenish white	glabrous	Broad
S. glandulosus	*glandulosus*	rose, lavender, violet, purple	pubescent - hispid	Weak
S. glandulosus	*josephinensis*	white, creamy white	pubescent - hispid	Strict

Caulanthus amplexicaulis (Pepper and Norwood, 2001)

Species	Variety	Flower Color		Serpentine Endemicity*
C. amplexicaulis	*amplexicaulis*	purple		Weak
C. amplexicaulis	*barbarae*	yellow, creamy white		Strict

SOURCE: *Safford et al., 2005.

nonserpentine populations of *S. glandulosus* showed that the serpentine populations were entirely serpentine tolerant, whereas the nonserpentine populations did not display any tolerance (Kruckeberg, 1950). Response of the serpentine and nonserpentine populations to Ca-reconstituted serpentine soils with increasing Ca:Mg ratios, showed that the serpentine populations were significantly more tolerant of lower soil Ca:Mg ratios than the nonserpentine populations (Kruckeberg, 1950).

Hybrid cross and fertility analysis studies of the subspecies revealed that most subspecies crosses had low interfertility (Kruckeberg, 1957). Only crosses of subspecies *raichei* and *arkii* (both formerly *S. glandulosus* ssp. *glandulosus*) with *secundus* and *pulchellus* generally resulted in relatively high interfertility. Within subspecies crosses, interfertility was inversely correlated with distance, with geographically closer populations having greater interfertility than populations further from each other. Phylogenetic analysis of *S. glandulosus* subspecies support the findings of hybrid crosses, with genetic identity being correlated with geographical distance between populations (Mayer et al., 1994). Furthermore, serpentine populations of ssp. *raichei, arkii,* and *glandulosus* were found to have greater genetic identity with nonserpentine populations found within the same region than more distant regions (Mayer and Soltis, 1994, 1999). It appears that serpentine tolerance has either evolved numerous times within the genus or been repeatedly lost among the descendants of a serpentine tolerant ancestor (Mayer and Soltis, 1994).

Caulanthus and closely related *Streptanthus* are both members of the Streptanthoid complex. *Caulanthus amplexicaulis* consists of two varieties including the serpentine endemic var. *barbarae* and the nonserpentine var. *amplexicaulis* (Pepper and Norwood, 2001). *C. amplexicaulis* var. *barbarae* has yellow flowers and is restricted to serpentine in the far southern Coast Range of California. *C. amplexicaulis* var. *amplexicaulis* has purple flowers and is found exclusively on nonserpentine soils at least 75 km away to the east from the nearest var. *barbarae* populations. Although the two varieties are ecologically and geographically isolated, they are completely interfertile (Pepper and Norwood, 2001). Phylogenetic analysis of populations of the two *C. amplexicaulis* varieties revealed that the varieties are distinct and monophyletic. Analysis of other species of *Caulanthus* and *Streptanthus* included in the study, however, revealed that the two genera are not monophyletic. This result is supported by chemotaxonomic studies of both genera (Rodman et al., 1981). Additionally, serpentine endemism is not monophyletic, suggesting that serpentine tolerance and subsequent endemism has arisen independently several times in the Streptanthoid complex (Rodman et al., 1981; Pepper and Norwood, 2001).

Many reassigned species of genus *Thlaspi* (tribes Thlaspidae and Noccaeeae; Brassicaceae) are Ni hyperaccumulators and serpentine endemics (Brooks, 1998;

Peer et al., 2003). Recent phylogenetic studies of genus *Thlaspi* and closely related genera revealed that *Thlaspi* is not monophyletic (Koch and Mummenhoff, 2001; Koch and Al-Shehbaz, 2004). All New World and most Old World *Thlaspi* have since been reassigned to genus *Noccaea* (Koch and Al-Shehbaz, 2004). Although not discussed by the authors, the phylogenies suggest either that serpentine toler-ance and heavy metal hyperaccumulation has evolved numerous times within tribe Noccaeeae, or that the ancestral state of Noccaeeae is presence of serpentine tolerance and heavy metal hyperaccumulation, with some derivative species hav-ing lost one or both of the traits.

Noccaea fendleri includes ssp. *glauca, idahoense, siskiyouense* (all formerly vari-eties of *Thlaspi montanum*), and *californica* (formerly *Thlaspi californicum*). *N. fendlerii* has a widespread western U.S. distribution. *N. fendlerii* ssp. *glauca* popu-lations occur on both serpentine and nonserpentine substrates in California, Or-egon, and Washington (Kruckeberg, 1967; Reeves et al., 1983; Boyd and Martens, 1994). *N. fendlerii* ssp. *siskiyouense* and *californica* are localized serpentine endem-ics in southwestern Oregon and northern California, respectively (U.S. Fish and Wildlife Service, 2003). Like most other *Noccaea* species, *N. fendleri* ssp. *glauca, siskiyouense,* and *californica* are all known to be Ni hyperaccumulators (Reeves et al., 1983).

Serpentine and nonserpentine populations of *N. fendleri* ssp. *glauca* from Washington exhibit ecotype differentiation with respect to serpentine soil toler-ance in both the glasshouse and the field (Kruckeberg, 1967). Comparison of the Ni uptake of serpentine and nonserpentine populations of *N. fendleri* ssp. *glauca* from California in the glasshouse showed that both serpentine and nonserpentine populations hyperaccumulate Ni (Boyd and Martens, 1998). Ni hyperaccumula-tion in the species has been shown to deter insect herbivory (Boyd and Martens, 1994). Field-collected plant tissue of *N. fendleri* ssp. *glauca* from serpentine (high Ni soil) and nonserpentine (very low Ni soil) populations were analyzed, and only plants from the serpentine population were found to have Ni concentrations high enough to indicate hyperaccumulation (Boyd and Martens, 1998). When plants of both serpentine and nonserpentine populations were grown on the same Ni-rich substrate in a glasshouse and the tissue was analyzed, however, plants of both pop-ulations exhibited Ni hyperaccumulation (Boyd and Martens, 1998; Peer et al., 2006). It has been suggested that the nonserpentine populations of *N. fendleri* ssp. *glauca* are effectively preadapted for serpentine tolerance, and Ni hyperaccumula-tion is a constitutive trait (Boyd and Martens, 1998). Similar results were found for *Noccaea goesingense* (formerly *Thlaspi goesingense;* Austria; Reeves and Baker, 1984) and *Noccaea cochleariforme* (formerly *Thlaspi japonicum;* Japan; Mizuno et al., 2003, 2005) serpentine and nonserpentine populations.

Noccaea caerulescens (formerly *Thlaspi caerulescens*) from Europe exhibits dif-ferentiation with serpentine ecotypes displaying significantly greater Ni tolerance and Ni uptake than nonserpentine ecotypes (Assunção et al. 2003, 2008). Significant variation in Ni uptake has also been found between serpentine populations of the serpentine endemic *Noccaea pindica* (formerly *Thlaspi pindicum*) from Greece (Taylor and Macnair, 2006). Some populations of the species display significantly greater Ni uptake and root-to-shoot translocation than other populations when grown in the same substrate (Taylor and Macnair, 2006).

Like *Noccaea*, many species of genus *Alyssum* (Brassicaceae) are also Ni hyperaccumulators and serpentine endemics (Brooks, 1998). Comparison of serpentine and nonserpentine populations of *A. murale* grown in culture solution with Ni showed that the serpentine population was significantly more tolerant of elevated Ni concentrations than the nonserpentine population (Asemaneh et al., 2006). Tissue Ni analysis of nine populations of the serpentine endemic *A. bertolonii* grown in culture solution containing Ni revealed ecotypic variation with respect to both Ni tolerance and uptake (Galardi et al., 2007), similar to *Noccaea pindica*. Phylogenetic analysis of species and subspecies in *Alyssum* has revealed that the Ni hyperaccumulation mechanism has been lost or acquired several times (Mengoni et al., 2003). Studies of serpentine and nonserpentine populations of *A. inflatum*, a Ni hyperaccumulator from Iran, showed that plants from the serpentine population were more tolerant of low solution Ca:Mg ratios than plants of the nonserpentine population when grown in culture solution (Ghasemi and Ghaderian, 2009). There was no difference between plants of the two populations with respect to Ni tolerance or uptake.

ADAPTATION TRENDS

As shown with the many diverse taxa, several recurring adaptations to serpentine soils are evident. These adaptations include (1) inherently slower growth rate; (2) high root:shoot mass ratios; (3) early flowering time; (4) regulation of Ca and Mg, including selective Ca uptake and enhanced root-to-shoot translocation and Mg exclusion at the root level or restricted root-to-shoot translocation; and (5) heavy metal regulation, including root-level exclusion or sequestration, and variation in heavy metal uptake and root-to-shoot translocation in heavy metal hyperaccumulators.

Inherently slower growth rate, high root:shoot mass ratios, and early flowering time appear to be strongly associated with drought tolerance, although slower growth rate and high root:shoot mass ratio may also help cope with low macronutrient (N, P, K) availability. Regulation of Ca and Mg uptake is associated with the maintenance of balanced tissue Ca and Mg concentrations in response to very low Ca and high Mg soil concentrations. Species of order Poales contain type II cell walls, which have much lower Ca concentrations than type I cell walls in most other vascular plant species. As a result, species of Poales are effectively preadapted to low substrate Ca concentrations and do not appear to display ecotypic differentiation with respect to Ca and Mg regulation as frequently as species in other plant families. Root-level heavy metal exclusion and sequestration are common tolerance mechanisms to the elevated bioavailable heavy metal concentrations in serpentine soils. Evolution of heavy metal hyperaccumulation has been demonstrated in ecotypes of a number of species. Studies have shown that metal hyperaccumulation confers protection from insect herbivory and serves a role in osmotic regulation and drought tolerance in some species.

Recurring evolutionary trends in selection for serpentine tolerance are evidence for preadaptation and ecotypic differentiation occurring by catastrophic selection. This mode of selection appears to be especially common for tolerance mechanisms involving Ca and Mg regulation. Convergent and parallel evolution of ecotypes numerous times within the same species also appears to be common and is represented by several species within diverse families. The role of parallel evolution in the origin of species (Levin, 2001) is at an early stage of discovery, and we believe that serpentine ecosystems can provide model settings for the study of parallel speciation (i.e., the parallel origins of traits that confer adaptations and reproductive isolation) in plants.

SUMMARY

Plant adaptation and evolutionary mechanisms on serpentine soils bear many similarities to other phytotoxic soils such as alkaline, saline soils and acidic, metal ore mine tailings. Unlike mine tailings and serpentine, saline soils can often have a broad, continuous, widespread distribution, especially in arid inland environments and coastlines (Läuchli and Lüttge 2002). In contrast, mine tailings and serpentine areas are limited and typically have a discontinuous, widespread distribution across the landscape, resembling island environments.

Although metal ore mine tailings globally have the same general conditions of low pH and high bioavailable metal concentrations, these conditions can be highly variable due to mine host rock type, ore type, and quantity of associated elemental sulfur and sulfate-bearing minerals (Quilty, 1975; Williamson et al., 1982; Tordoff et al., 2000). Compared to metal ores and their host rock types, serpentine has a relatively consistent mineral composition (Brady et al., 2005; Alexander et al., 2007; Kazakou et al., 2008). Regardless of climate, topography, and associated organisms, serpentine consistently weathers to produce shallow, rocky soils, low in bioavailable N, P, K, and Ca and high in bioavailable Mg, and often Ni, Cr, and Co. Low Ca:Mg ratio is the defining characteristic of serpentine soils worldwide.

Serpentine soils have a widespread distribution, occurring on every continent and found in virtually every type of climate (Kruckeberg, 1984; Brooks, 1987; Roberts and Proctor, 1992). Serpentine also supports a high diversity of plant species and vegetation types. Additionally, it is a natural phytotoxic environment as opposed to being an unnatural, anthropogenic, phytotoxic environment like mine tailings. Widespread, island-like distribution; consistency of adverse soil chemical conditions; diversity of climate, plant species, and vegetation type; and natural, relatively undisturbed conditions make serpentine soils ideal model ecosystems for studies of intraspecific variation and resulting plant evolution.

Acknowledgments

We thank Mike Mayer (University of San Diego) and Ihsan Al-Shehbaz (Missouri Botanical Garden) for taxonomic information regarding *Streptanthus* and *Noccaea*, Tanner Harris for assisting in the literature searches, Kelly Bougher for compiling literature cited, and Jessica Wright and Bob Boyd for useful feedback during manuscript preparation.

LITERATURE CITED

Adams, R. P., and Nguyen, S. (2007) Post-Pleistocene geographic variation in *Juniperus communis* in North America. *Phytologia*, 89, 43–57.

Adams, R. P. and Pandey, R. N. (2003) Analysis of *Juniperus communis* and its varieties based on DNA fingerprinting. *Biochemical Systematics and Ecology*, 31, 1271–78.

Adams, R. P., Pandey, R. N., Leverenz, J. W., Dignard, N., Hoegh, K., and Thorfinnsson, T. (2003) Pan-Arctic variation in *Juniperus communis*: Historical biogeography based on DNA fingerprinting. *Biochemical Systematics and Ecology*, 31, 181–92.

Alexander, E. B., Coleman, R. G., Keeler-Wolf, T., and Harrison, S. (2007) *Serpentine Geoecology of Western North America*. Oxford University Press, New York.

Al-Shehbaz, I. A., and Mayer, M. S. (2008) New or noteworthy *Streptanthus* (Brassicaceae) for the flora of North America. *Novon*, 18, 279–82.

Antonovics, J. (2006) Evolution in closely adjacent plant populations X: Long-term persistence of prereproductive isolation at a mine boundary. *Heredity*, 97, 33–37.

Asemaneh, T., Ghaderian, S. M., Crawford, S. A., Marshall, A. T., and Baker, A. J. M. (2006) Cellular and subcellular compartmentation of Ni in the Eurasian serpentine plants *Alyssum bracteatum*, *Alyssum murale* (Brassicaceae) and *Cleome heratensis* (Capparaceae). *Planta*, 225, 193–202.

Ashworth, V. E. T. M., O'Brien, B. C., and Friar, E. A. (2001) Survey of *Juniperus communis* (Cupressaceae) L. varieties from the western United States using RAPD fingerprints. *Madroño*, 48, 172–76.

Assunção, A. G. L., Bleeker, P., ten Bookum, W. M., Vooijs, R., and Schat, H. (2008) Intraspecific variation of metal preference patterns for hyperaccumulation in *Thlaspi caerulescens*: Evidence from binary metal exposures. *Plant and Soil*, 303, 289–99.

Assunção, A. G. L., ten Bookum, W. M., Nelissen, H. J. M., Vooijs, R., Schat, H., and Ernst, W. H. O. (2003) Differential metal-specific tolerance and accumulation patterns among *Thlaspi caerulescens* populations originating from different soil types. *New Phytologist*, 159, 411–19.

Baker, A. J. M. (1987) Metal tolerance. *New Phytologist*, 106, 93–111.

Bhatia, N. P., Baker, A. J. M., Walsh, K. B., and Midmore, D. J. (2005) A role for nickel in osmotic adjustment in drought-stressed plants of the nickel hyperaccumulator *Stackhousia tryonii* Bailey. *Planta*, 223, 134–39.

Boyd, R. S. (2007) The defense hypothesis of elemental hyperaccumulation: Status, challenges and new directions. *Plant and Soil*, 293, 153–76.

Boyd, R. S., Davis, M. A., and Balkwill, K. (2008) Elemental patterns in Ni hyperaccumulating and non-hyperaccumulating ultramafic soil populations of *Senecio coronatus*. *South Africa Journal of Botany*, 74, 158–62.

Boyd, R. S., and Martens, S. N. (1992) The raison d'être for metal hyperaccumulation by plants. In *The Vegetation of Ultramafic (Serpentine) Soils. Proceedings of the First International Conference on Serpentine Ecology* (eds. A. J. M. Baker, J. Proctor, and R. D. Reeves), pp. 279–89. Intercept, Andover.

Boyd, R. S., and Martens, S. N. (1994) Nickel hyperaccumulation by *Thlaspi montanum* var. *montanum* is acutely toxic to an insect herbivore. *Oikos*, 70, 21–25.

Boyd, R. S., and Martens, S. N. (1998) Nickel hyperaccumulation by *Thlaspi montanum* var. *montanum* (Brassicaceae): A constitutive trait. *American Journal of Botany*, 85, 259–65.

Boyd, R. S., Wall, M. A., Santos, S. R., and Davis, M. A. (2009) Variation of morphology and elemental concentrations in the California nickel hyperaccumulator *Streptanthus polygaloides* (Brassicaceae). *Northeastern Naturalist*, 16, 21–38.

Bradshaw H. D. Jr. (2005) Mutations in CAX1 produce phenotypes characteristic of plants tolerant to serpentine soils. *New Phytologist*, 167, 81–88.

Brady, K. U., Kruckeberg, A. R., and Bradshaw H. D. Jr. (2005) Evolutionary ecology of plant adaptation to serpentine soils. *Annual Review of Ecology, Evolution, and Systematics*, 36, 243–66.

Bratteler, M., Baltisberger, M., and Widmer, A. (2006a) QTL analysis of intraspecific differences between two *Silene vulgaris* ecotypes. *Annals of Botany*, 98, 411–19.

Bratteler, M., Lexer, C., and Widmer, A. (2006b) Genetic architecture of traits associated with serpentine adaptation of *Silene vulgaris*. *Journal of Evolutionary Biology*, 19, 1149–56.

Briggs, D., and Walters, S. M. (1997) *Plant Variation and Evolution*. Cambridge University Press, Cambridge.

Broadley, M. A., Bowen, H. C., Cotterill, H. L., Hammond, J. P., Meacham, M. C., Mead, A., and White, P. J. (2003) Variation in the shoot calcium content of angiosperms. *Journal of Experimental Botany*, 54, 1431–46.

Brooks, R. R. (1987) *Serpentine and Its Vegetation. A Multidisciplinary Approach*. Dioscorides Press, Portland, OR.

Brooks, R. R. (1998) *Plants that Hyperaccumulate Heavy Metals*. CAB International, Wallingford, Oxon, UK.

Burt, R., Fillmore, M., Wilson, M. A., Gross, E. R., Langridge, R. W., and Lammers, D. A. (2001) Soil properties of selected pedons on ultramafic rocks in Klamath Mountains, Oregon. *Communications in Soil Science and Plant Analysis*, 32, 2145–75.

Cheeseman, J. M. (1988) Mechanisms of salinity tolerance in plants. *Plant Physiology*, 87, 547–50.

Cheng, N., Pittman, J. K., Barkla, B. J., Shigaki, T., and Hirschi, K. D. (2003) The *Arabidopsis* cax1 mutant exhibits impaired ion homeostasis, development, and hormonal responses and reveals interplay among vacuolar transporters. *Plant Cell,* 15, 347–64.

Cooke, S. S. (1994) *The Edaphic Ecology of Two Western North American Composite Species.* Ph.D. dissertation. University of Washington, Seattle.

Demarty, M., Morvan, C., and Thellier, M. (1984) Calcium and the cell wall. *Plant, Cell, and Environment,* 7, 441–48.

Desrochers, A., and Bohm, B. A. (1995) Biosystematic study of *Lasthenia californica* (Asteraceae). *Systematic Botany,* 20, 65–84.

Edashige, Y., Ishii, T., Hiroi, T., and Tomoyuki, F. (1995) Cell-wall polysaccharides of cambial tissue: Structural analysis of the polysaccharides of the primary wall from xylem differentiating zone of *Cryptomeria japonica* D. Dom. *Holzforschung,* 49, 197–202.

Enberg, A., and Wu, L. (1995) Selenium assimilation and differential response to elevated sulfate and chloride salt concentrations in two saltgrass ecotypes. *Ecotoxicology and Environmental Safety,* 32, 171–78.

Espeland, E. K., and Rice, K. J. (2007) Facilitation across stress gradients: The importance of local adaptation. *Ecology,* 88, 2404–9.

Freeman, J. L., Lindblom, S. D., Quinn, C. F., Fakra, S., Marcus, M. S., and Pilon-Smits, E. A. H. (2007) Selenium accumulation protects plants from herbivory by Orthoptera via toxicity and deterrence. *New Phytologist,* 175, 490–500.

Freeman, J. L., Quinn, C. F., Lindblom, S. D., Klamper, E. M., and Pilon-Smits, E. A. H. (2009) Selenium protects the hyperaccumulators *Stanleya pinnata* against black-tailed prairie dog herbivory in native seleniferous habitats. *American Journal of Botany,* 96, 1075–85.

Freitas, H., and Mooney, H. (1996) Effects of water stress and soil texture on the performance of two *Bromus hordeaceus* ecotypes from sandstone and serpentine soils. *Acta Oecologia,* 17, 307–17.

Friedman, J., and Barrett, C. H. (2009) Wind of change: New insights on the ecology and evolution of pollination and mating in wind-pollinated plants. *Annals of Botany,* 103, 1515–27.

Furnier, G. R., and Adams, W. T. (1986) Geographic patterns of allozyme variation in Jeffrey pine. *American Journal of Botany,* 73, 1009–15.

Gabbrielli, R., Pandolfini, T., Vergnano, O., and Palandri, M. R. (1990) Comparison of two serpentine species with different nickel tolerance strategies. *Plant and Soil,* 122, 271–77.

Galardi, F., Corrales, I., Mengoni, A., Pucci, S., Barletti, L., Barzanti, R., Arnetoli, M., Gabbrielli, R., and Gonnelli, C. (2007) Intra-specific differences in nickel tolerance and accumulation in the Ni-hyperaccumulator *Alyssum bertolonii. Environmental and Experimental Botany,* 60, 377–84.

Gardner, M., and Macnair, M. (2000) Factors affecting the co-existence of the serpentine endemic *Mimulus nudatus* Curran and its presumed progenitor, *Mimulus guttatus* Fischer ex DC. *Biological Journal of the Linnean Society,* 69, 443–59.

Ghasemi, R., and Ghaderian, S. M. (2009) Responses of two populations of an Iranian nickel-hyperaccumulating serpentine plant, *Alyssum inflatum* Nyar., to substrate Ca/Mg quotient and nickel. *Environmental and Experimental Botany,* 67, 260–68.

Gonnelli, C., Galardi, F., and Gabbrielli, R. (2001) Nickel and copper tolerance and toxicity in three Tuscan populations of *Silene paradoxa*. *Physiologia Plantarum*, 113, 507–14.

Griffin, J. R. (1962) *Intraspecific Variation in* Pinus sabiniana Dougl. Ph.D. dissertation. University of California, Berkeley.

Hafrén, J., Daniel, G., and Westermark, U. (2000) The distribution of acidic and esterified pectin in cambium, developing xylem and mature xylem of *Pinus sylvestris*. *IAWA Journal* 21, 157–68.

Hall, J. L. (2002) Cellular mechanisms for heavy metal detoxification and tolerance. *Journal of Experimental Botany*, 53, 1–11.

Hanson, B., Lindblom, S. D., Loeffler, M. L., and Pilon-Smits, E. A. H. (2004) Selenium protects plants from phloem-feeding aphids due to both deterrence and toxicity. *New Phytologist*, 162, 655–62.

Harrison, S., Rice, K., and Maron, J. (2001) Habitat patchiness promotes invasion by alien grasses on serpentine soil. *Biological Conservation*, 100, 45–53.

He, B., Yang, X., Ni, W., Wei, Y., Long, X., and Ye, Z. (2002) *Sedum alfredii*: A new lead-accumulating ecotype. *Acta Botanica Sinica*, 44, 1365–70.

Hendrick, M. F. (2008) *Characterizing Serpentine Soil Tolerance in* Mimulus guttatus (*Phrymaceae*). B.A. thesis. Reed College, Portland, OR.

Higgins, S. S., and Mack, R. N. (1987) Comparative responses of *Achillea millefolium* ecotypes to competition and soil types. *Oecologia*, 73, 591–97.

Hughes, R., Bachmann, K., Smirnoff, N., and Macnair, M. R. (2001) The role of drought tolerance in serpentine tolerance in the *Mimulus guttatus* Fischer ex DC. complex. *South African Journal of Science*, 97, 581–86.

Jarvis, M. C. (1984) Structure and properties of pectin gels in plant cell walls. *Plant, Cell, and Environment*, 7, 153–64.

Jarvis, M. C., Briggs, S. P. H., and Knox, J. P. (2003) Intercellular adhesion and cell separation in plants. *Plant, Cell, and Environment*, 7, 153–64.

Jarvis, M. C., Forsyth, W., and Duncan, H. J. (1988) A survey of the pectic content of non-lignified monocot cell walls. *Plant Physiology*, 88, 309–14.

Jenkinson, J. L. (1974) *Ponderosa Pine Progenies: Differential Response to Ultramafic and Granitic Soils*. U.S. Forest Service, Berkeley, CA.

Jhee, E. M., Boyd, R. S., and Eubanks, M. D. (2005) Nickel hyperaccumulation as an elemental defense of *Streptanthus polygaloides* (Brassicaceae): Influence of herbivore feeding mode. *New Phytologist*, 168, 331–43.

Johnston, W. R., and Proctor, J. (1981) Growth of serpentine and non-serpentine races of *Festuca rubra* in solutions simulating the chemical conditions in a toxic serpentine soil. *Journal of Ecology*, 69, 855–69.

Jurjavcic, N. L., Harrison, S., and Wolf, A. T. (2002) Abiotic stress, competition, and the distribution of the native annual grass *Vulpia microstachys* in a mosaic environment. *Oecologia*, 130, 555–62.

Kawase, D., Yumoto, T., Hayashi, K., and Sato, K. (2007) Molecular phylogenetic analysis of the infraspecific taxa of *Erigeron thunbergii* A. Gray distributed in ultramafic rock sites. *Plant Species Biology*, 22, 107–15.

Kawase, D., Yumoto, T., and Sato, K. (2009) Phylogeography of a rare serpentine plant, *Arenaria katoana* Makino (Caryophyllaceae). *Acta Phytotaxonomica et Geobotanica*, 60, 19–25.

Kayama, M., Quoreshi, A. M., Uemura, S., and Koike T. (2005) Differences in growth characteristics and dynamics of elements absorbed in seedlings of three spruce species raised on serpentine soil in northern Japan. *Annals of Botany*, 95, 661–72.

Kazakou, E., Dimitrakopoulos, P. G., Baker, A. J. M., Reeves, R. D., and Troumbis, A. Y. (2008) Hypotheses, mechanisms and trade-offs of tolerance and adaptation to serpentine soils: From species to ecosystem level. *Biological Reviews*, 83, 495–508.

Koch, M., and Al-Shehbaz, I. A. (2004) Taxonomic and phylogenetic evaluation of the American *"Thlaspi"* species: Identity and relationship to the Eurasian genus *Noccaea* (Brassicaceae). *Systematic Botany*, 29, 375–84.

Koch, M., and Mummenhoff, K. (2001) *Thlaspi* s.str. (Brassicaceae) versus *Thlaspi* s.l.: Morphological and anatomical characters in the light of ITS nrDNA sequence data. *Plant Systematics and Evolution*, 227, 209–25.

Kruckeberg, A. R. (1950) *An Experimental Inquiry into the Nature of Endemism on Serpentine Soils*. Ph.D. dissertation. University of California, Berkeley.

Kruckeberg, A. R. (1957) Variation in fertility of hybrids between isolated populations of the serpentine species, *Streptanthus glandulosus* Hook. *Evolution*, 11, 185–211.

Kruckeberg, A. R. (1967) Ecotypic response to ultramafic soils by some plant species of northwestern United States. *Brittonia*, 19, 133–51.

Kruckeberg, A. R. (1984) *California Serpentines: Flora, Vegetation, Geology, Soils, and Management Problems*. University of California Press, Berkeley.

Kruckeberg, A. R. (1986) An essay: The stimulus of unusual geologies for plant speciation. *Systematic Biology*, 11, 455–63.

Kruckeberg, A. R. (1991) An essay: Geoedaphics and island biogeography for vascular plants. *Aliso*, 13, 225–38.

Läuchli, A., and Lüttge, U. (2002) *Salinity: Environment-Plants-Molecules*. Kluwer Academic Publishers, Dordrecht.

Lee, W. G., Littlejohn, R. P., and Prema, P. G. (1991) Growth of *Pinus radiata* in relation to foliar element concentrations on ultramafic soil, New Zealand. *New Zealand Journal of Botany*, 29, 163–67.

Levin, D. A. (2000). *The Origin, Expansion, and Demise of Plant Species*. Oxford University Press, Oxford,.

Levin, D. A. (2001) The recurrent origin of plant races and species. *Systematic Botany*, 26, 197–204.

Lewis, H. (1962) Catastrophic selection as a factor in speciation. *Evolution*, 16, 257–71.

Löfgren, A. (2002) Effects of isolation on distribution, fecundity, and survival in the self-incompatible *Achillea millefolium* (L.). *Ecoscience*, 9, 503–8.

Lombini, A., Llugany, M., Poschenrieder, C., Dinelli, E., and Barcelo, J. (2003) Influence of the Ca/Mg ratio on Cu resistance in three *Silene armeria* ecotypes adapted to calcareous soil or to different, Ni- or Cu-enriched, serpentine sites. *Journal of Plant Physiology*, 160, 1451–56.

Lyons, K. G., Shapiro, A. M., and Schwartz, M. W. (In Press) Distribution and ecotypic variation of the invasive annual barbed goatgrass on serpentine soil. *Invasive Plant Science and Management*.

Macnair, M. R. (1983) The genetic control of copper tolerance in the yellow monkey flower, *Mimulus guttatus*. *Heredity*, 50, 283–93.

Macnair, M. R. (1989a) A new species of *Mimulus* endemic to copper mines in California. *Botanical Journal of the Linnean Society*, 100, 1–14.

Macnair, M. R. (1989b) The potential for rapid speciation in plants. *Genome*, 31, 203–10.

Macnair, M. R. (1992) Preliminary studies on the genetics and evolution of the serpentine endemic *Mimulus nudatus* Curran. In *The Vegetation of Ultramafic (Serpentine) Soils* (eds. A. J. M. Baker, J. Proctor, and R. D. Reeves), pp. 409–19. Intercept, Andover, UK.

Macnair, M. R., and Cumbes, Q. J. (1989) The genetic architecture of interspecific variation in *Mimulus*. *Genetics*, 122, 211–22.

Macnair, M.R., and Gardner, M. (1998) The evolution of edaphic endemics. In *Endless Forms: Species and Speciation* (eds. D. J. Howard and S. H. Berlocher), pp. 157–71. Oxford University Press, New York.

Macnair, M. R., Macnair, V. E., and Martin, B. E. (1989) Adaptive speciation in *Mimulus*: An ecological comparison of *M. cupriphilus* with its presumed progenitor, *M. guttatus*. *New Phytologist*, 112, 269–79.

Main, J. L. (1974) Differential responses to magnesium and calcium by native populations of *Agropyron spicatum*. *American Journal of Botany*, 61, 931–37.

Marschner, H. (2002) *Mineral Nutrition of Higher Plants*. Academic Press, San Diego.

Mason, H. L. (1946a) The edaphic factor in narrow endemism I. The nature of environmental influences. *Madroño*, 8, 209–40.

Mason, H. L. (1946b) The edaphic factor in narrow endemism II. The geographic occurrence of plants of highly restricted patterns of distribution. *Madroño*, 8, 241–72.

Mayer, M. S., and Beseda, L. (2010) Reconciling taxonomy and phylogeny in the *Streptanthus glandulosus* complex (Brassicaceae). *Annals of the Missouri Botanical Garden*, 97, 106–16.

Mayer, M. S., and Soltis, P. S. (1994) The evolution of serpentine endemics: A chloroplast DNA phylogeny of the *Streptanthus glandulosus* complex (Cruciferae). *Systematic Botany*, 19, 557–74.

Mayer, M. S., and Soltis, P. S. (1999) Intraspecific phylogeny analysis using ITS sequences: Insights from studies of the *Streptanthus glandulosus* complex (Cruciferae). *Systematic Biology*, 24, 47–61.

Mayer, M. S., Soltis, P. S., and Soltis, D. E. (1994) The evolution of the *Streptanthus glandulosus* complex (Cruciferae): Genetic divergence and gene flow in serpentine endemics. *American Journal of Botany*, 81, 1288–99.

McMillan, C. (1956) The edaphic restriction of *Cupressus* and *Pinus* in the coast ranges of central California. *Ecological Monographs*, 26, 177–212.

McNeilly, T., and Antonovics, J. (1968) Evolution in closely adjacent plant populations IV. Barriers to gene flow. *Heredity*, 23, 219–38.

Mead, D. J., and Gadgil R. L. (1978) Fertiliser use in established radiata pine stands in New Zealand. *New Zealand Journal of Forestry Science*, 8, 105–34.

Mengoni, A., Baker, A. J. M., Bazzicalupo, M., Reeves, R. D., Adigüzel, N., Chianni, E., Galardi, F., Gabbrielli, R., and Gonnelli, C. (2003) Evolutionary dynamics of nickel hyperaccumulation in *Alyssum* revealed by ITS nrDNA analysis. *New Phytologist,* 159, 691–99.

Mesjasz-Przybyłowicz, J., Barnabas, A., and Przybyłowicz, W. (2007) Comparison of cytology and distribution of nickel in roots of Ni-hyperaccumulating and non-hyperaccumulating genotypes of *Senecio coronatus. Plant and Soil,* 293, 61–78.

Miller, S. P., and Cumming, J. R. (2000) Effects of serpentine soil factors on Virginia pine (*Pinus virginiana*) seedlings. *Tree Physiology,* 20, 1129–35.

Mizuno, N., Nosaka, S., Mizuno, T., Horie, K., and Obata, H. (2003) Distribution of Ni and Zn in the leaves of *Thlaspi japonicum* growing on ultramafic soil. *Soil Science and Plant Nutrition,* 49, 93–97.

Mizuno, T., Obata, H., Horie, J., Nosaka, S., and Mizuno, N. (2005) Comparison of Ni/Zn accumulation ability of *Thlaspi japonicum* from three different areas in Hokkaido. *Soil Science and Plant Nutrition,* 51, 589–94.

Morrey D. R., Balkwill K., Balkwill M.-J., and Williamson, S. (1992) A review of some studies of the serpentine flora of southern Africa. In *The Vegetation of Ultramafic (Serpentine) Soils* (eds. A. J. M. Baker, J. Proctor, and R. D. Reeves), pp. 147–57. Intercept, Andover, UK.

Murren, C. J., Douglass, L., Gibson, A., and Dudash, M. R. (2006) Individual and combined effects of Ca/Mg ratio and water on trait expression in *Mimulus guttatus. Ecology,* 87, 2591–602.

Nagy, L., and Proctor, J. (2001) The effects of available Fe and Ni on ultramafic and nonultramafic *Cochlearia* spp. (Brassicaceae). *South African Journal of Science,* 97, 586–90.

Nordal, I., Haraldsen, K. B., Ergon, A., and Eriksen, A. B. (1999) Copper resistance and genetic diversity in *Lychnis alpina* (Caryophyllaceae) populations on mining sites. *Folia Geobotanica,* 34, 471–81.

Nyberg Berglund, A. B. (2005) *Postglacial Colonization and Parallel Evolution of Metal Tolerance in the Polyploid* Cerastium alpinum. Ph.D. dissertation. Mid Sweden University, Uppsala.

Nyberg Berglund, A. B., Dahlgren, S., and Westerbergh, A. (2003) Evidence for parallel evolution and site-specific selection of serpentine tolerance in *Cerastium alpinum* during the colonization of Scandinavia. *New Phytologist,* 161, 199–209.

Nyberg Berglund, A. B., and Westerbergh, A. (2001a) Two postglacial immigration lineages of the polyploidy *Cerastium alpinum* (Caryophyllaceae). *Hereditas,* 134, 171–83.

Nyberg Berglund, A. B., and Westerbergh, A. (2001b) Genetic differentiation of a polyploidy plant on ultramafic soils in Fennoscandia. *South African Journal of Science,* 97, 533–35.

O'Dell, R. E., and Claassen, V. P. (2006) Serpentine and nonserpentine *Achillea millefolium* accessions differ in serpentine substrate tolerance and response to organic and inorganic amendments. *Plant and Soil,* 279, 253–69.

O'Dell, R. E., and Claassen, V. P. (2008) Parallel evolution and convergent physiological tolerance mechanisms of serpentine-tolerant *Achillea millefolium* (Asteraceae) edaphic ecotypes. *Sixth International Conference on Serpentine Ecology meeting abstracts.* Bar Harbor, ME.

Oline, D. K., Mitton, J. B., and Grant, M. C. (2000) Population and subspecific genetic differentiation in the foxtail pine (*Pinus balfouriana*). *Evolution,* 54, 1813–19.

Pavlova, D. (2009) Morphological variation in *Teucrium chamaedrys* in serpentine and non-serpentine populations. *Northeastern Naturalist,* 16, 39–55.

Peer, W. A., Mahmoudian M., Freeman, J. L., Lahner, B., Richards, E. L., Reeves, R. D., Murphy, A. S., and Salt, D. E. (2006) Assessment of plants from the Brassicaceae family as genetic models for the study of nickel and zinc hyperaccumulation. *New Phytologist,* 172, 248–60.

Peer, W. A., Mamoudian, M., Lahner, B., Reeves, R. D., Murphy, A. S., and Salt, D. E. (2003) Identifying model metal hyperaccumulating plants: Germplasm analysis of 20 Brassicaceae accessions from a wide geographical area. *New Phytologist,* 159, 421–30.

Pepper, A. E., and Norwood, L. E. (2001) Evolution of *Caulanthus amplexicaulis* var. *barbarae* (Brassicaceae), a rare serpentine endemic plant: A molecular phylogenetic perspective. *American Journal of Botany,* 88, 1479–89.

Piro, G., and Dalessandro, G. (1998) Cell-wall biosynthesis in differentiating cells of pine root tips. *Phytochemistry,* 47, 1201–6.

Proctor, J. (1971) The plant ecology of serpentine II. Plant response to serpentine soils. *Journal of Ecology,* 59, 397–410.

Proctor, J., and Woodell, S. J. R. (1975) The ecology of serpentine soils. *Advances in Ecological Research,* 9, 256–366.

Quilty, J. A. (1975) Guidelines for rehabilitation of tailings dumps and open cuts. *Journal of the Soil Conservation Service of New South Wales,* 31, 95–107.

Quinn, C. F., Freeman, J. L., Galeas, M. L., Klamper, E. M., and Pilon-Smits, E. A. H. (2008) The role of selenium in protecting plants against prairie dog herbivory: Implications for the evolution of selenium hyperaccumulation. *Oecologia,* 155, 267–75.

Rajakaruna, N. (2004) The edaphic factor in the origin of plant species. *International Geology Review,* 46, 471–78.

Rajakaruna, N., Baldwin, B. G., Chan, R., Desrochers, A. M., Bohm, B. A., and Whitton, J. (2003a) Edaphic races and phylogenetic taxa in the *Lasthenia californica* complex (Asteraceae: Heliantheae): An hypothesis of parallel evolution. *Molecular Ecology,* 12, 1675–79.

Rajakaruna, N., and Bohm, B. A. (1999) The edaphic factor and patterns of variation in *Lasthenia californica* (Asteraceae). *American Journal of Botany,* 86, 1576–96.

Rajakaruna, N., and Boyd, R. S. (2008) The edaphic factor. In *The Encyclopedia of Ecology. Volume 2* (eds. S. E. Jorgensen and B. Fath), pp. 1201–7. Elsevier, Oxford.

Rajakaruna, N., Siddiqi, M. Y., Whitton, J., Bohm, B. A., and Glass, A. D. M. (2003b) Differential responses to Na^+/K^+ and Ca^{2+}/Mg^{2+} in two edaphic races of the *Lasthenia californica* (Asteraceae) complex: A case for parallel evolution of physiological traits. *New Phytologist,* 157, 93–103.

Reeves, R. D. (1992) The hyperaccumulation of nickel by serpentine plants. In *The Vegetation of Ultramafic (Serpentine) Soils. Proceedings of the First International Conference on Serpentine Ecology* (eds. A. J. M. Baker, J. Proctor, and R. D. Reeves), pp. 253–77. Intercept, Andover, UK.

Reeves, R. D., and Baker, A. J. M. (1984) Studies on metal uptake by plants from serpentine and non-serpentine populations of *Thlaspi goesingense* Hálácsy (Cruciferae). *New Phytologist*, 98, 191–204.

Reeves, R. D., Brooks, R. R., and Macfarlane, R. M. (1981) Nickel uptake by Californian *Streptanthus* and *Caulanthus* with particular reference to the hyperaccumulators *S. polygaloides* Gray (Brassicaceae). *American Journal of Botany*, 68, 708–12.

Reeves, R. D., Macfarlane, R. M., and Brooks, R. R. (1983) Accumulation of nickel and zinc by western North American genera containing serpentine-tolerant species. *American Journal of Botany*, 70, 1297–303.

Richau, K. H., and Schat, H. (2009) Intraspecific variation of nickel and zinc accumulation and tolerance in the hyperaccumulator *Thlaspi caerulescens*. *Plant and Soil*, 314, 253–62.

Roberts, B. A., and Proctor, J. (1992) *The Ecology of Areas with Serpentinized Rocks. A World View.* Kluwer Academic Press, Dordrecht.

Roche, B., Lloyd, M., and Raieta, C. (2004) Local adaptation of the lyre-leaved rock cress, *Arabidopsis lyrata*, in serpentine and non-serpentine habitats. *Ecological Society of America meeting abstracts.* Portland, OR.

Rodman, J. E., Kruckeberg, A. R., and Al-Shehbaz, I. A. (1981) Chemotaxonomic diversity and complexity in seed glucosinates of *Caulanthus* and *Streptanthus* (Cruciferae). *Systematic Botany*, 6, 197–222.

Rozema, J., Rozema-Dijst, E., Freijsen, A. H. J., and Huber, J. J. L. (1978) Population differentiation within *Festuca rubra* L. with regard to soil salinity and soil water. *Oecologia*, 34, 329–41.

Safford, H. D., Viers, J. H., and Harrison, S. P. (2005) Serpentine endemism in the California flora: A database of serpentine affinity. *Madroño*, 52, 222–57.

Sambatti, J. B. M., and Rice, K. J. (2006) Local adaptation, patterns of selection, and gene flow in the Californian serpentine sunflower (*Helianthus exilis*). *Evolution*, 60, 696–710.

Sambatti, J. B. M., and Rice, K. J. (2007) Functional ecology of ecotypic differentiation in the Californian serpentine sunflower (*Helianthus exilis*). *New Phytologist*, 175, 107–19.

Schmitt, J. (1983) Density-dependent pollinator foraging, flowering phenology, and temporal pollen dispersal patterns in *Linanthus bicolor*. *Evolution*, 37, 1247–57.

Selby, J. P., Wright, K. M., and Willis, J. H. (2008) Evolution of serpentine tolerance in the *Mimulus guttatus* species complex. *Sixth International Conference on Serpentine Ecology abstracts*, Bar Harbor, ME.

Shaw, A. J. (1990) *Heavy Metal Tolerance in Plants: Evolutionary Aspects.* CRC Press, Boca Raton, FL.

Shigaki, T., Rees, I., Nakhleh, L., and Hirschi, K. D. (2006) Identification of three distinct phylogenetic groups of CAX cation/proton antiporters. *Journal of Molecular Evolution*, 63, 815–25.

Silvertown, J., and Charlesworth, D. (2001) *Introduction to Plant Population Biology*, 4th ed. Blackwell Publishing, MA.

Simson, B. W., and Timell, T. E. (1978a) Polysaccharides in cambial tissues of *Populus tremuloides* and *Tilia americana*. I. Isolation, fractionation and chemical composition of the cambial tissues. *Cellulose Chemistry and Technology*, 12, 39–50.

Simson, B. W., and Timell, T. E. (1978b) Polysaccharides in cambial tissues of *Populus trem-uloides* and *Tilia americana*. IV. 4-O-methylglucuronoxylan and pectin. *Cellulose Chem-istry and Technology*, 12, 79–84.

Smith, R. A. H., and Bradshaw, A. D. (1979) The use of metal tolerant plant populations for the reclamation of metalliferous wastes. *Journal of Applied Ecology*, 16, 595–612.

Sors, T. G., Ellis, D. R., Na, G. N., Lahner, B., Lee, S., Leustek, T., Pickering, I. J., and Salt, D. E. (2005) Analysis of sulfur and selenium assimilation in *Astragalus* plants with vary-ing capacities to accumulate selenium. *Plant Journal*, 42, 785–97.

Spencer, S. C., and Porter, J. M. (1997) Evolutionary diversification and adaptation to novel environments in *Navarretia* (Polemoniaceae). *Systematic Botany*, 22, 649–68.

Staal, M., Maathuis, F., Elzenga, T., Overbeek, J., and Prins, H. (1991) Na^+/H^+ antiport ac-tivity in tonoplast vesicles from roots of the salt-tolerant *Plantago maritima* and the salt-sensitive *Plantago media*. *Physiologia Plantarum*, 82, 179–84.

Taylor, S. I., and Levy, F. (2002) Responses to soils and a test for preadaptation to serpentine in *Phacelia dubia* (Hydrophyllaceae). *New Phytologist*, 155, 437–47.

Taylor, S. I., and Macnair, M. R. (2006) Within and between population variation for zinc and nickel accumulation in two species of *Thlaspi* (Brassicaceae). *New Phytologist*, 169, 505–14.

Thomas, J. R., McNeil, M., Darvill, A. G., and Albersheim, P. (1987) Structure of plant cell walls. XIX. Isolation and characterization of wall polysaccharides from suspension-cultured Douglas fir cells. *Plant Physiology*, 83, 659–71.

Thomson, D. M. (2007) Do source-sink dynamics promote the spread of an invasive grass into a novel habitat? *Ecology*, 88, 3126–34.

Tibbetts, R. A., and Smith, J. A. C. (1992) Vacuolar accumulation of calcium and its interac-tion with magnesium availability. In *The Vegetation of Ultramafic (Serpentine) Soils. Pro-ceedings of the First International Conference on Serpentine Ecology* (eds. A. J. M. Baker, J. Proctor, and R. D. Reeves), pp. 367–73. Intercept, Andover, UK.

Tilstone, G. H., and Macnair, M. R. (1997) Nickel tolerance and copper–nickel co-tolerance in *Mimulus guttatus* from copper mine and serpentine habitats. *Plant and Soil*, 191, 173–80.

Tordoff, G. M., Baker, A. J. M., and Willis, A. J. (2000) Current approaches to the revegeta-tion and reclamation of metalliferous mine wastes. *Chemosphere*, 41, 219–28.

Turner, T. L., von Wettberg, E. J., and Nuzhdin, S. V. (2008) Genomic analysis of differentia-tion between soil types reveals candidate genes for local adaptation in *Arabidopsis lyrata*. *PLos ONE*, 3, 1–7.

Turner, T. L., Bourne, E. C., von Wettberg, E. J., Hu, T. T., and Nuzhdin, S. V. (2010) Popula-tion resequencing reveals local adaptation of *Arabidopsis lyrata* to serpentine soils. *Na-ture Genetics*, 42, 260–63.

U.S. Fish and Wildlife Service. (2003) *Recovery Plan for Kneeland Prairie Penny-Cress* (Thlaspi californicum). U.S. Fish and Wildlife Service, Portland, OR.

Vogel, J. (2008) Unique aspects of the grass cell wall. *Current Opinion in Plant Biology*, 11, 301–7.

Von Wettberg, E. J. (2008) Population differentiation in performance under heavy metal and low Ca:Mg stress in *Arabidopsis lyrata*. *Botany 2008 meeting abstracts*, Vancouver, British Columbia.

Westerbergh, A. (1994) Serpentine and non-serpentine *Silene dioica* plants do not differ in nickel tolerance. *Plant and Soil,* 167, 297–303.

Westerbergh, A., and Saura, A. (1992) The effect of serpentine on the population structure of *Silene dioica* (Caryophyllaceae). *Evolution,* 46, 1537–48.

Westerbergh, A., and Saura, A. (1994) Gene flow and pollinator behavior in *Silene dioica* populations. *Oikos,* 71, 215–24.

Williamson, N. A., Johnson, M. S., and Bradshaw, A. D. (1982) *Mine Wastes Reclamation.* Mining Journal Books, London.

Williamson, S. D., Robinson, E. R., and Balkwill, K. (1997) Evolution of two serpentine endemic taxa in Mpumalanga. *South African Journal of Botany,* 63, 507–13.

Woodell, S. R. J., Mooney, H. A., and Lewis, H. (1975) The adaptation to serpentine soils in California of the annual species *Linanthus androsaceus* (Polemoniaceae). *Bulletin of the Torrey Botanical Club,* 102, 232–38.

Wright, J. W. (2007) Local adaptation to serpentine soils in *Pinus ponderosa. Plant and Soil,* 293, 209–17.

Wright, J. W., and Stanton, M. L. (2007) *Collinsia sparsiflora* in serpentine and nonserpentine habitats: using F2 hybrids to detect the potential role of selection in ecotypic differentiation. *New Phytologist,* 173, 354–66.

Wright, J. W., Stanton, M. L., and Scherson, R. (2006) Local adaptation to serpentine and non-serpentine soils in *Collinsia sparsiflora. Evolutionary Ecology Research,* 8, 1–21.

Wu, C. A., Lowry, D. B., Cooley, A. M., Wright, K. M., Lee, Y. W., and Willis, J. H. (2008) *Mimulus* is an emerging model system for the integration of ecological and genomic species. *Heredity,* 100, 220–30.

Wu, L., Enberg, A. W., and Guo, X. (1997) Effects of elevated selenium and salinity concentrations in root zone on selenium and salt secretion in saltgrass (*Distichlis spicata* L.). *Ecotoxicology and Environmental Safety,* 37, 251–58.

Yang, X. E., Li, T. Q., Long, X. X., Xiong, Y. H., He, Z. L., and Stoffella, P. J. (2006) Dynamics of zinc uptake and accumulation in the hyperaccumulating and non-hyperaccumulating ecotypes of *Sedum alfredii* Hance. *Plant and Soil,* 284, 109–19.

Zhao, F., Hamon, R. E., Lombi, E., McLaughlin, M. J., and McGrath, S. P. (2002) Characteristics of cadmium uptake in two contrasting ecotypes of the hyperaccumulator *Thlaspi caerulescens. Journal of Experimental Botany,* 53, 535–43.

6

Genomic Approaches to Understanding Adaptation

Eric J. von Wettberg, *Florida International University and Fairchild Tropical Botanical Garden*
Jessica Wright, *Pacific Southwest Research Station, USDA Forest Service*

Advances in molecular biology technologies allow us to ask ecological and evolutionary questions in more elaborate and detailed ways in any species—not just a narrow set of model organisms. These technological advances allow us to probe the mechanistic and genetic basis for expressed traits, explore patterns of genetic variation in organisms for signs of selection and evidence of past population processes, and identify cryptic members of communities, such as soil microbes that are critical to both plant success and ecosystem processes. Here we illustrate the power of these tools with examples from serpentine systems.

Next-generation sequencing technologies (Shendure and Ji, 2008; von Bubnoff, 2008; Metzger, 2010) can determine all of the sequence differences (genomics) or all of the differences in gene expression (transcriptomics) between individuals, such as plants from serpentine and nonserpentine populations. In these approaches, either DNA is sequenced directly, or mRNA is converted to cDNA and then sequenced using next-generation sequencing technologies. One then assembles the thousands to millions of individual sequences generated from a sample into a genome or transcriptome either de novo from the sample or by assembling against a previously sequenced reference genome or transcriptome. Because assembly to an existing data set is much easier, particularly with genomic data, genomic technologies are more effective to use with model organisms or their relatives. However, in transcriptomics, where only coding sequence is used, de novo assembly creates higher quality alignments even with lower levels of coverage. Comparing all the DNA sequences (the genome) between, for example, serpentine and nonserpentine plants will identify genes that vary in the two habitats—potentially highlighting genes under differential selection in the two environments (e.g., Turner et al., 2010). In serpentine systems, a novel and informative approach will be the comparison of gene expression (the transcriptome) in adapted and nonadapted individuals on serpentine and nonserpentine substrates. Furthermore, this can be extended using highly controlled conditions to explore gene expression of plants growing under different components of serpentine soils (high Mg, low Ca, low nutrients, high metals, etc.) or with different microbial symbionts. This approach will greatly improve our understanding of plant responses to the complexities of serpentine soils.

Just as sequencing DNA and RNA is now possible on the scale of the entire genome, technologies are emerging that can characterize proteins and small metabolites and even the ions found in tissue extracts from nonmodel plants. Proteomics and metabolomics rely on high-throughput mass spectrometry to characterize, respectively, amino acids and small metabolites (i.e., the chemical signatures of specific cellular processes), and these techniques can characterize a large portion of the plant proteome (Rampitsch and Srinivasan, 2006; Huang and Xu, 2008) and metabolome (Schauer and Fernie, 2006; Allwood et al., 2008). Ionomics uses various forms of mass spectrometry to understand the distribution of ions in tissue (Salt et al., 2008). Examining protein and metabolite differences between serpentine and nonserpentine populations can provide crucial information about the mechanism of differentiation. If one has candidate genes from sequence differences (genomics) or expression differences in mRNA (transcriptomics), one can use protein and metabolite differences to verify that pathways influenced by candidate genes are in fact altered (Schauer and Fernie, 2006). If one lacks sequence or expression data, one can use protein, metabolite, or ion differences to narrow the search for candidate genes to a candidate pathway.

The plants (and animals, fungi, and microbes) of serpentine ecosystems make ideal study systems in which to implement these emerging methods. Serpentine ecosystems have several interesting and useful characteristics. In particular, they offer an opportunity for effective replication and a chance to examine biological changes along steep gradients. Because most serpentine belts occur as archipelagos of serpentine outcrops, there is the potential for multiple, independent colonization of outcrops by different taxa. As a result, there are often many independent populations/species with which to work. Because serpentines vary both on a fine scale within an outcrop and also substantially among outcrops, there is the potential to tease apart different aspects of a complex multifactorial tolerance mechanism. In addition, within a single outcrop and its surrounding nonserpentine matrix, there can be extremely steep gradients, providing an opportunity to study stark variation on a fine scale. Many plant taxa have multiple clades occurring across a serpentine belt, and these clades can be used as replicates to develop tests to determine the basis of tolerating serpentine.

We think serpentine systems are ideally suited for addressing several grand questions in evolutionary biology and ecology. We focus on two broad areas that we feel encompass many questions about the physiological mechanisms, evolutionary processes, and community dynamics that lead to unique serpentine communities. These questions could be asked in most systems, but are ready for study in serpentine systems:

1. What is the genetic and mechanistic basis of adaptive traits?
2. Are microbes associated with plants in different habitats, and do they contribute to plant adaptation?

We recast these queries as particular questions about serpentine-associated plant systems.

WHAT IS THE GENETIC AND MECHANISTIC BASIS OF SERPENTINE TOLERANCE?

There are two general approaches to determining the genetic basis of traits. One is to go from traits that differ back to genes, a "reverse ecological" approach (Friesen and von Wettberg, 2010). With this approach, one finds traits associated with performance in certain habitats, such as serpentine soils. Then one finds variation in the trait segregating in a cross (i.e., quantitative train locus [QTL] mapping) or across a set of populations (association mapping) and uses the segregating DNA sequence variation to infer the genes responsible. An alternative approach that has just become possible, thanks to next-generation sequencing technologies, is to let the genes tell which traits should differ among individuals, based on the nature and types of genes that differ between individuals from different habitats. This is moving from the genes to traits: "forward ecology." We have learned a great deal about the genes controlling ecologically important traits using reverse ecology, which was the only feasible approach until recently. However, we expect great gains with the development of forward ecological approaches, which can be used for organisms that are long-lived and difficult to manipulate.

To date the best model for understanding the genetic basis of tolerance of serpentine soils has been the study of heavy metal hyperaccumulation. Although most serpentine plants are not hyperaccumulators, hyperaccumulators must be tolerant of the metals they uptake. As potentially useful plants for phytoremediation, they have received considerable attention. As an example of how the reverse ecological approach works, we examine work on *Arabidopsis halleri,* the hyperaccumulator most closely related to the well-known model plant *A. thaliana.*

A. halleri is a zinc and cadmium hyperaccumulator from calamine soils in Europe, which are anthropogenically rich in heavy metals. With only about five millions years of separation from *A. thaliana*, transferring genetic tools to *A. halleri* has been relatively easy (Roosens et al., 2008b). One of the first studies of the genetic basis of metal accumulation in this species (Macnair et al., 1999) was based on a cross between heavy metal–accumulating (Zn, Cd) *A. halleri* and nonaccumulating *A. lyrata* var. *petraea*, the European subspecies of *A. lyrata*, formerly *A. petraea* (O'Kane and Al-Shehbaz, 2003). The F2s from this interspecific cross made uncovering the genetic basis possible because all surveyed populations of *A. halleri* had been found to hyperaccumulate heavy metals, whereas *A. l. petraea* has never been observed to hyperaccumulate. Initial work with this cross before a QTL map was developed suggested that zinc tolerance and zinc hyperaccumulation are independent traits (Macnair et al., 1999).

Filatov et al. (2006) used an *A. thaliana* cDNA microarray to find genes with expression differences between *A. halleri* and the more closely related *A. l. petraea*, as well as the extreme phenotypic outliers (high and low Zn accumulators) of the F3s of this interspecific cross. They found several candidate genes differing in expression. Others have compared gene expression in *A. halleri* directly to gene expression in *A. thaliana* without the use of crosses or mapping populations by hybridizing cDNA of both species to *A. thaliana* cDNA microarrays (Becher et al., 2004; Weber et al., 2004, 2006). These studies have found large numbers of differentially expressed genes between these relatively closely related species, although many of these differences are clearly not due to the hyperaccumulation trait.

To determine the physical location of the genes responsible for differences in hyperaccumulation phenotypes, subsequent work used segregating markers, primarily mapped microsatellites, and also single nucleotide polymorphisms (SNPs) and other marker types to genetically map Zn accumulation and tolerance QTLs. Filatov et al. (2007) used an F2 cross of *A. halleri* and *A. l. petraea* to map Zn accumulation, whereas Willems et al. (2007) used a back-cross from the same species (different populations) to examine Zn tolerance. Roosens et al. (2008a) have narrowed the QTL regions based on synteny with the *A. thaliana* genome sequence, which allows them to focus on particular candidate genes in their regions; particular loci have yet to be narrowed to actual sequence differences, although several likely candidate genes occur in these intervals. Using the same cross as Willems et al. (2007), Courbot et al. (2007) were able to localize a QTL for Cd accumulation, finding some overlap in QTL. It is possible that tolerance and hyperaccumulation of these two metals have a similar genetic basis, although the regions are too broad to be certain, and other work (Macnair, 1993) suggests a separate genetic basis for different metals.

To show that candidate genes for hyperaccumulation generated by expression surveys (Filatov et al., 2006) and QTL mapping (Courbot et al., 2007; Filatov et al.,

2007; Willems et al., 2007; Roosens et al., 2008a) actually generate phenotypic differences, functional characterization of candidate genes is still required. Hanikenne et al. (2008) were able to localize and show a functional role by using RNA interference (RNAi) on one of the candidates, *HMA4*, a heavy metal ATPase, that has three copies in *A. halleri* but only one in *A. thaliana* (a copy number variant). Reducing expression of this gene with RNAi reduced the level of Zn and Cd uptake and reduced whole plant tolerance. Furthermore, inserting the *A. halleri* copy into *A. thaliana* in both the coding region, which exists in triplicate in *A. halleri,* and cis-regulatory regions of this gene complex, causes *A. thaliana* to both tolerate and accumulate Zn (Hanikenne et al., 2008).

If generating a new mutation is not possible, another reverse ecological approach is to turn to mutants or transgenics in a model organism to look for particular candidate genes and characterize these mutations. This was the approach taken by Bradshaw (2005), who screened nearly one million mutant *Arabidopsis* plants in serpentine-like low Ca and high Mg hydroponic conditions. He found a mutation in the gene CAX1 (a calcium-proton antiporter) that allowed plants to perform well under low Ca:Mg conditions. This performance benefit in the mutant could be due to increased Ca storage in the vacuole (Cheng et al., 2003).

Other tools researchers have at their disposal for reverse ecology are field reciprocal transplant experiments and common garden experiments. Observed phenotypes can be associated with genetic or molecular variation in many ways, as described above. These experiments are discussed at length elsewhere in this book (see Chapter 7), so we do not discuss them here. Nevertheless, we make one other point related to reciprocal transplant experiments and common garden experiments. Performance of lineages in nonserpentine habitats should not be overlooked. Similarly, characterization of mechanisms should occur in both serpentine and nonserpentine environments. Understanding differences in performance in nonserpentine habitats can be key to understanding the basis of costs of tolerating serpentine soil.

In light of the genomics revolution, a forward ecological approach can now be taken to address questions about the genetic basis of serpentine tolerance in any species of interest. Accurate, high-throughput tools exist to characterize differences in serpentine and nonserpentine plants in ions, small primary and secondary metabolites, proteins, RNA, and DNA (Wright and von Wettberg, 2009). The genes that differ in sequence, transcripts that differ in expression level, or proteins or metabolites that differ in concentration that one finds using these approaches provide a list of candidate genes and pathways for serpentine tolerance. Based on what is known about the function of the gene, each candidate gene suggests a potential pathway for tolerance. A pathway allows us to explore the role of these candidate genes with related tools. For example, ecophysiological approaches can characterize hormonal activity or nutrient and water relations potentially altered by genetic variation in a pathway.

Forward ecological approaches go from genes to traits that are under selection. Genes will show different patterns of variation depending on the patterns of selection, drift, and demographic processes affecting them. As a consequence, patterns of genetic variation across habitats can shed light on the population processes operating across habitats and the manner in which serpentine tolerance evolves (see Chapter 5 in this volume). Population genetic analyses can reveal patterns of dispersal between patches and habitats, differences in population size, and the balance between drift and different types of selection, such as balancing and purifying selection. As sequencing technology advances, it becomes possible to use population genetic approaches on whole genome data sets for a forward ecological approach (Friesen and von Wettberg, 2010), find all the genes that differ between habitats, and build up from those to pathways and traits that differ between habitats.

Surveys of populations with molecular markers are a step toward a forward ecological approach, although many studies lack the high-density genomic coverage necessary to infer selection on different genes. Some surveys have examined variation across population in hyperaccumulating *A. halleri*. Meyer et al. (2009) looked at four populations, from soils differing in Zn levels. They developed 820 dominant amplified fragment length polymorphism (AFLP) loci, and calculated Fst, a measure of among-population differentiation, for each locus. AFLP loci with Fst values significantly higher than the average Fst for all loci are likely to be linked to regions that differ due to differential selection. The Fst-outlier approach (Beaumont, 2005) they employed can be a useful genomic method for identifying loci likely to be under selection when whole genome sequence information is lacking.

A handful of studies have used molecular markers to characterize population genetic processes operating in serpentine plants. Some have shown little molecular genetic variation between serpentine and nonserpentine populations (Chapter 7 in this volume). Others have shown some differentiation in native taxa. For example, AFLP markers show that the rare eastern North American serpentine aster, *Symphyotrichum depauperatum*, is distinct from the potential ancestral taxa *S. pilosum* and *S. parviceps*, which are common to the great plains of North America (Gustafson and Latham, 2005). Although the study was not designed to specifically address the question of how many times serpentine tolerance evolved, the low differentiation among serpentine populations in their AFLP data is consistent with a single evolution of serpentine tolerance and subsequent colonization of other patches, but this is not conclusive. Interestingly, the presence of another population on magnesium-rich diabase soil south of the central Atlantic North American serpentine belt can be interpreted to imply that diabase tolerance is a bridge to serpentine tolerance, although this hypothesis would require formal investigation. Further work, with more markers and greater examination of soil tolerance phenotypes, is needed to show more about the history of serpentine tolerance in this species. Allozymes showed that there are two clades of *Cerastium alpinum* that are postglacial colonists of serpentine in Scandinavia (Nyberg Berglund and Westerbergh, 2001) and that serpentine tolerance appears to be evolving separately in them (Nyberg Berglund et al., 2004). Unfortunately similar genetic information is lacking for most serpentine taxa, and several of the most species-rich serpentine regions, such as Cuba and New Caledonia, are almost entirely unstudied.

With the feasibility of genotyping individuals at large numbers of loci covering a large portion of the genome, it becomes possible to ask functional questions in addition to the demographic and historical questions one might ask with only small numbers of neutral markers. For example, one can look at candidate genes for tolerance of particular environmental variables. Eckert et al. (2009) sequenced 121 candidate genes for cold tolerance from Douglas-fir and found eight loci with patterns of variation consistent with natural selection after accounting for demography. As approaches for aligning massive amounts of sequence information improve, it will be increasingly feasible to do more of these types of studies with more nonmodel organisms (Wall et al., 2009).

Work in model plants can give a flavor for how we can eventually proceed toward genome-enabled population genetics of serpentine plants. Genome-wide sequence variation in 20 disparate genotypes of *A. thaliana* ecotypes from across Eurasia showed several different patterns of variation across the genome, with some chromosomal regions undergoing patterns of purifying selection (i.e., selection against uncommon deleterious alleles), and others showing higher levels of variability (Clark et al., 2007). The variation observed across the entire genome is an illustration of how traditional molecular markers, such as is common in genetic studies of nonmodel organisms, can miss interesting variation in patterns of selection across the genome.

More recently, Turner et al. (2010) have sequenced pooled samples of individuals of *A. lyrata* from serpentine habitats and nearby granitic outcrops and found vast numbers of genes differing between plants growing in the two habitats. The study shows the effectiveness of next-generation sequencing for characterizing genetic differences across habitats. Several of the candidate genes show population-level patterns of variation consistent with differential selection between soil types, both in Pennsylvania in the North American subspecies *A. lyrata lyrata* and in Scotland in the European subspecies *A. lyrata petraea*. That the same polymorphisms can be found segregating between soil types in both subspecies, separated by at least several thousand years, suggests not only that the variation being selected is ancient but also that particular alleles are regularly selected in certain soil types. Such variation is an obvious target for further investigation and perhaps eventually breeding germplasm suited to particular soil types. It also suggests that there is not extremely strong selection against serpentine alleles in nonserpentine habitats over much of the range of *A. lyrata* in northern Eurasia and eastern North America, where serpentine substrates are rare and most populations occur on other substrates. Either alleles that are beneficial on serpentinic substrates are not extremely costly in other habitats, or in some other stressful habitats there is also selection for the same alleles found at high frequencies on serpentine substrates.

The forward ecological approach soon could be the preferred approach in hard-to-cultivate, long-lived, or extremely rare serpentine plants. Although direct experimentation is preferred, in these cases it is not a realistic or even desirable option. For example, in long-lived perennials like *A. lyrata*, fitness cannot be estimated in a single season or even several. This makes the traditional selection analysis (Lande and Arnold, 1983; Rausher, 1992) difficult to implement with these species. Furthermore, in many situations, we may not want to do reciprocal transplant experiments for fear of contaminating local populations or destroying habitat for experimentation. Inferring selection based on genetic variation means that all we need are small amounts of tissue.

Many inferences coming out of population genetic analyses depend on comparing patterns of observed variation to models of different processes, such as drift, selection, and mutation, and how they impact genetic variation within and among genes and populations. Several tests of molecular evolution have emerged since the emergence of neutral theory in the 1970s, which guide the inferences that we draw from observed patterns of genetic variation. We direct readers to reviews of a variety of tests and the underlying understanding of selection upon which they are based (Nielsen, 2005; Hedrick, 2006). Applying these tests on a genomic scale has become increasingly feasible. As with all statistical tests, however, particular care is still required to understand their limitations, use them only as appropriate, and only interpret them with caution. Genes and traits evolve in light of population structure and historical contingency, and often results are only valid with reference to the populations one has sampled (Nordborg and Weigel, 2008; Nielsen, 2009).

Population genomic studies generally find many genomic regions experiencing many different patterns of selection—purifying selection, that is, removal of background deleterious alleles; selective sweeps that drive regions to fixation and variation in closely linked genetic regions; balancing selection, which generates regions with maintained polymorphisms; hitchhiking, where alleles are dragged toward fixation by nearby regions under selection; and regions of low recombination where mutations can accumulate (Kreitman and Akashi, 1995; Charlesworth et al., 2003; Nielsen, 2005; Charlesworth, 2009). All can be found in the model organisms examined to date. Applying such tests to serpentine taxa, with deep coverage of genomes, could be a step forward. Regions of the genome that show evidence for selective sweeps (i.e., regions with greatly reduced allelic diversity compared to the rest of the genome), associated with populations in serpentine habitats, are clear candidates for genes conferring serpentine tolerance and are worthy of further examination. Most of the genome is likely to not be under positive selection on serpentine sites. The balance of different processes can be used to understand the selective history of the species, to understand how it came to tolerate and potentially be restricted to serpentine soils.

To date, almost all applications of such tests on a genome-wide scale have been in organisms with recent population expansions (humans, fruit flies, weedy *Arabidopsis*, and domesticated plants and animals). Organisms that occur across heterogeneous habitats, such as serpentine bodenvags that tolerate serpentine but occur off of it as well, and rare organisms, such as the endemic flora of many serpentine regions that are fully restricted to serpentine substrates, are likely to show markedly different patterns of genomic variation.

The forward ecological approach has not been fully implemented in a serpentine system. The work of Turner et al. (2010) is the closest example, but the sample pooling approach taken in that study made it impossible to implement several of the tests described here. We await the implementation of this approach on a fully genomic scale in a serpentine plant.

WHICH MICROBES ARE ASSOCIATED WITH PLANTS IN SERPENTINE HABITATS, AND DO THEY CONTRIBUTE TO PLANT ADAPTATION?

Plants are ubiquitously associated with microbial partners and antagonists. For example, mycorrhizae are crucial for plant P uptake and may affect the balance of other elements (Smith et al., 2010). Endophytic fungi and bacteria can provide plants with novel defensive chemicals (Rodriguez et al., 2008). Rhizobia are a major source of input of atmospheric N into terrestrial ecosystems (Gage, 2004). Finally, all plant hormones can be produced by soil microbes (Frankenberger and Arshad, 1995). These examples leave open the possibility that plant phenotypes are widely modified by endophytes and soil symbionts. To understand microbial contributions to plant phenotypes and community dynamics, it is essential to know what the microbes are and how abundant they are. Because many soil microbes are difficult to impossible to culture, and few make structures visible without extensive microscopy, characterization of microbial communities remains incomplete.

Just as next-generation sequencing revolutionizes how we can approach genetic variation, it also changes how we look at microbes associated with a plant community. The examination of microbes with new sequencing technologies has become the emerging field of meta-genomics (Handelsman et al., 1998; Handelsman,

2004, 2005; Hugenholtz and Tyson, 2008). In meta-genomics, DNA is extracted from environmental samples—plant tissue or even soils—and directly sequenced. One can either take the raw DNA and sequence it or amplify conserved regions of particular taxa. Search tools can then be used to find sequences in plant or soil samples that are fungal, bacterial, or viral. These sequences indicate the presence of partners and/or antagonists, and the number of times a particular microbial sequence appears in a sequencing batch indicates the microbe's abundance. These sequences can give unprecedented insight into microbial diversity and abundance. Furthermore, beyond simply identifying taxa, they can allow inferences to be made about function. By sequencing entire samples, rather than simply conserved genes used for taxonomic assessment, one can infer microbial function, such as nitrogen or phosphorus metabolism, or heavy metal chelation.

Now that we have tools with the power to identify low abundance or otherwise cryptic microbes, we can ask how essential microbes are for tolerance of serpentine soil (see Chapter 8 in this volume). It is possible that an underappreciated source of serpentine adaptation in plants is the presence of a particular microbe that detoxifies serpentine or ameliorates the shortage of particular nutrients. If this is the case, the important aspects of genetic differentiation between adapted and nonadapted plants may be genes promoting the interaction with the microbe, not actually tolerating the abiotic stress itself.

A large body of work already exists characterizing a portion of the microbial portion of serpentine systems. Mycorrhizae are particularly well studied. Many studies have used morphological characters in place of molecular methods in mycorrhizae of serpentine barrens in the mid-Atlantic region of the United States (Casper and Castelli, 2007; Cumming and Kelly, 2007; Casper et al., 2008). Morphological approaches can work well for taxa with well-characterized spores and can give an accurate estimate of abundance in the soil. They are also useful for surveying large amounts of soil for the presence of mycorrhizae. But they are much weaker for species and subspecies variation within mycorrhizal taxa, as only variation in spore structure is used.

Approaches that employ molecular genetic variation in place of or in addition to morphological characters can unveil mycorrhizal and microbial diversity not found in morphological characters. A RAPD (random amplification of polymorphic DNA) survey of microbial diversity in New Caledonian serpentine plant communities sites found substantial differences in microbial community profiles across soil types (Lenczewski et al., 2009), although culturing uncovered additional taxa. These approaches can uncover portions of the community that cannot be cultured or identified visually, although they also may miss taxa.

By targeting mycorrhizae-colonized root tips or nodules, it is possible to sample mycorrhizae and rhizobia directly. Primers designed to target conserved genes across a broad taxonomic group, such as fungi, are particularly useful because they allow for the amplification of genes from previously uncultivated microorganisms and allow the recovery of genes from organisms where genomic information is not available. By using a primer combination designed to exclude plant sequences but target mycorrhizae, Schechter and Bruns (2008) were able to amplify and sequence 1950 clones from 24 root samples of *Collinsia sparsiflora* from three serpentine and three nonserpentine populations in the Coast Range of northern California. Using this approach they found 19 mycorrhizal operational taxonomic units (OTUs) belonging to 6 genera, with a single OTU restricted to serpentine soils. Based on clone numbers found in different soils, abundance of fungal OTU also varied substantially across soil types, with an *Acaulospora* OTU-dominating serpentine and a *Glomus* OTU dominating nonserpentine sites. This approach is quite powerful in that it can characterize community composition and abundance and can resolve intraspecific variation if enough sequences are generated. The difficulty has been the expense and time of doing this with traditional sequencing.

Meta-genomic approaches have the power to change this. For example, a recent study used one type of next-generation sequencing, a massively parallel sequencing-by-synthesis approach, to characterize the mycorrhizal community of a boreal forest and found as many mycorrhizal species in 10 square meters as there were tree species present in the entire forest (Opik et al., 2009). These authors first narrowed their community sequencing to mycorrhizae by amplifying a mycorrhizal-specific gene. As next-generation sequencing technologies continue to develop, sequencing entire samples of dirt or root tips will be increasingly feasible. Sequencing unamplified samples, rather than those enriched for a particular conserved gene, will allow assessments of functional diversity and abundance to be made based on the types of genes found across taxa and kingdoms. We await major advances in the coming years in our understanding of the cryptic microbial members of plant communities, and we believe serpentine soils are a promising community for this. We hope these approaches will complement existing approaches in the future, both in widely profiling communities by sequencing DNA extracted from samples and in a functionally targeted fashion, such as by sequencing from mycorrhizal roots and nodules.

CONCLUSIONS

We can envision emerging tools allowing a broad research program examining the genetic basis of tolerance of serpentine soils. A genome-enabled research program might have the following components: first, looking for RNA/protein/metabolite/ion differences to discover pathways suggested by candidate genes uncovered in genomic and transcriptomic scans; and second, profiling communities to uncover the identity, abundance, and function of microbial members of serpentine communities. Ultimately we advocate a synthetic approach, using molecular genetic, population genetic, and community genetic approaches to characterizing how plants tolerate serpentine soils and the evolutionary and community implications of this tolerance. By incorporating tools from across biology, we can do much more to understand what is inside the black box of organisms on serpentine systems.

Acknowledgments

We thank the editors for the opportunity to put together our thoughts on genome-enabled evolutionary ecology for serpentine systems; Maren Friesen, Sergey Nuzhdin, Stephanie Porter, and Tom Turner for helpful discussion; and Mary Knapp, the editors, and two reviewers for careful comments on drafts of this chapter. Eric von Wettberg's research on serpentinomics has been supported by the National Science Foundation Plant Genome Research Program #0820846, the National Science Foundation Office of International Science and Engineering #0751073, and National Institute of Health National Research Service Award #F32ES015443.

LITERATURE CITED

Allwood, J. W., Ellis, D. I., and Goodacre, R. (2008) Metabolomic technologies and their application to the study of plants and plant-host interactions. *Physiologia Plantarum,* 132, 117–35.

Beaumont, M. A. (2005) Adaptation and speciation: What can Fst tell us? *Trends in Ecology and Evolution,* 20, 435–40.

Becher, M., Talke, I., Krall, L., and Kramer, U. (2004) Cross-species microarray transcript profiling reveals high constitutive expression of metal homeostasis genes in shoots of the zinc hyperaccumulator *Arabidopsis halleri. Plant Journal,* 37, 251–68.

Bradshaw, H. D. (2005) Mutations in CAX1 produce phenotypes characteristic of plants tolerant to serpentine soils. *New Phytologist,* 167, 81–88.

Casper, B., Bentivenga, S., Ji, B., Doherty, J., Edenborn, H., and Gustafson, D. (2008) Plant-soil feedback: Testing the generality with the same grasses in serpentine and prairie soils. *Ecology,* 89, 2154–64.

Casper, B., and Castelli, J. (2007) Evaluating plant-soil feedback together with competition in a serpentine grassland. *Ecology Letters,* 10, 394–400.

Charlesworth, B. (2009) Effective population size and patterns of molecular evolution and variation. *Nature Reviews Genetics,* 10, 195–205.

Charlesworth, B., Charlesworth, D., and Barton, N. (2003) The effects of genetic and geographic structure on neutral variation. *Annual Review of Ecology, Evolution, and Systematics,* 34, 99–125.

Cheng, N., Pittman, J., Barkla, B., Shigaki, T., and Hirschi, K. (2003) The *Arabidopsis* cax1 mutant exhibits impaired ion homeostasis, development, and hormonal responses and reveals interplay among vacuolar transporters. *Plant Cell,* 15, 347–64.

Clark, R., Schweikert, G., Toomajian, C., Ossowski, S., Zeller, G., Shinn, P., Warthmann, N., Hu, T., Fu, G., Hinds, D., Chen, H., Frazer, K., Huson, D., Schoelkopf, B., Nordborg, M.,

Raetsch, G., Ecker, J., and Weigel, D. (2007) Common sequence polymorphisms shaping genetic diversity in *Arabidopsis thaliana*. *Science*, 317, 338–42.

Courbot, M., Willems, G., Motte, P., Arvidsson, S., Roosens, N., Saumitou-Laprade, P. and Verbruggen, N. (2007) A major quantitative trait locus for cadmium tolerance in *Arabidopsis halleri* colocalizes with HMA4, a gene encoding a heavy metal ATPase. *Plant Physiology*, 144, 1052–65.

Cumming, J., and Kelly, C. (2007) *Pinus virginiana* invasion influences soils and arbuscular mycorrhizae of a serpentine grassland. *Journal of the Torrey Botanical Society*, 134, 63–73.

Eckert, A., Bower, A., Wegrzyn, J., Pande, B., Jermstad, K., Krutovsky, K., Clair, J., and Neale, D. (2009) Association genetics of coastal Douglas fir (*Pseudotsuga menziesii* var. *menziesii*, Pinaceae). I. Cold-hardiness related traits. *Genetics*, 182, 1289–302.

Filatov, V., Dowdle, J., Smirnoff, N., Ford-Lloyd, B., Newbury, H. J., and Macnair, M. R. (2006) Comparison of gene expression in segregating families identifies genes and genomic regions involved in a novel adaptation, zinc hyperaccumulation. *Molecular Ecology*, 15, 3045–59.

Filatov, V., Dowdle, J., Smirnoff, N., Ford-Lloyd, B., Newbury, H. J., and Macnair, M. R. (2007) A quantitative trait loci analysis of zinc hyperaccumulation in *Arabidopsis halleri*. *New Phytologist*, 174, 580–90.

Frankenberger, W. T., and Arshad, M. (1995) *Phytohormones in Soils: Microbial Production and Function*. Marcel Dekker, New York.

Friesen, M. L., and von Wettberg, E. J. (2010) Adapting genomics to the study the ecology and evolution of agricultural systems. *Current Opinion in Plant Biology*, 13, 119–215.

Gage, D. (2004) Infection and invasion of roots by symbiotic, nitrogen-fixing rhizobia during nodulation of temperate legumes. *Microbiology and Molecular Biology Reviews*, 68, 280–95.

Gustafson, D., and Latham, R. (2005) Is the serpentine aster, *Symphyotrichum depauperatum* (Fern.) Nesom, a valid species and actually endemic to eastern serpentine barrens? *Biodiversity and Conservation*, 14, 1445–52.

Handelsman, J. (2004) Metagenomics: Application of genomics to uncultured microorganisms. *Microbiology and Molecular Biology Reviews*, 68, 669–85.

Handelsman, J. (2005) Sorting out metagenomes. *Nature Biotechnology*, 23, 38–39.

Handelsman, J., Rondon, M., Brady, S., Clardy, J., and Goodman, R. (1998) Molecular biological access to the chemistry of unknown soil microbes: A new frontier for natural products. *Chemistry and Biology*, 5, R245–49.

Hanikenne, M., Talke, I., Haydon, M., Lanz, C., Nolte, A., Motte, P., Kroymann, J., Weigel, D., and Kramer, U. (2008) Evolution of metal hyperaccumulation required cis-regulatory changes and triplication of HMA4. *Nature*, 453, 391–95.

Hedrick, P. W. (2006) Genetic polymorphism in heterogeneous environments: The age of genomics. *Annual Review of Ecology Evolution and Systematics*, 37, 67–93.

Huang, B., and Xu, C. (2008) Identification and characterization of proteins associated with plant tolerance to heat stress. *Journal of Integrative Plant Biology*, 50, 1230–37.

Hugenholtz, P., and Tyson, G. (2008) Microbiology: Metagenomics. *Nature*, 455, 481–83.

Kreitman, M., and Akashi, H. (1995) Molecular evidence for natural-selection. *Annual Review of Ecology and Systematics,* 26, 403–22.

Lande, R., and Arnold, S. (1983) The measurement of selection on correlated characters. *Evolution,* 37, 1210–26.

Lenczewski, M., Rigg, L., Enright, N., Jaffre, T., and Kelly, H. (2009) Microbial communities of ultramafic soils in maquis and rainforest at Mont Do, New Caledonia. *Austral Ecology,* 34, 567–76.

Macnair, M. R. (1993) The genetics of metal tolerance in vascular plants. *New Phytologist,* 124, 541–59.

Macnair, M. R., Bert, V., Huitson, S. B., Saumitou-Laprade, P., and Petit, D. (1999) Zinc tolerance and hyperaccumulation are genetically independent characters. *Proceedings of the Royal Society of London Series B: Biological Sciences,* 266, 2175–79.

Metzger, M. L. (2010) Sequencing technologies—the next generation. *Nature Reviews Genetics,* 11, 31–46.

Meyer, C., Vitalis, R., Saumitou-Laprade, P., and Castric, V. (2009) Genomic pattern of adaptive divergence in *Arabidopsis halleri,* a model species for tolerance to heavy metal. *Molecular Ecology,* 18, 2050–62.

Nielsen, R. (2005) Molecular signatures of natural selection. *Annual Review of Genetics,* 39, 197–218.

Nielsen, R. (2009) Adaptitionism—30 years after Gould and Lewontin. *Evolution,* 63, 2487–90.

Nordborg, M., and Weigel, D. (2008) Next-generation genetics in plants. *Nature,* 456, 720–23.

Nyberg Berglund, A. B., Dahlgren, S., and Westerbergh, A. (2004) Evidence for parallel evolution and site-specific selection of serpentine tolerance in *Cerastium alpinum* during the colonization of Scandinavia. *New Phytologist,* 161, 199–209.

Nyberg Berglund, A. B., and Westerbergh, A. (2001) Two postglacial immigration lineages of polyploid *Cerastium alpinum* (Caryophyllaceae). *Hereditas,* 134, 171–83.

O'Kane, S., and Al-Shehbaz, I. (2003) Phylogenetic position and generic limits of *Arabidopsis* (Brassicaceae) based on sequences of nuclear ribosomal DNA. *Annal of the Missouri Botanical Garden,* 90, 603–12.

Opik, M., Metsis, M., Daniell, T., Zobel, M., and Moora, M. (2009) Large-scale parallel 454 sequencing reveals host ecological group specificity of arbuscular mycorrhizal fungi in a boreonemoral forest. *New Phytologist,* 184, 424–37.

Rampitsch, C., and Srinivasan, M. (2006) The application of proteomics to plant biology: A review. *Canadian Journal of Botany,* 84, 883–92.

Rausher, M. D. (1992) The measurement of selection on quantitative traits: Biases due to environmental covariances between traits and fitness. *Evolution,* 46, 616–26.

Rodriguez, R., Henson, J., Van Volkenburgh, E., Hoy, M., Wright, L., Beckwith, F., Kim, Y., and Redman, R. (2008) Stress tolerance in plants via habitat-adapted symbiosis. *ISME Journal,* 2, 404–16.

Roosens, N., Willems, G., Gode, C., Courseaux, A., and Saumitou-Laprade, P. (2008a) The use of comparative genome analysis and syntenic relationships allows extrapolating the position of Zn tolerance QTL regions from *Arabidopsis halleri* into *Arabidopsis thaliana*. *Plant and Soil*, 306, 105–16.

Roosens, N., Willems, G., and Saumitou-Laprade, P. (2008b) Using *Arabidopsis* to explore zinc tolerance and hyperaccumulation. *Trends in Plant Science*, 13, 208–215.

Salt, D. E., Baxter, I., and Lahner, B. (2008) Ionomics and the study of the plant ionome. *Annual Review of Plant Biology*, 59, 709–33.

Schauer, N., and Fernie, A. R. (2006) Plant metabolomics: Towards biological function and mechanism. *Trends in Plant Science*, 11, 508–16.

Schechter, S., and Bruns, T. (2008) Serpentine and non-serpentine ecotypes of *Collinsia sparsiflora* associate with distinct arbuscular mycorrhizal fungal assemblages. *Molecular Ecology*, 17, 3198–210.

Shendure, J., and Ji, H. (2008) Next-generation DNA sequencing. *Nature Biotechnology*, 26, 1135–45.

Smith, S., Facelli, E., Pope, S. and Smith, F. (2010) Plant performance in stressful environments: interpreting new and established knowledge of the roles of arbuscular mycorrhizas. *Plant and Soil*, 326, 3–20.

Turner, T., Bourne, E. C., von Wettberg, E. J., Hu, T. T., and Nuzhdin, S. V. (2010) Population resequencing reveals local adaptation of *Arabidopsis lyrata* to serpentine soils. *Nature Genetics*, 42, 260–63.

von Bubnoff, A. (2008) Next-generation sequencing: The race is on. *Cell*, 132, 721–23.

Wall, P. K., Leebens-Mack, J., Chanderbali, A. S., Barakat, A., Wolcott, E., Liang, H., Landherr, L., Tomsho, L. P., Hu, Y., Carlson, J. E., Ma, H., Schuster, S. C., Soltis, D. E., Soltis, P. S., Altman, N., and dePamphilis, C. W. (2009) Comparison of next generation sequencing technologies for transcriptome characterization. *BMC Genomics*, 10, 347.

Weber, M., Harada, E., Vess, C., von Roepenack-Lahaye, E., and Clemens, S. (2004) Comparative microarray analysis of *Arabidopsis thaliana* and *Arabidopsis halleri* roots identifies nicotianamine synthase, a ZIP transporter and other genes as potential metal hyperaccumulation factors. *Plant Journal*, 37, 269–81.

Weber, M., Trampczynska, A., and Clemens, S. (2006) Comparative transcriptome analysis of toxic metal responses in *Arabidopsis thaliana* and the Cd^{2+}-hypertolerant facultative metallophyte *Arabidopsis halleri*. *Plant, Cell, and Environment*, 29, 950–63.

Willems, G., Drager, D., Courbot, M., Gode, C., Verbruggen, N., and Saumitou-Laprade, P. (2007) The genetic basis of zinc tolerance in the metallophyte *Arabidopsis halleri* ssp *halleri* (Brassicaceae): An analysis of quantitative trait loci. *Genetics*, 176, 659–74.

Wright, J. W., and von Wettberg, E. J. (2009) Serpentinomics—an emerging new field of study. *Northeastern Naturalist*, 16, 285–96.

7

Local Adaptation in Heterogeneous Landscapes

Reciprocal Transplant Experiments and Beyond

Jessica W. Wright, *Pacific Southwest Research Station, USDA Forest Service*
Maureen L. Stanton, *University of California, Davis*

Adaptation to different selection pressures (i.e., adaptive differentiation) is thought to be a major driver of diversification and accordingly has been a focus of evolutionary biology since the early work of Turesson (1922) and Clausen, Keck, and Hiesey (1940, 1948). The classic work of Clausen's team revealed the power of reciprocal transplant experiments to determine whether populations are locally adapted to their home environment, characterize phenotypes that differentiate locally adapted ecotypes, and, if sufficiently well replicated, identify axes of environmental variation along which adaptive differentiation has occurred. Their extensive experiments and observations set a new standard for characterizing ecologically relevant variation within species and testing adaptive hypotheses and have had an enormous impact on fields as diverse as forestry, crop breeding, and restoration ecology. In an advisory letter written to F. Austen at the Institute of Forest Genetics, Placerville, California (dated April 16, 1938), William Hiesey outlined some of the major challenges to understanding adaptive variation:

> Our experience with different herbaceous species also teaches us that it is impossible to predict just how a given race from one elevation will react when brought to another altitude. Sometimes the expected reaction of a given race at a given station may be actually reversed when the test is applied. As far as we know, only experimental plantings, such as you plan at your transect gardens can determine this point.

Clearly, we have learned an enormous amount about evolutionary adaptation since Hiesey's time. In addition to great theoretical advances in the field, empirical studies indicate that local adaptation is a common response to heterogeneous environments. Leimu and Fischer (2008) conducted a meta-analysis of reciprocal transplant experiments and found that 71% of transplant sites showed evidence for local adaptation. Nonetheless, many challenges face those who seek a predictive understanding of adaptive differentiation and its mechanisms. In this chapter, written more than 70 years after Hiesey's letter, we argue, just as Hiesey did, that reciprocal transplant experiments remain the cornerstone technique for studies of plant adaptation. We also discuss how this classic approach can be improved and informed by recent advances in quantitative genetics, statistics, and genomics.

LOCAL ADAPTATION VERSUS ECOTYPIC DIFFERENTIATION

Before discussing details of experimental design for studying plant adaptation, we feel compelled to recommend that investigators be much more careful in distinguishing between local adaptation and ecotypic differentiation. Both processes are the result of adaptive differentiation, and both are detected by comparing the performance of individuals moved back into their native environment (as transplantation controls) to the performance of individuals moved there from a different environment. However, because these two types of adaptive differentiation have very different consequences for population management and conservation, care needs to be taken to use the correct term when describing experimental results.

Local adaptation improves performance in a specific home site. In its most basic (and most common) form, a reciprocal transplant experiment tests for the existence of local adaptation between two populations. In this approach, lineages from two sites are reciprocally transplanted into both sites as experimental destinations (Waser and Price, 1985; Galen et al., 1991; Nagy and Rice, 1997; Hufford et al., 2008). Significantly better performance exhibited by lineages moved back into their "home" sites, relative to lineages originating from other sites, demonstrates local adaptation to specific home-site attributes. That said, without doing further manipulations, this experimental design does not identify the environmental axes most critical to local selection.

In contrast, an ecotype is comprised of a set of populations adapted to an identifiable *type* of habitat and a common set of environmental conditions (e.g., serpentine soils). To achieve the more ambitious goal of testing for ecotypic differentiation, it is critically important to conduct a field transplant experiment in which genotypes from multiple populations occupying each habitat type are moved into multiple destinations belonging to each habitat type (Sambatti and Rice, 2006; Wright et al., 2006a; Lowry et al., 2008; Raabová et al., 2008; Gonzalo-Turpin and

Hazard, 2009). Ecotypic differentiation is demonstrated when genotypes moved into their native habitat type, even when it is not their home site, outperform genotypes transplanted from a contrasting habitat type. These two kinds of adaptive differentiation are not mutually exclusive—even among populations of a given ecotype, there may be additional local adaptation to home site characteristics.

Goals for This Chapter

In this chapter, we explore the use of reciprocal transplant and common garden experiments as tools for documenting patterns of plant adaptation within heterogeneous environments, with a particular emphasis on serpentine habitats. We begin by evaluating alternative criteria for documenting local adaptation and/or ecotypic differentiation using reciprocal transplant experiments and then review field common garden and reciprocal transplant experiments that have focused on plant species that grow both on and off serpentine soils. We then discuss some important issues that must be considered when designing and analyzing such experiments:

1. the difficulties associated with estimating plant fitness in field trials;
2. the role of maternal environmental effects in experimental design and interpretation;
3. how transplant experiments can be designed to take advantage of naturally occurring environmental heterogeneity to understand key selective factors associated with adaptive differentiation;
4. the importance of reciprocal transplant and common garden experiments for testing adaptive hypotheses generated from molecular genetic data; and
5. additional experimental methods that can be used to identify the role of natural selection in shaping adaptation in serpentine and nonserpentine habitats.

DOCUMENTING ADAPTIVE DIFFERENTIATION

There are varying opinions about the specific patterns of transplant performance that demonstrate the existence of adaptive differentiation between focal populations. In their comprehensive and influential review, Kawecki and Ebert (2004) argued that local adaptation is demonstrated between two populations when their fitness reaction norms cross, such that genotypes from each local habitat perform better in their home site than do transplants from the alternative habitat. By extension, according to these criteria, ecotypic differentiation between two habitat types would be shown when lineages transplanted into their original habitat type consistently outperform genotypes derived from the alternative habitat type, even when the destination is not the specific home site. This pattern of crossing fitness reaction norms is indicative of *reciprocal* adaptation, such that both focal populations (or ecotypes) are adapted to their native environment and are poorly adapted for the alternative. The kind of performance trade-off shown by crossing fitness reaction norms is a prerequisite for spatial heterogeneity in selection pressures to maintain genetic variation, but here we argue that there are some situations in which adaptation to alternative sites or habitat types may not occur reciprocally.

Depending on the extent to which contrasting habitats are connected by gene flow, selection can increase the frequencies of alleles conferring locally greater fitness in some habitats but not in others. Such asymmetry may be common when interconnected populations occupy habitats that differ in overall quality. A number of theoretical treatments suggest that source-to-sink gene flow can allow adaptation to high-quality or abundant habitats while preventing adaptation to contrasting selection regimes in marginal environments (Holt and Gaines, 1992; García-Ramos and Kirkpatrick, 1997; Kirkpatrick and Barton, 1997; reviewed in Bridle and Vines, 2007). There are a few empirical explorations of this idea as well (e.g., Stanton and Galen, 1997; Emery, 2006). Similarly, some populations may be more likely to adapt to a local habitat if there are other factors operating (such as self-fertilization or shifts in flowering time or pollinators) that reduce genetic input from alternative selection regimes (Antonovics and Bradshaw, 1970; Emery, 2009; reviewed in Levin, 2009). Alternatively, if gene flow among populations is rare, locally low levels of genetic variation may preclude local adaptation in some populations, for example, those in isolated sites colonized by small founder groups.

STUDIES OF ADAPTATION IN SERPENTINE SYSTEMS

Kruckeberg was an early pioneer in using common garden experiments to compare responses of plant populations to serpentine soils (1951, 1967). His experiments, conducted mostly under greenhouse conditions but also in serpentine field sites, were principally designed to determine whether populations of bodenvag (i.e., soil-indifferent) species found on serpentine soils are more tolerant of that soil type than are nonserpentine populations. Kruckeberg's experiments paved the way for subsequent investigations that have used reciprocal transplant experiments in the field to test for adaptive differentiation between serpentine and nonserpentine plant populations.

Soil-focused common garden studies conducted under controlled conditions have often indicated that serpentine plant populations are tolerant of the chemical properties of serpentine soil (reviewed in Brady et al., 2005; see Chapter 5 in this volume). However, only experiments conducted under field conditions can test whether serpentine and nonserpentine populations are adapted to the totality of the serpentine or nonserpentine habitat. There have been a limited number of field reciprocal transplant experiments on serpentine and nonserpentine sites since Kruckeberg's early studies. Jurjavcic and colleagues (2002) studied the native annual grass *Vulpia microstachys* growing on serpentine rocky slopes, serpentine meadows, and nonserpentine meadows. They conducted a reciprocal transplant experiment by planting out field-collected seeds from each of the three habitats using a split-split plot design (using five blocks that each contained adjacent areas of all three habitat types). They found that plants from the serpentine rocky slopes performed best in that habitat, whereas there was no significant variation in performance among the three sources in the two meadow habitats. Sambatti and Rice (2006) studied annual serpentine sunflowers (*Helianthus exilis*; Asteraceae) growing on serpentine and nonserpentine seeps. They reciprocally planted field-collected seeds among four sites (two serpentine and two nonserpentine) and found strong evidence for a home-site advantage (local adaptation)—plants growing in their home site performed better than those from other seed sources. In addition, they found evidence for adaptation to soil type, indicating differentiation into serpentine versus nonserpentine ecotypes. Hufford et al. (2008) transplanted two native bunchgrass species between a serpentine and nonserpentine site separated by approximately 50 km. They found evidence for a home-site advantage, with plants from both species performing best when transplanted back to their home environment. Wright et al. (2006b) conducted reciprocal transplant experiments using greenhouse-reared seeds from six focal populations (three serpentine and three nonserpentine) of the annual plant *Collinsia sparsiflora* and demonstrated reciprocal ecotypic adaptation to both serpentine and nonserpentine habitats. All of the previous field experiments used annual plants as subjects for experiments lasting less than a year. A rare exception is provided by Wright (2007)—in a common garden experiment run for 36 years, *Pinus ponderosa* trees from a serpentine site performed better on serpentine soil than did trees from a nonserpentine source population.

GETTING THE MOST FROM FIELD TRANSPLANT EXPERIMENTS: FIVE DECISIONS EVERY RESEARCHER MUST MAKE

For the remainder of this chapter, we use several examples from our own work on *C. sparsiflora* and *P. ponderosa* to illustrate some of the lessons learned from field transplant experiments aimed at understanding patterns of plant adaptation to heterogeneous serpentine/nonserpentine landscapes. These two study systems and our experimental methods have been described in detail elsewhere (Wright et al., 2006a, 2006b; Wright and Stanton, 2007; Wright, 2007; Lau et al., 2008).

1. How Is the Performance of Transplants Going to Be Assessed?

Implicit in this whole discussion is the assumption that we can determine which plants perform "best" in a transplant experiment. How to estimate plant fitness is a very complex topic that we can only touch on here, but we point out a few of the most common problems faced when conducting transplant studies on species with different life histories.

Unsurprisingly, long-lived plants such as woody shrubs and trees pose a huge challenge for experimental fitness comparisons, as lifetime fertility is usually impossible to measure. This is not only due to individual longevity per se but to the fact that such species partition resources among alternative functions within and between successive life stages. Annual seed production can fluctuate dramatically, especially in masting species, and taking fertility measurements for two to three years may not approximate lifetime performance. Many investigators working on long-lived perennials estimate fitness by taking size or growth measurements at specific life stages, based on the assumption that larger individuals will produce more seeds. This is generally a sound approach. However, in some stressful habitats, including serpentine, there may be selection for increased investment in below-ground biomass (Rajakaruna and Bohm, 1999; Rajakaruna et al., 2003), which could result in periods of reduced above-ground growth and reproduction. Selection may also favor investment in other maintenance or defense functions in such situations (see Chapter 8 in this volume). For example, in a reciprocal transplant between nutrient-rich clay soils and nutrient-poor sandy soils, Fine et al. (2004) found that trees from clay soils grew larger in both habitats—but only when herbivores were excluded. When herbivores were allowed access, there was evidence for adaptive differentiation between the habitat types, driven in part by a trade-off between rapid growth and herbivore defense. Demonstrations of genetically based adaptive differentiation in long-lived perennials are hampered not only by problems with lifetime fitness estimation but also by the difficulty of controlling for effects of parental environment on performance (see later discussion).

For annual plants or short-lived perennials, a convenient estimate of lifetime fitness is the number or weight of all seeds produced, but there are limitations for these measures. Seed dormancy is often extensive in short-lived plant species, raising the question of how to interpret the performance of experimentally sown seeds that do not emerge as seedlings. Various approaches to this problem, all imperfect, include: (1) transplanting seedlings instead of sowing seeds (a procedure that ignores potential selection during germination and emergence); (2) assigning unemerged seeds a fitness of zero and including them in the analysis (this ignores dormancy, which may be ecotype- and habitat-specific, and also generates statistically problematic, non-normal fitness distributions); (3) including only emergent plants in the analysis of fitness (again, sidestepping the messy issue of selection at the seed stage); and (4) estimating performance for successive, multiplicative life cycle stages, as well as for cumulative fitness across all life stages. This latter approach still requires assumptions about the fates of unemerged seeds, but analyzing each life stage separately can provide insights into when differentiating selection is potentially acting (e.g., Kalisz and McPeek, 1992; Wright and Stanton, 2007).

2. How Are Maternal Effects Going to Be Considered in the Experimental Design?

Common sense and experience tell us that plants growing in nutrient-rich sites, with plenty of water and sunlight, are going to grow large compared to plants growing in nutrient-poor sites, and they are going to produce bigger seeds. Those big seeds are, in turn, more likely to produce vigorous offspring, which themselves are likely to grow large (e.g., Stanton, 1984). This is an example of a simple, yet pervasive effect of the maternal plant's environment on phenotype and fitness in the next generation (reviewed in Roach and Wulff, 1987; Donohue, 2009). The existence of such cross-generational plasticity has significant implications for plant population responses to heterogeneous habitats (e.g., Stanton and Galen, 1997; Turelli, 1997; Galloway, 2005; Galloway and Etterson, 2007) and can also complicate the interpretation of experiments aiming to measure adaptive genetic variation between habitat types.

In serpentine systems, for example, the nutrient-poor substrate can result in fewer resources being available for developing seeds (Wright et al., 2006b). Indeed, we believe it is common in serpentine reciprocal transplant experiments to observe greater success of seeds collected from nonserpentine soils than those from serpentine, even when seeds collected from both habitat types are grown on serpentine soils. How often such results are published is unclear (but see Branco, 2009). In a greenhouse experiment conducted with field-collected seeds of *C. sparsiflora* collected from both serpentine and nonserpentine sites, grown on and off of serpentine soils (Wright and Stanton, unpublished data), plants grown from seeds collected in nonserpentine sites were larger than those from serpentine sites, regardless of the soil into which they were transplanted (Figure 7.1), a result from which one might reasonably conclude that there is no evidence for local adaptation to soil type in this system. However, other plants from these same populations were grown on standard greenhouse soils to reduce potential maternal effects on offspring performance and then crossed within each population to produce experimental seeds that were planted into the field in a reciprocal transplant design. The results of the field experiment unequivocally demonstrated strong, genetically based ecotypic differentiation to both serpentine and nonserpentine soils, with plants from replicate populations performing best in their native soil habitat (Figure 7.1; Wright et al., 2006b).

These contrasting results lead to the following: question: under what circumstances should we conduct experiments that either limit or incorporate parental environmental effects on offspring performance? The answer rests on the research

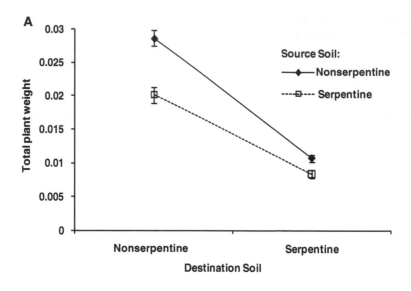

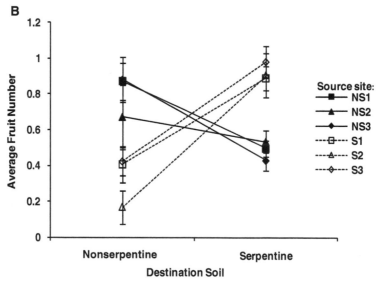

FIGURE 7.1. A: Results of greenhouse experiment using field-collected seeds, showing no evidence for reciprocal local adaptation. B: Results of field reciprocal transplant experiment using seeds produced through crosses in the greenhouse.

question of greatest interest to the investigator. If the primary aim is to measure the degree of adaptive genetic differentiation between populations occupying contrasting environments (e.g., soil types), then it is appropriate to minimize the effects of those different parental environments on offspring fitness or performance. Alternatively, if the primary aim is to document the consequences of seed or pollen dispersal between different populations or habitats, then the use of field-collected seeds for experimental transplantation is most appropriate.

When experimental seeds or cuttings are collected from multiple source populations for use in experimental studies, genetic variation within that collection is completely confounded with variation caused by contrasting parental environments. There are several useful approaches for minimizing such cross-generational environmental effects.

The most common approach used in short-lived species is to collect seeds from natural populations; grow them in a common growth chamber, greenhouse, or garden environment; and then conduct crosses to produce experimental seeds for use in reciprocal experiments (e.g., reviewed in Roach and Wulff, 1987; Kawecki and Ebert 2004; Hereford and Winn, 2008; Lowry et al., 2008). We mention three caveats about this time-honored approach. First, although this methodology minimizes maternal effects on offspring phenotype when properly applied, there may still be effects carrying over from the contrasting grand-maternal environments from which field-collected seeds were sampled (reviewed in Roach and Wulff, 1987). Therefore, parental environmental effects can never be said to be eliminated—they can only be reduced or minimized. Second, given that growth chamber and greenhouse environments are far from uniform (Potvin and Tardif, 1988), the correct application of this approach requires full randomization and/or rotation of plants from different lineages across bench or garden locations. Finally, when using seeds or seedlings produced in common garden settings, it is important to remember that the observed attributes of the experimental plants are likely to be specific to the common garden conditions experienced by the parental generation. To cite just one example, the maternal plant's environment is known to influence the timing of seedling emergence (Platenkamp and Shaw, 1993; Galloway, 2002; Donohue et al., 2005a), and emergence timing, in turn, can significantly alter patterns of genetic variance and covariance for ecologically important traits and for fitness (Donohue et al., 2005b; Brock et al., 2009). It follows that the conditions prevailing within any given common garden environment will have lasting effects on phenotype on the experimental generation.

Long-lived species pose a special problem for investigators wishing to reduce maternal effects. Growing trees in a common garden and conducting crosses is less practical over typical funding cycles, and in many cases is not possible (but see Dormling and Johnsen, 1992, who used clone banks, collections of grafted trees, often grown for conservation and seed production). A partial solution to this problem is to run an experiment long enough so that the early effects of maternal environment on offspring phenotype are diminished in magnitude, relative to the life-long effects of genotype and the current experimental environment (reviewed in Roach and Wulff, 1987; for a recent case study, see Donohue et al., 2005a). For example, a reciprocal transplant experiment in *Pinus ponderosa* showed that trees grown from seeds collected in serpentine sites grew larger on serpentine soils than nonserpentine trees, but this critical difference was statistically significant only after 20 years of growth (Wright, 2007). Another approach that can reduce some effects of variation in quality of the maternal environment is to weigh the seeds used in the experiment and then include seed weight as a covariate in subsequent analyses (reviewed in Roach and Wulff, 1987; but see Hereford and Moriuchi, 2005). This method accounts for some differences in maternal investment. Another approach was used by Branco (2009), who showed that acorns collected from nonserpentine sites produced larger seedlings in both serpentine and nonserpentine soils. She attempted to reduce the effects of habitat-specific maternal provisioning by removing the acorns early in the growth of seedlings, before acorns were depleted of nutrients.

Parental environmental effects may be very important in determining the performance of genotypes dispersing across heterogeneous landscapes, and so are worthy of careful study and analysis in their own right (e.g., Dormling and Johnsen, 1992). Better provisioned, immigrant seeds produced in favorable micro-habitats may outcompete lower quality seeds that are produced locally within more stressful locations, even if the smaller seeds are genetically more adapted to their home site (Stanton and Galen, 1997; Turelli, 1997). Similarly, pollen produced by plants growing in favorable conditions may experience a competitive advantage during ovule fertilization (Young and Stanton, 1990; Delph et al., 1997), and local soil conditions may also change the stylar environment of the maternal parent in a way that alters the outcome of competition between local and immigrant pollen grains (Searcy and Mulcahy, 1985). Such cross-generational environmental effects are likely to be especially important to consider in habitat restoration. For example, Gordon and Rice (1998) studied the consequences of seed collection source for efforts to restore different populations of wiregrass (*Aristida beyrichiana*). They used field-collected seeds in a reciprocal transplant experiment, because they were most interested in the ecological outcome of seed additions, rather than in the genetic characterization of their source populations. Using field-collected seeds from varying parental environments can also provide insight into what happens in nature when animals, wind, humans, or other dispersal vectors move seeds or pollen from one habitat to another.

In summary, we recommend that researchers make informed decisions about whether to minimize, account for, or manipulate maternal effects based on the questions of primary interest. Explicit consideration of these priorities will allow each investigator to design the most appropriate experiment. Ultimately, our understanding of how parental environment and dispersal jointly affect adaptive differentiation in heterogeneous environments will be advanced by experiments in which genetic origin and parental environment are independently manipulated within the context of factorial reciprocal transplant designs.

3. How Is the Experiment Going to Be Replicated?

Experimental site replication within serpentine and nonserpentine "soil types" is critical for assessing adaptation to soil type and can provide clues into the drivers of diversifying selection.

In studies of plant evolution and ecology in contrasting edaphic environments, we are often interested in acquiring general insights into adaptation to different types of soils (ecotypic differentiation), rather than in reaching specific conclusions about two or more particular locations (local adaptation). A reciprocal transplant experiment study involving just two destination sites (one on serpentine and the other not) does not allow the investigator to distinguish site-specific adaptation from adaptation to soil type. To demonstrate adaptation to soil type, a reciprocal transplant experiment must be designed so that more than one site representing each soil type is represented, and lineages from all sites are moved into all other sites. A number of reciprocal transplant experiments have used such a design in nonserpentine systems (Lowry et al., 2008; Raabová et al., 2008; Gonzalo-Turpin and Hazard, 2009). One of the scientifically useful features of serpentine systems is that the patchy nature of serpentine outcroppings often makes it possible to establish replicate study sites at experimentally tractable spatial scales, within the same bioregion, elevation zone, and climatic regime, thus minimizing the impacts of site-to-site variation that is not due to variation in the soil substrate (e.g., Sambatti and Rice, 2006; Wright et al., 2006a).

To conduct a rigorous test for adaptation to soil type, a researcher can include a fixed main effect of "soil type" in the statistical analysis of a replicated transplant experiment, including the random effect of "site" (nested within soil type). Box 7.1 shows the SAS code used in Wright et al. (2006b), using such an analysis. With this approach, ecotypic adaptation to "soil type" is demonstrated by a significant interaction between source soil type and destination soil type, in which transplants perform better in their native soil type than do transplants originating from the alternative soil type. Planned contrasts or a second, more focused model conducted within each soil type can be used to test for specific home-site adaptation, that is, to test whether transplants returned to their home location outperform transplants from other locations belonging to the same soil type. Given that distinct

BOX 7.1. SAS CODES FROM RT EXPERIMENTS

This is the SAS code used to analyze local adaptation in Wright et al. (2006b). Variables are defined below.

```
proc sort data=one; by site dummy;
proc means n mean stderr std min max; var flnum frnum;
by site dummy;
output out=mlocal mean= mflnum mfrnum
std=vflnum vfrnum;

data rel1;
merge one mlocal; by site dummy;

rfrnum=(frnum-mfrnum)/vfrnum;
rflnum=(flnum-mflnum)/vflnum;

proc mixed;
class ssoil dsoil site momorig plot;
model rfrnum=ssoil dsoil ssoil*dsoil/ ddfm=kenwardroger;
random momorig(ssoil) momorig*site(ssoil*dsoil)
dsoil*momorig(ssoil) ssoil*site(dsoil) plot(dsoil*site);
weight invfruits;
lsmeans ssoil*dsoil/pdiff adjust=tukey;

Run;

Quit;
```

Data sets

One- data set that contains all the raw data from the experiment, including fruit number and flower number.

Variables (in the order in which they appear)

Site- The name of the field site where each plant was grown (6 total)

dummy- A dummy variable to help with the Merge function (dummy = 1)

flnum- The number of flowers recorded on each plant

frnum- the number of fruits recorded on each plant

ssoil- the source soil type (serpentine/ non-serpentine) each plant came from

dsoil- the destination soil type each plant was grown in

momorig- the name of the field site where each plant came from (the origin of the mom of each seed).

plot- the number of the plot within each site were the plant was grown (5 plots per site)

invfruits- This weighting factor was used to account for uneven variances. It equals the inverse of the variance of the number of fruits on each plant, calculated by plot.

population sites will vary for multiple soil factors and other environmental attributes, this nested analysis of variance approach provides a stringent test for adaptation to soil type. Without replication, soil type (serpentine versus nonserpentine) is completely confounded with home site.

If true replication of sites within soil type is not possible, common garden experiments can be designed to reduce the confounding of idiosyncratic site variation with differences between soil types. For example, the long-term transplant experiment on *Pinus ponderosa* described in Wright (2007) was begun in 1967, when the original planners did not consider replication of soil type important. Seeds were collected from a serpentine site and a nonserpentine site in the Sierra Nevada foothills. They were germinated in the greenhouse, on site-appropriate soils, and planted out into two planting sites, one serpentine and one nonserpentine. One feature of this experiment, which reduces the confounding of soil type adaptation and adaptation to a home site that would result from an unreplicated reciprocal transplant experiment, is that seedlings were planted into sites 1–2 km away from where the seeds themselves were collected—no tree was grown in its "home" environment. Other studies have used a similar "common garden" approach (Lowry et al., 2008), but we stress again that a stringent test for adaptation to soil type requires true site replication.

In weighing the advantages and disadvantages of alternative experimental designs, investigators clearly need to prioritize the potential questions they hope to answer. In some restoration applications, it is most important to determine which specific seed sources will be most effective at particular sites of interest. Alternatively, if the goal is to understand a general pattern of adaptation, or to generate methods for predicting which seed sources are likely to be successful in restoring sites of certain attributes, then replication of experiments across multiple sites is much more important and will require the allocation of more resources.

In addition to providing stringent tests for ecotypic differentiation, replication of transplant studies across well-characterized sites and microsites can potentially reveal the axes of soil variation that are most critical to adaptive differentiation. All researchers in the serpentine field are keenly aware that both serpentine and nonserpentine sites are extremely variable with respect to many environmental factors, including soil chemistry and texture. Such site-to-site heterogeneity can be seen as something of a nuisance in that it may dilute the signal of adaptation to alternative soil type categories. An alternative approach is to use multivariate heterogeneity among experimental sites to explain variation in the relative performance of experimental plant lineages from different origins. This can yield new insights into the important environmental axes along which adaptive differentiation has occurred, as well as into the specific site attributes that drive site-specific variation in fitness.

An example of this approach comes from a large collaborative experiment conducted mostly on serpentine soils at the McLaughlin Natural Reserve in the California Coast Range (Wright et al., 2006a; Lau et al., 2008), in which niche modeling based on very local edaphic heterogeneity was shown to predict the performance of a serpentine ecotype of *C. sparsiflora* transplanted into 100 well-characterized microsites. Here we demonstrate another way to gain insights into how both serpentine and nonserpentine ecotypes are influenced by specific axes of soil heterogeneity. At each test plot location, a soil sample was analyzed for 26 chemical and physical attributes, along with slope, aspect, and soil depth. First, we use a principal components analysis (PCA) to characterize axes of variation, and find that the first principal component (PC1) describes 30% of the measured variation among test plots. Plots with high values of PC1 were characterized by more soil hydrogen and lower pH (greater acidity) and by higher levels of soil potassium, whereas plots with higher values of PC2 (accounting for another 22% of among-microsite variation) had lower soil calcium and magnesium. Also loading strongly onto PC2 are heavy metals [cobalt (+), copper (−), and iron (+)], cation exchange capacity (−) and soil texture (+ for % sand).

Serpentine and nonserpentine lineages of *C. sparsiflora* responded very differently to microsite variation along these two principal component axes (Figure 7.2). Plants from serpentine had poorer survival in test plots with higher values for PC1and more negative values of PC2. In contrast, the few plots in which the nonserpentine lineages performed well cluster near mean values for PC1 and PC2. As the study area is mostly mixed serpentine grassland, the noticeably lower survival of transplants from nonserpentine lineages is consistent with our previous finding of ecotypic differentiation to soil type in these populations (Wright et al., 2006b). Many soil attributes are highly intercorrelated, and therefore, correlations between specific soil attributes (e.g., Ca:Mg ratio) and growth performance only suggest hypotheses that can be tested through further experimentation—for example, by manipulations of soil chemistry.

As with any study, alternative experimental designs need to be evaluated with respect to the questions of greatest interest to the investigator—is the researcher interested in general effects of serpentine versus nonserpentine soils, or would it be beneficial to know more about which aspects of the underlying (and always complex) soil variation impact plant performance? Including many plots and locally measured environmental variables in an experimental design does allow for greater sampling of edaphic variation, but a larger number of plots will often limit the number of individuals or population sources represented per plot. For example, Wright et al. (2006a) monitored just 20 seedlings per plot, but the overall experimental design required the planting of 4000 seeds. This latter design constraint limits the ability to measure the effects of soil heterogeneity on genetic variation in performance unless very large experiments can be conducted.

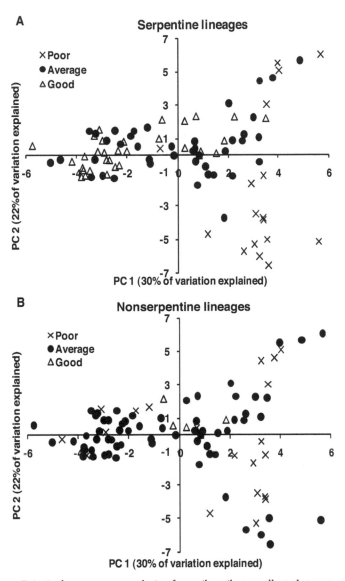

FIGURE 7.2. Principal component analysis of 22 soil attributes collected at 97 test plots used in a large transplant experiment. There was a maximum of 10 seedlings surviving to flowering in each plot for each source soil type (20 total seedlings per plot). A: performance of serpentine lineages; B: performance of nonserpentine lineages. For lineages from each soil type, performance within a plot was categorized as follows: Poor: 0/10 seedlings survived to flowering; Average: 1–5/10 seedlings survived to flowering; Good: ≥6/10 seedlings survived to flowering.

4. How Can My Reciprocal Transplant or Common Garden
Experiment Be Integrated with Analyses of Molecular Genetic
Variation to Better Understand Differentiation
among Plant Populations?

Molecular genetic approaches that sample loci such as allozymes, microsatellites, random amplifications of polymorphic DNA (RAPDs), and amplified fragment length polymorphisms (AFLPs) can offer important insights into the evolutionary history of populations (Wang and Szmidt, 2001). Each of these genetic markers allows investigators to partition genetic variation within and among populations and habitat types using hierarchical *F*-statistics (Wright, 1969) or multilocus analogs thereof (Weir, 1996). For any two locally adapted populations there will be differences in the frequencies of alleles that influence habitat-specific performance due to natural selection. However, if there is limited gene flow between those habitats, differences in neutral allele or marker frequencies will also accumulate due to the random effects of genetic drift. Especially when relatively few major genes contribute heavily to adaptive traits (see one possible serpentine example in Bradshaw, 2005), there may be little correspondence between adaptive differentiation and neutral genetic divergence.

To the extent that the sampled loci are selectively neutral, as is often assumed, a spatially structured analysis of allele frequencies reflects the net effects of realized gene flow at the different spatial and/or ecological scales included in the analysis (Weir, 1996). If differences have arisen purely from neutral genetic drift, one should see similar patterns of allele frequency variation at all polymorphic loci. For example, if different soil types cause differences in flowering time between populations at nearby sites, this could result in decreased gene flow and correspondingly faster neutral genetic divergence between those populations (Brady et al., 2005; Wright et al., 2006b). In this case, even strong molecular genetic differentiation between populations would not necessarily indicate adaptive differentiation, because even neutral loci will differentiate between the populations due to drift. Conversely, a number of recent studies have shown that plant populations displaying marked phenotypic differences show little divergence at neutral markers (e.g., Merilä and Crnokrak, 2001; McKay and Latta, 2002). This pattern suggests that individuals or gametes are moving between contrasting selection regimes, but there is sufficient genetic recombination to allow differentiation to persist specifically at loci under strong local selection.

Our own work illustrates how combining field experiments with genetic marker analysis can advance the assessment of serpentine adaptation. In both *P. ponderosa* and *C. sparsiflora*, transplant experiments showed that serpentine lineages outperformed nonserpentine lineages on serpentine soil (Wright et al., 2006b; Wright, 2007). However, in both cases we also found few differences in putatively neutral allozyme allele frequencies between the serpentine and nonserpentine populations (Table 7.1; Wright, 2007). Importantly, researchers using only molecular genetic approaches would not have found evidence for critical genetic differences between populations on the two soil ecotypes, as was revealed by transplant experiments. In situations like this, the use of molecular markers alone to make management decisions could lead to incorrect decisions or restoration practices.

In contrast to variants at neutral genetic loci, alleles that increase performance under specific conditions are expected to differ in frequency between environmental selection regimes. This occurs because individuals bearing the locally fit allele will have a greater chance of surviving and reproducing in that habitat. When screening genetic marker variation across heterogeneous landscapes, we very occasionally sample loci that are either under divergent selection or are closely linked to loci that are. In these cases, environment-based genetic differentiation at such loci should be substantially greater than that seen at neutral loci (e.g., Clegg and Allard, 1972; Hamrick and Allard, 1972). Although such "outlier" genetic differences provide no conclusive proof of local adaptation, their discovery may provide important clues into mechanisms of adaptation (reviewed in

TABLE 7.1   Analysis of Molecular Variation for Allozyme Marker Data for *Pinus ponderosa* and *Collinsia sparsiflora*

| | | *Pinus ponderosa* | | |
	df	Sum of Squares	Variance Component	% of Variation
Among groups (S vs. NS source populations)	1	1.534	−0.00185	−0.24
Among maternal half-sibships within groups	22	39.159	0.07841	10.01[a]
Within maternal half-sibships	306	216.262	0.70674	90.23[a]

| | | *Collinsia sparsiflora* | | |
	df	Sum of Squares	Variance Component	% of Variation
Among soil types (S vs. NS)	1	0.451	0.00051	0.64
Among populations within soil type	4	1.097	0.00154	1.97[a]
Within populations	796	60.753	0.07632	97.39[a]

NOTE: Seven polymorphic loci were scored for ponderosa pine, while four were scored for *C. sparsiflora*. Analyses were conducted in Arelquin (Excoffier et al., 2005).

[a]Significant at $p < 0.05$.

Storz, 2005). In our own work on *C. sparsiflora*, for example, allele frequency variation at one locus, malic enzyme (*Me*), contrasted sharply with that at four other polymorphic allozyme loci surveyed, in showing dramatic allele frequency differences between McLaughlin Reserve populations on serpentine and nonserpentine soils (three populations on each soil type were included in the sample; Wright and Stanton, unpublished data). Similarly, Oline and colleagues (2000) surveyed three polymorphic allozyme loci in foxtail pines from multiple California populations on both serpentine and nonserpentine. They found that allele frequencies at glucose-6-phosphate isomerase (*Gpi*) did not differ between serpentine and nonserpentine sites, whereas soil type was associated with significant differentiation at fluorescent esterase (*Fest1*) and marginally significant allele frequency differences at the malate dehydrogenase (*Mdh1*) locus. Strong allele frequency differences like these can be used as a starting point for further genetic analyses and experiments aimed at identifying the genes at which adaptive differentiation has occurred.

Although they provide definitive evidence for (or against) local adaptation, reciprocal transplant experiments are difficult and labor-intensive to conduct, even in tractable short-lived species. Accordingly, it is important to develop more broadly applicable techniques that can indicate whether adaptive differentiation is likely to have occurred and can point the investigator toward potential adaptive mechanisms. Specifically, those interested in soil-based ecotypes are keen to understand which genes are important for enhancing survival, growth, and reproduction on serpentine soils.

One promising approach is to screen replicate populations from contrasting habitat types and then use molecular techniques to identify loci that show unusually high levels of habitat-specific allele frequency differences (e.g., "genome scans," reviewed by Storz, 2005). Though still in its infancy, great promise is shown by the emerging field of ecological genomics, or ecogenomics (Kammenga et al., 2007; Ouborg and Vriezen, 2007; see also Chapter 6 in this volume). Essentially, genomic screening methods aim to compare and contrast the *entire* genome for individuals from contrasting selection regimes of interest, for example, from serpentine and nonserpentine sites. Consistent genomic differences that are seen in multiple populations from different habitat types point to genes or genomic regions that are likely to be under divergent selection. Another approach is to compare the transcriptome (all of the expressed mRNA) of serpentine and nonserpentine adapted plants (see Chapter 6 in this volume) to find differences in gene expression that may indicate mechanisms for serpentine tolerance. These techniques are currently costly and labor-intensive, but rapid technical and bioinformatic advances mean that population genomics will soon be a widely available approach, even for nonmodel systems (Wright and von Wettberg, 2009).

5. Which Genotypes Should Be Transplanted?

Simple transplants of parental types are of limited use in identifying traits that are under diversifying selection across soil types.

Natural selection causes local adaptation, but patterns of diversifying selection on specific genes or traits are not revealed in standard reciprocal transplant experiments, which test a limited number of genotypes in a small number of environments (for example, Sambatti and Rice, 2006; Wright et al., 2006b; Wright, 2007). Except when there is extensive gene flow between habitats with strongly divergent selection regimes, "parental" genotypes adapted to those different environments are likely to be mosaics of genetic loci that have differentiated due to direct selection, linkage to directly selected alleles and/or random drift. When such nonrecombinant genotypes are used in reciprocal or common garden transplant experiments, these historical genetic and phenotypic associations make it difficult to ascertain which of the differentiated traits do in fact represent adaptations to the different habitats.

For example, in our work on *C. sparsiflora* growing at the McLaughlin Natural Reserve, we find predominantly white-flowered individuals growing on nonserpentine soils and purple-flowered individuals on serpentine. (We have subsequently found that this trend is not true throughout the range of *C. sparsiflora*.) Our reciprocal transplant experiment clearly demonstrated the existence of serpentine and nonserpentine ecotypes that are adapted to each soil type, but how can we determine whether purple flowers increase fitness on serpentine soils and/or whether white flowers confer an advantage on nonserpentine soils? In other words, can a plant can be serpentine-tolerant and have white flowers? Because individuals with this recombinant phenotype do not exist in our study populations, we had to create them through a set of crosses, following a method pioneered by Jordan (1991) for studying mechanisms of local adaptation in plants (Wright and Stanton, 2007).

First, we reared plants from field-collected seeds in the greenhouse. We then cross-pollinated individuals from four different populations. The F1 offspring produced from these crosses were heterozygous at most genetic loci (e.g., they all had pink flowers), and so would be of limited use for detecting selection on individual traits. These F1 plants were grown under greenhouse conditions and were then crossed with each other to produce a highly recombinant F2 generation. To maximize the resulting phenotype combinations, we selected F1 plants to cross such that each F2 lineage had grandparents from four different populations—two serpentine and two nonserpentine. The processes of independent assortment and recombination reshuffled the serpentine and nonserpentine genomes, producing a wide array of unique recombinant phenotypes. Petal color varied from pure white to dark purple among F2 individuals, and petal color should have become disassociated from allelic variation at most other loci.

We then planted out these F2 seeds into six different sites—three serpentine, and three nonserpentine. For each emergent seedling, we measured survival and the number of fruits and flowers as estimates of lifetime fitness. We recorded a series of phenotypic traits for each individual (including emergence date, cotyledon size, flowering date, flower color, and flower size) and then used multivariate selection gradient analysis (Lande and Arnold, 1983) to determine which traits influenced fitness in serpentine and nonserpentine sites.

Our results were counterintuitive, but still informative. In both serpentine and nonserpentine destinations, individual plants with darker flowers tended to have higher fitness. This was expected in serpentine habitats, where we usually find purple-flowered plants, but we did not see the expected pattern of plants with white flowers having higher fitness in nonserpentine habitats. This result indicates that petal color is not under divergent selection between serpentine and nonserpentine habitats, at least under our experimental conditions. It is important to note that F2 transplants were only allowed to self-fertilize in this study. To avoid genetic contamination of the study sites, naturally occurring pollinators were excluded from experimental plots once transplants came into flower. This did not allow us to test the idea that pollinator-mediated selection may favor different petal pigments in the two soil habitats. We also found that other traits were under selection, but there was no evidence for diversifying selection between the two soil types. How can this be? A reciprocal transplant experiment being conducted at the same time in the same sites, with similar pollinator exclusion, provided clear evidence for local adaptation to serpentine and nonserpentine soils (Wright et al., 2006b), so there must have been diversifying selection acting between the soil types during the course of the F2 transplant experiment. What we failed to show was that this differential selection was in fact acting on the traits we measured. We draw two conclusions and lessons from this hybrid transplant experiment.

First, our results demonstrate that soil-based selection must be acting on traits that we did *not* measure. As just one example, root physiological traits are rarely included in multivariate analyses of selection on plants (but see Volis et al., 2004), but are likely to be of critical importance in adaptation to soil type (Sambatti and Rice, 2007). Assessing selection on traits that can only be measured destructively presents a great challenge to field biologists, one that can potentially be overcome in tractable species by the use of recombinant inbred experimental lines. If root characters vary much more among lineages than within lineages, then it should be possible to assess the lifetime fitness on some individuals, while harvesting related individuals to estimate the root traits of their intact relatives. Measuring selection in this way requires large experiments, because the analysis is essentially conducted on family trait means or breeding values (Rausher, 1992).

Second, given the fact that we enclosed our flowering transplants within mesh cages to prevent genetic contamination of our study sites, we cannot exclude the possibility that pollinators or seed predators may exert selection for the two different petal colors in serpentine versus nonserpentine habitats at the McLaughlin Reserve. Indeed, concern about genetic contamination of natural populations can impose significant constraints on the design of field transplant experiments. Although maladapted alleles are likely to be removed from populations over time through locally purifying selection, genetic contamination from transplant studies could change patterns of spatial genetic variation at neutral loci, potentially hindering subsequent genetic analyses. Especially in heavily used research areas, care must be taken to minimize such negative interactions between different studies.

CONCLUSIONS

Reciprocal transplant experiments in general, and those involving serpentine systems in particular, are important tools for understanding plant adaptation to heterogeneous environments. In this chapter, we discussed several important factors that should be considered when designing reciprocal transplant or common garden experiments. We stressed that proper replication of such experiments across multiple sites, using experimental lineages (potentially including recombinant lines) in combination with detailed edaphic measurements and/or manipulations, can help us learn about the specific traits that may be under diversifying selection across soil types and the axes of environmental variation that are most critical to that diversification. We have made substantial progress in understanding serpentine adaptation since Kruckeberg's seminal studies, but there is much yet to be learned about adaptation to heterogeneous environments, including those that are mosaics of serpentine and non-serpentine soils. Reciprocal transplant experiments, paired with emerging technologies such as genomics and transcriptomics (see Chapter 6 in this volume), will continue to be important tools in the future.

Acknowledgments

The authors thank the editors for their tireless work to make this volume a success. We also received very helpful comments from two anonymous reviewers. The research on *C. sparsiflora* described here was funded through grants from the Packard Foundation and the Andrew Mellon Foundation. Many members of the Stanton lab, along with friends and family members, generously helped plant out the field experiments, and we especially want to thank Dr. Rosa Scherson for her assistance with many phases of the project. Cathy Koehler and Paul Aigner at the Donald and Sylva McLaughlin Reserve also provided critical logistical support. The ponderosa pine plantations were established and supported for all 38 years of their existence by the Institute of Forest Genetics, Pacific Southwest Research Station, USDA Forest Service.

LITERATURE CITED

Antonovics, J., and Bradshaw, A. D. (1970) Evolution in closely adjacent plant populations VIII. Clinal patterns at a mine boundary. *Heredity*, 25, 349–62.

Bradshaw, H. D. Jr. (2005) Mutations in *CAX*1 produce phenotypes characteristic of plants tolerant to serpentine soils. *New Phytologist*, 167, 81–88.

Brady, K. U., Kruckeberg, A. R., and Bradshaw, H. D. Jr. (2005) Evolutionary ecology of plant adaptation to serpentine soils. *Annual Review of Ecology, Evolution, and Systematics*, 36, 243–66.

Branco, S. (2009) Are oaks locally adapted to serpentine soils? *Northeastern Naturalist*, 16 (Special Issue 5), 329–40.

Bridle, J. R., and Vines, T. H. (2007) Limits to evolution at range margins: When and why does adaptation fail? *Trends in Ecology and Evolution*, 22, 140–47.

Brock, M. T., Stinchcombe, J. R., and Weinig, C. (2009) Indirect effects of FRIGIDA: Floral trait (co)variances are altered by seasonally variable abiotic factors associated with flowering time. *Journal of Evolutionary Biology*, 22, 1826–38.

Clausen, J., Keck, D. D., and Hiesey, W. M. (1940) *Experimental Studies on the Nature of Species I. Effect of Varied Environments on Western North American Plants*. Carnegie Institution of Washington, Washington, DC.

Clausen, J., Keck, D. D., and Hiesey, W. M. (1948) *Experimental Studies on the Nature of Species III. Environmental Responses of Climatic Races of* Achillea. Carnegie Institution of Washington, Washington DC.

Clegg, M. T., and Allard, R. W. (1972) Patterns of genetic differentiation in the slender wild oat species *Avena barbata*. *Proceedings of the National Academy of Sciences, USA*, 69, 1820–24.

Delph, L. F., Johannsson, M. H., and Stephenson, A. G. (1997) How environmental factors affect pollen performance: Ecological and evolutionary perspectives. *Ecology*, 78, 1632–39.

Donohue, K. (2009) Completing the cycle: Maternal effects as the missing link in plant life histories. *Philosophical Transactions of the Royal Society B-Biological Sciences*, 364, 1059–74.

Donohue, K., Dorn, L., Griffith, C., Kim, E., Aguilera, A., Polisetty, C. R., and Schmitt, J. (2005a) Niche construction through germination cueing: Life-history responses to timing of germination in *Arabidopsis thaliana*. *Evolution*, 59, 771–85.

Donohue, K., Dorn, L., Griffith, C., Kim, E., Aguilera, A., Polisetty, C. R., and Schmitt, J. (2005b) Environmental and genetic influences on the germination of *Arabidopsis thaliana* in the field. *Evolution*, 59, 740–57.

Dormling, I., and Johnsen, Ø. (1992) Effects of the parental environment on full-sib families of *Pinus sylvestris*. *Canadian Journal of Forestry Research*, 22, 88–100.

Emery, N. C. (2006) *Ecology and evolution of distribution patterns in vernal pool annual plant populations*. Ph.D. thesis, University of California, Davis.

Emery, N.C. (2009) Ecological limits and fitness consequences of cross-gradient pollen movement in *Lasthenia fremontii*. *American Naturalist*, 174, 221–35.

Excoffier, L., Laval, G., and Schneider, S. (2005) Arlequin ver. 3.0: An integrated software package for population genetics data analysis. *Evolutionary Bioinformatics Online*, 1, 47–50.

Fine, P. V. A, Mesones, I., and Coley, P. D. (2004) Herbivores promote habitat specialization by trees in Amazonian forests. *Science*, 305, 663–65.

Galen, C., Shore, J. S., and Deyoe, H. (1991) Ecotypic divergence in alpine *Polemonium viscosum*: Genetic structure, quantitative variation, and local adaptation. *Evolution*, 45, 1218–28.

Galloway, L. F. (2002) The effect of maternal phenology on offspring characters in the herbaceous plant *Campanula americana*. *Journal of Ecology*, 90, 851–58.

Galloway, L. F. (2005) Maternal effects provide phenotypic adaptation to local environmental conditions. *New Phytologist*, 166, 93–99.

Galloway, L. F., and Etterson J. R. (2007) Transgenerational plasticity is adaptive in the wild. *Science*, 318, 1134–36.

García-Ramos, G., and Kirkpatrick, M. (1997) Genetic models of adaptation and gene flow in peripheral populations. *Evolution*, 51, 21–28.

Gonzalo-Turpin, H., and Hazard, L. (2009) Local adaptation occurs along altitudinal gradient despite the existence of gene flow in the alpine plant species *Festuca eskia*. *Journal of Ecology*, 97, 742–51.

Gordon, D. R., and Rice, K. J. (1998) Patterns of differentiation in wiregrass (*Aristida beyrichiana*): Implications for restoration efforts. *Restoration Ecology*, 6, 166–74.

Hamrick, J. L., and Allard, R. W. (1972) Microgeographical variation in allozyme frequencies in *Avena barbata*. *Proceedings of the National Academy of Sciences. USA*, 69, 2100–2104.

Hereford, J., and Moriuchi, K. S. (2005) Variation among populations of *Diodia teres* (Rubiaceae) in environmental maternal effects. *Journal of Evolutionary Biology*, 18, 124–31.

Hereford, J., and Winn, A. A. (2008) Limits to local adaptation in six populations of the annual plant *Diodia teres*. *New Phytologist*, 178, 888–96.

Holt, R. D., and Gaines, M. S. (1992) Analysis of adaptation in heterogeneous landscapes: Implications for the evolution of fundamental niches. *Evolutionary Ecology*, 6, 433–47.

Hufford, K. M., Mazer, S. J., and Camara, M. D. (2008) Local adaptation and effects of grazing among seedlings of two native California bunchgrass species: Implications for restoration. *Restoration Ecology*, 16, 59–69.

Jordan, N. (1991) Multivariate analysis of selection in experimental plant populations derived from hybridization of two ecotypes of the annual plant *Diodia teres* W. (Rubiaceae). *Evolution*, 45, 1760–72.

Jurjavcic, N. L., Harrison, S., and Wolf, A. T. (2002) Abiotic stress, competition, and the distribution of the native annual grass *Vulpia microstachys* in a mosaic environment. *Oecologia*, 130, 555–62.

Kalisz, S., and McPeek, M. A. (1992) Demography of an age-structured annual: Resampled projection matrices, elasticity analyses, and seed bank effects. *Ecology*, 73, 1082–93.

Kammenga, J. E., Herman, M. A., Ouborg, N. J., Johnson, L., and Breitling, R. (2007) Microarray challenges in ecology. *Trends in Ecology and Evolution*, 22, 273–79.

Kawecki, T. J., and Ebert, T. (2004) Conceptual issues in local adaptation. *Ecology Letters*, 7, 1225–41.

Kirkpatrick, M., and Barton, N. H. (1997) Evolution of a species' range. *American Naturalist*, 150, 1–23.

Kruckeberg, A. R. (1951) Intraspecific variability in the response of certain native plant species to serpentine soil. *American Journal of Botany*, 38, 408–19.

Kruckeberg, A. R. (1967) Ecotypic response to ultramafic soils by some plant species of northwestern United States. *Brittonia*, 19, 133–51.

Lande, R., and Arnold, S. J. (1983) The measurement of selection on correlated characters. *Evolution*, 37, 1210–26.

Lau, J. A., McCall, A. C., Davies, K. F., McKay, J. K., and Wright, J. W. (2008) Herbivores and edaphic factors constrain the realized niche of a native plant. *Ecology*, 89, 754–62.

Leimu, R., and Fischer, M. (2008) A meta-analysis of local adaptation in plants. *PLoS ONE*, 12, e4010.

Levin, D. A. (2009) Flowering-time plasticity facilitates niche shifts in adjacent populations. *New Phytologist*, 183, 661–66.

Lowry, D. B., Rockwood, R. C., and Willis, J. H. (2008) Ecological reproductive isolation of coast and inland races of *Mimulus guttatus*. *Evolution*, 62, 2196–214.

McKay, J. K., and Latta, R. G. (2002) Adaptive population divergence: Markers, QTL and traits. *Trends in Ecology and Evolution*, 17, 285–91.

Merilä, J., and Crnokrak, P. (2001) Comparison of genetic differentiation at marker loci and quantitative traits. *Journal of Evolutionary Biology*, 14, 892–903.

Nagy, E. S., and Rice, K. J. (1997) Local adaptation in two subspecies of an annual plant: Implications for migration and gene flow. *Evolution*, 51, 1079–89.

Oline, D. K., Mitton, J. B., and Grant, M. C. (2000) Population and subspecific genetic differentiation in the foxtail pine (*Pinus balfouriana*). *Evolution*, 54, 1813–19.

Ouborg, N. J., and Vriezen, W. H. (2007) An ecologist's guide to ecogenomics. *Journal of Ecology*, 95, 8–16.

Platenkamp, G. A. J., and Shaw, R. G. (1993) Environmental and genetic maternal effects on seed characters in *Nemophila menziesii*. *Evolution*, 47, 540–55.

Potvin, C., and Tardif, S. (1988) Sources of variability and experimental designs in growth chambers. *Functional Ecology*, 2, 123–30.

Raabová, J., Fischer, M., and Münzbergová, Z. (2008) Niche differentiation between diploid and hexaploid *Aster amellus*. *Oecologia*, 158, 463–72.

Rajakaruna, N., and Bohm, B. (1999) The edaphic factor and patterns of variation in *Lasthenia californica* (Asteraceae). *American Journal of Botany*, 86, 1576–96.

Rajakaruna, N., Siddiqi, M. Y., Whitton, J., Bohm, B., and Glass, A. D. M. (2003) Differential responses to Na^+/K^+ and Ca^{2+}/Mg^{2+} in two edaphic races of the *Lasthenia californica* (Asteraceae) complex: A case for parallel evolution of physiological traits. *New Phytologist*, 157, 93–103.

Rausher, M. D. (1992) The measurement of selection on quantitative traits: Biases due to environmental covariances between traits and fitness. *Evolution*, 46, 616–26.

Roach, D. A., and Wulff, R. D. (1987) Maternal effects in plants. *Annual Review of Ecology and Systematics*, 18, 209–35.

Sambatti, J. B. M., and Rice, K. J. (2006) Local adaptation, patterns of selection, and gene flow in the Californian serpentine sunflower (*Helianthus exilis*). *Evolution*, 60, 696–710.

Sambatti, J. B. M., and Rice, K. J. (2007) Functional ecology of ecotypic differentiation in the Californian serpentine sunflower (*Helianthus exilis*). *New Phytologist*, 175, 107–19.

Searcy, K. B., and Mulcahy, D. L. (1985) Pollen tube competition and selection for metal tolerance in *Silene dioica* (Caryophyllaceae) and *Mimulus guttatus* (Scrophulariaceae). *American Journal of Botany*, 72, 1695–99.

Stanton, M. L. (1984) Seed variation in wild radish: Effect of seed size on components of seedling and adult fitness. *Ecology*, 65, 1105–112.

Stanton, M. L., and Galen, C. (1997) Life on the edge: Adaptation versus environmentally mediated gene flow in the snow buttercup, *Ranunculus adoneus*. *American Naturalist*, 150, 143–78.

Storz, J. F. (2005) Using genome scans of DNA polymorphism to infer adaptive population divergence. *Molecular Ecology*, 14, 671–88.

Turelli, M. (1997) Appendix: Environmental heterogeneity, maternal effects and spatial patterns of genetic variation. *Evolution*, 51, 93–94.

Turesson, G. (1922) The genotypical response of plant species to the habitat. *Hereditas*, 3, 211–350.

Volis, S., Verhoeven, K. J. F., Mendlinger, S., and Ward, D. (2004) Phenotypic selection and regulation of reproduction in different environments in wild barley. *Journal of Evolutionary Biology*, 17, 1121–31.

Wang, X.-R., and Szmidt, A. E. (2001) Molecular markers in population genetics of forest trees. *Scandinavian Journal of Forest Research*, 16, 199–220.

Waser, N. M., and Price, M. V. (1985) Reciprocal transplant experiments with *Delphinium nelsonii* (Ranunculaceae): Evidence for local adaptation. *American Journal of Botany*, 72, 1726–32.

Weir, B. S. (1996) *Genetic Data Analysis II. Methods for Discrete Population Genetic Data*. Sinaur Associates, Sunderland, MA.

Wright, J. W. (2007) Local adaptation to serpentine soils in *Pinus ponderosa*. *Plant and Soil*, 293, 209–17.

Wright, J. W., Davies, K. F., Lau, J. A., McCall, A. C., and McKay, J. K. (2006a) Experimental verification of ecological niche modeling in a heterogeneous environment. *Ecology*, 87, 2433–39.

Wright, J. W., and Stanton, M. L. (2007) *Collinsia sparsiflora* on serpentine and non-serpentine soils: Using F2 hybrids to detect the potential role of selection in ecotypic differentiation. *New Phytologist*, 173, 354–66.

Wright, J. W., Stanton, M. L., and Scherson, R. (2006b) Local adaptation to serpentine soils in *Collinsia sparsiflora*. *Evolutionary Ecology Research*, 8, 1–21.

Wright, J. W., and von Wettberg, E. J. (2009) Serpentinomics—an emerging new field of study. *Northeastern Naturalist*, 16, 285–96.

Wright, S. (1969) *Evolution and the Genetics of Populations. Volume 2. The Theory of Gene Frequencies*. University of Chicago Press, Chicago.

Young, H. J., and Stanton, M. L. (1990) Influence of environmental quality on pollen competitive ability in wild radish. *Science*, 248, 1631–33.

8

Herbivory and Other Cross-Kingdom Interactions on Harsh Soils

Sharon Y. Strauss, *University of California, Davis*
Robert S. Boyd, *Auburn University*

Edaphically stressful substrates, like serpentine, present plants with challenges that differ from other substrates. Stressful substrates often require plant adaptations to toxicity stressors like heavy metals, as well as to low nutrient concentrations or abnormal ratios of necessary elements (Proctor and Woodell, 1975; Ellis and Weis, 2006). Pressures from enemies may also be greater on edaphically stressful substrates than on normal soils (Fine et al., 2005). On the other hand, substrates with high concentrations of heavy metals (like serpentine) may provide plants with opportunities for elemental defense, such as heavy metal accumulation (Martens and Boyd, 1994; Mroz, 2008). We describe some of the aspects of harsh substrates, paying particular attention to serpentine substrates, that might affect plant interactions with enemies and may be central to edaphic specialization. We also briefly consider interactions with pathogens and pollinator and dispersal mutualists.

HERBIVORES AND EDAPHIC SPECIALIZATION

There is growing evidence that interactions with enemies may be a key element to understanding the traits associated with edaphic specialization. Fine et al. (2005) showed that tropical tree white-sand edaphic specialists were more resistant to herbivory on white sand than were their relatives from richer red-clay soils.

White-sand specialists invested more in phenolic and tannin defenses in their leaves than did their red-clay congeners. Whereas white-sand specialists were more defended, they were also more vulnerable to herbivory and, once damaged, had a lower capacity to regrow or survive relative to their red-clay congeners (Fine et al., 2005). In fact, white-sand specialists only performed better than their red-clay congeners on white sand and in the presence of enemies. These authors argue that herbivores and stressful soil environments combine to prevent invasion of white-sand habitats by red-clay relatives.

In another case where herbivores may mediate edaphic specialization, Lau et al. (2008) planted serpentine and nonserpentine ecotypes of *Collinsia sparsiflora* into microhabitats of known soil chemical composition inside and outside the niche of the serpentine ecotype, where niche was defined using a niche model with multiple soil variables. Herbivory on serpentine (and nonserpentine) ecotypes was greater in areas outside the serpentine ecotype niche. In addition, serpentine ecotypes, like the white-sand tree species described, were less tolerant of herbivory than were their nonserpentine ecotype counterparts. In this case, herbivory levels explained a significant amount of variation in the niche of ecotypes that was not explained by soil traits (Lau et al., 2008). Together, Fine et al. (2005) and Lau et al. (2008) provide mounting evidence that interactions with enemies may be a key feature of edaphic adaptation.

Plants growing on serpentine and other edaphically stressful substrates may experience increased selective pressures from herbivory arising from several different aspects of edaphically stressful substrates. These are listed and discussed in turn.

Harsh Substrates May Be Challenging for Plants and May Increase Costs of Herbivore Damage

Although serpentine soils are known for their high Ni content and low Ca:Mg ratio as major plant stressors, they are also low in P and other key nutrients, as well as water content in hot Mediterranean climates (Sambatti and Rice, 2007; Baythavong et al., 2009). In field nutrient additions, separately and in combination, N and P were found to be more limiting than Ca in reducing growth of two native grasses on serpentine (Turitzin, 1982), although Ca regulation above and beyond N and P additions appears to be a key aspect of serpentine adaptation in evergreen shrubs (O'Dell et al., 2006). Harsh substrates like serpentine also have low vegetation cover: a fact that, in turn, may slow the build-up of organic matter in soils. Proctor and Woodell (1975) indicate that few surveys have actually measured percent cover on bare serpentine substrate, but estimates from five different investigators ranged from less than 1% to 16% cover. Low cover not only reduces nutrient build-up but may increase drought stress through lack of shading of the soil and reduced opportunities for facilitation (Maestre et al., 2009).

When environments are stressful, plants may experience greater impacts of herbivory than when they grow in richer substrates. The resource availability hypothesis of plant defense posed by Coley and colleagues (1985) suggests that costs of losing tissues in low resource environments (resources could be nutrients or water) should select for plants that invest constitutively in defenses. Thus, even though attack rates may be similar across substrate types, the same amount of herbivory may be more costly in terms of plant fitness when it occurs on resource-poor substrates. Because the ability to regrow after damage is also closely linked to resource availability (reviewed in Strauss and Agrawal, 1999), tolerance to herbivory is also expected to be low on serpentine and other edaphically stressful substrates. Results from both Fine et al. (2005) and Lau et al. (2008) support this prediction.

There are many avenues for further investigation. Experiments that control damage levels across substrate types would address whether serpentine-adapted species are more or less tolerant to damage. An intriguing recent paper by Palomino et al. (2007) found that nickel soil additions increased tolerance to heavy damage in the nickel-hyperaccumulating mustard *Thlaspi montanum*. Nutrient and water additions combined with experimental herbivore damage might indicate whether tolerance ability of plants on serpentine is due to low nutrients, low water, or other attributes of serpentine soils or soil biota.

Lower Local Diversity and Density of Plants on Harsh Substrates Can Alter Microclimate and Conspicuousness of Plants to Herbivores

In many areas, one can find reduced above-ground biomass on edaphically stressful substrates compared to adjacent habitats. Habitats with more bare ground can lead to hotter, brighter, and often drier microclimates than those found in normal soils. Herbivores are generally more abundant in high light environments in the tropics (Coley, 1983; Aide and Zimmerman, 1990; Louda and Rodman, 1996; Basset et al., 2001) and other areas (Louda and Rodman, 1996; Niesenbaum and Kluger, 2006). *Cardamine cordifolia* plants growing in sunny habitats in Colorado were routinely more heavily attacked by butterfly larvae than were conspecifics growing in shade nearby (Louda and Rodman, 1996). This preference for high light environments may arise from thermal needs of insect herbivores for feeding or oviposition (e.g., Sherman and Watt, 1973; Moore et al., 1988) and/or from differences in nutritional quality between sun and shade leaves (Louda and Rodman, 1996; Osier and Jennings, 2007). California serpentine habitats are often high light environments consisting of dry scree slopes. These microclimates might favor some types of herbivory. For example, Lau et al. (2008) found that growing on a slope was associated with higher levels of herbivory on *Collinsia sparsiflora* plants.

In addition to creating warmer environments, sparser, less diverse communities on edaphically stressful substrates like serpentine may render plants more conspicuous to both olfactorally and visually searching herbivores. Root (1973) long ago identified the importance of complex plant communities in determining pest loads and damage to plants. Associational resistance occurs when heterospecific neighbors decrease the damage level experienced by another plant species. Simplified environments may make plants more easily found by searching herbivores through reduction of this associational resistance. Although initially most of the action of neighboring plants was thought to be due to olfactory masking of host odors by neighbors, more recently several studies have shown that olfactory cues may work at long distances for orientation to patches but visual cues are very important for insects in landing or oviposition success (Finch and Collier, 2000; Campbell and Borden, 2006; Stenberg and Ericson, 2007). Host searching can be disrupted by surrounding host plants with plant models made from green paper or simply with sheets of green paper (Kostal and Finch, 1994), neither of which release plant olfactory cues. In fact, for eight different cruciferous-feeding herbivores from a range of orders, plants growing in experimentally manipulated backgrounds of green nonhost leaves, dry (brown) nonhost leaves, green paper, black plastic, and bare ground, plants in the bare ground treatment routinely had 15% to >90% higher levels of herbivory (Stenberg and Ericson, 2007). Moreover, plants offered to herbivores in clear plastic bags where no olfactory cues were available were routinely found by herbivores (Doring et al., 2004; Couty et al., 2006; Stenberg and Ericson, 2007). Consistent with these data is the fact that leaf outlines are often very important cues for host finding (Rausher, 1978; Karban and Courtney, 1987; Mackay and Jones, 1989; Hirota and Kato, 2001), and leaf outlines may be more conspicuous on sparsely vegetated substrates.

Visual cues may also be important for vertebrate herbivores. Recent studies by Fadzly et al. (2009) and one of these authors (Strauss, unpublished data) suggest that leaf coloration may serve to hide plants from visually searching herbivores. Functions of leaf coloration may not be limited to photosynthetic ability: they may serve a role in crypsis or aposematism (Lev-Yadun et al., 2004; Lev-Yadun, 2006). Leaf coloration of lancewood (*Pseudopanax crassifolius*) in New Zealand has been proposed as a defense against moa grazing (Fadzly et al., 2009). Leaves at moa browsing height have patterns that strongly resemble large thorns. Several serpentine scree-adapted species in the clade *Streptanthus* (Brassicaceae; also thought to include genera *Caulanthus*, *Streptanthella*, *Guillenia*; Mayer, personal communication) have brown mottled leaves that appear to match the color of the serpentine scree substrate. *Streptanthus* species routinely experience damage levels in excess of 20–30% leaf area loss (Shapiro, 1981; Strauss, unpublished data). This loss is even higher (close to 100%) along deer trails, near bushes, and near rodent holes (Strauss, unpublished data). There are at least six instances of cryptic (brown) leaves in the *Streptanthus* clade of ~50 taxa; of these, all are serpentine-growing species, and there appear to be at least two independent origins of this trait. A search of the Web and literature shows that plants with leaves that appear to match the substrate can be found on screes of New Zealand (*Notothlaspi* and *Ranunculus*), serpentine slopes of California (*Lotus*, *Streptanthus*), sand dunes of Israel (Lev-Yadun, 2006), and famously, stone plants of South Africa (Ellis and Weis, 2006). Edaphically stressful, barren substrates, or sparsely vegetated substrates in general, may select for leaf crypsis in palatable species to hide leaves from visually searching herbivores.

The alternative hypothesis that sparser plant densities would reduce herbivore damage should also be considered. Dense plant communities may attract greater numbers of herbivores, or may cause them reside there for longer periods of time, than less dense communities (e.g., Root, 1973). Surveys of plant density and damage levels across habitat types would shed light on these questions. Manipulations of plant densities or conspicuousness across habitat types would also be informative with respect to damage levels.

Unusual Substrates May Provide Opportunities for Elemental Defenses

Most plant defenses are organic chemicals (Hay, 1996; Dearing et al., 2005), and these have been extensively studied over the past 50 years. Organic defenses include groups of chemicals such as polyphenols, alkaloids, terpenes, and terpenoids. These chemical defenses are synthesized in plants primarily from photosynthate, using various metabolic pathways. More recently, elemental plant defenses have been described (Martens and Boyd, 1994), in which the defensive chemical is an element taken up from the soil. Some plants appear to take up compounds in relation to their concentration in the soil, including metals (Reeves et al., 1981; Mroz, 2008). In many cases of elemental defenses, a plant takes up an element to an extraordinary concentration: plants that do this are called hyperaccumulators (Reeves and Baker, 2000).

Serpentines have been an important focus of elemental defense studies because the majority of hyperaccumulator species grow on serpentine soils. More than 400 hyperaccumulators of various elements have been identified, but most (75%) hyperaccumulate Ni (Reeves and Baker, 2000), and most Ni hyperaccumulators (85–90%) are endemic to serpentine soils (Reeves and Adigüzel, 2008). Tests of elemental defense (reviewed by Boyd, 2007) have reported defensive effects (in some but not all cases) for a number of elements (As, Cd, Ni, Se, and Zn), most of which are heavy metals. Again, the most explored element is Ni, and serpentine Ni hyperaccumulators have been the focus of those experimental explorations. Boyd (2007) pointed out that there are many questions about elemental defenses in need of experimental investigation, and serpentine metal hyperaccumulators and their habitats can feature prominently in future work. For example, very few experiments regarding elemental defenses have been field-based. Also, more exploration of the lower limits of elemental defenses is needed: metal levels below the hyperaccumulator level may be defensively effective against some plant natural enemies (e.g., Coleman et al., 2005), suggesting that elemental defenses may be more widespread among serpentine plants than is currently recognized. In addition, as mentioned earlier, Ni may also serve to increase plant tolerance to herbivore damage in Ni-adapted plants (Palomino et al., 2007).

Hyperaccumulator plants also have the potential to become model systems for experimental exploration of elemental defense. Reverse genetics, the use of gene silencing techniques to produce a specific phenotype, has been of great value to physiologists who study gene functions, but the technique has been used much less by ecologists/evolutionary biologists (Schwachtje et al., 2008). Recent uses of reverse genetics to study organic plant defense chemicals (e.g., Steppuhn and Baldwin, 2007) have shown how natural enemies respond to phenotypes that lack particular defensive chemicals. These organic plant chemical studies provide models for experimental investigations of elemental defenses, if gene silencing can be used to convert a hyperaccumulator phenotype into a nonhyperaccumulator one. Past tests of elemental defenses often have manipulated soil metal concentration to manipulate metal accumulation level in plants, and this has been recognized as an advantage for study of elemental plant defenses (Boyd, 1998a; Pollard, 2000). Boyd (2007) pointed out that this technique has experimental limitations because it changes multiple plant qualities and therefore does not isolate plant metal concentration as a sole experimental factor. Fortunately, we are gaining considerable knowledge of the molecular mechanisms underlying hyperaccumulation (e.g., Verbruggen et al., 2009), so reverse genetic approaches soon should be available for experimental systems that include elemental plant defenses.

Because of their unusual response to certain elements, there has been considerable interest in hyperaccumulator species (e.g., Brooks, 1998), including how hyperaccumulators evolved. Elemental defense may have been a selective force underlying the evolution of metal hyperaccumulation in at least some cases. This connection between elemental defense and hyperaccumulation evolution, called the defensive enhancement hypothesis by Boyd (2007), suggests that relatively low levels of metal accumulation may have provided an initial defensive benefit for a plant species against some natural enemies. From this initial level of metal and its defensive benefit, increased metal uptake and sequestration ability may have magnified the defensive benefit and provided an evolutionary pathway that eventually culminated in hyperaccumulation ability.

The elemental defense hypothesis is also associated with another related hypothesis: the joint effects hypothesis (Boyd, 2007). Investigators of plant chemical defenses are realizing that mixtures of defenses have important effects that are not revealed in studies of individual compounds. In general, combinations of potentially toxic materials may act additively, synergistically, or antagonistically. For example, additive effects occur when one chemical's effects are independent of another's. Jhee et al. (2006) reported the effectiveness of combinations of some elemental defenses toward an herbivorous moth (*Plutella xylostella*) using artificial diet amended with metals. When Zn was combined pairwise with Cd, Ni, or Pb, toxic effects were additive, and the combination treatments caused greater mortality than single metals. Synergy occurs when chemicals interact in a way that increases their joint toxicity beyond that expected if their effects were additive. As an example, Jensen and colleagues (2006) reported synergy of methylmercury and selenate in a study of effects of these chemicals on an insect detritivore (the fly *Megaselia scalaris*). Mixtures of as little as the LC5 (lethal concentration 5%) of both chemicals resulted in 100% larval mortality. Finally, antagonism occurs when effects of one chemical reduce the negative impact of another. The joint effects hypothesis (Boyd, 2007) suggests that mixtures of multiple elemental defenses or elemental and organic defenses may exhibit either additive or synergistic effects, so a relatively low concentration of a metal may have a defensive effect. If this is correct, then the least amount of metal that begins to provide a defensive effect may be relatively small, allowing defensive enhancement to begin to occur at low levels of metal accumulation in plant tissues. This phenomenon could have provided selective support for the evolution of metal accumulation and eventually hyperaccumulation in plants.

Elemental defenses may not be limited to plants. Elements (including heavy metals) found at relatively high concentrations in herbivores (or higher trophic levels) may have defensive benefits (Boyd and Martens, 1998) but have rarely been investigated. Nickel-hyperaccumulating plants in serpentine ecosystems can be fed on by specialized herbivores that themselves have high whole-body Ni levels (Boyd, 2009). If high Ni concentrations have defensive effects, these "high-Ni insects" (Boyd, 2009), defined as those containing >500 mg Ni/kg (on a whole-body, dry mass basis), would be good choices for experimental tests of elemental defense against predators or pathogens. An initial investigation by Boyd and Wall (2001), using the high-Ni mirid heteropteran *Melanotrichus boydi* (Figure 8.1), found that mortality of crab spiders (*Misumena vatia*) was significantly greater when the spiders were fed the high-Ni insect compared to those fed low-Ni prey, providing some evidence for a defensive effect. No significant mortality effect was found for similar experiments using two other predators (Boyd and Wall, 2001) or for experiments (Boyd, 2002) testing the effectiveness of several pathogens against *Melanotrichus boydi* compared to a low-Ni mirid (*Lygus hesperus*). As a consequence of these mixed results, more investigation is needed to determine whether elemental defenses are effective for high-Ni insects.

Surveys of serpentine insect faunas are rare, so it is likely that many more high-Ni insects await discovery. A recent review (Boyd, 2009) reported that 14 high-Ni insect species have been discovered to date. Most of these (~75%) are true bugs (Heteroptera), such as *M. boydi* illustrated in Figure 8.1, along with a few

FIGURE 8.1. *Melanotrichus boydi.*

beetles (Coleoptera) and one grasshopper (Orthoptera). The maximum whole-body Ni concentration reported to date is 3500 µg Ni/g (Boyd et al., 2006) for the South African grasshopper (*Stenoscepa* sp. nov.). All of these insects are herbivores collected from Ni-hyperaccumulating species, with the exception of one (*Rhinocoris neavei*). This latter insect is thought to include in its diet a high-Ni insect (the beetle *Chrysolina pardalina*, which feeds on the Ni hyperaccumulator *Berkheya coddii*) and thus obtains its Ni from the plant indirectly (Migula et al., 2007). This newly discovered category of insects provides a unique opportunity to study metal tolerance mechanisms in obviously highly metal-tolerant species. A few such studies have begun (e.g., Przybylowicz et al., 2003; Augustyniak et al., 2008), but the potential for discovery of innovative adaptations in these insects is great.

Plant defense compounds may affect cross-kingdom relationships other than those between plants and herbivores. This is an important area of ecological research regarding organic defense compounds. For example, plant organic defenses may negatively affect detritivores or decomposers that feed on senescent plant matter, leading to impacts on community and ecosystem ecology. Elemental defenses of serpentine plants may have similar effects on cross-kingdom interactions, but these are little studied. Boyd and Martens (1998) suggested that metal hyperaccumulation could reduce the activity of detritivores and decomposers, basing this suggestion on studies of organic defense chemicals that decrease activity of these organisms. A few pioneering studies show that these effects of metals may occur and thus deserve further exploration. For example, Gonçalves et al. (2007) showed that the isopod detritivore *Porcellio dilatatus* had greater mortality and consumed less leaf material when fed leaves of a Ni-hyperaccumulating species compared to isopods fed leaves of three nonhyperaccumulator species. Boyd et al. (2008) conducted a field study on decomposition rate of leaves of the South African plant species *Senecio coronatus*. They took advantage of an unusual feature of this species: some serpentine populations of this plant hyperaccumulate Ni, whereas others do not. Thus, they could compare decomposition rates of high- and low-Ni field-collected leaves, placing samples in decomposition bags in the field. To their surprise, they found that only very high Ni levels affected leaf decomposition rate. Leaves from a population containing 15,000 mg Ni/kg decomposed more slowly than nonhyperaccumulator leaves, but leaves from a population containing 9200 mg Ni/kg did not. Thus, their study indicated that only extremely high Ni levels—more than an order of magnitude greater than the minimum level used to define Ni hyperaccumulation (1000 mg Ni/kg)—impacted decomposition.

MUTUALISTS AND EDAPHIC SPECIALIZATION

Animals and plants interact in the largely mutualistic interactions of biotic pollination and biotic seed dispersal. These interactions, especially biotic pollination, have been important to the evolution and diversification of plants. An important factor that can influence pollination and seed dispersal success is fragmentation of a plant's habitat (Harris and Johnson, 2004). Human-caused habitat fragmentation is a major concern of conservation biology and an important threat to planetary biodiversity (Fisher and Lindenmayer, 2007). Experimental studies of the effects of habitat fragmentation on pollination and seed dispersal are numerous, but the habitat patchiness of many serpentine areas provides an ideal system in which to study various aspects of habitat fragmentation. Serpentine systems lend themselves to studies of effects of habitat patch size (or a potential correlate, population size), distance from other patches, and so on, on pollination/seed dispersal (see Chapter 13 in this volume for pollination review). For example, Wolf (2001) used California serpentine sites to study effects of outcrop size on cross-kingdom species interactions using the perennial herb *Calystegia collina*. She reported that pollinators and herbivores were equally active on both large and small outcrops, but reproductive output on small outcrops was limited by availability of compatible pollen (the species is self-incompatible). The reduced seed output of small outcrop populations did not decrease their genetic diversity compared to large outcrop populations, however, because the species reproduces vegetatively.

In terms of seed dispersal, the patchiness of serpentine habitats provides a model system useful for testing ecological questions about the role of spatial patch dynamics in population biology. For example, habitat patchiness can influence population viability (Turner, 2005) as well as the roles of local adaptation and phenotypic plasticity in population evolution (Sultan and Spencer, 2002). These questions can be particularly important when trying to understand species invasions (see Chapter 9 in this volume), including invasions into serpentine habitats, which are often rich in endemic species and thus are of heightened conservation interest (Harrison et al., 2001). Thomson (2007) illustrated the utility of serpentine habitats for invasive species studies by examining the performance of *Aegilops triuncialis* (goatgrass) at the edge of extreme serpentine habitats (relatively rocky and barren sites) and in adjacent serpentine grassland habitats. Her comparison of the species in these two subdivisions of the serpentine habitat showed greater dispersal ability of the invader in the extreme edge areas, as well as differences in some other population biology parameters. Although probably wind-dispersed on her study areas, she noted that sites with animal dispersers (the species is also dispersed via ectozoochory) would probably have more long-distance dispersal events, and this feature also would impact invasion by this species.

Serpentine habitats also present opportunities to study conflicts between plant chemical defenses and plant mutualists. In many cases of pollen/seed dispersal, plants offer nutritional rewards (pollen, nectar, fruit pulp) to mutualist animal dispersers. If rewards offered by hyperaccumulator plants are high in metals (elemental defenses), the metals may affect the suitability of the reward and thus the interaction. For example, secondary chemicals in nectar can negatively affect pollinators, and there has been a great deal of interest in understanding why the nectar of some plant species is toxic to certain floral visitors (e.g., Adler, 2001). Hypotheses for functions of toxic nectar include protection against nectar robbers, protection against microbial decomposers, and effects on pollinator behavior, although Adler (2001) points out that toxic nectar may have no function or even be detrimental to a plant. If nectars contain elemental defenses, they may have similar important ecological consequences for both plants and their mutualists. Similarly, secondary chemicals can be found in fruits, and these can affect both mutualists and antagonists. For example, chilies produce capsaicinoids in their fruits, and these influence whether the fruits are eaten by mammals (which can taste capsaicinoids and act as seed predators) or birds (which cannot taste them and act as seed dispersers) and thus influence seed dispersal (Tewksbury and Nabhan, 2001). Capsaicinoids also provide a protective effect against fungal attack of seeds (Tewksbury et al., 2008), illustrating the complex effects of defensive chemicals on cross-kingdom interactions.

Unfortunately, little information is available regarding the concentrations of metals in nectar or pollen produced by hyperaccumulators; to our knowledge there are no experimental studies of the impacts of these rewards on mutualists. Nectar and pollen are often produced in small quantities and thus pose technical difficulties for metal quantification (Boyd, 1998b). But some studies have compared whole-body Ni levels of floral visitors collected from hyperaccumulator versus non-hyperaccumulator flowers. For example, Boyd et al. (2006) compared Ni levels of honeybees (*Apis mellifera*) collected from *Hybanthus austrocaledonicus* (a Ni hyperaccumulator) or *Myodocarpus fraxinifolius* (a nonhyperaccumulator) flowers at a serpentine site in New Caledonia, finding eightfold more Ni in bees from the hyperaccumulator flowers. A study from California compared Ni levels in two bee species collected from flowers of a Ni hyperaccumulator on a serpentine site versus those collected from a nonhyperaccumulator on a nearby nonserpentine site (Wall and Boyd, 2002). The bees collected from hyperaccumulator flowers contained about threefold more Ni, again suggesting that floral rewards of Ni hyperaccumulator plants contain more Ni than those from other species.

Fruit pulp may also be a reward for mutualists, in this case seed dispersers, and there is some information regarding metal concentrations of those tissues that are produced by metal hyperaccumulators. Fruit pulp of some fleshy-fruited Ni hyperaccumulators is relatively high in Ni. For example, the New Caledonian tree *Sebertia acuminata* produces fruits with as much as 6900 mg Ni/kg in their pulp (Boyd et al., 2006). To our knowledge, effects of such high-element tissues on mutualist seed dispersers have not yet been explored.

Hypotheses about the impact of elemental defenses on mutualists can be tested in two ways: directly and indirectly. Direct tests of the toxicity of hyperaccumulator plant rewards include testing them on generalist pollinators or seed dispersers, but to our knowledge these have not yet been attempted. Another approach, albeit an indirect one, is to use phylogenetically controlled independent contrasts (Felsenstein, 1985; Garland et al., 1992) to investigate whether hyperaccumulators are less likely than nonhyperaccumulator serpentine plants to be pollinated or have their seeds dispersed via animal mutualists. Phylogenies that include hyperaccumulator species are being developed (e.g., Jansen et al., 2002), but ecological studies of seed dispersal of serpentine soil plant species are few. Progress in both of these areas will be required to apply this approach to these questions.

Finally, because many serpentine habitats are isolated and favor traits that are markedly divergent from those adaptive in surrounding habitats (see Chapters 4–6 in this volume), selection for reduced seed dispersal and/or vegetative reproduction could accompany serpentine adaptation. Reduced seed dispersal could also bypass conflicting selection between fruit defense and dispersal mutualists. To support this hypothesis, *Crepis sancta* (Asteraceae) grows in fragmented urban habitats, many of which are surrounded by concrete; it also grows in unfragmented larger weedy areas. There has been strong selection for and rapid evolution of increased investment in pappus-free seeds relative to wind-dispersed seeds with pappus when *C. sancta* grows in urban concrete islands relative to unfragmented areas (Cheptou et al., 2009). Again, a comparative approach using phylogenetic independent contrasts could evaluate whether there has been a shift toward decreased dispersal distance of seeds or increased vegetative reproduction on serpentine or other edaphically isolated taxa by comparing related taxa growing on and off serpentine and changes in reproductive traits.

FUNGAL MUTUALISTS AND PATHOGENS AND EDAPHIC SPECIALIZATION

Because serpentine soils are low in phosphorus and often dry, it is possible that arbuscular mycorrhizal fungal (AMF) root associates—fungi that help plants procure phosphorus, resist drought, and play other key roles—could be important mutualists enabling plant growth on serpentine (Batten et al., 2006; Casper et al., 2008; Doherty et al., 2008; Schechter and Bruns, 2008). *Collinsia sparsiflora* ecotypes associate with distinct AMF assemblages across a serpentine gradient—serpentine ecotypes associate with AMF in the genus *Acaulospora*, whereas nonserpentine ecotypes associate more with *Glomus* (Schechter and Bruns, 2008). Richness and evenness of AMF communities were also significantly higher in serpentine assemblages. Because AMF assemblages were also influenced by soil composition, it is not clear whether serpentine soils directly affect AMF communities or whether plants in serpentine selectively associate with *Acaulospora* in serpentine habitats. In Pennsylvania, there are also differences in the AMF communities with which the grass *Sorghastrum nutans* associates in serpentine and nonserpentine areas. In these cases, serpentine AMF do not seem to confer benefits to plant hosts (Doherty et al., 2008). Plants were inoculated with AMF inocula or whole-soil microbial communities from either prairie or serpentine soils in the presence of nickel. Nickel reduced AMF root colonization in both serpentine and nonserpentine grass ecotypes. In addition, serpentine AMF and whole microbial community treatments also decreased plant biomass relative to uninoculated plants, whereas AMF inocula from prairie soils had no effect on plant performance. AMF from serpentine did not improve Ni tolerance, nor did they improve plant performance. In general, more work needs to be done to understand whether AMF provide benefits in serpentine soils. Given that serpentine habitats can be drier, it will also be important in the future to test the possible advantages of AMF associates under drought conditions, where they may be important mutualists (Marulanda et al., 2009).

Even less information is available on the relationships between plants and their pathogens in serpentine soils. In early work, researchers observed that some serpentine taxa were more susceptible to attack by pathogenic fungi when planted on normal (nonmineralized) soil (Tadros, 1957; Brooks, 1987), although in some cases no differences were observed (e.g., Reeves and Baker, 1984). More recent investigations have shown that Ca metabolism is important in plant resistance to pathogens, and Ca additions to soil have resulted in lower levels of disease across a wide group of pathogens and their taxonomically diverse hosts. Because serpentine is characterized by low Ca:Mg ratios, serpentine taxa may be particularly vulnerable to pathogen attack. Using experimental Ca additions to serpentine soils, as well as surveys of field populations, Springer et al. (2007) and Springer (2009a) showed that the serpentine flax *Hesperolinon californicum* experienced greater levels of infection by the fungal rust *Melampsora lini* in serpentine, low-Ca environments relative to nonserpentine soils. Moreover, *H. californicum* tissue Ca levels were closely correlated with soil calcium availability. It therefore appears that low Ca concentrations of serpentine soils may compromise *H. californicum*'s ability to resist infection. A wider investigation of pathogen pressure experienced by *Hesperolinon* species associated with soils of varying serpentine qualities (Springer, 2009b) revealed that rust disease was less frequent and less severe in more extreme (lower Ca) soils, a conclusion opposite to that from studies of *H. californicum*. Clearly, much remains to be learned regarding the influence of edaphic factors on plant disease in serpentine soils.

Pathogens are also a natural enemy that may have stimulated evolution of hyperaccumulation as an elemental defense on serpentine soils. For example, infection levels by some fungal and bacterial natural enemies are lessened when plants are protected by hyperaccumulated Ni or Zn (see review in Boyd, 2007). Freeman et al. (2005) reported an interesting connection between pathogen defense, Ni tolerance, and Ni hyperaccumulation in several species of *Thlaspi*. Among these species, Ni tolerance increases with salicylic acid level, and this in turn stimulates production of glutathione, which protects cells from oxidative stress. Salicylic acid is also an important signal molecule that initiates pathogen resistance responses in these species. In the Ni hyperaccumulator *Thlaspi goesingense*, a species with high salicylic acid levels and therefore high Ni tolerance, infection by powdery mildew fails to change salicylic acid levels and trigger protective responses to the pathogen. Thus, the species is hypersensitive to this pathogen when grown in low-Ni media. Freeman et al. (2005) concluded that the pathogen protective effect of Ni hyperaccumulation in *T. goesingense* replaces protection provided in other species by signaling pathways that use elevated levels of salicylic acid to stimulate other pathogen defense mechanisms.

CROSS-KINGDOM INTERACTIONS MAY BE KEY FEATURES OF EDAPHIC ADAPTATION

In conclusion, there is growing evidence that adaptation to harsh edaphic conditions is inextricably linked to plant interactions with other species from other kingdoms, ranging from microbes to animals. We find selection for increased defense against herbivores, reduced tolerance to leaf damage and changes in susceptibility to pathogens on serpentine and other edaphically stressful substrates. Moreover, substrate composition appears to directly affect traits through nutrient availability: through availability of metals or other elements that can be coopted into a defensive role, and through the lack of key elements important to immunofunction. Soil substrates create both stoichiometric challenges and opportunities for edaphic specialists. Divergent selective pressures across substrate types, coupled with constraints caused by stressful substrates, may ultimately drive speciation and may explain why the source of much biodiversity is associated with edaphic specialization (see Chapters 4, 5 in this volume).

The information presented herein also illustrates the usefulness of serpentine habitats for studies of speciation and adaptation in edaphically stressful situations. Important antagonistic and mutualistic cross-kingdom interactions, such as herbivory, biotic seed/pollen dispersal, and mycorrhiza formation, are influenced by the unique suite of habitat features that characterize serpentine environments. Thus, studies of serpentine communities and comparison of serpentine and non-serpentine species and communities can yield important insights into fundamental biological questions like the generation of biodiversity hot spots and the speciation process.

Acknowledgments

This work was supported by the Alabama Agricultural Experiment Station (RSB), the UC Davis CAES Agricultural Experiment Station (SYS), and by NSF DEB-0919559 (SYS).

LITERATURE CITED

Adler, L. S. (2001) The ecological significance of toxic nectar. *Oikos,* 91, 409–20.

Aide, T. M., and Zimmerman, J. K. (1990) Patterns of insect herbivory, growth, and survivorship in juveniles of a neotropical liana. *Ecology,* 71, 1412–21.

Augustyniak, M., Przybylowicz, W., Mesjasz-Przybylowicz, J., Migula, P., Tarnawska, M., Glowacka, E., and Babczynska, A. (2008) Nuclear microprobe studies of grasshopper feeding on nickel hyperaccumulating plants. *X-Ray Spectrometry,* 37, 142–45.

Basset, Y., Charles, E., Hammond, D. S., and Brown, V. K. (2001) Short-term effects of canopy openness on insect herbivores in a rain forest in Guyana. *Journal of Applied Ecology,* 3, 1045–58.

Batten, K. M., Scow, K. M., Davies, K. F., and Harrison, S. P. (2006) Two invasive plants alter soil microbial community composition in serpentine grasslands. *Biological Invasions,* 8, 217–30.

Baythavong, B. S., Stanton, M. L., and Rice, K. J. (2009) Understanding the consequences of seed dispersal in a heterogeneous environment. *Ecology,* 90, 2118–28.

Boyd, R. S. (1998a) Hyperaccumulation as a plant defensive strategy. In *Plants that Hyperaccumulate Heavy Metals* (ed. R. R. Brooks), pp. 181–201. CAB International, Oxford.

Boyd, R. S. (1998b) The significance of metal hyperaccumulation for biotic interactions. *Chemoecology*, 8, 1–7.

Boyd, R. S. (2002) Does elevated body Ni concentration protect insects against pathogens? A test using *Melanotrichus boydi* (Heteroptera: Miridae). *American Midland Naturalist*, 147, 225–36.

Boyd, R. S. (2007) The defense hypothesis of elemental hyperaccumulation: Status, challenges and new directions. *Plant and Soil*, 293, 153–76.

Boyd, R. S. (2009) High-nickel insects and nickel hyperaccumulator plants: A review. *Insect Science*, 16, 19–31.

Boyd, R. S., Davis, M. A., and Balkwill K. (2008) Does hyperaccumulated nickel affect leaf decomposition? A field test using *Senecio coronatus* (Asteraceae) in South Africa. *Chemoecology*, 18, 1–9.

Boyd, R. S., Davis, M. A., Wall, M. A., and Balkwill, K. (2006) Metal concentrations of insects associated with the South African Ni hyperaccumulator *Berkheya coddii* (Asteraceae). *Insect Science*, 13, 85–102.

Boyd, R. S., and Martens, S. N. (1998) The significance of metal hyperaccumulation for biotic interactions. *Chemoecology*, 8, 1–7.

Boyd, R. S., and Wall, M. A. (2001) Responses of generalist predators fed high-Ni *Melanotrichus boydi* (Heteroptera: Miridae): Elemental defense against the third trophic level. *American Midland Naturalist*, 146, 186–98.

Boyd, R. S., Wall, M. A., and Jaffré, T. (2006) Nickel levels in arthropods associated with Ni hyperaccumulator plants from an ultramafic site in New Caledonia. *Insect Science*, 13, 271–77.

Brooks, R. R. (1987) *Serpentine and Its Vegetation: A Multidisciplinary Approach*. Dioscorides Press, Portland.

Brooks, R. R., ed. (1998) *Plants that Hyperaccumulate Heavy Metals: Their Role in Phytoremediation, Microbiology, Archaeology, Mineral Exploration and Phytomining*. CAB International, Wallingford.

Campbell, S. A., and Borden, J. H. (2006) Integration of visual and olfactory cues of hosts and non-hosts by three bark beetles (Coleoptera: Scolytidae). *Ecological Entomology*, 31, 437–49.

Casper, B. B., Bentivenga, S. P., Ji, B. M., Doherty, J. H., Edenborn, H. M., and Gustafson, D. J. (2008) Plant-soil feedback: Testing the generality with the same grasses in serpentine and prairie soils. *Ecology*, 89, 2154–64.

Cheptou, P.-O., Carrue, O., Rouifed, S., and Cantarel, A. (2008) Rapid evolution of seed dispersal in an urban environment in the weed *Crepis sancta*. *Proceedings of the National Academy of Sciences, USA*, 105, 3796–99.

Coleman, C. M., Boyd, R. S., and Eubanks, M. D. (2005) Extending the elemental defense hypothesis: Dietary metal concentrations below hyperaccumulator levels could harm herbivores. *Journal of Chemical Ecology*, 31, 1669–81.

Coley, P. D. (1983) Herbivory and defensive characteristics of tree species in a lowland tropical forest. *Ecological Monographs*, 53, 209–33.

Coley, P. D., Bryant, J. P., and Chapin, F. S. (1985) Resource availability and plant antiherbivore defense. *Science*, 230, 895–99.

Couty, A., Van Emden, H., Perry, J. N., Hardie, J., Pickett, J. A., and Wadhams, L. J. (2006) The roles of olfaction and vision in host-plant finding by the diamondback moth, *Plutella xylostella*. *Physiological Entomology*, 31, 134–45.

Dearing, M. D., Foley, W. J., and McLean, S. (2005) The influence of plant secondary metabolites on the nutritional ecology of herbivorous terrestrial vertebrates. *Annual Review of Ecology, Evolution and Systematics*, 36, 169–89.

Doherty, J. H., Ji, B. M., and Casper, B. B. (2008) Testing nickel tolerance of *Sorghastrum nutans* and its associated soil microbial community from serpentine and prairie soils. *Environmental Pollution*, 151, 593–98.

Doring, T. F., Kirchner, S. M., Kuhne, S., and Saucke, H. (2004) Response of alate aphids to green targets on coloured backgrounds. *Entomologia Experimentalis et Applicata*, 113, 53–61.

Ellis, A. G., and Weis, A. E. (2006) Coexistence and differentiation of "flowering stones": The role of local adaptation to soil microenvironment. *Journal of Ecology*, 94, 322–35.

Fadzly, N., Jack, C., Schaefer, H. M., and Burns, K. C. (2009) Ontogenetic colour changes in an insular tree species: Signalling to extinct browsing birds? *New Phytologist*, 184, 495–501.

Felsenstein, J. (1985) Phylogenies and the comparative method. *American Naturalist*, 125, 1–15.

Finch, S., and Collier, R. H. (2000) Host-plant selection by insects—a theory based on "appropriate/inappropriate landings" by pest insects of cruciferous plants. *Entomologia Experimentalis et Applicata*, 96, 91–102.

Fine, P. V. A., Daly, D. C., Munoz, G. V., Mesones, I., and Cameron, K. M. (2005) The contribution of edaphic heterogeneity to the evolution and diversity of Burseraceae trees in the western Amazon. *Evolution*, 59, 1464–78.

Fisher, J., and Lindenmayer, D. B. (2007) Landscape modification and habitat fragmentation: A synthesis. *Global Ecology and Biogeography*, 16, 265–80.

Freeman, J. L., Garcia, D., Kim, D., Hopf, A., and Salt, D. E. (2005) Constitutively elevated salicylic acid signals glutathione-mediated nickel tolerance in *Thlaspi* nickel hyperaccumulators. *Plant Physiology*, 137, 1082–91.

Garland, T., Harvey, P. H., and Ives, A. R. (1992) Procedures for the analysis of comparative data using phylogenetically independent contrasts. *Systematic Biology*, 41, 18–32.

Gonçalves, M. T., Gonçalves, S. C., Portugal, A., Silva, S., Sousa, J. P., and Freitas, H. (2007) Effects of nickel hyperaccumulation in *Alyssum pintodasilvae* on model arthropods representatives of two trophic levels. *Plant and Soil*, 293, 177–88.

Harris, L. F., and Johnson, S. D. (2004) The consequences of habitat fragmentation for plant-pollinator mutualisms. *International Journal of Tropical Insect Science*, 24, 29–43.

Harrison, S., Rice, K., and Maron, J. (2001) Habitat patchiness promotes invasion by alien grasses on serpentine soil. *Biological Conservation*, 100, 45–53.

Hay, M. E. (1996) Marine chemical ecology: What's known and what's next? *Journal of Experimental Marine Biology and Ecology*, 200, 103–34.

Hirota, T., and Kato, Y. (2001) Influence of visual stimuli on host location in the butterfly, *Eurema hecabe*. *Entomologia Experimentalis et Applicata*, 101, 199–206.

Jansen, S., Broadley, M. R., Robbrecht, E., and Smets E. (2002) Aluminum hyperaccumulation in angiosperms: A review of its phylogenetic significance. *Botanical Review*, 68, 235–69.

Jensen, P. D., Johnson, L. R., and Trumble, J. T. (2006) Individual and joint actions of selenate and methylmercury on the development and survival of insect detritivore *Megaselia scalaris* (Diptera: Phoridae). *Archives of Environmental Contamination and Toxicology*, 50, 523–30.

Jhee, E. M., Boyd, R. S., and Eubanks, M. D. (2006) Effectiveness of metal-metal and metal-organic compound combinations against *Plutella xylostella*: Implications for plant elemental defense. *Journal of Chemical Ecology*, 32, 239–59.

Karban, R., and Courtney, S. (1987) Intraspecific host plant choice: Lack of consequences for *Streptanthus tortuosus* (Cruciferae) and *Euchloe hyantis* (Lepidoptera, Pieridae). *Oikos*, 48, 243–48.

Kostal, V., and Finch, S. (1994) Influence of background on host-plant selection and subsequent oviposition by the cabbage root fly (*Delia radicum*). *Entomologia Experimentalis et Applicata*, 70, 153–63.

Lau, J. A., McCall, A. C., Davies, K. F., McKay, J. K., and Wright, J. W. (2008) Herbivores and edaphic factors constrain the realized niche of a native plant. *Ecology*, 89, 754–62.

Lev-Yadun, S. 2006. Defensive functions of white coloration in coastal and dune plants. *Israel Journal of Plant Sciences*, 54, 317–25.

Lev-Yadun, S., Dafni, A., Flaishman, M. A., Inbar, M., Izhaki, I., Katzir, G., and Ne'eman, G. (2004) Plant coloration undermines herbivorous insect camouflage. *Bioessays*, 26, 1126–30.

Louda, S. M., and Rodman, J. E. (1996) Insect herbivory as a major factor in the shade distribution of a native crucifer (*Cardamine cordifolia* A. Gray, bittercress). *Journal of Ecology*, 84, 229–37.

Mackay, D. A., and Jones, R. E. (1989) Leaf shape and the host-finding behaviour of two ovipositing monophagous butterfly species. *Ecological Entomology*, 14, 423–31.

Maestre, F. T., Callaway, R. M., Valladares, F., and Lortie, C. J. (2009) Refining the stress-gradient hypothesis for competition and facilitation in plant communities. *Journal of Ecology*, 97, 199–205.

Martens, S. N., and Boyd, R. S. (1994) The ecological significance of nickel hyperaccumulation: A plant chemical defense. *Oecologia*, 98, 379–84.

Marulanda, A., Barea, J. M., and Azcon, R. (2009) Stimulation of plant growth and drought tolerance by native microorganisms (AM fungi and bacteria) from dry environments: Mechanisms related to bacterial effectiveness. *Journal of Plant Growth Regulation*, 28, 115–24.

Migula, P., Przybylowicz, W. J., Mesjasz-Przybylowicz, J., Augustyniak, M., Nakonieczny, M., Glowacka, E., and Tarnawska, M. (2007) Micro-PIXE studies of elemental distribution in sap-feeding insects associated with Ni hyperaccumulator, *Berkheya coddii*. *Plant and Soil*, 293, 197–207.

Moore, L. V., Myers, J. H., and Eng, R. (1988) Western tent caterpillars prefer the sunny side of the tree, but why? *Oikos*, 51, 321–26.

Mroz, L. (2008) Between-population variation in plant performance traits and elemental composition of *Colchicum autumnale* L. and its relation to edaphic environments. *Acta Societatis Botanicorum Poloniae*, 77, 229–39.

Niesenbaum, R. A., and Kluger, E. C. (2006) When studying the effects of light on herbivory, should one consider temperature? The case of *Epimecis hortaria* F. (Lepidoptera: Geometridae) feeding on *Lindera benzoin* L. (Lauraceae). *Environmental Entomology*, 35, 600–606.

O'Dell, R. E., James, J. J., and Richards, J. H. (2006) Congeneric serpentine and nonserpentine shrubs differ more in leaf Ca:Mg than in tolerance of low N, low P, or heavy metals. *Plant and Soil*, 280, 49–64.

Osier, T. L., and Jennings, S. M. (2007) Variability in host-plant quality for the larvae of a polyphagous insect folivore in midseason: The impact of light on three deciduous sapling species. *Entomologia Experimentalis et Applicata*, 123, 159–66.

Palomino, M., Kennedy, P. G., and Simms, E. L. (2007) Nickel hyperaccumulation as an anti-herbivore trait: Considering the role of tolerance to damage. *Plant and Soil*, 293, 189–95.

Pollard, A. J. (2000) Metal hyperaccumulation: A model system for coevolutionary studies. *New Phytologist*, 146, 179–81.

Proctor, J., and Woodell, S. R. J. (1975) The ecology of serpentine soils. *Advances in Ecological Research*, 9, 356–66.

Przybylowicz, W. J., Mesjasz-Przybylowicz, J., Migula, P., Glowacka, E., Nakonieczny, M., and Augustyniak, M. (2003) Functional analysis of metals distribution in organs of the beetle *Chrysolina pardalina* exposed to excess of nickel by Micro-PIXE. *Nuclear Instruments and Methods in Physics Research B*, 210, 343–48.

Rausher, M. D. (1978) Search image for leaf shape in a butterfly. *Science*, 200, 1071–73.

Reeves, R. D., and Adigüzel, N. (2008) The nickel hyperaccumulating plants of Turkey and adjacent areas: A review with new data. *Turkish Journal of Biology*, 32, 143–53.

Reeves, R. D., and Baker, A. J. M. (1984) Studies on metal uptake by plants from serpentine and non-serpentine populations of *Thlaspi goesingense* Halacsy (Cruciferae). *New Phytologist*, 98, 191–204.

Reeves, R. D., and Baker, A. J. M. (2000) Metal-accumulating plants. In *Phytoremediation of Toxic Metals: Using Plants to Clean Up the Environment* (eds. I. Raskin and B. D. Ensley), pp. 193–229. Wiley, New York.

Reeves, R. D., Brooks R. R., and Macfarlane, R. M. (1981) Nickel uptake by Californian *Streptanthus* and *Caulanthus* with particular reference to the hyperaccumulator *S. polygaloides* Gray (Brassicaceae). *American Journal of Botany*, 68, 708–12.

Root, R. B. (1973) Organization of plant-arthropod association in simple and diverse habitats: fauna of collards (*Brassica oleracea*). *Ecological Monographs*, 43, 95–120.

Sambatti, J. B. M., and Rice, K. J. (2007) Functional ecology of ecotypic differentiation in the Californian serpentine sunflower (*Helianthus exilis*). *New Phytologist*, 175, 107–19.

Schechter, S. P., and Bruns, T. D. (2008) Serpentine and non-serpentine ecotypes of *Collinsia sparsiflora* associate with distinct arbuscular mycorrhizal fungal assemblages. *Molecular Ecology*, 17, 3198–210.

Schwachtje, J., Kutschbach, S., and Baldwin, I. T. (2008) Reverse genetics in ecological research. PLoS ONE 3(2): e1543. DOI: 10.1371/journal.pone.0001543.

Shapiro, A. M. (1981) Egg-mimics of *Strepthanthus* (Cruciferae) deter oviposition by *Pieris sisymbrii* (Lepidoptera: Pieridae). *Oecologia*, 48, 142–43.

Sherman, P. W., and Watt, W. B. (1973) Thermal ecology of some *Colias* butterfly larvae. *Journal of Comparative Physiology*, 83, 25–40.

Springer, Y. P. (2009a) Edaphic quality and plant-pathogen interactions: Effects of soil calcium on fungal infection of a serpentine flax. *Ecology*, 90, 1852–62.

Springer, Y. P. (2009b) Do extreme environments provide a refuge from pathogens? A phylogenetic test using serpentine flax. *American Journal of Botany*, 96, 2010–21.

Springer, Y. P., Hardcastle, B. A., and Gilbert, G. S. (2007) Soil calcium and plant disease in serpentine ecosystems: A test of the pathogen refuge hypothesis. *Oecologia*, 151, 10–21.

Stenberg, J. A., and Ericson, L. (2007) Visual cues override olfactory cues in the host-finding process of the monophagous leaf beetle *Altica engstroemi*. *Entomologia Experimentalis et Applicata*, 125, 81–88.

Steppuhn, A., and Baldwin, I. T. (2007) Resistance management in a native plant: Nicotine prevents herbivores from compensating for plant protease inhibitors. *Ecology Letters*, 10, 499–511.

Strauss, S. Y., and Agrawal, A. A. (1999) The ecology and evolution of plant tolerance to herbivory. *Trends in Ecology and Evolution*, 14, 179–85.

Sultan, S. E., and Spencer, H. G. (2002) Metapopulation structure favors plasticity over local adaptation. *American Naturalist*, 160, 271–83.

Tadros, T. M. (1957) Evidence of the presence of an edapho-biotic factor in the problem of serpentine tolerance. *Ecology*, 38, 14–23.

Tewksbury, J. J., and Nabhan, G. P. 2001. Seed dispersal: Directed deterrence by capsaicin in chillies. *Nature*, 412, 403–4.

Tewksbury, J. J., Reagan, K. M., Machnicki, N. J., Carlo, T. A., Haak, D. C., Peñaloza, A. L. C., and Levey, D. J. (2008) Evolutionary ecology of pungency in wild chilies. *Proceedings of the National Academy of Sciences, USA*, 105, 11808–11.

Thomson, D. M. (2007) Do source-sink dynamics promote the spread of an invasive grass into a novel habitat? *Ecology*, 88, 3126–34.

Turitzin, S. N. (1982) Nutrient limitations to plant growth in a California serpentine grassland. *American Midland Naturalist*, 107, 95–99.

Turner, M. G. (2005) Landscape ecology: What is the state of the science? *Annual Review of Ecology, Evolution and Systematics*, 36, 319–44.

Verbruggen, N., Hermans, C., and Schat, H. (2009) Molecular mechanisms of metal hyperaccumulation in plants. *New Phytologist*, 181, 759–76.

Wall, M. A., and Boyd, R. S. (2002) Nickel accumulation in serpentine arthropods from the Red Hills, California. *Pan-Pacific Entomologist*, 78, 168–76.

Wolf, A. (2001) Conservation of endemic plants in serpentine habitats. *Biological Conservation*, 100, 35–44.

9

Invasions and the Evolution of Range Limits

Diane M. Thomson, *Claremont Colleges*
Brooke S. Baythavong and Kevin J. Rice,
University of California, Davis

Understanding how species are sometimes able to expand their range boundaries into a novel environment is key to managing biological invasions, a major component of global change and threat to biodiversity (Hastings et al., 2005). Paradoxically, although invasions are a critical conservation problem, they also can potentially contribute to the development of solutions by creating opportunities to study how and why range limits form or change. Ecology and evolution are rich in theory and models explaining the development and maintenance of range limits, so there is no lack of hypotheses about these processes (Lenormand, 2002; Kawecki, 2008; Gaston, 2009). However, testing range limits models in the field has proven to be very difficult, and there is a real imbalance between the number of ideas emerging from theory and the amount of data available to evaluate them (Sexton et al., 2009). Invasions provide some of the best chances to study shifts in range boundaries as they happen and improve our fundamental understanding of the processes that control species distributions (Sexton et al., 2009). As a result, research on invasions can also contribute to better predictions of how species ranges might respond to other aspects of global change, such as shifts in climate (see Chapter 17 in this volume).

In this chapter, we argue that serpentine habitats, particularly invasive plants in serpentine, are extremely useful systems for testing hypotheses about evolution and spatial spread at species range boundaries. First, we briefly outline the broad

Serpentine: The Evolution and Ecology of a Model System, edited by Susan Harrison and Nishanta Rajakaruna. Copyright © by The Regents of the University of California. All rights of reproduction in any form reserved.

framework shared by many theoretical models of range limits and some of the major predictions they generate. Then we discuss the challenges inherent in testing these predictions on the ground. With this background in mind, we describe how past and ongoing research in serpentine has addressed some of those challenges and sum up what this work tells us about range limits. Finally, we suggest some potential directions for new work.

MODELING RANGE LIMITS

A variety of model structures have been used to develop ideas about range limits and local adaptation; nevertheless these diverse approaches share the same basic underlying framework (e.g., Kirkpatrick and Barton, 1997; Bridle and Vines, 2007; Kawecki, 2008). In the absence of a physical barrier, distribution boundaries can be explained by a change in net population growth rates from zero or positive near the center of species ranges to negative near the margin. These spatial differences in population growth presumably correspond with a gradient in mean individual fitness, creating selection at the margin. This raises the question of why populations at the margin do not adapt to local conditions over time, enabling further spread. As a result, the challenge is seemingly to understand why species are not more often able to overcome distributional limitations through evolutionary change (Bridle and Vines, 2007).

Theoretical explanations of this tendency toward "niche conservatism" (Holt and Gaines, 1992) hinge on the fact that dispersal and gene flow can both promote and oppose adaptation at the range edge. First, populations at range margins where growth rates are lower may be smaller and more fragmented, and this can lead to asymmetric gene flow, with more individuals moving from core habitats into the margin than vice versa. If dispersal rates are too high, this creates genetic swamping that works against local adaptation to the marginal environment even in the presence of strong selection (Kawecki, 2008; Gaston, 2009). Many theoretical studies model such dynamics by simplifying continuous spatial variation along the gradient between range centers and margins into two discrete categories of habitat patches: source, or "core," near the range center, and sink, or "margin" at the distribution edge (Holt and Gaines, 1992; Kawecki 1995, 2000; Kawecki and Holt, 2002; Sultan and Spencer, 2002). In this framework, high dispersal rates from core to marginal patches prevent local adaptation at the range boundary.

However, low dispersal can also act against adaptation for both demographic and genetic reasons. With limited immigration, populations at the margin may become extinct due to low growth rates or Allee effects before adaptation occurs. In addition, genetic drift in small, marginal populations could lead to a lack of variation and low evolutionary potential (Garant et al., 2007; but see Goodnight, 1988, for a contrasting view on the potential for founder events to increase additive genetic variance because of epistasis). As a result of this balancing act between the positive and negative effects of migration and gene flow on adaptation, some have argued that we should expect evolution at range boundaries to occur relatively rarely, under a narrow range of dispersal rates (Holt and Gaines, 1992).

In addition to migration and gene flow, phenotypic plasticity is an important part of the picture emerging from this theoretical work. Interestingly, some models suggest that high rates of migration may work to favor plastic phenotypes rather than locally adapted ones, because greater dispersal raises the probability that individuals will encounter different selective environments than their parents (Sultan and Spencer, 2002). Alternatively, some have argued that plasticity may increase evolutionary potential by allowing species to colonize new environments more effectively and maintain populations at the margin with limited immigration (Crispo, 2008). These differences in theory have not yet resolved into clear predictions about how plasticity might influence local adaptation or range margins (Ghalambor et al., 2007).

TAKING MODELS TO THE FIELD

In contrast to the abundance of theoretical work summarized in the previous section, empirical field tests of these models are relatively scarce (Diffendorfer, 1998; Gaston, 2009). This is not surprising, given the challenges involved in putting all the important pieces together: an understanding of how selection varies spatially with habitat heterogeneity, estimates of migration rates and their effects on population demography and gene flow, and information on plasticity in traits that may be under selection. Translating the inevitably simplified structure of a model to more realistic field conditions is also problematic in some basic ways. For example, the majority of range limit models assume either a patchy spatial structure (core versus marginal) or a gradual selection gradient. In real populations, spatial heterogeneity is generally much more complicated. For some species, even definitively identifying which populations in a system of patches have higher fitness and what environmental gradient drives those differences can be extremely difficult (Diffendorfer, 1998).

Given these challenges, most empirical studies testing predictions from range limit models focus relatively narrowly on only part of the bigger picture. For example, a number of studies have tested for either dispersal limitation (e.g., Primack and Miao, 1992), lack of genetic variation (e.g., Hoffman et al., 2003), or reduced survivorship and reproduction (e.g., Angert, 2006) at range boundaries. However, very few studies measure more than one of these key range limit model parameters (Gaston, 2009). Alternatively, controlled lab model systems, such as bacteria, have proved more tractable for exploring how multiple factors interact in determining range limits (Forde et al., 2004). This work provides an important way to isolate and test some of the underlying mechanisms in range models, but the results can be difficult to extrapolate directly into natural populations.

The study of invasive species offers a potentially powerful approach for testing range limit models under more realistic field conditions. Invasions allow for both direct observation of range expansions as they occur and comparisons between species that do and do not achieve rapid spread once established in a new geographic region. Along these lines, studies relating the distributions of invasive species in the same genera have been used to identify the physiological traits that most limit the introduced ranges (Dangles et al., 2008). Traditionally, invasive spread has been attributed mostly to broad environmental tolerance and adaptations for high rates of dispersal in some introduced species (Sakai et al., 2001). However, in the past few years a number of studies have generated greater interest in the potential role of rapid evolutionary change (e.g., Maron et al., 2004; Bossdorf et al., 2005; Lee, 2002; Leger and Rice, 2007). Work on the genetics of invading species also helps shed light on processes facilitating range expansion. For example, recent studies have demonstrated that hybridization (e.g., Ainouche et al., 2009) and multiple introductions from different source populations (e.g., Kolbe et al., 2007) may be particularly important to the creation of genetic variation in small founder populations, potentially enabling adaptation to novel conditions.

Serpentine environments are especially useful systems for applying this kind of approach. The patchy spatial structure of serpentine soils and the strong environmental gradients in these habitats match up well with the structure of many theoretical models. Although species distributions in any environment are influenced by multiple interacting factors, the key role played in serpentine by particular soil traits such as heavy metal concentrations helps simplify the problem of measuring gradients in selection. By analogy, some of the most important field studies on local adaptation and parapatric speciation have used habitats contaminated with metals by mining as model systems (Antonovics, 2006). Finally, many plant invasions of serpentine are ongoing or in relatively early stages; in fact, until recently, serpentine environments were thought to be largely uninvasible because of their extreme abiotic conditions (Huenneke et al., 1990; Harrison et al., 2006). In contrast, many introduced species became established in comparable nonserpentine habitats long enough ago that reconstructing the process of range expansion is very difficult. For example, in nonserpentine California grasslands, the spread of introduced species that now dominate the ecosystem primarily occurred in the nineteenth century, complicating the problem of identifying the causal mechanisms.

In the rest of this chapter, we focus on recent studies of serpentine plant invasions that highlight the potential of this system for addressing basic questions about the evolution of range limits. First, we summarize what is known about the spatial distribution of invasive species in and near serpentine, and the hypotheses these patterns suggest about range expansions into this habitat type. This section is followed by a description of studies that have gone beyond characterizing spatial patterns into quantifying the processes that shape evolutionary change at range margins: local dispersal and spatial gradients in selection. This work focuses on two invasive species, barbed goatgrass (*Aegilops triuncialis*) and *Erodium cicutarium*. Finally, we review studies in serpentine that tackle different hypotheses for range expansion from a genetic perspective and specifically research on the roles of polyploidy, genetic bottlenecks, and phenotypic plasticity in goatgrass invasions.

INVASIVE PLANTS IN SERPENTINE

Spatial Distributions

Until recently, serpentine habitats in California have generally been viewed as a refuge habitat where natives predominate because they are relatively free of competition from invasive species (Harrison, 1999a, 1999b). Invasive plants are less abundant on serpentine than in adjacent nonserpentine habitats, and within serpentine lower soil $Ca^{2+}:Mg^{2+}$ ratios are associated with reduced numbers of invasive species (Huenneke et al., 1990; Harrison 1999a, 1999b). This pattern supports the idea that an inability to tolerate the harsher soil conditions characterized by low $Ca^{2+}:Mg^{2+}$ ratios explains the scarcity of invasive plants in serpentine patches.

However, a number of invasive species common in nonserpentine matrix habitat are also found on serpentine. These primarily consist of introduced European annual grasses, such as *Bromus* and *Avena* spp., and also a few common exotic forbs, for example, *Erodium cicutarium* (Harrison et al., 2001). Over the past decade, several studies on the spatial distributions of these species suggest hypotheses about what enables them to establish and persist on serpentine, even if at low abundances. Harrison (1999a) showed that small serpentine patches were much more likely to contain introduced species than large, continuous areas, but even in small patches only those with relatively high Ca^{2+} levels sustained invasive plants. Along the same lines, Harrison et al. (2001) found that for large serpentine patches, two common invasive grasses were more abundant within 50 m of the edge than in the interior, although soil conditions did not vary across the same gradient. One possible explanation is source-sink dynamics, with high dispersal rates from the matrix maintaining populations of introduced plants in small patches that have a larger edge-to-interior ratio. However, if conditions in the sink habitat are too harsh, even high dispersal rates might not compensate demographically for lower fitness there, explaining the complete absence of introduced species in patches with very low calcium levels. Alternatively, we would also predict a greater diversity of invasive plants in small patches if these habitats were not true sinks but greater propagule pressure increased the chances of successful initial establishment. Interestingly, some invasive grasses from small serpentine patches perform better on their home soils than conspecifics from nonserpentine soils (Harrison et al., 2001). This suggests that populations of invasive species persisting in small patches of serpentine may serve as a springboard to local adaptation.

Dispersal Dynamics, Demography, and Spread: Barbed Goatgrass

Although most invasive grasses are relatively sparse in serpentine environments, at least one introduced plant has begun spreading aggressively onto even very harsh serpentine soils over the past 15–20 years: *A. triuncialis* (barbed goatgrass) (Meimberg et al., 2006; Thomson, 2007). This invasion is an obvious cause for concern from a conservation perspective, given the importance of serpentine patches as islands of endemic plant diversity and refuges for many natives from the impacts of other introduced grasses. Goatgrass invasion has profound effects on both the structure and functioning of the serpentine ecosystem. Batten et al. (2006) documented that both arbuscular mycorrhizal soil communities and abundances of bacterial groups associated with sulfur cycling differ markedly in areas recently invaded by goatgrass compared to native communities. These changes reduce the performance of an important serpentine native species (*Lasthenia californica*) (Batten et al., 2008). Likewise, both lignin:N and C:N ratios in above- and below-ground litter are higher in serpentine grasslands dominated by goatgrass, slowing litter decomposition and rates of carbon and nutrient cycling (Drenovsky and Batten, 2007). These changes in nutrient cycling and the soil microbial community may tip the competitive balance in favor of goatgrass relative to native plants and increase its rate of spread into serpentine meadows (Batten et al., 2008).

However, the arrival of goatgrass in serpentine has also raised a lot of interesting questions about what enables this particular introduced species to thrive both in serpentine and nonserpentine habitats. Given the long lag time between initial introduction of goatgrass into California around 1914 and the observed recent spread into serpentine (Meimberg et al., 2006), one hypothesis is that evolutionary change has enabled local adaptation to this novel environment. Alternatively, returning to a broader range limit model framework, goatgrass could achieve large population sizes on some serpentine patches because of very high dispersal rates from surrounding areas where dense stands occur. Finally, goatgrass may possess preadaptations or phenotypic plasticity in certain traits that allow for positive population growth in both serpentine and nonserpentine environments. Testing these alternative hypotheses requires integration of both demographic measures and genetic information. We start by discussing work on the demographic side and turn to the genetic perspective in a subsequent section.

The demographic data necessary to distinguish among these hypotheses include measures of population growth in both serpentine and nonserpentine patches, and, critically, dispersal between them. As noted earlier, under field conditions, characterizing dispersal dynamics in particular is very challenging. In many cases, plant dispersal rates can only be inferred indirectly through genetic

data (e.g., Stanton and Galen. 1997) or by using seed traps; the latter method re-
moves seeds from the environment before their fitness can be measured (e.g.,
Clark et al., 1998). In contrast, goatgrass dispersal ecology facilitates estimation of
both movement rates and their consequences for fitness and population growth,
key for constructing a full picture of dynamics at the range margin. The unit of
dispersal is a large woody spike, consisting of multiple fruits (spikelets), each containing up to two seeds. Spikes can be readily marked and relocated after dispersal
and germination, making this species especially well suited for testing range limit
models (Dyer, 2004; Thomson, 2007).

Thomson (2007) used this strategy to measure movement of goatgrass spikes
across environmental gradients and track their subsequent demographic fates.
The study site (McLaughlin Reserve, California) consists of a matrix "core" habitat,
where soils are less serpentinite in nature and goatgrass occurs in very high densities as a near monoculture, and adjacent patches of marginal "edge" consisting of
harsher serpentine outcrops with low $Ca^{2+}:Mg^{2+}$ ratios and plant cover. Two results of this work indicate that high dispersal is unlikely to explain the ability of
goatgrass to persist in serpentine outcrops. First, predicted population growth
rates were strongly positive in both edge and core habitats. In fact, edge growth
rates were higher than those observed in the core, although density dependence
and intraspecific competition appear to account for most of this difference. Second, in spite of much higher densities in the core, mean dispersal distances were
very short, with little evidence of net migration into the edge. Instead, low plant
cover and greater exposure to wind appear to move edge spikes longer distances
than those in the more densely vegetated core. This is an interesting result because
most range limit models assume migration is predominantly from core to margin.
However, in this system movement rates are asymmetric in the opposite direction.
Habitat and density-dependent dispersal have been identified as potentially important in animal population dynamics at range edges; for example, mobile organisms may leave lower quality habitats or patches that have reached carrying capacity at a higher rate (Diffendorfer, 1998). However, the potential for complicated
dispersal dynamics has not been explored much for plants.

Given that these results do not support the presence of source-sink dynamics,
the observed differences in individual demographic rates between the core and
edge habitats may help identify some of the life history traits where plasticity or
local adaptation could facilitate persistence at the margin. In particular, Thomson
(2007) generally observed much lower germination rates on the edge. Edge populations also showed much higher germination rates from "second seeds," which
tend to remain dormant until the year after their sibling from the same spikelet
has emerged (Dyer, 2004). One hypothesis is that plants on the edge may change
their maternal allocation strategies and the division of resources between first
and second seeds in ways that help buffer against the harsher and more variable

microclimates. There is some evidence that plants growing in edge habitats allocate more to second seeds than do those in the core (Thomson, unpublished data). However, the extent to which these differences influence germination probabilities and seedling survivorship or growth is still unclear.

Rapid Evolution of Seed Dispersal Distance: Erodium cicutarium

The work on barbed goatgrass described in the previous section illustrates how serpentine environments can be used to explore the consequences of demographic variation and dispersal patterns for population growth at range boundaries. However, as discussed earlier, dispersal influences not just population dynamics but the selection environment experienced by progeny; for this reason, evolution of traits related to dispersal may be especially important in establishing range limits. For example, some models of invasive spread predict strong selection for increased dispersal ability in species expanding their ranges (Travis and Dytham, 2002). Furthermore, a number of empirical studies support the importance of evolution in dispersal traits to geographic range expansion, including examples from weedy plants (Cody and Overton. 1996) to bush crickets (Thomas et al., 2001) to cane toads (Phillips et al., 2006).

Dispersal distance controls the level of environmental heterogeneity experienced by progeny within a lineage across generations. Environmental heterogeneity determines patterns of selection within and among populations, so selection on dispersal distance may vary between populations that differ in the spatial scale of heterogeneity. If the strength and direction of selection on seed dispersal distance is sufficiently different across space, this could lead to genetic divergence in traits mediating seed dispersal distance among populations occurring in different habitat types.

To characterize how selection on dispersal varies across the distribution of a species, it is necessary to quantify patterns of spatial variation in environmental factors for naturally occurring populations on scales relevant to that species; this is a very challenging problem. Recent research by Baythavong et al. (2009) on the invasion of *Erodium cicutarium* into serpentine soils examines how patterns of selection on seed dispersal distance may vary across the invasive range by comparing populations on nonserpentine grasslands and those in adjacent serpentine soils at the edge of the species distribution. This work provides another very nice example of how contrasts between invasive populations on adjacent serpentine and nonserpentine habitats can be used to test ideas about range limits.

E. cicutarium is an introduced Mediterranean annual that has successfully invaded both nonserpentine and serpentine grasslands throughout California, in some cases achieving high densities even on serpentine (Mensing and Byrne, 1998; Baythavong et al., 2009). As with barbed goatgrass, quantifying the relationship between offspring fitness and seed dispersal distance is possible in part because the seed dispersal biology of *E. cicutarium* is well characterized. These seeds are ballistically dispersed from the maternal plant in a diaspore composed of a coiled awn and a carpel surrounding the seed. Seeds can disperse up to 1 m but travel about 50 cm on average (Stamp, 1989b). Following the primary ballistic dispersal, seeds may travel further when the coiled awn undergoes hygroscopic coiling and uncoiling in response to wetting and drying cycles. Under field conditions, seeds of *E. moschatum*, a species that is morphologically almost indistinguishable from *E. cicutarium*, travel on average 7.8 cm following primary ballistic seed dispersal via hygroscopic crawling (Stamp, 1989a).

Although the seed dispersal biology of *E. cicutarium* is well understood, until recently very little was known about the pattern of microenvironmental heterogeneity encountered by dispersing seeds within invasive populations and the fitness consequences of variation in dispersal distance. Baythavong et al. (2009) quantified the grain of environmental variation associated with different dispersal distances, and the corresponding patterns of variation in progeny fitness, with field experiments on serpentine and nonserpentine patches at the McLaughlin Reserve in northern California. In both habitats, seeds produced within dense patches of *E. cicutarium* were collected and dispersed to one of four distances (0 m, 0.5 m, 1 m, and 10 m) from their original location (Baythavong et al., 2009). The experimental distances were chosen to represent plausible natural dispersal distances (0–1 m) and rarer long-distance seed dispersal facilitated by ants (Harmon and Stamp, 1992) or rodents (Stamp, 1989a). Survival and lifetime fitness of all experimental plants were quantified, as well as environmental factors (the number of conspecifics, above-ground grass and forb biomass, soil water content, and soil availability), enabling a comprehensive analysis of how the spatial scale of environmental heterogeneity influences offspring fitness for serpentine and nonserpentine populations.

A multivariate analysis of variance indicated distinct patterns of fine scale spatial variation on serpentine compared to nonserpentine soils (Baythavong et al., 2009). No significant variation between the different dispersal distances was detected for any of the environmental variables on nonserpentine soils, whereas soil water content and above-ground grass biomass varied significantly among distances on serpentine soils. These results indicate that dispersing progeny of *E. cicutarium* experience different levels of microenvironmental variation on serpentine and nonserpentine soils and provide a preliminary indication that patterns of selection on seed dispersal distance could differ between these habitat types.

To fully understand variation in selection on seed dispersal distance among populations of *E. cicutarium*, it is important to compare the relationships between microenvironmental variation and offspring fitness on each soil type. In addition to the soil environment, the distribution and abundance of conspecifics, as well as of grasses and other forbs, are all likely to affect the survival and lifetime fitness of individual *E. cicutarium.* As a result, both the direct effects of variation in soil on *E. cicutarium* progeny fitness and the indirect effects mediated by intra- and inter-specific competition must be considered. To address this issue, Baythavong et al. (2009) used structural equation modeling to assess the multivariate direct and indirect causal relationships among progeny fitness and edaphic and biotic factors. Multigroup analysis revealed significant differences between serpentine and non-serpentine sites. In general, direct and indirect effects of edaphic variation on biotic heterogeneity were stronger on serpentine soils than on nonserpentine soils. This pattern is consistent with Kruckeberg's (1984) assertion that the harsh edaphic environment on serpentine soils is the primary driver of the distribution and abundance of plant populations growing there.

Consistent with these differences in habitat heterogeneity, spatial variation in lifetime fitness of *E. cicutarium* progeny dispersed from their maternal home sites differed among soil types. Progeny had equivalent lifetime fitness at all dispersal distances on nonserpentine patches, whereas progeny dispersed 0.5 m from their home site on serpentine soils had higher fitness than seeds dispersed 10 m away (Baythavong et al., 2009). The contrasting patterns of spatial variation in progeny fitness on the two soil types indicate that selection likely favors reduced dispersal distances on serpentine. This could result in selection for reduced fruit length, a trait important in determining ballistic seed dispersal distance (Stamp, 1989b), at the range margin. Considered together, these results indicate that localized selection gradients, or internal range limits, may have a strong influence on the evolution of dispersal; in turn, these changes in dispersal patterns could influence spread at the range margin.

GENETIC VARIATION, PLASTICITY, AND INVASION: BARBED GOATGRASS

The studies discussed so far test hypotheses about range expansions through characterizing spatial distributions, demography, selection gradients, and dispersal patterns, but none explicitly include genetics. Developing a full picture of evolutionary processes at population margins, however, also depends on a better understanding of both patterns in genetic variation across species ranges and the role of adaptive plasticity in determining distributions. In this section, we return to barbed goatgrass (*A. triuncialis*) and discuss how recent work on population genetics and plasticity in this species provides further insight into serpentine invasions specifically and range expansions in general.

Going back to a theoretical point of view, many important predictions in range limit models clearly depend on the amount of genetic variation present at the margin of a species distribution. Theory and some empirical evidence predict that lowered genetic variation due to bottlenecks should significantly reduce the evolutionary capacity of a species to adapt to new conditions and expand its range (Barrett and Kohn, 1991; Ghalambor et al., 2007; Olivieri, 2009). Rapid range expansion by invasive species is all the more interesting in light of this prediction, given that introduced species in their new ranges may bring only a fraction of the genetic variation present within populations in the home ranges. Invasive plant species often experience significant genetic bottlenecks on introduction (Barrett and Kohn, 1991; Dlugosch and Parker, 2008a).

Goatgrass is no exception in this regard. Molecular evidence indicates that this grass underwent an extreme genetic bottleneck during its colonization of California (Meimberg et al., 2006). A comparison of genetic variation between 11 California goatgrass populations and 57 Eurasian accessions indicates that goatgrass experienced a reduction in genetic variation of over an order of magnitude during the invasion of California. In addition, evidence supports only a small number of introduction events, suggesting that the capacity of goatgrass to adapt and spread into serpentine is not the result of multiple introductions (Meimberg et al., 2006). When coupled with the fact that goatgrass is highly selfing (i.e., essentially homozygous for all loci tested), the ability of this species to spread rapidly in both serpentine and nonserpentine grasslands is intriguing, to say the least, and reflects a broader problem: although theory predicts that extremely low genetic variation at the range margin should prevent spread, a significant number of introduced plant species still spread quickly into new environments, despite reduced genetic variation resulting from demographic bottlenecks. This pattern may be the result of adaptive plasticity (Williams et al., 1995) and/or reflect the fact that these genetically depauperate populations still exhibit the evolutionary potential to evolve rapidly in their new ranges (Dlugosch and Parker, 2008a, 2008b).

Molecular genetic work also provides some potential answers to this key problem. The study by Meimberg et al. (2006) revealed that goatgrass in its home range exhibits very little genetic structure, with highly mixed genotypes dispersed over broad geographical scales. This pattern suggests that goatgrass is a weedy colonizing species in its home range. Thus, goatgrass may be preadapted for its role as a major invader in California, given a coevolutionary past with both human-caused disturbance and human-facilitated dispersal (e.g., as a crop contaminant and a hitchhiker on livestock). These results support the idea that broad environmental tolerance (rather than rapid evolution) explains the success of goatgrass as an invasive species.

Although most of the molecular genetic evidence for serpentine goatgrass populations indicates genetic uniformity (Meimberg et al., 2006), the fact that this species is an allotetraploid raises the possibility that fixed heterozygosity resulting from genome duplication may increase its ecological amplitude and tolerance (Ramsey and Schemske, 2002). The potential importance of hybridization and polyploidy in facilitating plant invasions has been noted elsewhere, especially where polyploid formation occurs in the new range (Ellstrand and Schierenbeck, 2000). However, the adaptive importance of recurrent or multiple origins of allopolyploidy has received much less attention for polyploids in general and invasive polyploids in particular (Soltis and Soltis, 1999; Soltis et al., 2003). It is not well recognized that a major genetic bottleneck occurs when single diploid genotypes from two species hybridize to form an allotetraploid that is then reproductively isolated from its diploid progenitors. Multiple diploid origins of invading polyploid lineages could reduce genetic bottlenecks during invasion events because each lineage would have an independent and unique set of diploid progenitors. This diversification of introduced lineages would then provide a greater evolutionary potential for the invader in its new range. At the genus level in goatgrass, evidence suggests that *Aegilops* allopolyploid species with a greater number of independent origins exhibit greater ecological amplitude. Meimberg et al. (2009) found a positive correlation between the number of allopolyploid origins for a species and both the total range and tendency to be common.

More directly relevant to serpentine invasions were the results of this study specific to barbed goatgrass. Meimberg et al. (2009) examined phenotypic variation in seed mass, flowering time, and awn length of five different cpDNA haplotypes that represented distinct allopolyploid origins. There are significant differences in all three traits, suggesting that multiple origins may have increased genetic diversity and thus decreased bottlenecking effects in this species during its introduction to California. Of particular interest were results showing significant variation in flowering time. Field measurements in serpentine barbed goatgrass populations suggest that variation in soil moisture exerts strong selection on flowering time (McKay and Rice, unpublished data). Genetic variation in flowering time resulting from multiple allopolyploid origins may thus facilitate goatgrass adaptation to moisture stress in serpentine soils.

The ability of goatgrass to spread in California despite an extreme bottleneck in genetic variability also suggests that phenotypic plasticity might play an important role in facilitating the colonization of this species into edaphically stressful serpentine habitats. Although the importance of within-generation plasticity has long been suggested as a mechanism for adaptation in invading species (Baker, 1965; Bradshaw, 1965; Rice and Mack, 1991; Williams and Black, 1996; Sakai et al., 2001), the role of maternal environmental effects or transgenerational plasticity in adaptation is largely unknown for invasive plants (but see Miao et al., 1991). Adaptive within-generation plasticity represents the adjustment of phenotypic expression of an individual to current environmental conditions in a way that increases individual fitness. Adaptive transgenerational plasticity, on the other hand, occurs when individuals respond to environmental cues by adjusting the phenotypes of their progeny in a way that increases progeny fitness (Donohue and Schmitt, 1998). Historically, transgenerational plasticity (i.e., maternal environmental effects) has been considered a nuisance factor when trying to obtain good estimates of genetic parameters; more recently the potential importance of this type of plasticity in adaptation has been recognized for both plants and animals (Mousseau and Fox, 1998; Galloway, 2005; Galloway and Etterson, 2007). Transgenerational plasticity represents a predictive cueing mechanism by which parental responses to current conditions may adaptively precondition offspring for the environment that they are most likely to encounter (Donohue and Schmitt, 1998; Galloway, 2005).

Recent work suggests that this form of plasticity may play an important role in allowing goatgrass to cope with the edaphic stress of serpentine habitats (Dyer et al., 2010). Measurements of seasonal variation in soil moisture indicate that serpentine patches dry out more rapidly in the spring than nonserpentine grassland sites (Sambatti and Rice, 2006), and phenotypic selection analyses suggest there is strong selection for earlier flowering in serpentine goatgrass populations (McKay and Rice, unpublished data). Given a fitness advantage for earlier flowering phenotypes on serpentine, shortening the flowering time of progeny would represent an adaptive transgenerational plastic response for maternal plants growing on serpentine. We think this shift in flowering phenology reflects aspects of a stress avoidance strategy that contrasts with the stress tolerance syndrome discussed shortly; in essence, goatgrass may be pursuing a mixed strategy of both stress tolerance and avoidance. By using highly inbred families of goatgrass to grow "replicate" maternal genotypes in both serpentine and nonserpentine soil, Dyer et al. (2010) found that goatgrass mothers growing in serpentine soils produced offspring with shortened flowering time, the predicted adaptive response.

To examine the potential role of stress tolerance in goatgrass adaptation to harsh serpentine soils, Dyer et al. (2010) also examined photosynthetic rates of progeny produced from maternal genotypes replicated across both serpentine and nonserpentine soil environments. Using both clones and genetically uniform split families, they found a significant decrease in photosynthetic rates in progeny produced by mothers growing in serpentine soils (Dyer et al., 2010). Despite this reduction, or down-regulation, in photosynthesis, split family progeny from mothers grown on serpentine were larger in size at the end of the experiment. The observation that goatgrass fitness increases when maternal environmental cues lead to lowered photosynthetic rates supports the notion that reducing photosynthesis and metabolic activity may represent an adaptive response for plants growing in harsh edaphic environments; this idea has been characterized as a stress tolerance syndrome (Chapin et al., 1993). Support for the direct adaptive role of the metabolic shifts associated with this stress tolerance syndrome is provided by studies that document the complex interactions among changes in hormonal balance, allocation patterns, nutrient use, and water relations (summarized in Chapin, 1991). Together, these results for goatgrass indicate that the predictive cuing capacity (Donohue and Schmitt, 1998) of transgenerational plasticity may play a key adaptive role in facilitating the spread of goatgrass into serpentine, a habitat characterized by a complex mosaic of edaphic patches with varying levels of abiotic stress.

CONCLUSIONS AND FUTURE DIRECTIONS

As a whole, what do these recent studies of serpentine invasions in California tell us about the problem of predicting changes in species range limits and the ability of existing theory to explain patterns for real populations in the field? We close by first considering some of the lessons from this work to date and what they demonstrate about the value of serpentine habitats as study systems. We then suggest some potentially fruitful new areas of research on range limits in serpentine invasions.

First, although the results of this work reinforce the findings of many previous studies, they also highlight some potentially important mechanisms that have not received as much attention. For example, from a demographic perspective, many predictions of range limit models are driven by asymmetric dispersal. In goatgrass, there is indeed evidence of asymmetrical migration, but unexpectedly both density and habitat-dependent dispersal promote a much higher per capita migration rate at the distribution edge. This may mean that dispersal from the center to the margin is not in fact dramatically higher than movement in the opposite direction (Thomson, 2007). These results suggests that both theoretical explorations of more complicated dispersal dynamics and further empirical work to characterize real dispersal patterns at range edges would improve our understanding of species distribution limits.

From a genetic standpoint, studies on barbed goatgrass support the long-held idea that invasive populations tend to be highly bottlenecked and lacking in heterozygosity (Meimberg et al., 2006). However, they also suggest that a less appreciated source of genetic diversity—multiple allopolyploid origins—may provide enough phenotypic variation to enable goatgrass spread across a range of environments that differ strongly in selection pressures. This work also points to a specific trait, flowering time, that may play a particularly critical role in allowing goatgrass on serpentine to avoid drought stress (Meimberg et al., 2009). Research on goatgrass in general also provides evidence for the long-standing hypothesis that broad environmental tolerance and phenotypic plasticity are important adaptive mechanisms for invasive spread. However, this work focuses attention on a mechanism that has not been explored much in the context of invasive spread or range limits: adaptive transgenerational plasticity (Dyer et al., 2010). Whether the result of adaptive plasticity or of genetic differentiation associated with ecotype formation, these results emphasize the need to better understand the evolutionary dynamics of populations at the margins of species distributions. In particular, "evolutionary experiments," such as reciprocal transplants and phenotypic selection analyses, are needed to examine local adaptation along range margins. In addition, the use of quantitative and molecular analyses of genetic variation under selection along both internal and geographic range margins can provide important information on the capacity for further evolutionary response at distribution boundaries.

Finally, previous range limit models have explored the balance between selection differentials and dispersal in a number of ways, but little of this work has emphasized the role of selection on dispersal traits themselves in determining range limits. In particular, there is a real opportunity to use experimental studies of local selection gradients (i.e., internal range limits) to better understand mechanisms underlying geographic range limits (Sexton et al., 2009). Research on *E. cicutarium* clearly illustrates the value of this approach by demonstrating how localized changes in edaphic conditions across a species distribution can lead to shifts in the relationship between dispersal distance and offspring fitness, thus creating a strong selection gradient on dispersal behavior (Baythavong et al., 2009).

Additional field studies involving a greater diversity of introduced species would help greatly in determining whether these same mechanisms are generally important in other populations. One interesting avenue of future research would be to apply similar methods for some invasive species that are abundant adjacent to and occasionally found on serpentine soils but have not successfully established large populations in serpentine habitats. This ability to carry out comparative studies on multiple species encountering the same range boundary and responding in different ways could potentially prove to be a powerful approach for testing the generality of model predictions. However, even though most of the studies reviewed here focus on two species, they suggest some productive new directions for modeling efforts. In particular, the use of demographic, genetic, and evolutionary approaches on the same species opens up the possibility of fully parameterizing a range limit model with empirical data, a new and exciting direction for the study of species distributions.

Acknowledgments

The authors thank Susan Harrison and Nishanta Rajakaruna for giving us the opportunity to contribute this chapter and for all their assistance with the editing process, as well as Katrina Dlugosch and an anonymous reviewer for a number of helpful comments.

LITERATURE CITED

Ainouche, M. L., Fortune, P. M., Salmon, A., Parisod, C., Grandbastien, M. A., Fukunaga, K., Ricou, M., and Misset, M. T. (2009) Hybridization, polyploidy and invasion: Lessons from *Spartina* (Poaceae). *Biological Invasions*, 11, 1159–73.

Angert, A. A. (2006) Demography of central and marginal populations of monkeyflowers (*Mimulus cardinalis* and *M. lewisii*). *Ecology*, 87, 2014–25.

Antonovics, J. (2006) Evolution in closely adjacent plant populations X: Long-term persistence of prereproductive isolation at a mine boundary. *Heredity*, 97, 33–37.

Baker, H.G. (1965) Characteristics and modes of origin of weeds. In *The Genetics of Colonizing Species* (eds. H.G. Baker and G.L. Stebbins), pp. 147–72. Academic Press, New York.

Barrett S.C.H., and Kohn, J.R. (1991) Genetic and evolutionary consequences of small population size in plants: Implications for conservation. In *Genetics and Conservation of Rare Plants* (eds. D.A. Falk and K.E. Holsinger), pp. 3–30. Oxford University Press, New York.

Batten, K.M., Scow, K.M., and Espeland, E.K. (2008) Soil microbial community associated with an invasive grass differentially impacts native plant performance. *Microbial Ecology*, 55, 220–28.

Batten, K.M., Scow, K.M., Davies, K.F., and Harrison, S.P. (2006) Two invasive plants alter soil microbial community composition in serpentine grasslands. *Biological Invasions*, 8, 217–30.

Baythavong, B.S., Stanton, M.L. and Rice, K.J. (2009) Understanding the consequences of seed dispersal in a heterogeneous environment. *Ecology*, 90, 2118–28.

Bossdorf, O., Auge, H., Lafuma, L., Rogers, W.E., Sieman, E., and Prati, D. (2005) Phenotypic and genetic differentiation between native and introduced plant populations. *Oecologia*, 144, 1–11.

Bradshaw, A.D. (1965) Evolutionary significance of phenotypic plasticity in plants. *Advances in Genetics*, 13, 115–55.

Bridle, J.R., and Vines, T.H. (2007) Limits to evolution at range margins: When and why does adaptation fail? *Trends in Ecology and Evolution*, 22, 140–47.

Chapin, F.S. III. (1991) Integrated responses of plants to stress. *Bioscience*, 41, 29–36.

Chapin, F.S. III, Autumn, K., and Pugnaire, F. (1993) Evolution of suites of traits in response to environmental stress. *American Naturalist*, 142, S78–92.

Clark, J.S., Macklin, E., and Wood, L. (1998) Stages and spatial scales of recruitment limitation in southern Appalachian forests. *Ecological Monographs*, 68, 213–35.

Cody, M.L., and Overton, J.M.C. (1996) Short-term evolution of reduced dispersal in island plant populations. *Journal of Ecology*, 84, 53–61.

Crispo, E. (2008) Modifying effects of phenotypic plasticity on interactions among natural selection, adaptation and gene flow. *Journal of Evolutionary Biology*, 21, 1460–69.

Dangles, O., Carpio, C., Barragan, A.R., Zeddam, J.L., and Silvain, J.F. (2008) Temperature as a key driver of ecological sorting among invasive pest species in the tropical Andes. *Ecological Applications*, 18, 1795–809.

Diffendorfer, J.E. (1998) Testing models of source-sink dynamics and balanced dispersal. *Oikos*, 81, 417–33.

Dlugosch, K.M., and Parker, I.M. (2008a) Founding events in species invasions: Genetic variation, adaptive evolution, and the role of multiple introductions. *Molecular Ecology*, 17, 431–49.

Dlugosch, K.M., and Parker, I.M. (2008b) Invading populations of an ornamental shrub show rapid life history evolution despite genetic bottlenecks. *Ecology Letters*, 11, 701–9.

Donohue, K., and Schmitt, J. (1998) Maternal environmental effects: Adaptive plasticity? In *Maternal Effects as Adaptations* (eds. A. Mousseau and C. W. Fox), pp. 137–58. Oxford University Press, Oxford.

Drenovsky, R. E., and Batten, K. M. (2007) Invasion by *Aegilops triuncialis* (barb goatgrass) slows carbon and nutrient cycling in a serpentine grassland. *Biological Invasions*, 9, 107–16.

Dyer, A. R. (2004) Maternal and sibling factors induce dormancy in dimorphic seed pairs of *Aegilops triuncialis. Plant Ecology*, 172, 211–18.

Dyer, A. R., Brown, C. S., Espeland, E. K., McKay, J. M., Meimberg, H., and Rice, K. J. (2010) The role of adaptive trans-generational plasticity in biological invasions of plants. *Evolutionary Applications*, 3, 179–92.

Ellstrand, N. C., and Schierenbeck, K. A. (2000) Hybridization as a stimulus for the evolution of invasiveness in plants? *Proceedings of the National Academy of Sciences, USA*, 97, 7043–50.

Forde, S. E., Thompson, J. N., and Bohannan, B. J. M. (2004) Adaptation varies through space and time in a coevolving host-parasitoid interaction. *Nature*, 431, 841–44.

Galloway, L. F. (2005) Maternal effects provide phenotypic adaptation to local environmental conditions. *New Phytologist*, 166, 93–100.

Galloway, L. F., and Etterson, J. R. (2007) Transgenerational plasticity is adaptive in the wild. *Science*, 318, 1134–36.

Garant, D., Forde, S. E., and Hendry, A. P. (2007) The multifarious effects of dispersal and gene flow on contemporary adaptation. *Functional Ecology*, 21, 434–43.

Gaston, K. J. (2009) Geographic range limits: Achieving synthesis. *Proceedings of the Royal Society B-Biological Sciences*, 276, 1395–406.

Ghalambor, C., McKay, J. K., Carroll, S., and Reznick, D. (2007) Adaptive versus non-adaptive phenotypic plasticity and the potential for contemporary adaptation in new environments. *Functional Ecology*, 21, 394–407.

Goodnight, C. J. (1988) Epistasis and the effect of founder events on the additive genetic variance. *Evolution*, 42, 441–54.

Harmon, G. D., and Stamp, N. E. (1992) Effects of postdispersal seed predation on spatial inequality and size variability in an annual plant, *Erodium cicutarium* (Geraniaceae). *American Journal of Botany*, 79, 300–305.

Harrison, S. (1999a) Local and regional diversity in a patchy landscape: Native, alien, and endemic herbs on serpentine. *Ecology*, 80, 70–80.

Harrison, S. (1999b) Native and alien species diversity at the local and regional scales in a grazed California grassland. *Oecologia*, 121, 99–106.

Harrison, S., Grace, J. B., Davies, K. F., Safford, H. D., and Viers, J. H. (2006) Invasion in a diversity hotspot: Exotic cover and native richness in the Californian serpentine flora. *Ecology*, 87, 695–703.

Harrison, S., Rice, K., and Maron, J. (2001) Habitat patchiness promotes invasion by alien grasses on serpentine soil. *Biological Conservation*, 100, 45–53.

Hastings, A., Cuddington, K., Davies, K. F., Dugaw, C. J., Elmendorf, S., Freestone, A., Harrison, S., Holland, M., Lambrinos, J., Malvadkar, U., Melbourne, B. A., Moore, K., Taylor,

C., and Thomson, D. (2005) The spatial spread of invasions: New developments in the-ory and evidence. *Ecology Letters*, 8, 91–101.

Hoffmann, A. A., Hallas, R. J., Dean, J. A., and Schiffer, M. (2003) Low potential for climatic stress adaptation in a rainforest *Drosophila* species. *Science*, 301, 100–102.

Holt, R. D., and Gaines, M. S. (1992) Analysis of adaptation in heterogeneous landscapes—implications for the evolution of fundamental niches. *Evolutionary Ecology*, 6, 433–47.

Huenneke, L. F., Hamburg, S. P., Koide, R., Mooney, H. A., and Vitousek, P. M. (1990) Effects of soil resources on plant invasion and community structure in Californian serpentine grassland. *Ecology*, 71, 478–91.

Kawecki, T. J. (1995) Demography of source-sink populations and the evolution of ecologi-cal niches. *Evolutionary Ecology*, 9, 38–44.

Kawecki, T. J. (2000) Adaptation to marginal habitats: contrasting influence of the dispersal rate on the fate of alleles with small and large effects. *Proceedings of the Royal Society of London Series B-Biological Sciences*, 267, 1315–20.

Kawecki, T. J. (2008) Adaptation to marginal habitats. *Annual Review of Ecology Evolution and Systematics*, 39, 321–42.

Kawecki, T. J., and Holt, R. D. (2002) Evolutionary consequences of asymmetric dispersal rates. *American Naturalist*, 160, 333–47.

Kirkpatrick, M., and Barton, N. H. (1997) Evolution of a species' range. *American Natural-ist*, 150, 1–23.

Kolbe, J. J., Glor, R. E., Schettino, L. R., Lara, A. C., Larson, A., and Losos, J. B. (2007) Mul-tiple sources, admixture, and genetic variation in introduced *Anolis* lizard populations. *Conservation Biology*, 21, 1612–25.

Kruckeberg, A. R. (1984) *California Serpentines: Flora, Vegetation, Geology, Soils, and Man-agement Problems*. University of California Press, Berkeley.

Lee, C. E. (2002) Evolutionary genetics of invasive species. *Trends in Ecology and Evolution*, 17, 386–91.

Leger, E. A., and Rice, K. J. (2007) Assessing the speed and predictability of local adaptation in invasive California poppies (*Eschscholzia californica*). *Journal of Evolutionary Biology*, 20, 1090–103.

Lenormand, T. (2002) Gene flow and the limits to natural selection. *Trends in Ecology and Evolution*, 17, 183–89.

Maron, J. L., Vila, M., Bommarco, R., Elmendorf, S., and Beardsley, P. (2004) Rapid evolu-tion of an invasive plant. *Ecological Monographs*, 74, 261–80.

Meimberg, H., Hammond, J. I., Jorgensen, C. M., Park, T. W., Gerlach, J. D., Rice, K. J., and McKay, J. K. (2006) Molecular evidence for an extreme genetic bottleneck during intro-duction of an invading grass to California. *Biological Invasions*, 8, 1355–66.

Meimberg, H., Rice, K. J.,Milan, N. F, Njoku, C. C., and McKay, J. K. (2009) Multiple origins promote the ecological amplitude of allopolyploid *Aegilops* (Poaceae). *American Journal of Botany*, 96, 1262–73.

Mensing, S. A., and Bryne, R. (1998). Pre-mission invasion of *Erodium cicutarium* in the California grassland. *Journal of Biogeography*, 25, 757–62.

Miao, S. L., Bazzaz, F. A., and Primack, R. A. (1991) Effects of maternal nutrient pulse on reproduction of two colonizing *Plantago* species. *Ecology*, 72, 586–96.

Mousseau, T. A., and Fox, C. W. (1998). *Maternal Effects as Adaptations*. Oxford University Press, Oxford.

Phillips, B. L., Brown, G. P., Webb, J. K., and Shine, R. (2006) Invasion and the evolution of speed in toads. *Nature*, 439, 803.

Primack, R. B., and Miao, S. L. (1992) Dispersal can limit local plant distribution. *Conservation Biology*, 6, 513–19.

Olivieri, I. (2009) Alternative mechanisms of range expansion are associated with different changes of evolutionary potential. *Trends in Ecology and Evolution*, 24, 289–92.

Ramsey, J., and Schemske, D. W. (2002) Neopolyploidy in flowering plants. *Annual Review of Ecology and Systematics*, 33, 589–639.

Rice, K. J., and Mack, R. N. (1991) Ecological genetics of *Bromus tectorum*. II. Intraspecific variation in phenotypic plasticity. *Oecologia*, 88, 84–90.

Sakai, A. K., Allendorf, F. W., Holt, J. S., Lodge, D. M., Molofsky, J., With, K. A., Baughman, S., Cabin, R. J., Cohen, J. E., Ellstrand, N. C., McCauley, D. E., O'Neil, P., Parker, I. M., Thompson, J. N., and Weller, S. G. (2001) The population biology of invasive species. *Annual Review of Ecology and Systematics*, 32, 305–32.

Sambatti, J. B. M., and Rice, K. J. (2006) Local adaptation, patterns of selection, and gene flow in the Californian serpentine sunflower (*Helianthus exilis*). *Evolution*, 60, 696–710.

Sexton, J. P., McIntyre, P. J., Angert, A. L., and Rice, K. J. (2009) Evolution and ecology of species range limits. *Annual Review of Ecology, Evolution and Systematics*, 40, 415–36.

Soltis, D. E., and Soltis, P. S. (1999) Polyploidy: Recurrent formation and genome evolution. *Trends in Ecology and Evolution*, 14, 348–52.

Soltis, D. E., Soltis, P. S., and Tate, J. A. (2003) Advances in the study of polyploidy since plant speciation. *New Phytologist*, 161, 173–91.

Stamp, N. E. (1989a) Efficacy of explosive vs hygroscopic seed dispersal by an annual grassland species. *American Journal of Botany*, 76, 555–61.

Stamp, N. E. (1989b) Seed dispersal of 4 sympatric grassland annual species of *Erodium*. *Journal of Ecology*, 77, 1005–20.

Stanton, M. L., and Galen, C. (1997) Life on the edge: Adaptation versus environmentally mediated gene flow in the snow buttercup, *Ranunculus adoneus*. *American Naturalist*, 150, 143–78.

Sultan, S. E., and Spencer, H. G. (2002) Metapopulation structure favors plasticity over local adaptation. *American Naturalist*, 160, 271–83.

Thomas, C. D., Bodsworth, E. J., Wilson, R. J., Simmons, A. D., Davies, Z. G., et al. (2001) Ecological and evolutionary processes at expanding range margins. *Nature*, 411, 577–81.

Thomson, D. M. (2007) Do source-sink dynamics promote the spread of an invasive grass into a novel habitat? *Ecology*, 88, 3126–34.

Travis, J. M. J., and Dytham, C. (2002) Dispersal evolution during invasions. *Evolutionary Ecology Research*, 4, 1119–29

Williams, D. G., and Black, R. A. (1996) Effects of nutrient amendment and environment on growth and gas exchange for introduced *Pennisetum setaceum* in Hawaii. *Canadian Journal of Botany* 74, 268–75.

Williams, D. G., Mack, R. N., and Black, R. A. (1995) Ecophysiology of introduced *Pennisetum setaceum* on Hawaii—the role of phenotypic plasticity. Ecology, 76, 1569–80.

PART TWO

Serpentine as a Model in Ecology and Conservation

10

Plant Competition and Facilitation in Systems with Strong Environmental Gradients

Kara A. Moore, *University of California, Davis*
Sarah C. Elmendorf, *University of British Columbia*

The distinctive plant communities and environmental gradients found in serpentine ecosystems provide a rich palette for testing fundamental questions about species interactions. Plant–plant interactions on serpentine soils differ from those on nonserpentine soils in the identity, abundance, and diversity of interacting plants. Interactions on serpentine also are distinct due to the high-stress, low-resource, and often extremely heterogeneous environment. One of the most striking elements of serpentine plant communities is the frequent dominance of native species in an otherwise exotic-dominated landscape. Serpentine soils are a significant refuge for native species diversity globally and characterized by high levels of species endemism (Harrison et al., 2006). Because competition is assumed to be a major determinant in the dominance of exotic species over natives, it comes as no surprise that the role of competition in shaping the often striking serpentine community boundaries has been a topic of frequent inquiry. Unique plant communities and soil conditions on serpentine often exist in close proximity to nonserpentine sites, providing an experimentally tractable opportunity to test basic questions of plant competition theory and invasion ecology (Figure 10.1).

In this chapter, we explore how plant interactions and their effects on community organization vary with stress, productivity, and environmental heterogeneity in serpentine ecosystems and along their boundaries with nonserpentine ecosystems. We begin with a brief overview of competition theory. We summarize

FIGURE 10.1. Contrasting levels in plant above-ground productivity, community composition, soil texture, and heterogeneity in (top) a harsh serpentine site and (bottom) an adjacent nonserpentine grassland (photos by K. Moore).

common experimental designs that have been employed to test for the effects of competition along gradients and differences in their interpretation. We follow with a review of special considerations for the study of interactions on serpentine soil, addressing the possibility of chemical amelioration of serpentine stress and the potential effects of environmental heterogeneity. We conclude with an evaluation of the benefits and complexities of serpentine as a model system for studying species interactions and suggestions for future study.

COMPETITION ALONG ENVIRONMENTAL GRADIENTS

There are two dominant and interrelated working hypotheses concerning the complex relationships between coexisting plants, environmental conditions, and species distributions. According to the stress gradient and productivity gradient hypotheses, individuals occupying low-productivity or high-stress sites will tend to have weak competitive interactions or even positive interactions. Serpentine soils tend to be both low in productivity and high in stress, which make them good candidates for testing these competition gradient theories. Serpentine sites generally have relatively shallow, gravelly soil, high levels of magnesium and possibly other metals, and low levels of plant available calcium, nitrogen, and phosphorus (Walker, 1954). They are often low in productivity, with plant density and size increasing toward the margins of the patches, as the effects of serpentine stress diminish. In addition, serpentine soils are often embedded in a matrix of nonserpentine soils. This provides an interesting model system for testing the predictions of competition gradient theories because variation in levels of stress and productivity and well-differentiated plant communities exist at scales ideal for empirical study.

Testing for Competition Differences along Gradients

The competition gradient hypothesis is typically tested using one of two types of experiments: either removing all naturally occurring neighbors from target individuals across an environmental gradient (neighbor removal experiments), or competing the same pair or suite of species across an environmental gradient (controlled identity competition experiments). Although their main objective is often similar—to evaluate effects of different habitat types on plant interactions—the two types of experiment differ both in specific focus of study and in interpretation, primarily in whether the same or different species are interacting across habitat types. We discuss the pros and cons of these types of experiments and then review findings from both serpentine and nonserpentine systems as they relate to competition gradient theory. A third and related type of competition experiment, adding resources to examine how amelioration of certain stressors alters species interactions among pairs or groups of species, has also been conducted on serpentine soils. We discuss such stress amelioration studies in a later section.

Neighbor removal experiments, the most common tests of the stress gradient theory, examine whether differences in environmental conditions alter the magnitude or direction of the plant neighborhood effect on a single target species. In these studies, performance of a target species, or phytometer, is compared across a range of environmental conditions with and without the neighboring plants that naturally occur there. Community composition generally covaries with environmental gradients. Thus, measured variation in the effect of neighbor removal across a gradient is due to the combination of abiotic and biotic differences. For this reason, it is typically not possible to differentiate the influence of varied identity of competitors from differences in resource or stress levels on plant interactions. This point is important to the interpretation of neighbor removal experiment results and warrants additional attention in the literature. An advantage of neighbor removal experiments is that they mimic a natural competitive milieu, where competitors likely differ along gradients.

Controlled identity experiments test whether environmental conditions alter the magnitude or direction of competitive hierarchies. In these studies, also called pair-wise experiments (see Maestre et al., 2009, for an excellent synthesis), the same target species (two or more) compete across a range of environmental conditions. Interpretation of results is more straightforward, because any difference in interaction strength will be due to environmental drivers rather than differences in competitor identity. However, such experiments may not successfully represent the forces that naturally drive competition in the field, where competitor identity and environment often change along the same gradient. Additionally, it may be impossible to measure interaction strength between the same species across an entire environmental gradient, because some regions may be outside the fundamental niche of one or both species. Because it is not practical to conduct competition experiments between every combination of species in a community, these experiments are an inadequate test of the multitudinous and diffuse interactions found throughout natural communities (Wilson and Keddy, 1986a). Notably, some meta-analyses of competition gradient studies have not made the distinction between neighbor removal and identity controlled experiments (e.g., Maestre et al., 2005, 2006; Lortie and Calloway, 2006), thus the interpretation of their results is at times not straightforward (but see Goldberg and Barton, 1992; Maestre et al., 2009).

Neighbor Removal Experiments

Neighbor removal studies in nonserpentine systems have demonstrated that the direction of plant interactions often switch from negative to neutral to positive across gradients of decreasing productivity (Kadmon, 1995; Sammul et al., 2000; Foster, 2002), increasing elevation (Choler et al., 2001; Callaway et al., 2002) and decreasing nutrients (Reader et al., 1994). The competition gradient hypothesis is somewhat supported by neighbor removal studies in serpentine systems. Proctor and Woodell (1975) found seeding of a serpentine-tolerant *Plantago* species was successful off serpentine soils only when neighbors were removed. Neighbor removal enhanced all demographic parameters of a native serpentine-tolerant grass only in productive habitats (Jurjavcic et al., 2002). In contrast, Williamson and Harrison (2002) found competitor removal benefited establishment of an invasive species both on and off serpentine soil.

Evaluation of how the performance of a target species covaries with neighbor community composition requires botanical surveys of the plant interaction neighborhood, in addition to more commonly measured with environmental gradients. We used this approach to determine whether community composition could accurately predict competitive intensity over multiple gradients in a heterogeneous serpentine and nonserpentine grassland in northern California (Elmendorf and Moore, 2008). Known gradients at the study area include soil transitions from serpentine to nonserpentine, variation in slope, aspect and soil depth (affecting water availability), nutrients, and dramatic differences in community composition. We tested for change in the competitive effect exerted by the neighboring plant community on six native and non-native grasses and forbs planted across these multiple gradients. Specifically, we asked whether the composition of the community itself predictably affected competitive intensity. We found that a community composition gradient, as described by a nonmetric multidimensional scaling ordination of species presences, was in fact the best correlate with competitive intensity at this site, outperforming a suite of other environmental gradients.

We postulate that species composition might frequently be a better predictor of competitive intensity than the measurement of soil chemistry or other typical environmental factors in neighbor removal studies. Species composition can affect a number of ecosystem processes, such as decomposition rates, disturbance, and nutrient cycling (Wardle et al., 1998; Eviner and Chapin, 2003). Whether the driving factor for the relationship between competitive intensity and composition is the underlying environmental gradient or the physiological properties of the various competitors requires additional studies. For example, a factorial study in which competitor identity is varied independently of environmental conditions would allow for the evaluation of both the effects of neighbor identity and environmental conditions on target performance.

Controlled Identity Competition Experiments

Interactions between the same pair of competitors across different environments have commonly been used to test the competition gradient hypothesis across gradients of nutrient, moisture, elevation, salinity, and productivity. Notably, many of these same gradients are found in serpentine systems, although few controlled identity competition experiments have been conducted on serpentine. The outcome of tests of the competition gradient hypothesis in nonserpentine systems has been mixed. Although some nonserpentine studies have found that plant interactions switch from negative to neutral to positive across gradients of increasing environmental stress (e.g., soil flooding, Brose and Tielborger, 2005; and elevation, Callaway, 1998) the results of many other studies fail to support the gradient hypotheses. To note a few examples, Boyden et al. (2005) found that competition was greater in high-nutrient conditions for only one of two species examined; La Peyre et al. (2001) found no consistent effects of salt stress on competition; Liancourt et al. (2005) found that the same species were strong and weak competitors regardless of soil nutrients or water availability; and Peltzer et al. (1998) found that competition levels were constant across a productivity gradient when species identities were fixed.

To our knowledge, a single pair-wise controlled identity plant interaction experiment has been conducted along a serpentine gradient under natural field conditions. Freestone (2006) examined differences in interactions between a serpentine seep specialist, *Delphinium uliginosum,* and an associated moss across a moisture gradient bordering a serpentine seep. Moss removal reduced the emergence of *Delphinium* across its range, indicating facilitation during the emergence stage. Contrary to the stress gradient hypothesis, there was no difference in the direction or intensity of interactions with soil moisture. Additional controlled identity experiments on serpentine gradients could be useful in examining how interactions between species shift over serpentine gradients and the boundaries of serpentine patches. In particular, studies that pair a native species found both off and on serpentine with an exotic species that has some serpentine tolerance would be helpful in elucidating the effects of invasion on persistence of the native species.

SERPENTINE PLANT COMMUNITY ORGANIZATION: FROM ENDEMICS TO EXOTICS

Serpentine communities are a distinctive collection of endemic species coexisting with native and non-native serpentine tolerators and often surrounded by nontolerant communities (see Chapter 11 in this volume). Physiological trade-offs may prevent a single species from being both stress-tolerant and a good competitor. Thus, where environmental stresses vary spatially, species may be distributed according to their respective tolerances for stress and competition. In a classic wetland study, Wilson and Keddy (1986b) found that species niches along a nutrient gradient marsh were related to their competitive ability, with more competitive species occupying more nutrient-rich areas. More generally, core–periphery or centrifugal theories of community organization posit that there is a single "core" habitat in which performance is best in the absence of competitors (Rosenswieg and Abramsky, 1986; Wisheu and Keddy, 1992). Due to trade-offs in tolerance for competition and stress, only superior competitors persist in core habitat, whereas inferior competitors are relegated to inferior habitats. Such a model has long been thought to apply to the distribution of endemics, tolerators, and exotic species in serpentine systems.

The diminutive appearance of plants growing on serpentine, and the often open or barren appearance of communities on serpentine in contrast to adjacent non-serpentine communities, has led to a common perception that serpentine tolerant or endemic species are poor competitors compared with their more robust non-serpentine neighbors and relegated to serpentine for this reason, rather than by physiological preference (Kruckeberg, 1954; Walker, 1954; Proctor and Woodell, 1971). Furthermore, we know from work in other stressful ecosystems that apparent endemism can be caused by competitive exclusion (MacDougall and Turkington, 2005; Veblen and Young, 2009). Although his work predated the popularization of core–periphery and centrifugal organization theories in the literature, Kruckeberg's (1954) experimentation with plant community organization on serpentine tested precisely these theories. In a now classic greenhouse experiment, Kruckeberg (1954) found that an endemic serpentine *Streptanthus* species was able to grow on nonserpentine soils in the absence of competition, supporting the idea that exclusion from the core habitat is not based solely on abiotic preferences. He then sowed a mixture of weedy exotics and a serpentine endemic *Streptanthus* species into two trays, one containing loamy soil and a second containing serpentine soil. On nonserpentine soils, *Streptanthus* germinated but was rapidly outcompeted by the exotics, whereas on serpentine soils, all the weedy exotics died due to toxicity, leaving only the stress-tolerating *Streptanthus*. Kruckeberg concluded that plant community differentiation across serpentine boundaries results from the different plant species' respective intolerances of competition or serpentine soils, with competitively dominant exotics thriving off serpentine, and endemics relegated to serpentine soils where exotics could not invade. In contrast, a greenhouse study by Powell and Knight (2009) found that a trade-off between stress tolerance and competition was unlikely to account for the range restriction of serpentine endemic *Cirsium fontinale*. It is also important to note that greenhouse studies on serpentine effects may not generalize to natural systems, because serpentine soil has been observed by many researchers to change dramatically in structure and texture when removed from the field and kept in a greenhouse. More empirical evidence is needed, particularly from field experimentation, on the role that competitive exclusion might play in driving serpentine endemism.

SPECIAL CONSIDERATIONS FOR THE STUDY OF COMPETITION ALONG SERPENTINE GRADIENTS

Serpentine gradients have both strengths and weaknesses as systems for studying plant interactions. On one hand, the stressors in serpentine systems, in particular the high ratio of magnesium to calcium and low nutrient availability, are relatively amenable to manipulation. These characteristics provide a wide range of possibilities for stress amelioration experiments. On the other hand, contrasting conditions between serpentine and nonserpentine soils include varying degrees of within-site environmental heterogeneity in addition to varying degrees of environmental stress. These may have confounding effects on plant interactions.

Amelioration of Serpentine Stress

Attempts to neutralize serpentine conditions and monitor the resulting competitive dynamics have been numerous, with varied results. Kruckeberg (1954) added calcium to serpentine soils and monitored outcome of competition between a weedy exotic mixture and the endemic *Streptanthus*. Results on serpentine soils with supplemental Ca mimicked those on nonserpentine soils—exotic species outcompeted the serpentine endemic. However, low Ca:Mg ratios are just one component of serpentine stressors, and growth on serpentine soils is not always Ca-limited. Huenneke et al. (1990) found that instead of Ca amelioration, adding macronutrients to serpentine soils seemed to reverse species dominance patterns. In this study, addition of nitrogen and phosphorus promoted invasion and increased dominance of exotic species in an otherwise native forb–dominated serpentine grassland. Similarly, Abraham et al. (2009) found that nitrogen addition increased the growth of an exotic perennial grass but not the productivity of native perennials, suggesting that nitrogen deposition might disproportionately benefit some exotic invaders by increasing their competitive effect. Evidence from these studies and others suggests that amelioration of serpentine stress may allow the colonization of serpentine sites by competitively dominant but less stress-tolerant species. This possibility raises a number of conservation concerns, especially in light of potential climate change and nitrogen deposition, because serpentine soils harbor a plethora of rare species and have thus far been resistant to takeover by exotic species (Proctor, 1999).

Environmental Heterogeneity and Effects on Competition

One of the compelling features of serpentine influenced landscapes is the relatively high level of heterogeneity in soil and biotic communities. This spatial heterogeneity in habitat characteristics promotes the existence of elevated diversity within and among serpentine patches. Environmental heterogeneity is central to most contemporary theories of competitive coexistence, such as spatial or temporal niche differentiation (Amarasekare, 2003; Levine and Rees, 2004), stochastic niche theory (Tilman, 2004), and the spatial storage effect (Chesson et al., 2004; see also Chapter 11 in this volume). Serpentine systems provide an opportunity to test these alternative theories for how spatial heterogeneity affects species coexistence, including whether its effects are deterministic or stochastic (e.g., Harrison et al., in press). Although the most frequently discussed scale of habitat heterogeneity is the contrast between serpentine and nonserpentine sites, it is important to note that within-habitat heterogeneity also varies between these systems. Microsite variation within serpentine patches is high (Baythavong, unpublished data), as are differences in soil and biotic characteristics among serpentine sites (Proctor, 1971; Proctor and Woodell, 1971). Serpentine substrates have higher rates of patchily distributed disturbance than do nonserpentine soils. With gravelly soils and frequent steep slopes, some serpentine sites are characterized by frequent erosion. Exposure of open soil by the Western pocket gopher (*Thomomys bottae*) has been found to be higher at some serpentine sites than in adjacent nonserpentine sites (Hobbs and Mooney, 1995). More research is needed to detail precisely the effect of such heterogeneity on serpentine plant communities and their interactions. Fine-scale spatial heterogeneity might contribute to species coexistence if different species are competitively favored in different adjacent microsites (Pacala and Tilman, 1994; Reynolds et al., 1997). Based on the stress gradient hypothesis, it is tempting to attribute the spectacular diversity that is seen on serpentine soils to unique, low competitive conditions. However, the availability of a multitude of specialized niches, as contrasted with the more homogeneous conditions off serpentine, may be more responsible for these contrasts in diversity and species dominance.

SYNTHESIS

In some respects, serpentine is an ideal system for the study of competition and stress gradients: high contrast in soil regimes and plant communities found often within tens of meters of each other is a methodological boon for field study. Dramatic differences between serpentine and adjacent nonserpentine sites ensure that differences are tested over a range of environmental conditions with clear importance to plants. In addition, spatial habitat heterogeneity at multiple spatial scales provides an opportunity to test, with care, questions about how heterogeneity may influence plant interactions.

As noted, serpentine ecosystems are characterized by the high degree of variation in their soils, productivity, disturbance level, and even the size of patches (Proctor and Woodell, 1971). For these reasons, it should not come as a surprise that results of studies on competition and stress in serpentine systems are somewhat variable. Intersite comparisons indicated that serpentine and their adjacent nonserpentine soils do not differ consistently in the degree of difference in soil toxicity, nutrient levels, productivity, or community composition. To compare the results of competition studies across serpentine sites or other ecosystem types, researchers need to very carefully describe the location and conditions in which species interactions are tested. In particular, rather than testing for differences along a stress gradient" studies might in some cases be effectively testing for differences along a microsite "heterogeneity gradient." This point warrants careful deliberation in the planning of field studies and the exploration of data and statistical analyses.

In addition, there are other factors that merit consideration in the study of competition and stress, on serpentine or other systems, which are rarely incorporated into empirical study. For example, although magnitude and direction of interactions may change over the course of plant life history (Grubb, 1977; Miriti, 2006), most studies focus on the growth and reproduction of competitors, often initiating treatments only at later life stage, rather than monitoring competition throughout the entire life of a plant. Increasingly, attention has been given to the emergence and seedling stage. For example, Abraham et al. (2009) found that the order of emergence affected competition between native perennial and exotic annual grasses on serpentine, and Baack et al. (2006) found that competition at the emergence stage was influential in limiting a native forb to serpentine soils. Another concern is that below-ground interactions are rarely included in experiments that examine competition along serpentine gradients (but see Hooper, 1998; Espeland and Rice, 2007). This is particularly relevant on serpentine soils, where light tends to be abundant and below-ground resources thought to be limiting. Others (Tilman, 1988) have pointed out that below-ground interactions might actually increase in importance in stressful environments. Morphological characteristics of serpentine plants suggest that this may apply here as well: plants growing on serpentine soils tend to have better developed root systems than those growing off serpentine (Brady et al., 2005), as do serpentine ecotypes of species found on and off serpentine (Sambatti and Rice, 2007). Explicit examination of below-ground interactions is generally challenging, and the difficulty of their exploration on serpentine soils is increased by the hard, gravelly texture. Any comparison of root growth on and off of serpentine is likely to be complicated by differences in the difficulty of separating roots from soil across this stress gradient.

Serpentine systems differ from many of the classically studied stress gradients (i.e., elevation, drought, salinity) in that with their complex geology, they contain multiple stressors rather than a single dominant limiting condition. The strong environmental gradients on serpentine may or may not be spatially correlated, and the forces they exert on plants can interact in complex ways. As a result, findings from competition studies on serpentine may not always generalize to other, simpler systems, but may instead reflect the complexity that is characteristic of many other natural systems. Serpentine offers a unique opportunity to study in an environment with dramatic variation in species composition, environmental conditions, and obvious spatial heterogeneity. Although they are perhaps more challenging to study than the single-gradient systems typically addressed in competition theory, careful study of serpentine systems may teach us volumes about how species interactions vary in complex heterogeneous systems.

Acknowledgments

We are grateful to Barbara Going, Laura Feinstein, and Kendi Davies for their insightful comments and discussion on this chapter.

LITERATURE CITED

Abraham, J. K., Corbin, J. D., and D'Antonio, C. M. (2009) California native and exotic perennial grasses differ in their response to soil nitrogen, exotic annual grass density, and order of emergence. *Plant Ecology*, 201, 445–56.

Amarasekare, P. (2003) Competitive coexistence in spatially structured environments: A synthesis. *Ecology Letters*, 6, 1109–22.

Baack, E. J., Emery, N. C., and Stanton, M. L. (2006) Ecological factors limiting the distribution of *Gilia tricolor* in a California grassland mosaic. *Ecology*, 87, 2736–45.

Boyden, S., Binkley, D., and Senock, R. (2005) Competition and facilitation between *Eucalyptus* and nitrogen-fixing *Falcataria* in relation to soil fertility. *Ecology*, 86, 992–1001.

Brady, K. U., Kruckeberg, A. R., and Bradshaw, H. D. (2005) Evolutionary ecology of plant adaptation to serpentine soils. *Annual Review of Ecology Evolution and Systematics*, 36, 243–66.

Brose, U., and Tielborger, K. (2005) Subtle differences in environmental stress along a flooding gradient affect the importance of inter-specific competition in an annual plant community. *Plant Ecology*, 178, 51–59.

Callaway, R. M. (1998) Competition and facilitation on elevation gradients in subalpine forests of the northern Rocky Mountains, USA. *Oikos*, 82, 561–73.

Callaway, R. M., Brooker, R. W., Choler, P., Kikvidze, Z., Lortie, C. J., Michalet, R., Paolini, L., Pugnaire, F. I., Newingham, B., Aschehoug, E. T., Armas, C., Kikodze, D., and Cook, B. J. (2002) Positive interactions among alpine plants increase with stress. *Nature*, 417, 844–48.

Chesson, P., Gebauer, R. L. E., Schwinning, S., Huntly, N., Wiegand, K., Ernest, M. S. K., Sher, A., Novoplansky, A., and Weltzin, J. F. (2004) Resource pulses, species interactions, and diversity maintenance in arid and semi-arid environments. *Oecologia*, 141, 236–53.

Choler, P., Michalet, R., and Callaway, R. M. (2001) Facilitation and competition on gradients in alpine plant communities. *Ecology*, 82, 3295–308.

Elmendorf, S., and Moore, K. (2008) Community composition data outperforms environmental data in predicting species' fecundity and abundance. *Conservation Biology*, 22, 1523–32.

Espeland, E. K., and Rice, K. J. (2007) Facilitation across stress gradients: The importance of local adaptation. *Ecology*, 88, 2404–9.

Eviner, V. T., and Chapin, F. S. (2003) Functional matrix: A conceptual framework for predicting multiple plant effects on ecosystem processes. *Annual Review of Ecology Evolution and Systematics*, 34, 455–85.

Foster, B. L. (2002) Competition, facilitation, and the distribution of *Schizachyrium sco-parium* along a topographic-productivity gradient. *Ecoscience*, 9, 355–63.

Freestone, A. L. (2006) Facilitation drives local abundance and regional distribution of a rare plant in a harsh environment. *Ecology*, 87, 2728–35.

Goldberg, D. E., and Barton, A. M. (1992) Patterns and consequences of interspecific competition in natural communities—a review of field experiments with plants. *American Naturalist*, 139, 771–801.

Grubb, P. J. (1977) Maintenance of species-richness in plant communities—importance of regeneration niche. *Biological Reviews of the Cambridge Philosophical Society*, 52, 107–45.

Harrison, S., Cornell, H. V., and Moore, K.A. (In Press) Spatial niches and coexistence: Testing theory with tarweeds. *Ecology*.

Harrison, S., Safford, H. D., Grace, J. B., Viers, J. H. and Davies, K. F. (2006) Regional and local species richness in an insular environment: Serpentine plants in California. *Ecological Monographs*, 76, 41–56.

Hobbs, R. J., and Mooney, H. A. (1995) Spatial and temporal variability in California annual grassland—results from a long-term study. *Journal of Vegetation Science*, 6, 43–56.

Hooper, D. U. (1998) The role of complementarity and competition in ecosystem responses to variation in plant diversity. *Ecology*, 79, 704–19.

Huenneke, L. F., Hamburg, S. P., Koide, R., Mooney, H. A.. and Vitousek, P. M. (1990) Effects of soil resources on plant invasion and community structure in Californian serpentine grassland. *Ecology*, 71, 478–91.

Jurjavcic, N. L., Harrison, S., and Wolf, A. T. (2002) Abiotic stress, competition, and the distribution of the native annual grass *Vulpia microstachys* in a mosaic environment. *Oecologia*, 130, 555–62.

Kadmon, R. (1995) Plant competition along soil-moisture gradients—a field experiment with the desert annual *Stipa capensis*. *Journal of Ecology*, 83, 253–62.

Kruckeberg, A. R. (1954) The ecology of serpentine soils. 3. Plant species in relation to serpentine soils. *Ecology*, 35, 267–74.

La Peyre, M. K. G., Grace, J. B., Hahn, E., and Mendelssohn, I. A. (2001) The importance of competition in regulating plant species abundance along a salinity gradient. *Ecology*, 82, 62–69.

Levine, J. M., and Rees, M. (2004) Effects of temporal variability on rare plant persistence in annual systems. *American Naturalist*, 164, 350–63.

Liancourt, P., Callaway, R. M., and Michalet, R. (2005) Stress tolerance and competitive-response ability determine the outcome of biotic interactions. *Ecology*, 86, 1611–18.

Lortie, C. J., and Callaway, R. M. (2006) Re-analysis of meta-analysis: Support for the stress-gradient hypothesis. *Journal of Ecology*, 94, 7–16.

MacDougall, A. S., and Turkington, R. (2005) Are invasive species the drivers or passengers of change in degraded ecosystems? *Ecology*, 86, 42–55.

Maestre, F. T., Callaway, R. M., Valladares, F., and Lortie, C. J. (2009) Refining the stress-gradient hypothesis for competition and facilitation in plant communities. *Journal of Ecology*, 97, 199–205.

Maestre, F. T., Valladares, F., and Reynolds, J. F. (2005) Is the change of plant-plant interactions with abiotic stress predictable? A meta-analysis of field results in arid environments. *Journal of Ecology*, 93, 748–57.

Maestre, F. T., Valladares, F., and Reynolds, J. F. (2006) The stress-gradient hypothesis does not fit all relationships between plant-plant interactions and abiotic stress: Further insights from arid environments. *Journal of Ecology*, 94, 17–22.

Miriti, M. N. (2006) Ontogenetic shift from facilitation to competition in a desert shrub. *Journal of Ecology*, 94, 973–79.

Pacala, S. W., and Tilman, D. (1994) Limiting similarity in mechanistic and spatial models of plant competition in heterogeneous environments. *American Naturalist*, 143, 222–57.

Peltzer, D. A., Wilson, S. D., and Gerry, A. K. (1998) Competition intensity along a productivity gradient in a low-diversity grassland. *American Naturalist*, 151, 465–76.

Powell, K. I., and Knight, T. M. (2009) Effects of nutrient addition and competition on biomass of five *Cirsium* species (Asteraceae), including a serpentine endemic. *International Journal of Plant Sciences*, 170, 918–25.

Proctor, J. (1971) Plant ecology of serpentine. 2. Plant response to serpentine soils. *Journal of Ecology*, 59, 397.

Proctor, J. (1999) Toxins, nutrient shortages and droughts: The serpentine challenge. *Trends in Ecology and Evolution*, 14, 334–35.

Proctor, J., and Woodell, S. R. J. (1971) Plant ecology of serpentine. 1. Serpentine vegetation of England and Scotland. *Journal of Ecology*, 59, 375–96.

Proctor, J., and Woodell, S. R. J. (1975) The ecology of serpentine soils. *Advances in Ecological Research*, 9, 255–366.

Reader, R. J., Wilson, S. D., Belcher, J. W., Wisheu, I., Keddy, P. A., Tilman, D., Morris, E. C., Grace, J. B., McGraw, J. B., Olff, H., Turkington, R., Klein, E., Leung, Y., Shipley, B., Vanhulst, R., Johansson, M. E., Nilsson, C., Gurevitch, J., Grigulis, K., and Beisner, B. E. (1994) Plant competition in relation to neighbor biomass—an Intercontinental study with *Poa pratensis*. *Ecology*, 75, 1753–60.

Reynolds, H. L., Hungate, B. A., Chapin, F. S., and Dantonio, C. M. (1997) Soil heterogeneity and plant competition in an annual grassland. *Ecology*, 78, 2076–90.

Rosenzweig, M. L., and Abramsky, A. (1986) Centrifugal community organization. *Oikos*, 46, 339–45.

Sambatti, J. B. M., and Rice, K. J. (2007) Functional ecology of ecotypic differentiation in the Californian serpentine sunflower (*Helianthus exilis*). *New Phytologist*, 175, 107–19.

Sammul, M., Kull, K., Oksanen, L., and Veromann, P. (2000) Competition intensity and its importance: Results of field experiments with *Anthoxanthum odoratum*. *Oecologia*, 125, 18–25.

Tilman, D. (1988) *Plant Strategies and the Dynamics and Structure of Plant Communities*. Princeton University Press, Princeton, NJ.

Tilman, D. (2004) Niche tradeoffs, neutrality, and community structure: A stochastic theory of resource competition, invasion, and community assembly. *Proceedings of the National Academy of Sciences, USA*, 101, 10854–61.

Veblen, K. E., and Young, T. P. (2009) A California grasslands alkali specialist, *Hemizonia pungens* ssp *pungens*, prefers non-alkali soil. *Journal of Vegetation Science*, 20, 170–76.

Walker, R. B. (1954) The ecology of serpentine soils 2. Factors affecting plant growth on serpentine soils. *Ecology,* 35, 259–66.

Wardle, D. A., Barker, G. M., Bonner, K. I., and Nicholson, K. S. (1998) Can comparative approaches based on plant ecophysiological traits predict the nature of biotic interactions and individual plant species effects in ecosystems? *Journal of Ecology,* 86, 405–20.

Williamson, J., and Harrison, S. (2002) Biotic and abiotic limits to the spread of exotic revegetation species. *Ecological Applications,* 12, 40–51.

Wilson, S. D., and Keddy, P. A. (1986a) Measuring diffuse competition along an environmental gradient—results from a shoreline plant community. *American Naturalist,* 127, 862–69.

Wilson, S. D., and Keddy, P. A. (1986b) Species competitive ability and position along a natural stress disturbance gradient. *Ecology,* 67, 1236–42.

Wisheu, I. C., and Keddy, P. A. (1992) Competition and centrifugal organization of plant communities: Theory and tests. *Journal of Vegetation Science,* 3, 147–56.

11

Community Invasibility

Spatial Heterogeneity, Spatial Scale, and Productivity

Kendi F. Davies, *University of Colorado, Boulder*

INTRODUCTION: WHY ARE SOME COMMUNITIES MORE INVASIBLE THAN OTHERS?

Invasive species are one of the most significant threats to native species diversity, and identifying the factors that make places more or less invasible has been one of the most important issues in the study of invasions (Wilcove et al., 1998; Pimentel et al., 2000). From a theoretical perspective, the reasons some communities are more invasible than others is a question that continues to intrigue ecologists (Levine et al., 2004; Davies et al., 2005; Fridley et al., 2007; Stohlgren et al., 2008; Cadotte et al., 2009) because it underlies fundamental concepts in community ecology: species coexistence and assembly (Chesson, 2000; Tilman, 2004). Therefore, its exploration offers insights into why communities are structured as they are.

Serpentine systems have provided significant insight into the reasons some communities are more invasible than others because the environment within these systems is often extreme. Spatial heterogeneity, spatial scale, and productivity have all proven to be critical elements in understanding the invasibility of communities. Serpentine systems have allowed researchers to examine these elements in greater depth. First, serpentine systems are very spatially heterogeneous in soil chemistry, texture, rockiness, and toxicity. Importantly, this spatial heterogeneity can exist at very small scales, making it easier to examine the effects of it on communities. Second, because spatial heterogeneity can be at both small scales and large scales,

serpentine systems have allowed researchers to uncover the importance of processes operating at both small (local) and large (regional) scales and how these
processes relate to spatial heterogeneity. Finally, serpentine systems can range
from very harsh and unproductive to quite productive. Thus, they provide an unusually wide range of productivity, uncovering processes that might not be possible to see in other systems. In what follows, I take each of these elements—spatial
heterogeneity, scale, and productivity—and contrast the contributions of studies
in serpentine systems compared with studies in systems that are not serpentine.

SPATIAL SCALE, SPATIAL HETEROGENEITY, AND THE INVASIBILITY OF COMMUNITIES

Elton (1958) first proposed that a high richness of native species armors sites
against invasion by making resources less available to newly arriving species. This
idea has been supported by many empirical studies that detected negative relationships between native and exotic diversity at small spatial scales—the scale of interaction between individuals (Elton, 1958; Turelli, 1981; Case, 1990; Tilman, 1997;
Knops et al., 1999; Stachowicz et al., 1999; Levine, 2000; Naeem et al., 2000; Lyons
and Schwartz, 2001; Brown and Peet, 2003; Levine et al., 2004). Thus, the idea that
competitive exclusion (Grime, 1973; Tilman, 1999) is in place at small scales, leading to resistance to invasion, is well established and has been confirmed experimentally: a recent meta-analysis of biotic resistance to invasion, incorporating 65
experiments from 24 different studies, revealed that competition from resident
species has strong and significant effects on both establishment and performance
of invaders (Levine et al., 2004).

In the late 1990s, Stohlgren et al. (1999) and Lonsdale (1999) showed that native and exotic diversity could be positively correlated when scales larger than that
of local interaction were considered. Their results first raised the possibility that
the processes determining invasibility and its relationship to diversity depend on
spatial scale. Since then, many other studies have reported positive relationships
between native and exotic diversity at large scales and furthermore that the most
diverse regions are also often the most invaded, particularly for plant communities
(Levine, 2000; Davies et al., 2005, 2007; Richardson et al., 2005; Stohlgren et al.,
2005; Harrison et al., 2006).

Since the studies of Stohlgren et al. (1999) and others, the potential for scale
dependence of the native–exotic relationship has stimulated much research, and
studies have reported both negative relationships at small scales and positive relationships at large scales within the same system (Levine, 2000; Brown and Peet,
2003; Davies et al., 2005; Knight and Reich, 2005).

Why native and exotic diversity should be positively correlated at large scales so
that diverse regions are also the most invaded regions has puzzled ecologists. Two

hypotheses could explain the positive correlation of native and exotic diversity at large scales. First, the environmental favorability hypothesis suggests that sites or landscapes with favorable environmental conditions for native species also have favorable conditions for exotic species (Levine and D'Antonio, 1999; Stohlgren et al., 1999; Levine, 2000; Brown and Peet, 2003). In other words, factors such as soil fertility or propagule supply rates vary between sites and cause between-site variation in both native and exotic diversity. Shea and Chesson (2002) generalized this idea and showed how a positive relationship at large spatial scales can arise by combining data from a series of negative relationships at smaller scales, where differences in diversity at larger scales were caused by environmental differences in the mean conditions between sites. However, their model only accounts for patterns in the mean diversity (alpha diversity) rather than the cumulative diversity (gamma diversity) of communities (Davies et al., 2005). Second, the environmental heterogeneity hypothesis suggests that not only variation in mean conditions between sites (as in the environmental favorability hypothesis) but also heterogeneity of conditions within sites can contribute to the positive relationship of native and exotic diversity at large scales.

What causes the relationship between native and exotic diversity to change from negative at small scales to positive at large scales? Davies et al. (2005) suggested, in line with evidence just discussed, that negative relationships tend to be detected at scales at which the environment and resources are relatively homogeneous and classic niche partitioning (and competitive exclusion) dominate (Grime, 1973; Tilman, 1999). They suggested that the relationship between native and exotic diversity becomes positive at scales at which spatial heterogeneity in the environment is such that coexistence mechanisms that depend on heterogeneity become dominant, resulting in communities that could be considered unsaturated. A comprehensive review of the diversity-invasibility paradox (Fridley et al., 2007) came up with a similar framework built on the shift from biotic to abiotic drivers as scale increases. Eight processes that could generate either negative or positive relationships were identified but all could be fitted within Fridley and colleagues' (2007) framework.

Recently, Stohlgren et al. (2008) suggested that invasion, rather than diminishing the diversity of native species in a community, has the overall effect of increasing the diversity of communities by adding exotic species. They based their conclusion on evidence that communities are unsaturated at large spatial scales, and, at least for plants, the lack of known species extinctions driven by competitive interactions. Consequently, we should generally find positive relationships between native and exotic diversity at large scales. Stohlgren and colleagues (2008) suggest that ultimately competition appears to play no role in structuring communities at large scales, even if it is important in structuring communities at small scales. With further investigation, however, this final point is proving to be incorrect (see also Harrison, 2008). Recent evidence from phylogenetic relatedness data suggests that competition still structures communities at large scales, in particular in determining which species can be present in a community, despite positive relationships between native and exotic diversity (Davies et al., submitted). I expand on the evidence from phylogenetic studies shortly.

Insights from Serpentine Systems

In a nested data set for grassland plants in Californian serpentine grasslands in the University of California McLaughlin Reserve, Davies et al. (2005) detected negative relationships between native and exotic diversity at small scales and positive relationships at large scales. Furthermore, both native and exotic diversity were positively correlated with spatial heterogeneity in abiotic conditions (variance of soil depth, soil nitrogen, and aspect) but were uncorrelated with average abiotic conditions, supporting the spatial heterogeneity hypothesis but not the favorable conditions hypothesis. Thus, Davies et al. (2005) demonstrated that the observed relationship between native and exotic diversity flipped from negative to positive at scales at which spatial heterogeneity in the environment came into play (correlated with native and exotic diversity and beta diversity).

These serpentine studies were able to show that spatial heterogeneity in species composition (beta diversity) and spatial environmental heterogeneity within metacommunities drove the positive relationship between native and exotic diversity at large scales (supporting the environmental heterogeneity hypothesis), rather than differences in mean (extrinsic) conditions between metacommunities (environmental favorability hypothesis). These observations are consistent with invasion and coexistence theories in heterogeneous environments. Habitat heterogeneity increases the number of both native and exotic species in metacommunities by allowing more species to invade while placing the resident native species at lower risk of extinction because of the presence of more niche opportunities for both natives and exotics in the presence of heterogeneity (Shea and Chesson, 2002; Pauchard and Shea, 2006).

Recently, Davies and colleagues (submitted) further tested these hypotheses using the phylogenetic relatedness of natives and exotics in this same system. They tested two hypotheses. First, if competitive exclusion and classic niche partitioning dominates, successful invaders should have niches that are different from those of the natives present. Thus, native and exotic species should be more distantly related than expected from a random assemblage model. Second, given the role of coexistence mechanisms dependent on environmental heterogeneity, native and exotic species should not be more distantly related than expected, and may be more closely related than expected if the environment filters membership of communities at large scales.

Davies et al. (submitted) found strong support for the first hypothesis, providing further evidence that competitive exclusion dominates at small scales. However, the second hypothesis was rejected: native and exotic species were more distantly related at large scales than expected. However, importantly, natives and exotics were significantly more distantly related at small scales than they were at large scales. These results suggest that as we transition from small to large scales, the effect of biotic resistance exclusion is relaxed but still present. In other words, although communities were saturated at small scales, they were not saturated at larger scales: more species could enter the community, thus increasing species richness. However, species did not invade indiscriminately at large scales; which species successfully invaded was determined in part by competitive exclusion. At large scales, communities were still resistant to invasion by particular invaders. Exotic species that were closely related to the species already established in the community were excluded.

The extreme and small scale of heterogeneity in abiotic conditions was critical for the foregoing insights. In particular variability in soil depth, soil nitrogen and aspect were important. Typical of serpentine systems, the underlying bedrock and the rockiness of the soil itself was very heterogeneous in space and contrasted strongly with the more "normal" soils that intersperse the serpentine soils. Similarly, the microtopography was very heterogeneous in space so that many microaspects were represented within a small area. Because of the mixture of serpentine and nonserpentine soils, the range of soil nitrogen was large. These extremes in heterogeneity likely made the effects of heterogeneity easier to detect. Similarly, the small scale of heterogeneity likely made it possible to contrast the effects of small and large spatial scales over a relatively small (and therefore intensely studied) area.

PRODUCTIVITY/HARSHNESS, SCALE, AND THE INVASIBILITY OF COMMUNITIES

Site productivity or harshness has also been used to try to understand why communities differ in their invasibility. Harsh conditions can alter the invasibility of communities (Cleland et al., 2004; Beisner et al., 2006; Davies et al., 2007; Perelman et al., 2007). I define harsh conditions to mean those that result from physiological or resource stress, resulting in low population densities (Chesson and Huntly, 1997). Although harshness alone cannot lead to lack of saturation (Chesson and Huntly, 1997), it could reduce saturation by interacting with other processes. Then the invasibility of communities may vary with productivity, even at the small spatial scales where competitive exclusion is expected to be the dominant mechanism. These ideas can essentially be traced back to Elton's (1958) hypothesis that increased diversity should lead to increased resource capture by the community but leave fewer resources available to potential invaders (e.g., Tilman, 1982).

Three hypotheses explain how productivity interacts with community invasibility: the environmental heterogeneity, environmental favorability, and the facilitation hypothesis. The environmental heterogeneity hypothesis proposes that harsh sites are more internally heterogeneous at small scales than productive sites are, so that in harsh sites, local scale plots with greater heterogeneity have more niches for both natives and exotics, leading to coexistence and a positive relationship between native and exotic diversity (Davies et al., 2005; Melbourne et al., 2007). In comparison, at productive sites, homogeneity of resources would lead to competitive exclusion of exotics by natives and a negative relationship. First, harsh sites might simply have greater environmental heterogeneity than productive sites at small scales. Sources of heterogeneity within sites could include rockiness, soil depth, and soil composition. A second possibility, suggested by Tilman (1988, 1987), is that competition occurs predominantly above ground at productive sites, where species compete for light, a single limiting resource, and shifts below ground at unproductive sites, where species compete for below-ground resources that tend to be more heterogeneous (e.g., Wilson and Tilman, 1991, 1993, 1995). Both possibilities lead to positive relationships between native and exotic diversity in harsh sites, even at very small spatial scales—the scale of interaction between individuals.

The environmental favorability hypothesis covers coexistence mechanisms that depend on mean conditions rather than heterogeneity in conditions (as already discussed). This hypothesis suggests that native and exotic diversity are positively correlated at large spatial scales because sites with favorable conditions for natives also have favorable conditions for exotics. Although this hypothesis was developed to explain large-scale positive relationships, it can also be applied at small spatial scales if communities are unsaturated so that there are weaker effects of resident species on the ability of new species to invade. This results in positive relationships between native and exotic diversity even at small spatial scales.

In the context of community invasibility and its relationship to productivity, most research examines the environmental favorability hypothesis, and researchers have found evidence to support it. An analysis of four long-term observational data sets for herb communities from long-term ecological research sites found that invaders were less likely to establish in high-productivity sites. The probability of invasion was reduced both within and following years with high productivity, except at a desert grassland site where high productivity was associated with increased invasion (Cleland et al., 2004). In harsh years, community saturation was reduced, changing the dynamic from one of competitive exclusion of invaders to coexistence. Finally, in Flooding Pampas of Argentina, sites that were harsher (more saline, shallower soils) were less invaded, but native and exotic diversity were still positively correlated across sites at large spatial scales (Perelman et al., 2007). Thus, results are somewhat mixed. These studies focus on how diversity and invasibility are related at large scales, but to truly understand the effect of productivity, we need to study its effects at small scales, the scale of local species interaction. This has usually been done only in the context of an experimental manipulation that looks at a single invader. For example, in Jamaican rock pools, high productivity (through experimentally boosted resources) increased the success rate of establishment of an invasive crustacean (Beisner et al., 2006). Local-scale, community-level analyses are needed.

The third hypothesis concerns facilitation, which can be more likely to occur at harsh, unproductive sites (Bertness and Callaway, 1994; Callaway, 1998; Choler et al., 2001; Elmendorf and Moore, 2007). Theory connecting facilitation to community invasibility is undeveloped, so it is unclear how facilitation would be expected to alter the diversity–invasibility relationship in a metacommunity context (but for invader facilitating invader. see Simberloff, 1999, 2006). However, facilitative interactions may be more common in general and more common than competitive interactions under harsh conditions resulting in different patterns of invasibility of communities at harsh and unproductive sites. This area warrants further investigation.

Insights from Serpentine Systems

Davies et al. (2007) examined California statewide scale data set for serpentine herbs, composed of 109 sites. They investigated the native–exotic diversity relationship at three scales (range: 1 m²–4000 km²) in a system with a wide range in the productivity of sites—communities were established in conditions that ranged from very harsh to relatively productive. Native and exotic diversity were positively correlated at all spatial scales, even the smallest ones, at which individuals were expected to interact directly (1 m²). Positive relationships at large scales are expected and common in empirical studies; however, detecting a positive relationship at the small scales of interaction between individuals is rare. In contrast, results from the statewide data set suggest that sites were not saturated, even at small scales (1 m²).

Davies et al. (2007) predicted that along with spatial scale, site productivity likely affected the invasibility of communities and thus the relationship between native and exotic diversity, especially at small scales, where competitive exclusion potentially varied with site productivity. They found that, at small-scale (1 m²), high-productivity sites, native and exotic diversity tended to be negatively correlated, whereas at unproductive sites, native and exotic diversity tended to be positively correlated, resulting in a significant relationship between the slope of the native exotic relationship and productivity. Also, because the majority of sites were less productive, the average relationship at small spatial scales for all sites considered together was positive. Furthermore, when productive and unproductive sites were considered separately, productive sites had a common positive relationship between native and exotic diversity, whereas unproductive sites had a common negative relationship.

Like most other studies concerning community invasibility and productivity, the environmental favorability hypothesis, likely explains Davies et al.'s (2007) finding. Recall that the environmental favorability hypothesis incorporates coexistence mechanisms that depend on mean conditions rather than heterogeneity in conditions. Davies and colleagues hypothesized that in their serpentine system, the majority of sites, except for the small subset with the highest herb cover, are probably unsaturated so that there are weaker effects of resident species on the ability of new species to invade. The positive effects of shared responses to the environment (i.e., shared tolerances) more easily overwhelm the negative effects of competition, resulting in a small-scale positive relationship between native and exotic diversity at harsh sites. For example, the availability of critical nutrients may vary between 1 m² quadrats so that quadrats with better conditions harbor both more natives and exotics. In contrast, at productive sites, competitive exclusion results in negative relationships between native and exotic diversity.

The environmental heterogeneity hypothesis did not explain the observed relationships between native and exotic diversity because if heterogeneity was behind the positive relationships detected, the presence of greater heterogeneity at unproductive sites should have resulted in greater diversity of both native and exotic species at harsh sites, whereas the opposite was the case. The least productive sites also had the fewest species, making heterogeneity-dependent coexistence an unlikely driver of the change in slope between productive and less productive sites at small spatial scales. Finally, Davies et al. (2007) could not discount the facilitation hypothesis, but neither did they have evidence to support it. The presence of facilitative (rather than competitive) local interactions at harsh compared to benign sites could have resulted in the relationships that were observed. Facilitation is more likely to occur at harsh, unproductive sites (Bertness and Callaway, 1994), but because theory connecting facilitation to community invasibility is undeveloped, it is not clear how facilitation would alter community invasibility in a metacommunity context.

Critically, these findings illustrate that the mechanisms that determine the invasibility of communities, in this case measured as the change in slope of the relationship between native and exotic diversity, do not depend on spatial scale per se but can change whenever or wherever environmental conditions change to promote species coexistence rather than competitive exclusion. In the 2007 serpentine study, this occurred within a single small spatial scale when the environment shifted from being locally productive (competitive exclusion) to unproductive (coexistence). It is very likely that the extreme range of productivity values at sites in this serpentine study system was critical to this discovery.

SYNTHESIS

Three unique abiotic features of serpentine systems make them superior models for studying community invasibility. First, the environment can be heterogeneous at very small spatial scales, making it easier to measure the effects of spatial heterogeneity on community processes. It can be easier to set up strategies that sample large amounts of heterogeneity. It can also be easier to sample at multiple spatial scales that incorporate significant spatial heterogeneity in the environment (even small scales). For example, at the McLaughlin Reserve grid, variables like soil depth and soil nutrient concentrations, which turned out to be critical to community structure and invasibility, can vary significantly within 1–2 m.

Second, not only can the environment be variable at very small spatial scales, the range of values encountered within any given environmental variable may be more extreme than in other systems, making it easier to detect the effects of spatial heterogeneity on community invasibility. For example, the range of soil nitrogen values between serpentine and interspersed nonserpentine soils can be large. Similarly, microtopography may be very heterogeneous in space so that many "micro" aspects are represented within a small area—the range of aspects present at small scales is larger than in other systems.

Third, following from the last point, productivity can vary widely within serpentine systems from sites that are barren and rocky to sites that are productive. Again, this provides researchers with better opportunities for determining the importance of site productivity on community structure and community invasibility. A study system with a smaller range of productivity values may not have allowed researchers to detect the trends described herein.

LITERATURE CITED

Beisner, B. E., Hovius, J., Hayward, A., Kolasa, J., and Romanuk, T. N. (2006) Environmental productivity and biodiversity effects on invertebrate community invasibility. *Biological Invasions*, 8, 655–64.

Bertness, M. D., and Callaway, R. (1994) Positive interactions in communities. *Trends in Ecology and Evolution*, 9, 191–93.

Brown, R. L., and Peet, R. K. (2003) Diversity and invasibility of southern Appalachian plant communities. *Ecology*, 84, 32–39.

Cadotte, M. W., Hamilton, M. A., and Murray, B. R. (2009) Phylogenetic relatedness and plant invader success across two spatial scales. *Diversity and Distributions*, 15, 481–88.

Callaway, R. M. (1998) Competition and facilitation on elevation gradients in subalpine forests of the northern Rocky Mountains, USA. *Oikos*, 82, 561–73.

Case, T. J. (1990) Invasion resistance arises in strongly interacting species-rich model competition communities. *Proceedings of the National Academy of Sciences, USA*, 87, 9610–14.

Chesson, P. (2000) Mechanisms of maintenance of species diversity. *Annual Review of Ecology and Systematics*, 31, 343–66.

Chesson, P., and Huntly, N. (1997) The roles of harsh and fluctuating conditions in the dynamics of ecological communities. *American Naturalist*, 150, 519–53.

Choler, P., Michalet, R., and Callaway, R. M. (2001) Facilitation and competition on gradients in alpine plant communities. *Ecology*, 82, 3295–308.

Cleland, E. E., Smith, M. D., Andelman, S. J., Bowles, C., Carney, K. M., Horner-Devine, M. C., Drake, J. M., Emery, S. M., Gramling, J. M., and Vandermast, D. B. (2004) Invasion in space and time: Non-native species richness and relative abundance respond to interannual variation in productivity and diversity. *Ecology Letters*, 7, 947–7.

Davies, K. F., Cavender-Bares, J., and Deacon, N. (Submitted) Native communities determine the identity of exotic invaders even at scales at which communities are unsaturated.

Davies, K. F., Chesson, P., Harrison, S., Inouye, B. D., Melbourne, B. A., and Rice, K. J. (2005) Spatial heterogeneity explains the scale dependence of the native-exotic diversity relationship. *Ecology*, 86, 1602–10.

Davies, K. F., Harrison, S., Safford, H. D., and Viers, J. H. (2007) Productivity alters the scale dependence of the diversity-invasibility relationship. *Ecology*, 1940–47.

Elmendorf, S. C., and Moore, K. A. (2007) Plant competition varies with community composition in an edaphically complex landscape. *Ecology*, 88, 2640–50.

Elton, C. S. (1958) *The Ecology of Invasions*. Methuen, London.

Fridley, J. D., Stachowicz, J. J., Naeem, S., Sax, D. F., Seabloom, E. W., Smith, M. D., Stohlgren, T. J., Tilman, D., and Von Holle, B. (2007) The invasion paradox: Reconciling pattern and process in species invasion. *Ecology*, 88, 3–17.

Grime, J. P. (1973) Control of species density in herbaceous vegetation. *Journal of Environmental Management*, 1, 151–67.

Harrison, S. (2008) Commentary on Stohlgren et al. (2008): The myth of plant species saturation. *Ecology Letters*, 11, 322–24.

Harrison, S., Grace, J. B., Davies, K. F., Safford, H. D., and Viers, J. H. (2006) Invasion in a diversity hotspot: Exotic cover and native richness in the Californian serpentine flora. *Ecology*, 87, 695–703.

Knight, K. S., and Reich, P. B. (2005) Opposite relationships between invasibility and native species richness at patch versus landscape scales. *Oikos*, 109, 81–88.

Knops, J. M. H., Tilman, D., Haddad, N. M., Naeem, S., Mitchell, C. E., Haarstad, J., Ritchie, M. E., Howe, K. M., Reich, P. B., Siemann, E., and Groth, J. (1999) Effects of plant species richness on invasion dynamics, disease outbreaks, insect abundances, and diversity. *Ecology Letters*, 2, 286–93.

Levine, J. M. (2000) Species diversity and biological invasions: Relating local process to community pattern. *Science*, 288, 852–54.

Levine, J. M., Adler, P. B., and Yelenik, S. G. (2004) A meta-analysis of biotic resistance to exotic plant invasions. *Ecology Letters*, 7, 975–89.

Levine, J. M., and D'Antonio, C. M. (1999) Elton revisited: A review of evidence linking diversity and invasibility. *Oikos*, 87, 15–26.

Lonsdale, W. M. (1999) Global patterns of plant invasions and the concept of invasibility. *Ecology*, 80, 1522–36.

Lyons, K.G., and Schwartz, M.W. (2001) Rare species loss alters ecosystem function—invasion resistance. *Ecology Letters*, 4, 358–65.

Melbourne, B.A., Cornell, H.V., Davies, K.F., Dugaw, C.J., Elmendorf, S., Freestone, A.L., Hall, R.J., Harrison, S., Hastings, A., Holland, M., Holyoak, M., Lambrinos, J., Moore, K., and Yokomizo, H. (2007) Invasion in a heterogeneous world: Resistance, coexistence or hostile takeover? *Ecology Letters*, 10, 77–94.

Naeem, S., Knops, J.M.H., Tilman, D., Howe, K.M., Kennedy, T., and Gale, S. (2000) Plant diversity increases resistance to invasion in the absence of covarying extrinsic factors. *Oikos*, 91, 97–108.

Pauchard, A., and Shea, K. (2006) Integrating the study of non-native plant invasions across spatial scales. *Biological Invasions*, 8, 399–413.

Perelman, S.B., Chaneton, E.J., Batista, W.B., Burkart, S.E., and Leon, R.J.C. (2007) Habitat stress, species pool size and biotic resistance influence exotic plant richness in the Flooding Pampa grasslands. *Journal of Ecology*, 95, 662–73.

Pimentel, D., Lach, L., Zuniga, R., and Morrison, D. (2000) Environmental and economic costs of nonindigenous species in the United States. *Bioscience*, 50, 53–65.

Richardson, D.M., Rouget, M., Ralston, S.J., Cowling, R.M., Van Rensburg, B.J., and Thuiller, W. (2005) Species richness of alien plants in South Africa: Environmental correlates and the relationship with indigenous plant species richness. *Ecoscience*, 12, 391–402.

Shea, K., and Chesson, P. (2002) Community ecology theory as a framework for biological invasions. *Trends in Ecology and Evolution*, 17, 170–76.

Simberloff, D. (1999) Positive interactions of nonindigenous species: Invasional meltdown? *Biological Invasions*, 1, 21–32.

Simberloff, D. (2006) Invasional meltdown 6 years later: Important phenomenon, unfortunate metaphor, or both? *Ecology Letters*, 9, 912–19.

Stachowicz, J.J., Whitlatch, R.B., and Osman, R.W. (1999) Species diversity and invasion resistance in a marine ecosystem. *Science*, 286, 1577–79.

Stohlgren, T.J., Barnett, D., Flather, C., Kartesz, J., and Peterjohn, B. (2005) Plant species invasions along the latitudinal gradient in the United States. *Ecology*, 86, 2298–309.

Stohlgren, T.J., Barnett, D.T., Jarnevich, C.S., Flather, C., and Kartesz, C. (2008) The myth of plant species saturation. *Ecology Letters*, 11, 315–26.

Stohlgren, T.J., Binkley, D., Chong, G.W., Kalkhan, M.A., Schell, L.D., Bull, K.A., Otsuki, Y., Newman, G., Bashkin, M., and Son, Y. (1999) Exotic plant species invade hot spots of native plant diversity. *Ecological Monographs*, 69, 25–46.

Tilman, D. (1982) *Resource Competition and Community Structure*. Princeton University Press, Princeton, NJ.

Tilman, D. (1987) On the meaning of competition and the mechanisms of competitive superiority. *Functional Ecology*, 1, 304–15.

Tilman, D. (1988) *Plant Strategies and the Dynamics and Structure of Plant Communities*. Princeton University Press, Princeton, NJ.

Tilman, D. (1997) Community invasibility, recruitment limitation, and grassland biodiversity. *Ecology*, 78, 81–92.

Tilman, D. (1999) The ecological consequences of changes in biodiversity: A search for general principles. *Ecology*, 80, 1455–74.

Tilman, D. (2004) Niche tradeoffs, neutrality, and community structure: A stochastic theory of resource competition, invasion, and community assembly. *Proceedings of the National Academy of Sciences, USA,* 101, 10854–61.

Turelli, M. (1981) Niche overlap and invasion of competitors in random environments. 1. Models without demographic stochasticity. *Theoretical Population Biology,* 20, 1–56.

Wilcove, D.S., Rothstein, D., Dubow, J., Phillips, A., and Losos, E. (1998) Quantifying threats to imperiled species in the United States. *Bioscience,* 48, 607–15.

Wilson, S.D., and Tilman, D. (1991) Components of plant competition along an experimental gradient of nitrogen availability. *Ecology,* 72, 1050–65.

Wilson, S.D., and Tilman, D. (1993) Plant competition and resource availability in response to disturbance and fertilization. *Ecology,* 74, 599–611.

Wilson, S.D., and Tilman, D. (1995) Competitive responses of 8 old-field plant-species in 4 environments. *Ecology,* 76, 1169–80.

12

Disturbance and Diversity in Low-Productivity Ecosystems

Hugh D. Safford, *Pacific Southwest Research Station,*
USDA Forest Service
Chris R. Mallek, *University of California, Davis*

Ecological theory predicts that the amount of resources available in an ecosystem should affect its response and sensitivity to ecological disturbances like fire, herbivory, and soil disturbance. Plant stature and life form, biomass, rates of growth, and plant palatability are all influenced by habitat productivity, and these factors play key roles in determining disturbance frequencies and intensities (Pickett and White, 1985; Bond and van Wilgen, 1996; Grime, 2001; Table 12.1). Theory and empirical investigations find that site quality is often related to rates of competitive displacement, with relatively unproductive, less competitive environments less reliant on disturbance for diversity regulation (Huston, 1994; Grime, 2001). Because the most significant direct effect of disturbance on vegetation is to increase available space and light (Grace, 1999), the effects of disturbance on plant diversity should correlate positively with productivity, since more productive plant communities are more limited by above-ground competition (Tilman, 1982). Ecologically, a disturbance of a given intensity will cause more change in space and light availability in dense vegetation than in open habitats, where these resources are already more abundant. On evolutionary time scales, species are less likely to specialize on regenerating after disturbance in communities where space and light are less limiting (Grubb, 1977; Bond and van Wilgen, 1996).

Given the great differences in productivity between serpentine and nonserpentine habitats, coupled with the heightened presence of endemics and species of

TABLE 12.1 "Contextual" Features of Serpentine Soil and Vegetation That Influence the Relationship between Disturbance and Plant Diversity

Feature	Serpentine Effect
Ecosystem structure	Stunted vegetation; relatively sparse, heterogeneous plant cover; low stem density and woody plant cover; relatively low biomass; significant areas of exposed soil; low canopy height; greater development of understory vegetation.
Resource base	Infertile soils with high Mg; low Ca, N, P, and K; high heavy metals (Cr, Ni). Available water capacity (AWC) can be higher or lower than normal soils, but many authors refer to low soil moisture of serpentine soils.
Species traits	Relatively slow growth rates, compared to conspecifics or congeners on more fertile soils. Many endemic plant species. Serpentine plants often exhibit xeromorphic traits traits such as small, thick, hairy, and/or evergreen foliage. In some regions (mostly humid, warm climates), some species hyperaccumulate heavy metals (Cr, Ni, Co), which may act as an herbivore defense. In California, some evidence of *higher* palatability of serpentine versus nonserpentine grasses. Root systems often deeper and better-developed than in more fertile soils. Often higher dominance of perennial forbs and grasses than on more fertile soils. In California, woody plants with postfire seeding strategy (serotiny and obligate seeders) are relatively more abundant in serpentine chaparral than in nonserpentine chaparral. Fire-dependent herbs (e.g., fire-cued germination) less common in serpentine than nonserpentine vegetation.
Landscape characteristics	Soils sometimes bare and rocky, often with relatively low litter cover. In humid climates, serpentine areas often support lower topographic relief than other substrate types. In drier climates, serpentine areas can be steep and rocky. Soil nutrient effect of serpentine can diminish effects of slope and aspect on vegetation development, leading to landscape vegetation pattern that is less patchy at the coarse scale. Relatively lower levels of tree and shrub cover on serpentine soils can enhance effects of wind and sun on soil and fuel drying and heating, and fire propagation.

NOTE: Also see Figure 12.5.

SOURCES: Kruckeberg 1984, 2004; Brooks 1987; Baker et al., 1992; Safford & Harrison, 2004; Alexander et al., 2007.

conservation concern in the former, it would seem that ecologists interested in studying the factors regulating diversity would gravitate to field and laboratory comparisons across this natural productivity gradient. However, such comparative studies remain rare, even in places where serpentine soils are a common part of the landscape. In the past two decades or so, the value of the serpentine-nonserpentine productivity gradient as a model system has become more apparent to community and landscape ecologists, and the number of studies investigating classic questions relating to biodiversity regulation has increased, especially in California. In this chapter, we carry out the first review of ecological studies of disturbance in serpentine vegetation, focusing on general patterns in the responses of plant species diversity to large ungulate herbivory, fire, and soil disturbance.

THE INTERACTING ROLES OF DISTURBANCE AND PRODUCTIVITY IN BIODIVERSITY REGULATION: THE THEORETICAL BACKGROUND

The modern synthesis of equilibrial and nonequilibrial theories for biodiversity regulation began in the mid-1970s, with recognition that competitive exclusion is often prevented in natural communities by factors like predation and harsh physical conditions. Connell's intermediate disturbance hypothesis (IDH) proposed that community structure is the product of three factors: disturbance; environmental heterogeneity, which is created and modified by disturbance; and recruitment, which is dependent on both of the former (Connell, 1978; Petraitis et al., 1989; Reice, 1994). The chief prediction of the IDH is that there is a unimodal relationship between diversity and disturbance, such that diversity is high at "intermediate" levels of disturbance (size or intensity or frequency) and lower at both lower and higher rates of disturbance (see Figures 12.1 and 12.2). At low levels of disturbance, equilibrial processes dominate community interactions, and superior competitors reduce diversity by eliminating inferior species. At high levels of disturbance, diversity drops as species with low recruitment rates disappear from the community.

Many studies have sought to evaluate the universality of the IDH, with varying results. Some authors have underscored the fact that different types of disturbance are not equivalent, and in many systems different disturbances interact to generate complex outcomes (Collins, 1987; Noy-Meir, 1995). Others have found that the unimodal disturbance–diversity relationship is dependent on, among other things, the type of disturbance, the sample size area, and productivity of the study system (Huston, 1994; Mackey and Currie, 2000, 2001). Proulx and Mazumder (1998) reviewed herbivory studies and found that all the studies from nutrient-poor ecosystems showed declines in species richness under heavy grazing, whereas 60% of

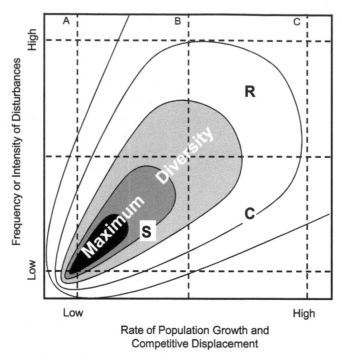

FIGURE 12.1. Effects of productivity (growth rate) and disturbance on local plant diversity, as predicted by the dynamic equilbrium model (DEM; Huston, 1994). Transects A through C represent cross-sections across the disturbance gradient at low, intermediate, and high productivity. The centroids of Grime's (2001) fundamental plant strategies (C, S, R) are also plotted. Figure altered and redrawn from Huston (1994).

studies from nutrient-rich systems showed increases in richness under heavy grazing.

The relationship between ecosystem productivity and species diversity has itself been the subject of intense debate for many years. Primary productivity—our focus—is the rate of change in plant biomass per unit area over time (Barbour et al., 1987). Numerous researchers have found broadly unimodal patterns relating species diversity to different (imperfect) surrogates for primary productivity (e.g., biomass, soil resource availability, precipitation) where diversity is maximized at intermediate productivity (Huston, 1994; Grace, 1999; Grime, 2001). It is generally thought that environments are too nutrient-poor at the low end of the productivity gradient for most species to survive. The often observed decline in diversity at the high end of the productivity gradient may be driven by local biotic processes like competitive exclusion, regional and historical processes, or both

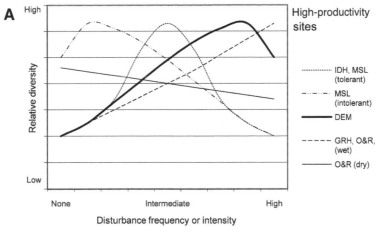

A   High ┐   High-productivity
                                               sites

                                               ············ IDH, MSL
                                                            (tolerant)

                                               —·—·— MSL
                                                            (intolerant)

                                               ——— DEM

                                               ----- GRH, O&R,
                                                            (wet)

                                               ——— O&R (dry)

     Low

        None   Intermediate   High
        Disturbance frequency or intensity

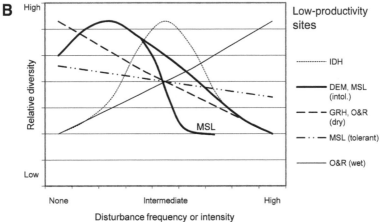

B   High ┐   Low-productivity
                                               sites

                                               ············ IDH

                                               ——— DEM, MSL
                                                            (intol.)

                                               ---- GRH, O&R
                                                            (dry)

                                               —··— MSL (tolerant)

                                               ——— O&R (wet)

     Low

        None   Intermediate   High
        Disturbance frequency or intensity

FIGURE 12.2. Theoretical disturbance-diversity curves for (A) high-productivity sites (see transect C in Figure 12.1); (B) low-productivity sites (see transect A in Figure 12.1). Diversity curves refer to relative (not absolute) diversity, and diversity levels should not be compared between curves. IDH = intermediate disturbance hypothesis (Connell, 1978); MSL = Milchunas, Sala, and Lauenroth (2003) (for disturbance-tolerant and -intolerant species); DEM = dynamic equilibrium model (Huston, 1994); GRH = grazing reversal hypothesis (Proulx and Mazumder, 1998); O&R = Olff and Ritchie (1998) (for wet and dry climates, large herbivores only). In (B), MSL differs from DEM in predicting that intolerant species in low-productivity sites cannot support high disturbance intensities or frequencies.

(Ricklefs and Schluter, 1993). Like the disturbance–diversity relationship, the unimodal productivity–diversity relationship may not be universal, and theoretical and empirical studies have also found support for positive and negative relationships (Rosenzweig and Abramsky, 1993; Abrams, 1995; Mittelbach et al., 2001).

A number of authors have suggested that species diversity patterns are best understood when the disturbance–diversity and productivity–diversity relationships are jointly considered as part of the same theoretical framework (e.g., Grime, 1977; Huston, 1994). Grime's 1979 CSR theory (Grime, 2001) and Huston's (1994) dynamic equilibrium model (DEM) are both based on a disturbance-by-productivity matrix (Figure 12.1). Grime (2001) described three primary plant strategies (C = competitors, S = stress tolerators, R = ruderal species) that result as evolutionary solutions to different permutations of the matrix, reasoning that a fourth strategy (for the extreme low-productivity/high-disturbance condition) was not viable in nature. Huston (1994) retained the fourth ecosystem condition, but otherwise CSR and the DEM make many of the same predictions. In this contribution, we focus our consideration on the DEM, because its retention of separate axes of productivity and disturbance (which are not truly independent) makes comparison of disturbance effects on serpentine and nonserpentine ecosystems more tractable (Figure 12.1).

Because increased ecosystem productivity usually leads to increased growth rates of individuals and populations, the DEM (and CSR) predicts that more frequent or severe disturbance is required to prevent competitive exclusion in highly productive habitats; highly productive, highly disturbed sites are dominated by "weedy" ruderal species, often at high species densities. At low productivity, high disturbance creates conditions where few species can survive. The DEM predicts maximal diversity where disturbance and productivity are balanced (Figure 12.1), because in these conditions the roles of local processes that drive mortality and competitive exclusion are either minimized (low ends of both gradients) or they offset one another (high ends), and larger scale factors that influence the size of the regional species pool (landscape heterogeneity, rates of immigration and speciation, etc.) come into play (Huston, 1994). Because conditions of low productivity and low disturbance allow the coexistence of many life history strategies, the DEM expects the highest diversity to occur in this region of Figure 12.1. Under high productivity, the DEM expects more or less monotonic increases in diversity as disturbance levels rise, whereas under low productivity, the model predicts diversity to generally decrease (Figure 12.2). Unimodal relationships between diversity and disturbance are only expected under intermediate conditions of productivity.

In their grazer reversal hypothesis (GRH), Proulx and Mazumder (2001) make broadly similar predictions to the DEM, with herbivore effects expected to increase diversity monotonically in high-productivity ecosystems and decrease diversity in low-productivity ecosystems (Figure 12.2). The GRH suggests that resource availability on its own is largely sufficient to explain these patterns, with some contribution from evolved grazing tolerance. Olff and Ritchie (1998; O&R) posit that precipitation and herbivore size interact with productivity to drive plant diversity responses to disturbance. For large ungulates, O&R predict a negligible to moderate loss in diversity in dry sites but a moderate to major gain in diversity for wet sites, depending on nutrient availability. The strongest disturbance-driven decrease in diversity is expected for low-fertility soils in dry climates, and the strongest increase for low-fertility soils in wet climates (Figure 12.2).

A number of authors have noted that changes in species diversity due to disturbance may depend on the historic relationship between the studied ecosystem, its component species, and the disturbance in question (Denslow, 1985; Milchunas et al., 1988; Huston, 1994; Olff and Ritchie, 1998). Organisms that have long been subject to a particular disturbance regime (e.g., certain patterns in frequency, size, or intensity) will adapt to that regime or leave the ecosystem. Milchunas, Sala, and Lauenroth (1988; MSL) developed a theoretical framework for understanding the responses of grasslands to grazing based on their evolutionary exposure to herbivory. In floras long associated with herding ungulates, the interaction between moderate grazing and high community productivity is likely to induce strong increases in diversity and large changes in composition; in floras lacking this long association with large herbivores, however, diversity may drop strongly (Milchunas et al., 1988; Figure 12.2). A similar example can be drawn for fire. Ecosystems that rarely experience wildfire are unlikely to support many taxa that possess adaptations for rapid recolonization of burned areas, whereas ecosystems that do experience frequent fire will have many niches for postfire specialists (Grubb, 1977; Denslow, 1985).

All of these predictions concern local-scale diversity, but disturbance–diversity theory also makes predictions about regional diversity. If we assume a landscape is homogeneous (in soil, weather, topography, etc.), then theory predicts that a given disturbance regime will generate a certain range of species diversity at the local scale. If the entire landscape is disturbed at the same time by a disturbance of homogeneous intensity, as it will be if conditions are perfectly homogeneous, then the diversity of the region will be equal to the diversity of the local site. Subjecting a homogeneous landscape to a series of spatially and temporally discontinuous disturbances would generate a patchwork of different successional stages, which might lower the average local diversity (depending, e.g., on productivity) but increase beta diversity among sites and therefore regional diversity as well. Still higher regional diversities would be generated in complex landscapes, where localities of differing productivities are subjected to differing regimes of disturbance (Ricklefs and Schluter, 1993; Huston, 1994; Grime, 2001).

SERPENTINE ECOSYSTEMS AND DISTURBANCE:
A REVIEW

Although ecological disturbances are ubiquitous (Pickett and White, 1985; Huston, 1994), and some serpentine ecosystems are clearly structured by disturbance (e.g., Safford and Harrison, 2004; Grace et al., 2007), research into the effects of disturbance on plant diversity in serpentine ecosystems remains rare. In our literature review, we found fewer than 50 published studies containing information relating to the diversity–disturbance relationship in serpentine vegetation, with few of these studies actually focused on disturbance itself and even fewer treating the subject quantitatively or statistically. To this point, disturbance research on serpentine habitats has focused on three types of ecological disturbance: herbivory, fire, and soil disturbances. For these three disturbance types, we summarize the results of our review.

Herbivory

Theory and empirical evidence suggest that grazing may either increase or decrease plant diversity, in the former case by reducing the dominance of superior competitors, in the latter by reducing the diversity of grazing-intolerant species (Collins et al., 1998; Gough and Grace, 1998; Olff et al., 1999; Grime, 2001). Because resource availability limits population growth, the direction of the plant diversity response to herbivory is widely thought to depend on ecosystem productivity (Huston, 1994; Olff et al., 1999; Grime, 2001). Generally speaking, grazing is expected to affect diversity positively in resource-rich ecosystems and negatively in resource-limited ecosystems, although evolutionary exposure to grazing and other factors may modify this relationship (Milchunas et al., 1988; Huston, 1994; Olff and Ritchie, 1998; Proulx and Mazumder, 1998; Figure 12.2).

Although many serpentine areas support grasslands, and although these grasslands are an important source of livestock forage in many parts of the world (Borhidi, 1988; Harrison, 1999; McCallum, 2006), there have been surprisingly few studies of the effects of grazing on serpentine plant diversity. Almost all of these have taken place in California, where research has been largely motivated by the desire to understand how herbivory and other disturbances relate to the persistence of a rapidly diminishing native grassland flora. In California, infertile substrates like serpentine are much less invaded than other substrates and provide a critical refuge for many native grassland species. Most non-native grassland species in California evolved in ecosystems subject to heavy pressure by herding ungulates; however, ungulate densities in California have been relatively low since the early Holocene extinctions, and grazing selection has likely been less important in the development of California's native grassland flora (Kimball and Schiffman,

2003). Serpentine/nonserpentine comparisons in California—and other similarly invaded regions—are thus uniquely well positioned to empirically examine the interactive effects of disturbance, productivity, soil type, and floral ancestry on plant diversity.

Herbivory Summary.   We found eight studies that statistically assessed the effects of ungulate grazing on species diversity in serpentine grassland (Huenneke, unpublished data, cited in Hobbs and Huenneke, 1992; Harrison, 1999; Safford and Harrison, 2001; Gelbard and Harrison, 2003; Harrison et al., 2003; McCallum, 2006; Niederer and Weiss, 2007; Safford, unpublished data), and one that had qualitatively assessed disturbance effects in a shrubland–grassland transition (Spence, 1957). Of the statistical assessments, seven conducted analysis at the local scale (<10 m^2), and all seven were in California. One study in South Africa evaluated grazing impacts at a broader spatial scale (1000 m^2; McCallum, 2006). Only two included analysis of the effects of varying disturbance levels (Safford and Harrison, 2001; Safford, unpublished data).

Our review suggests that low to moderate levels of ungulate (livestock) grazing in California annual grasslands tend to increase local species richness. On serpentine soils, the effect is primarily mediated through an increase in native species; richness increases on nonserpentine soils are more likely to be driven by increases in the non-native species component (Figure 12.3). Only two studies (Safford and Harrison, 2001; Safford, unpublished data) collected data from paired grazed–ungrazed plots, which allow unbiased measurement of grazing intensity and productivity. Both studies found evidence of decreased diversity at higher levels of grazing intensity (Figure 12.4). These studies also document interactions between productivity and grazing such that changes in diversity and composition due to grazing are greater in higher biomass sites.

In the reviewed studies, beta diversity in serpentine (and nonserpentine) grasslands was either unchanged or decreased by grazing; this appears to primarily be a result of floristic homogenization due to loss of grazing-intolerant taxa. Of four studies with explicit or inferred measurements of regional diversity in serpentine grassland, grazing drove an increase in one (Safford and Harrison, 2001), a decrease in one (McCallum, 2006), and no change in two (Gelbard and Harrison, 2003; Harrison et al., 2003). Three studies included regional measures from nonserpentine sites; two of them showed increases in regional diversity due to grazing (Safford and Harrison, 2001; Safford, unpublished data), one measured no change (Harrison, 1999).

The productivity of serpentine grassland is indisputably low on a global scale, but most of the reviewed studies find either a positive or negligible effect of grazing on local richness, which appears to contradict most theoretical expectations

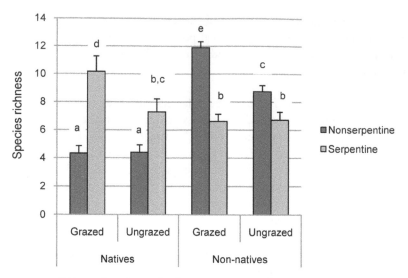

FIGURE 12.3. Effects of grazing and soil type on native and non-native species richness in serpentine and adjacent nonserpentine grasslands in northern California (Safford and Harrison, 2001).

(Figure 12.2). Most of the reviewed studies reported that they sampled areas subject to low to moderate grazing intensities (without providing actual measures of intensity). The two studies that included complete gradients of disturbance showed moderate increases in richness but negative effects on diversity as grazing intensity rose (Safford and Harrison, 2001; Safford, unpublished data). One study showed stronger losses in diversity on serpentine than on nonserpentine soils (Figure 12.4). Overall, the best theoretical match for these results is provided by the DEM and MSL (grazing-intolerant species) predictions for low-productivity soils, both of which predict an initial rise in diversity with disturbance, followed by a strong decrease at intermediate to high disturbance levels (Figure 12.2). At the moderately low levels of productivity that characterize serpentine grasslands in California (which is the only place where statistical assessment of grazing effects on plant diversity has been performed), low to moderate levels of herbivory reduce the biomass and thatch of the dominant species, which are often non-natives, providing competitive release for a suite of mostly native forbs and grasses (Hobbs and Huenneke, 1992; Weiss, 1999; Safford and Harrison, 2001; Harrison et al., 2003). At higher levels of herbivory, low levels of soil fertility and intrinsically slow growth rates of many serpentine species can result in their exclusion from the landscape. This pattern is exacerbated by the general grazing intolerance of many California species (Safford and Harrison, 2001; Kimball and Schiffman, 2003).

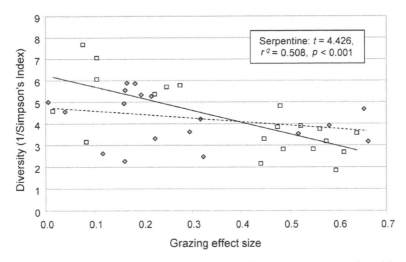

FIGURE 12.4. Local diversity versus grazing intensity for 18 nonserpentine (dotted line and diamonds) and 21 serpentine (solid line and squares) grassland sites across northern and central California. Diversity measured as 1/Simpson's Index, which measures the number of highly abundant species (lower diversity thus means that fewer species are dominating the plot and more species are becoming rare). Species richness is (weakly) lower under higher grazing intensities at serpentine sites but not at nonserpentine sites. With soil types pooled, $r^2 = 0.281$ ($p < 0.01$). Nine sites with <0 effect size are excluded from this graph (overall regression $r^2 = 0.278$ with them).

High-intensity, long-term grazing may be more likely to reduce diversity in serpentine grassland than in nonserpentine grassland (e.g., Dennis, 1989; Huntsinger et al., 1996), but species turnover due to grazing is likely to rise as productivity increases (Safford, unpublished data). The latter effect may be independent of soil type.

Community response to disturbance is highly dependent on the life histories of the species that are there to respond. In California, native species tend to respond most strongly to grazing on serpentine soils, whereas non-natives respond most strongly on nonserpentine soils (Safford and Harrison, 2001; Gelbard and Harrison, 2003; Harrison et al., 2003; Safford, unpublished data). The richness of non-natives appears to rarely (if ever) increase on serpentine soils as the result of low- to moderate-intensity grazing. Limited data from one ongoing California study suggest that grazing-tolerant species may be equally common in serpentine and nonserpentine grassland, with their contribution to biomass surprisingly high in the former (Safford, unpublished data).

The strength of the floristic homogenization effect of grazing may generally be stronger in more productive and nonserpentine sites, but some serpentine species may be resistant to herbivory for reasons not linked to herbivory itself (life form, magnesium content, leaf constitution). Herbivore pressure is not uniform across the landscape, and outcomes may differ based on whether grazing is light, moderate, or heavy. Although one study documented an increase in grazing effect with productivity (Safford, unpublished data), herbivore pressure may not necessarily be lower in serpentine grasslands. One California study found that the nutritional value of serpentine grass forage was actually higher than forage on adjacent nonserpentine soils (Rosiere and Vaughn, 1986).

Fire

For fire to occur, heat must be applied to fuel in the presence of oxygen. Once ignition has taken place, the behavior of fire (its intensity, rate of spread, etc.) is driven primarily by fuel characteristics, weather, and topography (Bond and van Wilgen, 1996; van Wagtendonk, 2007). The major difference between many serpentine habitats and nearby nonserpentine habitats is the amount and arrangement of biomass (fuel). In general, most serpentine vegetation is characterized by low rates of biomass accumulation. Compared to more fertile substrates, the vegetation of ultramafic soils is often of lower stature and cover, stands of woody plants are more open, soils are often rockier and frequently free of vegetation cover, and there is relatively high light and wind incidence at the ground (Table 12.1). Broad-scale weather patterns do not differ between serpentine and adjacent nonserpentine habitats, but the differing arrangements of biomass in the two habitats can lead to different effects of sun and wind on fuel drying and fire propagation. In some cases, topographic differences can also exist between serpentine areas and the nonserpentine matrix (Takaoka and Sasa, 1996).

In many parts of the globe, fire plays an important role in structuring ecosystems, but ecosystems themselves also play a role in structuring fire. In Mediterranean climate regions, islands of serpentine soils can support populations of regionally or locally rare plants that appear to be escaping competition and/or frequent fire (Vogl et al., 1977; Safford and Harrison, 2004). These taxa include species that are adapted to fire-driven reproductive cycles but require sufficient time between fires to generate adequate germplasm to guarantee population persistence (e.g., serotinous conifers). The ameliorating effect of open serpentine vegetation on wildfire spread is well known to firefighters in California, and forest fuel treatment plans sometimes incorporate natural serpentine fuel breaks. To this point, however, little is known about the role of fire in serpentine ecology or the relationship between fire and plant diversity in serpentine vegetation. Fire, fuel, and resource management in serpentine vegetation must be prosecuted almost entirely based on inference from contextual features, such as those in Table 12.1.

Fire Summary.   We reviewed the results of 21 fire studies in serpentine vegetation (Coombes and Forst, 1956; Wells, 1962; Miller, 1981; Knox, 1984; Borhidi, 1988; Parker, 1990; Tyndall, 1992, 1994; Takaoka and Sasa, 1996; McCoy et al., 1999; Arabas, 2000; Matos Mederos and Torres Bilbao, 2000; Chiarucci et al., 2001; Chiarucci, 2003; Harrison et al., 2003; Safford and Harrison, 2004; CAPO, 2007; Tolman, 2007; Safford, 2008; EBRPD, 2009; Reddy et al., 2009). Only four studies provided quantitative data on the effects of fire on local diversity (McCoy et al., 1999; Matos Mederos and Torres Bilbao, 2000; Harrison et al., 2003; Safford and Harrison, 2004), and only two carried out formal statistical analysis of those effects. Both of these latter studies were in California, and both investigated interactions with productivity and species provenance (Harrison et al., 2003; Safford and Harrison, 2004).

In California and other Mediterranean climate regions, fire has a strong positive effect on diversity in productive vegetation but less effect in unproductive (e.g., serpentine) sites, where space and light are less limiting in the undisturbed landscape (Harrison et al., 2003; Safford and Harrison, 2004, 2008). During postfire succession, diversity drops more rapidly in more productive sites (Figure 12.5). Safford and Harrison (2004) found an increase in local diversity in serpentine chaparral (sclerophyllous shrubland) due to fire (Figure 12.5), whereas Harrison and colleagues (2003) found no overall diversity effect of a late dry season fire in a serpentine grassland. In the latter, fire increased the number of native species in serpentine vegetation but not the number of non-natives. In both studies, the number of non-native weedy species in the postfire flora was disproportionately enhanced in nonserpentine versus serpentine chaparral. The effects of fire on beta and regional diversity were only statistically assessed in one study in California (Safford and Harrison, 2004). Beta diversity did not respond to fire in serpentine chaparral, but it rose in nonserpentine chaparral. Both chaparral types saw an increase in regional diversity due to the effects of fire.

Three publications from the wet tropics provided sufficient data to allow for a post facto evaluation of fire-diversity patterns (Borhidi, 1988; McCoy et al., 1999; Matos Mederos and Torres Bilbao, 2000). In Cuba and New Caledonia, local and regional diversity were both strongly reduced due to fire; in the latter, beta diversity among plots was enhanced in early successional habitats.

The effects of fire on diversity in serpentine ecosystems appear to be strongly dependent on ecological and evolutionary context. Data and anecdotes from areas of humid climate suggest that fire tends to have a negative effect on both local and regional diversity of serpentine vegetation (Coombes and Forst, 1956; Proctor and Woodell, 1971; Borhidi, 1988; McCoy et al., 1999; Matos Mederos and Torres Bilbao, 2000), although rare species may benefit from fire in some cases (Miller, 1981; Tyndall, 1992). Data from California and observations from Italy suggest that fires

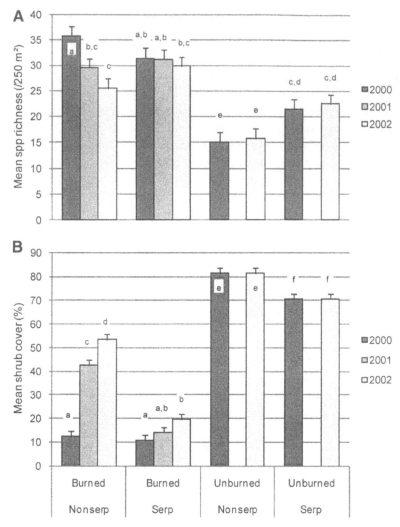

FIGURE 12.5. Postfire recovery of (A) local species richness and (B) woody cover on serpentine and sandstone chaparral sites burned in 1999 in northern California. Lowercase letters indicate significantly different means. Figure redrawn from Safford and Harrison (2004).

on serpentine in Mediterranean climate regions are more likely to increase local and regional diversity, although not as strongly as on more productive soils. Patterns in the reviewed studies support the existence of a peak in local diversity somewhere in the low to moderate disturbance intensity/frequency range. This would appear to best support the IDH, DEM, or MSL models (for disturbance-intolerant species). These patterns may be best explained by the relative rarity of natural (nonanthropogenic) fire in most humid landscapes, and its commonness in Mediterranean climates. Areas without evolutionary exposure to frequent fire are unlikely to evolve many species that fill fire-related niches (Grubb, 1977; Denslow, 1985), and modern increases in fire frequency due to human causes in these ecosystems are likely to have largely deleterious outcomes for plant diversity. On the other hand, the successful long-term suppression of fire in many semiarid areas of the planet has reduced plant diversity in many places (Hobbs and Huenneke, 1992; Sugihara et al., 2007), but this is probably not so much the case in vegetation on serpentine soils, where all signs point to a much looser relationship with fire than in most types of nonserpentine vegetation (Safford and Harrison, 2004, 2008; Safford, 2008).

The effects of fire on serpentine vegetation are longer lasting than they are on nonserpentine vegetation, probably due to lower productivity as well as slower intrinsic growth rates of many serpentine plant taxa (Proctor and Woodell, 1971; O'Dell et al., 2006). Postfire vegetation succession is a much slower process on serpentine soils, and therefore—even though the diversity effects of fire are not as strong—the vestiges of fire in serpentine community structure are longer lasting than on more productive soils.

Generally lower frequencies of fire in serpentine habitats have led to an association of fire-sensitive taxa with serpentine soils. In Mediterranean climate regions, woody species with fire-adapted regeneration strategies (serotiny, obligate seeding shrubs) that require longer fire-free periods (and less competition for resources) are disproportionately common on serpentine soils (Vogl et al., 1977; Safford and Harrison, 2004, 2008). In California, many of the herbaceous taxa requiring fire to cue germination are found in serpentine as well as nonserpentine sites, but their overall abundance and contribution to biomass is much lower in the serpentine vegetation (Safford and Harrison, 2004).

There are soil and geomorphology feedbacks between serpentine soils, climate, and disturbance. For example, more humid areas can break down more weatherable ultramafic rocks into less defined topography, which may lead to more rapid fire spread and more homogeneity in burn severity (Takaoka and Sasa, 1996). Drier climates leave much serpentine bedrock at the surface and increase burn heterogeneity and decrease fire spread (Safford and Harrison, 2004).

Compared with more productive soil types, vegetation on serpentine soils tends to support low fuel (biomass) loads and highly heterogeneous fuel structure. In drier regions or in areas with a pronounced dry season (e.g., Mediterranean climates), this leads to relatively low frequencies and intensities of fire in serpentine vegetation (Safford and Harrison, 2004; Safford, 2008). In wetter climatic regions, the lack of a dense woody overstory on serpentine soils may lead to the development of high biomass in the vegetation understory. In such regions, high levels of understory fuels, coupled with the enhanced influence of sun and wind on fuel moisture in the more open understory, may lead to higher flammability in serpentine vegetation (Proctor, 1999; Safford, 2008).

Soil Disturbance

Unlike herbivory and fire, there have been no grand holistic theoretical treatments of the relationship between soil disturbances and diversity. Short reviews of the effects of soil disturbance on diversity are provided by Hobbs and Huenneke (1992) and Huntly and Reichmann (1994). Considerations of the effects of soil disturbance on serpentine vegetation have focused on ablation by wind (Spence, 1957), frost heave (Rune, 1957; Dearden, 1979), soil movement due to shrink-swell clays (Alexander et al., 2007), precipitation and gravity (Oberhuber et al., 1997), and bioturbation by fossorial mammals (e.g., Hobbs and Mooney 1985, 1991, 1995; Hobbs et al., 1988, 2007). One group of researchers at Stanford University has been carrying out manipulative experiments of gopher disturbance effects on diversity for more than two decades (see Hobbs and Mooney, 1985; Hobbs et al., 2007). Their results underscore the important role that soil disturbance can play in maintaining habitat heterogeneity and regulating species diversity along both spatial and temporal gradients.

Soil Disturbance Summary.   We found more than a dozen studies that either qualitatively or quantitatively treated the subject of soil disturbance and diversity in serpentine vegetation (Rune, 1957; Spence, 1957; Dearden, 1979; Hobbs and Mooney, 1985, 1991, 1995; Hobbs et al., 1988, 2007; Safford and Harrison, 2001; Harrison et al., 2006a, 2006b; Davies et al., 2007; Safford, unpublished data). Only fossorial mammal disturbance was assessed statistically or experimentally, and then only in studies carried out in California. Only one study explicitly compared serpentine and nonserpentine soils (Safford and Harrison, 2001).

Our review suggests that the effects of soil bioturbation on local richness in herbaceous serpentine vegetation are nearly uniformly negative, although one set of studies found a unimodal response (Harrison et al., 2006a, 2006b; Davies et al., 2007; Safford, unpublished data). Beta diversity among plots appears primarily to be enhanced by soil disturbance (e.g., Hobbs and Mooney, 1985, 1991; Safford, unpublished data). The summed effect on regional diversity may be positive or negligible. These results more or less conform to expectations of five of six of the disturbance–diversity theories. This includes the unimodal response seen in one of the California studies, which is predicted by the DEM and MSL. Only the IDH is unsupported by the serpentine soil disturbance results we reviewed. It has been hypothesized that low-fertility ecosystems are characterized by lower levels of competition and/or shifting of competition from shoots to roots because of the relative abundance of light and scarcity of soil nutrients (Tilman, 1982; Huston, 1994; Grime, 2001). If serpentine grassland plants compete primarily for below-ground resources, then disturbances like tunneling, mounding, and underground herbivory, which affect root systems, may be especially deleterious to local richness. At the same time, these locally destructive activities may promote beta diversity by increasing habitat heterogeneity and providing colonization sites for disturbance-tolerant species whose habitat requirements are not met in undisturbed vegetation (Hobbs and Mooney, 1985, 1991, 1995; Hobbs and Huenneke, 1992).

Tunneling and mound-building animals like gophers can greatly increase spatial and temporal heterogeneity in areas with sufficiently deep soil and herbaceous vegetation (Hobbs and Mooney, 1985, 1991). This creates niche opportunities for many relatively uncompetitive species. Locally, gopher disturbance may be an important driver of rare species richness and abundance on serpentine soils, but common species are better buffered against the stochasticities in population size caused by soil disturbance and may react more strongly to broad-scale drivers like variations in climate (Hobbs et al., 2007).

Depending on the species that can get propagules to the site, a disturbance may either expand or decrease niche space. There are important feedbacks between the life histories of resident species and the type of disturbance. The timing of the disturbance is of major importance. Climatic conditions before, during, and after a disturbance are critical to determining the ultimate outcome, as is the timing of the disturbance relative to plant life cycles. Species that disperse seeds at the time of the disturbance are more likely to benefit (Hobbs and Mooney, 1985, 1991, 1995).

Deeper serpentine soils often sustain a high fine clay fraction (Alexander et al., 2007), which may be easier for gophers and other tunneling animals to move through, at least in the growing season before the soil hardens. This high fine clay fraction is also highly erodible when exposed to precipitation on sloping ground (Oberhuber et al., 1997; Alexander et al., 2007). The high content of shrink-swell clay minerals in some deeper serpentine soils can result in high levels of soil movement, restricting the survival of woody plants (Alexander et al., 2007). Shallow soils on ultramafic bedrock in cold climates may be more susceptible to frost heave and similar periglacial phenomena than other soil types (Rune, 1953; Dearden, 1979).

DIVERSITY AND DISTURBANCE ON SERPENTINE
SOILS: A SYNTHESIS

Only four published studies, plus one in progress, included serpentine and non-serpentine sites, disturbed and undisturbed controls, and explicit measures and statistical assessment of the effects of productivity (independent of soil type) and disturbance gradients on species diversity (Harrison, 1999; Safford and Harrison, 2001, 2004; Harrison et al., 2003; Safford, unpublished data). The limited number of disturbance studies in serpentine vegetation makes general conclusions difficult, but the potential for important findings is high. We provide a first-cut synthesis of the extant research.

Disturbance–Diversity Theory and Empirical Evidence from Serpentine Ecosystems

Most ecological theory predicts that disturbance should play a less important role in diversity regulation in low- versus high-productivity ecosystems, and our review supports this prediction. Serpentine studies generally found a negligible to moderate rise in local diversity at lower levels of disturbance on low-productivity sites, with a drop in diversity as disturbance intensity increases, in accordance with the dynamic equilibrium model (Huston, 1994) and the model of Milchunas et al. (1998) for disturbance-tolerant species. Exceptions to this included the studies of soil disturbance and studies of fire in humid climates, which did not detect the initial increase in diversity; although the results were not always conclusive, these studies therefore may better support models predicting monotonic decreases in diversity with disturbance (e.g., Milchunas et al., 1988, for disturbance-intolerant species). In studies that either compared serpentine and nonserpentine sites or analyzed continuous gradients of productivity and disturbance intensity, the general result was that increases in local diversity following disturbance were positively correlated with productivity (e.g., Safford and Harrison, 2001, 2004; Safford, unpublished data). Increasing intensity of disturbance reduced diversity monotonically in the grazing studies but showed a unimodal response in the fire study. These studies clearly indicate the inadequacy of the intermediate disturbance hypothesis, which is insensitive to changes in ecosystem productivity.

Several studies of serpentine floras have examined beta diversity (species dissimilarity among plots and sites), which is affected by site heterogeneity, disturbance, and dispersal limitation (e.g., Whittaker, 1960; Harrison, 1997; Freestone and Inouye, 2006). Based on our review, disturbance probably plays a lesser role in determining beta diversity in serpentine than in nonserpentine vegetation. For example, grazing reduces the heterogeneity among grassland plots to a greater degree in more productive sites (Safford, unpublished data).

The Importance of Species Identity and History

Serpentine and nonserpentine soils tend to support markedly different floras and life form spectra, and these dissimilarities undoubtedly play a role in driving differing responses to disturbance and other ecological processes on the two soil types. For example, many studies have noted the higher abundances of prostrate annual forbs, perennial geophytes, and perennial grasses in serpentine versus nonserpentine grassland, and there are notable differences in dominant plant families and genera as well (Kruckeberg, 1984; Safford et al., 2005; Alexander et al., 2007). Such traits affect ecosystem properties such as resource availability, nutrient cycling rates, soil chemistry and texture, decomposition, invasion, and responses to disturbance (Eviner and Chapin, 2003).

The much lower prevalence of weedy non-native species on serpentine may underlie many of the differences in grazing response in Californian ecosystems, although serpentine tolerance is developing in some non-native species (Harrison et al., 2001), and others have arrived with preadaptation to serpentine (Meimberg et al., 2005). The studies we reviewed from California clearly show that native and non-native species have the capacity to respond differently to herbivory, as is also seen in studies from nonserpentine systems (Kimball and Schiffman, 2003). Natives tended to respond positively to low–moderate grazing intensities on serpentine soils, whereas non-natives were more likely to respond positively to low–moderate intensity grazing on nonserpentine. Floristic differences between serpentine and nonserpentine sites may explain studies that find soil type a better predictor of grazing response than productivity (Harrison, 1999; Safford and Harrison, 2001; Safford, unpublished data).

Species identity and history are also important in the responses of diversity and composition to fire. Herbaceous fire-following species that germinate in response to fire are much more important members of the chaparral flora on nonserpentine than serpentine soils (Safford and Harrison, 2004). This component of the chaparral flora comprises a large portion of the postfire explosion of diversity in herbaceous species and contributes significantly to the differences in relative diversity response between nonserpentine and serpentine soils. Serotinous conifers in fire-prone regions are largely restricted to infertile soils like serpentine and the proportion of obligate seeding shrubs tends to be much higher on serpentine soils than on nearby nonserpentine substrates (Vogl et al., 1977; Safford and Harrison, 2004). The proportion of perennial forb and grass species is also generally higher in the serpentine than the nonserpentine flora (Kruckeberg, 1984; Alexander et al., 2007), and perennials and annuals respond very differently to disturbance.

Differences between Disturbances

Soil disturbances that primarily remove above-ground plant biomass tend to decrease local diversity, increase beta diversity, and benefit ruderal annual species at the expense of perennials (Huston, 1994; McIntyre et al., 1995). However, soil disturbances with strong below-ground effects may have contrasting consequences on plant communities. For example, gopher activity has especially negative effects on geophytes (Hobbs and Mooney, 1985, 1995; Huntly and Reichman, 1994). Gophers and other fossorial mammals also indirectly influence above-ground community structure through mixing unexploited subsoil into upper soil horizons; normally this increases soil fertility, but in serpentine soils, it may have the opposite effect because subsoils may be low in macronutrients (Hobbs and Mooney, 1991).

Fire and ungulate grazing have many similar effects on plant communities, but there are also important differences. Grazing is always selective, and its effects on plant communities depend on the relative abundances of preferred food species (Huston, 1994; Crawley, 1997). Fire consumes both living and dead material (Bond and Keeley, 2005), and it may be completely unselective in some conditions (e.g., hot, dry, windy weather) and somewhat more selective in others (e.g., moderate weather in low fuel conditions) (Bond and van Wilgen, 1996). In grasslands, effects of recent fire on vegetation cover and diversity can be similar to the effects of grazing, but because grazing continues to selectively reduce living biomass through the growing season, its effects are usually strongly stronger and more pronounced (Collins, 1987; Noy-Meir, 1995; Harrison et al., 2003). Within-study comparisons of grazing and fire show that the two disturbances can differ in, for example, the quantity and quality of biomass removed (Collins and Barber, 1986), traits and species favored (Collins, 1987; Fuhlendorf and Engle, 2004), the effects on community structure and diversity (Collins, 1987; Noy-Meir, 1995; Harrison, et al., 2003), and the spatial and temporal scales of these effects (Collins and Smith, 2006).

The Serpentine "Context": All Serpentine Is Not Equal

Gradients in bedrock and soil can play a major role in modifying vegetation on serpentine soils, thus creating strong variation in disturbance regimes and their effects. Climate can also play a major role in modifying serpentine vegetation and its ecological relationships with disturbance (Kruckeberg, 2004; Grace et al., 2007). In very cold or arid climates, serpentine vegetation may differ insignificantly in structure or composition from the vegetation on neighboring normal soils (McNaughton, 1968; Kruckeberg, 2004; Alexander et al., 2007); in these situations, disturbance factors linked to productivity are unlikely to differ between serpentine and nonserpentine vegetation.

In some humid regions where serpentine supports shrublands and open forest (e.g., Hokkaido, Japan; the wet tropics; the United Kingdom; eastern United States; northwesternmost California), the relatively low woody biomass in the overstory allows for the growth of a dense understory of flammable subshrubs and grasses, and the open vegetation structure also promotes the drying of fuels and wind-driven fire spread. In these situations, fire may be naturally more frequent in serpentine than nonserpentine vegetation (Miller, 1981; McCoy et al., 1999; Proctor, 1999; Arabas, 2000), and managing for an appropriate fire regime to maintain the unique serpentine vegetation is a major concern.

The nutrient factor in serpentine soils can become so dominant in some circumstances that other important drivers of vegetation pattern, such as topography, are largely overridden. We found multiple references to the relatively minor effects that variations in slope and aspect have on serpentine vegetation, diversity, and disturbance relative to nearby nonserpentine sites (e.g., McNaughton, 1968; Harrison, 1999; Safford and Harrison, 2001). The serpentine soil factor can likewise overwhelm disturbance as a driver of vegetation community structure. A number of the studies we reviewed found that soil type was more important than disturbance in explaining plant diversity patterns (Coombes and Forst, 1956; Harrison, 1999; Safford and Harrison, 2001; Harrison et al., 2003; Safford, unpublished data). The dearth of scientific studies exploring the effects of natural disturbance on serpentine vegetation is itself probably an effect of the relatively low importance of disturbance to serpentine vegetation in many parts of the world.

How Useful Are Serpentine–Nonserpentine Comparisons for Studying Ecological Patterns and Processes Related to Disturbance?

We believe that serpentine–nonserpentine comparisons provide a powerful model system for studying ecological and evolutionary processes related to disturbance and productivity. Local studies using serpentine and nonserpentine soils can be designed to capture very broad gradients of both productivity and disturbance, perhaps to a greater degree than almost any other natural terrestrial system. In many places, the difficulty comes in finding sites of enough fertility or water availability to complete the high end of the productivity gradient. In California, chaparral productivity in immediate postfire years is often very high, but many nonserpentine grasslands are not particularly productive. Regions with relatively high but not excessive rainfall and warm growing season temperatures (e.g., coastal or montane Mediterranean regions, the humid temperate zone, parts of the tropics and subtropics) may provide the best locations for studies of disturbance effects on diversity, because they are more likely to capture a broad productivity gradient.

Does the unusual plant species composition found on serpentine soils undermine their usefulness for studies related to soil productivity? We would counter that gradients of soil productivity are strong selective forces and evolution will always mold the flora in accordance with the reigning environment. Plants in all truly low-productivity sites are characterized by similar suites of morphologies and life history strategies (Grime, 2001), and serpentine plants fall well within this general suite.

LITERATURE CITED

Abrams, P. A. (1995) Monotonic or unimodal diversity-productivity gradients: What does competition theory predict? *Ecology*, 76, 2019–27.

Alexander, E. B., Coleman, R. G., Keeler-Wolf, T., and Harrison, S. (2007) *Serpentine Geoecology of Western North America*. Oxford University Press, New York.

Arabas, K. B. (2000) Spatial and temporal relationships among fire frequency, vegetation, and soil depth in an eastern North American serpentine barren. *Journal of the Torrey Botanical Society*, 127, 51–65.

Baker, A. J. M., Proctor, J., and Reeves, R. D. (1992) *The Vegetation of Ultramafic Soils*. Intercept, Andover.

Barbour, M. G., Burk, J. H., and Pitts, W. D. (1987) *Terrestrial Plant Ecology*, 2nd ed. Benjamin/Cummings, Menlo Park, NY.

Bond, W. J., and Keeley, J. E. (2005) Fire as a global "herbivore": The ecology and evolution of flammable ecosystems. *Trends in Ecology and Evolution*, 20, 387–94.

Bond, W. J., and van Wilgen, B. W. (1996) *Fire and Plants*. Chapman and Hall, London.

Borhidi, A. (1988) Vegetation dynamics of the savannization process on Cuba. *Vegetatio*, 77, 177–83.

Brooks, R. R. (1987) *Serpentine and Its Vegetation: A Multidisciplinary Approach*. Dioscorides Press, Portland.

CAPO. (2007) Un viaggio nell-Italia delle "pietre Verdi." *Aree protette— flora e vegetazione*. Coordinamento Aree Protette Ofiolitiche, Siena.

Chiarucci, A. (2003) Vegetation ecology and conservation on Tuscan ultramafic soils. *Botanical Review*, 69, 252–68.

Chiarucci, A., Rocchini, D., Leonzio, C., and de Dominicis, V. (2001) A test of vegetation-environment relationship in serpentine soils of Tuscany, Italy. *Ecological Research*, 16, 627–39.

Collins, S. L. (1987) Interaction of disturbances in tallgrass prairie: A field experiment. *Ecology*, 68, 1243–50.

Collins, S. L., and Barber, S. C. (1986) Effects of disturbance on diversity in mixed-grass prairie. *Vegetatio*, 64, 87–94.

Collins, S. L., Knapp, A. K., Briggs, J. M., Blair, J. M., and Steinauer, E. M. (1998) Modulation of diversity by grazing and mowing in native tallgrass prairie. *Science*, 280, 745–47.

Collins, S. L., and Smith, M. D. (2006) Scale-dependent interaction of fire and grazing on community heterogeneity in tallgrass prarie. *Ecology*, 87, 2058–67.

Connell, J. H. (1978) Diversity in tropical rain forests and coral reefs. *Science*, 199, 1302–10.

Coombes, D. E., and Forst, L. C. (1956) The heaths of the Cornish serpentine. *Journal of Ecology*, 44, 226–56.

Crawley, M. J. (1997) *Plant Ecology*, 2nd ed. Blackwell Science, Oxford.

Davies, K. F., Harrison, S., Safford, H. D., and Viers, J. H. (2007) Productivity alters the scale dependence of the diversity-invasibility relationship. *Ecology*, 88, 1940–47.

Dearden, P. (1979) Some factors influencing the composition and location of plant communities on a serpentine bedrock. *Journal of Biogeography*, 6, 93–104.

Dennis, A. (1989) *Effects of Defoliation on Three Native Perennial Grasses in the California Annual Grassland*. PhD diss., University of California, Berkeley.

Denslow, J. S. (1985) Disturbance-mediated coexistence of species. In *The Ecology of Natural Disturbance and Patch Dynamics* (eds. S. T. A. Pickett and P. S. White), pp. 307–24. Academic Press, Orlando.

EBRPD. (2009) *Serpentine Prairie Restoration Plan, Redwood Regional Park.* East Bay Regional Park District, Planning Stewardship Department, Oakland, CA.

Eviner, V. T., and Chapin, F. S. (2003) Functional matrix: A conceptual framework for predicting multiple plant effects on ecosystem processes. *Annual Review of Ecology Evolution and Systematics*, 34, 455–85.

Freestone, A. L., and Inouye, B. D. (2006) Dispersal limitation and environmental heterogeneity shape scale-dependent diversity patterns in plant communities. *Ecology*, 87, 2425–32.

Fuhlendorf, S. D., and Engle, D. M. (2004) Application of fire-grazing interaction to restore a shifting mosaic on tallgrass prairie. *Journal of Applied Ecology*, 41, 604–14.

Gelbard, J. L., and Harrison, S. (2003) Roadless habitats as refuges for native plant diversity in California grassland landscapes. *Ecological Applications*, 13, 404–15.

Gough, L. C., and Grace, J. B. (1998) Herbivore effects on plant species density at varying productivity levels. *Ecology*, 79, 1586–94.

Grace, J. B. (1999) The factors controlling species density in herbaceous plant communities: An assessment. *Perspectives in Plant Ecology, Evolution and Systematics*, 2(1), 1–28.

Grace, J. B., Safford, H. D., and Harrison, S. (2007) Large-scale causes of variation in the serpentine vegetation of California. *Plant and Soil*, 293, 121–32.

Grime, J. P. (1977) Evidence for the existence of three primary strategies in plants and its relevance to ecological and evolutionary history. *American Naturalist*, 111, 1169–94.

Grime, J. P. (2001) *Plant Strategies, Vegetation Processes, and Ecosystem Properties*, 2nd ed. Wiley, New York.

Grubb, P. J. (1977) The maintenance of species richness in plant communities: The importance of the regeneration niche. *Biological Reviews*, 52, 107–45.

Harrison, S. (1997) How natural habitat patchiness affects the distribution of diversity in California serpentine chaparral. *Ecology*, 78, 1989–906.

Harrison, S. (1999) Native and alien species diversity at the local and regional scales in a grazed California grassland. *Oecologia*, 12, 99–106.

Harrison, S. P., Grace, J., Davies, K. F., Safford, H. D., and Viers, J. H. (2006a) Invasion in a diversity hotspot: Exotic cover and native richness in the California serpentine flora. *Ecology*, 87, 695–703.

Harrison, S. P., Inouye, B., and Safford, H. D. (2003) Ecological heterogeneity in the effects of grazing and fire on grassland diversity. *Conservation Biology*, 17, 837–45.

Harrison, S., Rice, K. J., and Maron, J. L. (2001) Habitat patchiness promotes invasions by alien grasses on serpentine soil. *Biological Conservation*, 100, 45–53.

Harrison, S. P., Safford, H. D., Grace, J., Viers, J. H., and Davies, K. F. (2006b) Regional and local species richness in an insular environment: Serpentine plants in California. *Ecological Monographs*, 76, 41–56.

Hobbs, R. J., Gulmon, S. L., Hobbs, V. J., and Mooney, H. A. (1988) Effects of fertiliser addition and subsequent gopher disturbance on a serpentine annual grassland community. *Oecologia*, 75, 291–95.

Hobbs, R. J., and Huenneke, L. (1992) Disturbance, diversity, and invasion: Implications for conservation. *Conservation Biology*, 6, 324–37.

Hobbs, R. J., and Mooney, H. A. (1985) Community and population dynamics of serpentine grassland annuals in relation to gopher disturbance. *Oecologia*, 67, 342–51.

Hobbs, R. J., and Mooney, H. A. (1991) Effects of rainfall variability and gopher disturbance on serpentine annual grassland dynamics. *Ecology*, 72, 59–68.

Hobbs, R. J., and Mooney, H. A. (1995) Spatial and temporal variability in California annual grassland: Results from a long-term study. *Journal of Vegetation Science*, 6, 43–57.

Hobbs, R. J., Yates, S., and Mooney, H. A. (2007) Long-term data reveal complex dynamics in grassland in relation to climate and disturbance. *Ecological Monographs*, 77, 545–68.

Huntly, N., and Reichman, O. J. (1994) Effects of subterranean mammalian herbivores on vegetation. *Journal of Mammalogy*, 75, 852–59.

Huntsinger, L., McClaran, M. P., Dennis, A., and Bartolome, J. W. (1996) Defoliation response and growth of *Nassella pulchra* (A. Hitchc.) Barkworth from serpentine and non-serpentine populations. *Madroño*, 43, 46–57.

Huston, M. A. (1994) *Biological Diversity: The Coexistence of Species*. Cambridge University Press, Cambridge.

Kimball, S., and Schiffman, P. M. (2003) Differing effects of cattle grazing on native and alien plants. *Conservation Biology*, 17, 1681–93.

Knox, R. G. (1984) Age structure of forests on Soldiers Delight, a Maryland serpentine area. *Bulletin of the Torrey Botanical Club*, 111, 498–501.

Kruckeberg, A. R. (1984) California serpentines: Flora, vegetation, geology, soils, and management problems. *University of California Publications in Botany*, 87, 1–180.

Kruckeberg, A. R. (2004) *Geology and Plant Life: The Effects of Landforms and Rock Types on Plants*. University of Washington Press, Seattle.

Mackey, R. L., and Currie, D. J. (2000) A re-examination of the expected effects of disturbance on diversity. *Oikos*, 88, 483–93.

Mackey, R. L., and Currie, D. J. (2001) The diversity-disturbance relationship: Is it generally strong and peaked? *Ecology*, 82, 3479–92.

Matos Mederos, J., and Torres Bilbao, A. (2000) Primeros estadíos sucesionales del Cuabal en las serpentinas de Santa Clara. *Revista del Jardín Botánico Nacional*, 21(2), 167–82.

McCallum, D. A. (2006) *Diversity and Conservation of Ultramafic Flora in Swaziland*. PhD diss., University of the Witwatersrand, Johannesburg.

McCoy, S., Jaffré, T., Rigault, F., and Ash, J. E. (1999) Fire and succession in the ultramafic maquis of New Caledonia. *Journal of Biogeography*, 26, 579–94.

McIntyre, S., Lavorel, S., and Tremont, R. M. (1995) Plant life-history attributes: Their relationship to disturbance response in herbaceous vegetation. *Journal of Ecology*, 83, 31–44.

McNaughton, S. J. (1968) Structure and function in California grasslands. *Ecology*, 49, 962–72.

Meimberg, H., Hammond, J. I., Jorgensen, C. M., Park, T. W., Gerlach, J. D., Rice, K. J., and McKay, J. K. (2005) Molecular evidence for an extreme genetic bottleneck during introduction of an invading grass to California. *Biological Invasions*, 8, 1355–66.

Milchunas, D. G., Sala, O. E., and Lauenroth, W. K. (1988) A generalized model of the effects of grazing by large herbivores on grassland community structure. *American Naturalist*, 132, 87–106.

Miller, G. L. (1981) Secondary succession following fire on a serpentine barren. *Proceedings of the Pennsylvania Academy of Sciences*, 55, 62–64.

Mittelbach, G. G., Steiner, C. F., Scheiner, S. M., Gross, K. L., Reynolds, H. L., Waide, R. B., Willig, M. R., Dodson, S. I., and Gough, L. (2001) What is the observed relationship between species richness and productivity? *Ecology*, 82, 2381–96.

Niederer, C., and Weiss, S. B. (2007) Ecological management of serpentine grassland in the face of nitrogen deposition and invasive grasses. In *Ecological Society of America/Society for Restoration Ecology Joint Meeting, Abstract of presentation*, pp. 59–71. San Jose, CA. Available online at http://eco.confex.com/eco/2007/techprogram/P3812.HTM.

Noy-Meir, I. (1995) Interactive effects of fire and grazing on structure and diversity of Mediterranean grasslands. *Journal of Vegetation Science*, 6, 701–10.

Oberhuber, W., Pagitz, K., and Nicolussi, K. (1997) Subalpine tree growth in serpentine soil: A dendroecological analysis. *Plant Ecology*, 130, 213–21.

O'Dell, R. E., James, J. J., and Richards, J. H. (2006) Congeneric serpentine and nonserpentine shrubs differ more in leaf Ca:Mg than in tolerance of low N, low P, or heavy metals. *Plant and Soil*, 280, 49–64.

Olff, H., Brown, V. K., and Drent, R. H. (eds.) (1999) *Herbivores: Between Plants and Predators*. Blackwell Science, Oxford.

Olff, H., and Ritchie, M. E. (1998) Effects of herbivores on grassland plant diversity. *Trends in Ecology and Evolution*, 13, 261–65.

Parker, V. T. (1990) Problems encountered while mimicking nature in vegetation management: An example from fire prone vegetation. In *Proceedings of the Fifteenth Annual Natural Areas Conference. Bulletin No. 471* (eds. R. S. Mitchell, C. J. Sheviak, and D. J. Leopold), pp. 231–34. New York State Museum, New York.

Petraitis, P. S., Latham, R. E., and Niesenbaum, R. A. (1989) The maintenance of species diversity by disturbance. *Quarterly Review of Biology*, 64, 393–418.

Pickett, S. T. A., and White, P. S. (1985) *The Ecology of Natural Disturbance and Patch Dynamics*. Academic Press, Orlando.

Proctor, J., and Woodell, S. R. J. (1971) The plant ecology of serpentine: I. serpentine vegetation of England and Scotland. *Journal of* Ecology, 59, 375–95.

Proulx, M., and Mazumder, A. (1998) Reversal of grazing impact on plant species richness in nutrient-poor vs. nutrient-rich ecosystems. *Ecology*, 79, 2581–92.

Reddy, R. A., Balkwill, K., and McLellan, T. (2009) Plant species richness and diversity of the serpentine areas on the Witwatersrand. *Plant Ecology*, 201, 365–81.

Reice, S. R. (1994) Nonequilibrium determinants of biological community structure. *American Scientist*, 82, 424–35.

Ricklefs, R. E., and Schluter, D. (1993) *Species Diversity in Ecological Communities*. University of Chicago Press, Chicago.

Rosiere, R. E., and Vaughn, C. E. (1986) Nutrient content of sheep diets on a serpentine barrens range site. *Journal of Range Management*, 39, 8–13.

Rosenzweig, M. L., and Abramsky, Z. (1993) How are diversity and productivity related? In *Species Diversity in Ecological Communities* (eds. R. E. Ricklefs and D. Schluter), pp. 52–65. University of Chicago Press, Chicago.

Rune, O. (1953) Plant life on serpentines and related rocks in the north of Sweden. *Acta Phytogeografia Suecica*, 31, 1–139.

Safford, H. D. (2008) Fire and ultramafic vegetation in northern California. In *6th International Conference on Serpentine Ecology, Program and Abstracts* (ed. N. Rajakaruna), p. 6. College of the Atlantic, Bar Harbor.

Safford, H. D., and Harrison, S. P. (2001) Grazing and substrate interact to affect native vs. exotic diversity in roadside grasslands. *Ecological Applications*, 11, 1112–22.

Safford, H. D., and Harrison, S. (2004) Fire effects on plant diversity in serpentine vs. nonserpentine chaparral. *Ecology*, 85, 539–48.

Safford, H. D., and Harrison, S. (2008) The effects of fire on serpentine vegetation and implications for management. In *Proceedings of the 2002 Fire Conference on Managing Fire and Fuels in the Remaining Wildlands and Open Spaces of the Southwestern United States* (ed. M. Narog), PSW-GTR-189, pp. 321–28. USDA Forest Service, PSW Research Station, Albany.

Safford, H. D., Viers, J. H., and Harrison S. P. (2005) Serpentine endemism in the California flora: A database of serpentine affinity. *Madroño*, 52, 222–57.

Spence, D. H. N. (1957) Studies on the vegetation of Shetland: I. The serpentine debris vegetation in Unst. *Journal of Ecology*, 45, 917–45.

Sugihara, N. G., van Wagtendonk, J. W., Shaffer, K. E., Fites-Kaufman, J., and Thode, A. E. (eds.) (2007) *Fire in California's Ecosystems*. University of California Press, Berkeley.

Takaoka, S., and Sasa, K. (1996) Landform effects on fire behavior and post-fire regeneration in the mixed forests of northern Japan. *Ecological Research*, 11, 339–49.

Tilman, D. (1982) *Resource Competition and Community Structure*. Princeton University Press, Princeton, NJ.

Tolman, D. A. (2007) Soil patterns in three *Darlingtonia* fens in southwestern Oregon. *Natural Areas Journal*, 27, 374–84.

Tyndall, R. W. (1992) Historical considerations of conifer expansion in Maryland serpentine "barrens." *Castanea*, 57, 123–31.

Tyndall, R. W. (1994) Conifer clearing and prescribed burning effects to herbaceous layer vegetation on a Maryland serpentine barrens. *Castanea*, 59, 255–73.

van Wagtendonk, J. W. (2007) Fire as a physical process. In *Fire in California's Ecosystems* (eds. N. G. Sugihara, J. W. van Wagtendonk, K. E. Shaffer, J. Fites-Kaufman, and A. E. Thode), pp. 38–57. University of California Press, Berkeley.

Vogl, R. J., Armstrong, W. P., White, K. L., and Cole, K. L. (1977) The closed-cone pines and cypress. In *Terrestrial Vegetation of California* (eds. M. G. Barbour and J. Major), pp. 295–358. Wiley, New York.

Weiss, S. (1999) Cars, cows, and checkerspot butterflies: Nitrogen deposition and management of nutrient-poor grassland for a threatened species. *Conservation Biology*, 13, 1476–86.

Wells, P. V. (1962) Vegetation in relation to geological substratum and fire in the San Luis Obispo quadrangle. *Ecological Monographs*, 32, 79–103.

Whittaker, R. H. (1960) Vegetation of the Siskiyou Mountains, Oregon and California. *Ecological Monographs*, 30, 279–338.

13

Plant–Pollinator Interactions in Naturally Fragmented Habitats

Amy T. Wolf, *University of Wisconsin, Green Bay*
Robbin Thorp, *University of California, Davis*

Serpentine landscapes appear to provide ideal systems for studying the effects of habitat isolation on species interactions, including the relationship between flowering plants and their pollinators. Serpentine plant assemblages are generally discrete (Harrison et al., 2006), highly endemic (Skinner and Pavlik, 1994; Safford et al., 2005), and physiologically specialized (Brooks, 1987). Maps from California (Kruckeberg, 1984), Cuba (Reeves et al., 1999), New Caledonia (Dawson, 1981), South Africa (Williamson and Balkwill, 2006), and elsewhere illustrate the highly insular nature of serpentine substrates, a pattern that is repeated at multiple geographic scales (Harrison et al., 2006). In this chapter, we explore the characteristics of pollinator networks in serpentine landscapes and ask whether these networks can help us understand plant–pollinator interactions elsewhere. The answer to this question is important because many natural habitats are becoming increasingly fragmented as a result of human activities (Young et al., 1996). Aizen and Feinsinger (1994), Aguilar et al. (2006), and others have provided evidence that habitat fragmentation affects plant–pollinator networks and might present a threat to plant reproduction in human-dominated landscapes. An understanding of ecological dynamics in naturally fragmented landscapes like serpentine may provide valuable insights into the validity of this warning and may help guide conservation strategies for minimizing the unwanted effects of habitat fragmentation on plant–pollinator interactions.

The effects of regional versus local processes on community structure (Harrison and Cornell, 2008) can be evaluated in landscapes where naturally isolated serpentine patches are embedded in a matrix of other habitat types. Interactions of species within this habitat mosaic expose the forces that maintain species composition and population dynamics, including the relationships between plants and their pollinators. Resistance of serpentine patches to invasive species, for example, has revealed important insights about the role of physical conditions on interspecific competition in natural communities (Williamson and Harrison, 2002). Pollination studies on serpentine are few and incomplete, but recent studies described in this chapter suggest that the unique physical and geographic attributes of serpentine landscapes create outstanding field laboratories for studying pollination and related ecological interactions.

PLANT-POLLINATOR MUTUALISMS ON SERPENTINE

Approximately 60–80% of the world's 250,000 flowering plant species are pollinated by animals, mostly insects (Kremen et al., 2007). These plant species have evolved a wide variety of floral structures and attractant chemicals to ensure that pollen is transferred effectively to a compatible stigma. The evolution of flowers invariably involves resource allocation trade-offs, however (Lloyd, 1989; Strauss and Whittall, 2006). Energetic and nutrient trade-offs might be especially critical in habitats like serpentine, where soil chemistry, drought, and temperature extremes create severe resource limitations (Proctor and Woodell, 1975; Brooks, 1987; Brady et al., 2005). Allocation of resources to showy, nectar-rich flowers might occur at great expense to plant survival on serpentine, at least in the short term.

Resource limitation seems to be reflected in the types of plants that are successful on serpentine. Families of plants that tend to be disproportionately well represented on serpentine in California include the typically small-flowered Caryophyllaceae, Brassicaceae, and Linaceae, whereas large- or complex-flowered families like Scrophulariaceae, Hydrophyllaceae, Onagraceae, and Polemoniaceae are disproportionately poorly represented (Safford et al., 2005). Although there are exceptions to this generalization (e.g., *Calystegia collina*, *Mimulus layneae*) and quantitative studies on resource availability are needed, the preponderance of small flowers and the undeniably harsh edaphic conditions (Kruckeberg, 1954) suggest that over an entire growing season, nectar and pollen resources may be in short supply on serpentine compared with other habitats. Galen (1999) has provided a more general argument that reduced flower size is advantageous under resource-poor conditions.

In addition to generally small flower size, most serpentine plants are characterized by low stature and relatively high root:shoot ratios (Brady et al., 2005). Overall cover of these plants is sparse (Kruckeberg, 2004), leading to limited availability of flowers for pollinators in serpentine environments. Brady et al. (2005) noted that serpentine plants typically flower and set seed earlier than related taxa on nonserpentine soils to avoid drought conditions. This adaptation constrains the accumulation of leaf biomass and results in fewer available resources for the flowers themselves. Flowers on serpentine patches can be spectacular, but floral rewards are usually available to pollinators for only a short period due to low moisture and depressed nutrient levels (Brady et al., 2005). The sparse temporal distribution and small stature of serpentine flowers make it relatively uneconomical for bees and other pollinators to specialize on serpentine plant species (Cane and Sipes, 2006; Minckley and Roulston, 2006). Stebbins (1970) argued that floral specialization in plants should favor the most effective or abundant pollinators; generalization is favored when availability of pollinators is unpredictable from year to year (Johnson and Steiner, 2000). Using the same logic for pollinators, bee species and other visitors are unlikely to specialize on flowering plant species that are sparse or available for only a short time. Indeed, pollinators are generalists under most circumstances (Waser et al., 1996; Cane and Sipes, 2006), so we would expect little if any specialization of animals on either pollen (oligolecty) or nectar in serpentine landscapes.

Not all patchy and ephemeral habitats lack specialist pollinators. Thorp and Leong (1998) and Thorp (2007) reported that four conspicuous flowering plant species of vernal pools in California are visited by native, solitary, specialist bees. The bee specialists (oligoleges) collect pollen only from preferred flower species, and seed set in at least one of the plant species is significantly reduced in the absence of its specialist bee. A major difference between vernal pools and serpentine substrates is the extended availability of water in vernal pools, which can sustain flowers with richer or more predictable nectar and pollen resources.

The severe serpentine environment might affect plant–pollinator mutualisms in other ways. Boyd (2009) has demonstrated that hyperaccumulation of nickel by serpentine plants can lead to unusually high nickel concentrations in insects that eat these plants. Although the highest concentrations of nickel were found in folivores, Wall and Boyd (2006) and Boyd et al. (2006) found elevated levels of nickel in bees that visit serpentine flowers in California and New Caledonia, respectively. The extent and consequences of nickel accumulation in serpentine pollinators is unknown, but nickel hyperaccumulation might have broader food web effects through plant–pollinator networks on serpentine (Boyd and Martens, 1998).

Perhaps more significantly, the severe conditions for plant growth on serpentine soils may affect the timing of plant–pollinator interactions and, ultimately, the evolution of reproductive isolation. As mentioned above, individual plants on serpentine often flower earlier than conspecifics in nonserpentine habitats to avoid drought stress (Brady et al., 2005). Gailing et al. (2004) demonstrated that the serpentine-tolerant *Microseris douglassii*, for example, flowers and sets seeds earlier than its nonserpentine congener, *M. bigelovii*. A similar shift in flowering has been reported for serpentine populations of *Collinsia sparsifolia* (Wright et al., 2005). Differences in flowering schedules might promote reproductive isolation among local populations of plants in serpentine landscapes. In other words, the timing of plant–pollinator interactions might represent an important mechanism for speciation, perhaps playing a role in the well-known diversification of endemic serpentine plant species.

SPATIAL ECOLOGY OF POLLINATION

By their very nature, plant–pollinator interactions have a spatial dimension. Pollinators typically move among many flowers, inflorescences, and flower patches; the spatial distribution of these resources may have a significant impact on pollinator behavior (Pyke, 1978; Sih and Baltus, 1987; Bronstein, 1995; and others). One of behavioral ecology's most intriguing questions has been, "What are the evolutionary forces that drive animal foraging patterns?" A rich body of literature during the 1970s and 1980s asked whether species move optimally with respect to the spatial distribution of resources (MacArthur and Pianka, 1966; Charnov, 1976; Pyke, 1984). Although the optimization approach has been severely criticized (Pierce and Ollason, 1987), spatial dimensions of foraging behavior remain a legitimate subject of evolutionary research (Stearns and Schmid-Hempel, 1987; Parker and Maynard Smith, 1990). Numerous studies have evaluated pollinator foraging behavior among different configurations of flowers (Beattie, 1976; Heinrich, 1976, 1979; Pleasants and Zimmerman, 1983; Pleasants, 1989), although most of these analyses have focused on very local scales. The discrete configuration of many serpentine outcrops provides a useful field setting for exploring the effects of patch size, patch density, and patch quality on the movement patterns of pollinators.

Pollinator movements, in turn, may affect plant population genetics. Richards et al. (1999) used genetic markers in the short-lived perennial *Silene alba* to examine pollinator effectiveness in experimental patches of different size and isolation. Rates of pollen transfer were nearly eight times greater between patches separated by 20 m than between patches separated by 80 m. Goulson (2000) studied the use of experimental patches of *Trifolium repens* by the bumblebee *Bombus lapidarius*. The proportion of inflorescences visited by bees was greater in small patches than in large patches, a pattern predicted by a simple behavioral departure rule. Patchy serpentine landscapes dominated by endemic, habitat-specific plant species provide an ideal landscape for studying these spatial dynamics involving pollinator behavior and plant population genetics. Our research, described below, suggests that pollinators can play an important role in maintaining genetic diversity of the serpentine endemic *Calystegia collina*.

The relationship between species richness and area is one of the best-known patterns in ecology (Lomolino, 2000). In serpentine landscapes of California (and others like it), this relationship is complicated by the fact that serpentine patches are embedded in a matrix of nonserpentine grassland, rocky meadows, chaparral, and other habitat types. On a landscape scale, different regions are characterized by different biogeographic pools of serpentine endemic species. Harrison et al. (2006) evaluated the influence of local factors (e.g., area of serpentine patch, soil attributes) and regional factors (e.g., regional species pool, total serpentine area) on species richness in serpentine outcrops of California. Their results revealed important insights about the scale of community dynamics. Perhaps most significantly, they showed that area effects on endemic species richness are manifest mainly at the regional level. This same approach can be applied to plant–pollinator interactions of serpentine plants. For example, what are the relative roles of local versus regional effects on pollination success of serpentine flowers? How does the spatial configuration of serpentine patches influence the availability and effectiveness of pollinators? How does the species composition of the habitat matrix affect pollination success on serpentine patches? Just as Harrison et al. (2006) and others have successfully explored species richness in complex serpentine landscapes, the spatial dimensions of plant–pollinator interactions are ripe for future study.

Recently, conservation biologists have explored the effects of anthropogenic habitat loss and fragmentation on plant–pollinator mutualisms. Aguilar et al. (2006) used a meta-analysis to summarize studies of pollination success in fragmented versus unfragmented habitats. They found a strong negative effect of reduced habitat size and isolation on plant reproduction and suggested that pollination limitation is a major cause of reproductive impairment in fragmented habitats. Kearns et al. (1998) and Aizen et al. (2002) cited other studies of reduced pollination effectiveness in fragmented habitats. Experimental studies of habitat isolation have demonstrated that barriers between plants as close as 5 m apart can affect patch visitation rates by bumblebees on calcareous grasslands (Goverde et al., 2002). Kolb (2008) demonstrated that reduced plant population size in fragmented habitats has a negative effect on pollination success but no effect on rates of herbivory. She argued that the effects of habitat fragmentation are complex and must take into account both mutualistic and antagonistic species interactions.

Not all studies have shown strong negative effects of habitat fragmentation on pollination mutualisms, however (Kearns et al., 1998). Cane (2001) noted that bees, far and away the most important pollinators outside of the tropics, have broadly general preferences among plant species for nectar but in some cases, more specific requirements for pollen. "Habitat" for bees includes these floral resources plus suitable nesting substrates. Defined in this way, habitats for bees are intrinsically patchy and are often ephemeral (Bowers, 1985). Nesting habitat is typically separated from food resources, requiring multiple daily commutes between disjunct areas. Habitat fragmentation therefore becomes exceedingly difficult to define and assess for plant–pollinator mutualisms. According to Cane, experimental studies of isolated flower arrays do not address the habitat in which bees live, and therefore shed little light on the overall ecological effects of habitat fragmentation on bee–plant mutualisms. The matrix surrounding isolated habitat remnants might provide suitable nesting habitat or nectar resources for bees even though the vegetation type is dramatically different from that in the fragmented habitat (Cane, 2001; Marlin, and LaBerge 2001). In short, the effects of habitat fragmentation on plant–pollinator mutualisms are not always straightforward; future studies will be needed to understand plant–pollinator interactions in both habitat remnants and the surrounding ecological matrix.

SPATIAL ECOLOGY OF POLLINATION ON CALIFORNIAN SERPENTINE

Few (if any) detailed studies of plant–pollinator mutualisms have been conducted on serpentine, but the reproductive ecologies of several serpentine plants have been investigated by Westerbergh and Saura (1994), Wolf and Harrison (2001), and Wright and Stanton (2007). We studied the spatial ecology and pollination of three serpentine plant species in the North Coast Ranges of California during the late 1990s (Wolf et al., 1999, 2000; Wolf and Harrison, 2001). These species are not necessarily representative of the serpentine flora in this region, let alone serpentine plants in general, but they provide a glimpse of plant–pollinator interactions in this naturally patchy landscape.

Calystegia collina (Greene) Brummitt (Convolvulaceae) is a long-lived perennial plant endemic to the Coast Ranges of California. Within our study area in Napa, Lake, and Sonoma Counties, *C. collina* is mostly restricted to serpentine substrates, where it grows in discrete patches within both large and small outcrops. Patchiness occurs at several scales; serpentine formations are discontinuous over large geographic areas in the Coast Ranges, groups of serpentine outcrops occur regionally within these areas, and discrete *C. collina* patches of several meters to several hundred meters occur locally.

We studied reproductive ecology and population genetics of 12 *C. collina* patches on small (<5 ha) outcrops and 18 patches within large (>300 ha) outcrops (Wolf and Harrison, 2001). Experimental exclusion/pollination analyses showed that *C. collina* is an obligate outcrosser. Pollen from flowers outside a given patch yielded significantly higher fruit set than pollen from flowers within the patch (Wolf et al., 2000). Under these circumstances, isolation from other patches can be a serious disadvantage. Without compatible pollen from nearby flowers or flower patches, seed production is likely to be reduced or absent altogether.

Our comparison of *C. collina* reproduction on large and small outcrops was consistent with the experimental pollination results (Wolf and Harrison, 2001). Flowers on large outcrops were significantly less isolated than those on small outcrops. Correspondingly, the proportion of marked flowers producing seeds was significantly higher on large outcrops. Overall, the numbers of *C. collina* flowers (including those within the same patch) and *C. collina* flower patches within 100 m were strong predictors of successful pollination. These findings suggest that cross-pollination is critical for *C. collina* reproduction on serpentine outcrops, and the spatial distribution of local flower patches plays an important role in the success of pollination at a given site.

The local bee fauna is a critical element of serpentine plant population dynamics. Like many flowering plants, *C. collina* depends on bees for pollination (Figure 13.1). To better understand the observed patterns of seed production, we collected bees in standard bowl traps at 19 *C. collina* patches in small outcrops and 20 patches in large outcrops. Results showed several clear patterns. The numbers of bees collected in bowl traps and the numbers of bees visiting *C. collina* flowers were highly correlated with the numbers of *C. collina* flowers in a patch. In other words, bees seemed to be attracted to patches where flowers were abundant. Taxonomic richness of bees (number of genera) followed a similar pattern. The most

FIGURE 13.1. *Bombus melanopygus* at *Calystegia collina* flower.

frequently encountered types of bees in both traps and observations belonged to the genera *Lasioglossum* (*Dialictus*), *L.* (*Evylaeus*), *Halictus*, *Andrena*, and *Bombus*, in decreasing order of abundance.

Interestingly, *C. collina* patches on large outcrops produced over four times as many flowers per patch than did patches on small outcrops, despite a smaller average patch area. As expected, bee numbers were significantly higher in *C. collina* patches on large outcrops. When the relationship between numbers of flowers and numbers of bees was taken into account, however, the difference in bee numbers between large and small outcrops was not significant (Wolf and Harrison, 2001). The numbers of bee visits per flower per minute also did not differ between patches on large versus small outcrops. The question of why more flowers are present in *C. collina* patches on large outcrops has not been answered conclusively (Wolf and Harrison, 2001), and our findings provided little evidence that bee faunas on large serpentine outcrops were fundamentally different than those on small outcrops, other than their access to higher numbers and densities of *C. collina* flowers.

The serpentine sunflower, *Helianthus exilis* Gray (Asteraceae), illustrates a rather different lifestyle than *C. collina*. *H. exilis* is a robust annual restricted to occasionally moist seeps and rocky ravines within large serpentine outcrops. These localized microhabitats support a distinct assemblage of plant species (Harrison et al., 2000a). Like *C. collina*, *H. exilis* is endemic to the Coast Ranges of California and is taxonomically similar to a more widespread California species, *H. bolanderi* (Hickman, 1993). Unlike *C. collina*, however, *H. exilis* rarely (if ever) occurs on small serpentine outcrops due to the absence of seep microhabitats.

H. exilis flowers are self-compatible, but flowers fertilized by pollen from other individuals produced significantly more seeds than self-pollinated flowers (Wolf et al., 2000). Our field experiments showed that individuals are moderately pollen-limited; flowers treated with extra pollen produced 12% more seeds on average than naturally pollinated controls. Bees were by far the most frequent natural pollinators, accounting for 94% of all observed visits to *H. exilis* flowers. Again, unlike our findings for *C. collina*, neither patch isolation nor local population size was significantly correlated with the frequency of pollinator visits. These results suggest that the effects of isolation might be partly mitigated by the greater self-compatability of *H. exilis* compared with strict outcrossers like *C. collina*.

Another uncommon California endemic, *Fritillaria purdyi* Eastwood (Liliaceae), occurs in rocky openings of serpentine chaparral dominated by leather oak (*Quercus durata*) and several other woody species. Serpentine chaparral covers thousands of square kilometers within the range of *F. purdyi* in northwestern California (Skinner and Pavlik, 1994). This perennial herb is found in dense but widely scattered populations of a few hundred to a few thousand plants. Although not considered endangered, *F. purdyi* has been designated a "species of limited distribution" by the California Native Plant Society (Skinner and Pavlik, 1994).

Along with collaborators Jennifer (Williamson) Burt and Susan Harrison at the University of California, Davis, we studied the possible roles of herbivory, pollination limitation, and environmental variation in maintaining the patchy distribution of *F. purdyi* in serpentine chaparral of Lake County, California. Because this work is unpublished, we include more detail here than for the previous two species. To determine the importance of pollinators for *F. purdyi*, we randomly chose 100 plants with flower buds and assigned 25 to each of four treatments: (1) bagged control, (2) addition of self pollen, (3) addition of mixture of pollen from three different individuals, and (4) unbagged control. Flowers in treatments 1–3 were covered with a fine mesh bag to prevent pollinators from visiting. We recorded whether each flower produced a capsule and collected all mature capsules to count the number of seeds produced.

More than half (68%) of unbagged control flowers produced seed capsules compared with 0% of the bagged control treatments, showing that insects or other animal vectors are needed for successful pollination. Seed capsules were produced by 20% of artificially self-pollinated flowers and 96% of artificially cross-pollinated flowers ($p < 0.01$, chi-square 2 × 2 test of association).

Herbivory of flowering plants can affect reproductive success, and we found that this factor was important during the flowering period of *F. purdyi* on serpentine. On average, 15% ($\pm$ 4.1 SE) of the naturally occurring plants in our study population exhibited vertebrate herbivory during 1999, compared with 10% ($\pm$ 2.6) exhibiting insect herbivory. The incidence of herbivory was higher during 2000; 23.3% ($\pm$ 4.7) of sampled plants showed evidence of vertebrate herbivory, whereas 25.8% ($\pm$ 5.6) showed evidence of insect herbivory. The proportion of plants exhibiting vertebrate or insect herbivory on leaves or flowers was not correlated with the density of plants in sample plots during either year ($p > 0.05$, Spearman rank correlation).

We also conducted a manipulative experiment (Figure 13.2) to examine the effects of plant isolation on survival and reproduction of *F. purdyi*. Six groups of three flowering plants were transplanted 25–35 m from the main study population in each of four cardinal directions ($n = 72$). Another six sets of three transplanted plants (controls) were planted within the main study population. We recorded herbivory and seed capsule production for each transplant and control plant in 1999 and 2000. Visits by potential pollinators were recorded during 30-min observation periods at all flowering transplants and control plants. Observations were paired, with one observer recording visitors at distant transplant flowers and another observer simultaneously recording visitors at control flowers. We also marked 20 naturally isolated plants >25 m from the nearest *F. purdyi* patch of 10 or more plants (not shown in Figure 13.2) and recorded seed set and herbivory at the same time measurements were made at the transplants. Among the experimental plants that were transplanted >25 m beyond the main population (distant

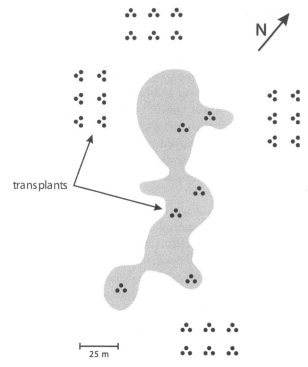

FIGURE 13.2. Map of *Fritillaria purdyi* study area. Experimental plants (•) were transplanted both within the existing population (shaded) and >25 m from the edge of the population.

transplants), 18% experienced insect herbivory on leaves, compared with 22% of transplants within the main population. This difference was not statistically significant ($p > 0.05$, logistic regression). Vertebrate leaf herbivory, on the other hand, was observed on 33% of the distant transplants but on 0% of the control transplants within the main population. The increased vertebrate leaf herbivory in distant plants compared with the main population was significant ($p = 0.024$, logistic regression). Flower herbivory followed a similar pattern to that of leaf herbivory; vertebrates ate 36% of the flowers in distant transplants, compared with 0% of the transplanted flowers within the main population ($p < 0.05$, logistic regression). Clearly, isolated plants experience elevated levels of vertebrate herbivory, probably by mule deer (*Odocoileus hemionus*) and brush rabbits (*Sylvilagus bachmani*). This negative density dependence is exactly opposite the pattern of herbivory on tropical rain forest seeds reported by Janzen (1970) and Connell (1971). A plausible explanation for this pattern is that existing *F. purdyi* patches tend to occur in large openings, whereas vertebrate activity is likely to be greater in areas with denser cover of *Q. durata* and other shrubs. Plants located far from *F. purdyi* patches are more likely to be located near shrubs and other cover for rabbits and deer.

Two native bee species were the primary visitors to *F. purdyi* flowers: *Andrena (Euandrena)* nr. *auricoma* Smith and *Andrena (Euandrena) microchlora* Cockerell. We occasionally saw honeybees (*Apis mellifera*) and bumblebees (*Bombus* spp.) visiting flowers. Average numbers of visits by potential pollinators during our 30-min observation periods were 1.77 ($\pm$ 0.44 SE) for the transplanted controls within the main *F. purdyi* population, and 0 for the distant transplant patches ($p <$ 0.01, Mann-Whitney U-test). Reduced numbers of visitors lead to reduced seed production in isolated flowers, a result that mirrors the Allee effect reported by Groom (1998) in the annual herb *Clarkia concinna*. None of the distant transplants produced seed capsules, and only 1 of the 18 transplant controls in the main population produced a capsule. The number of visitors within naturally occurring patches was positively correlated with the local (1 m^2) density of *F. purdyi* flowers ($r =$ 0.92, $p <$ 0.001). This result is similar to the density-dependent pollination rates found in *C. collina* (Wolf and Harrison, 2001).

Our experiments with *F. purdyi* demonstrated that the effects of pollination mutualisms are cast in the context of other biotic interactions. Insects attacked the foliage of a large proportion of *F. purdyi* individuals. Vertebrates damaged leaves and flowers in a smaller fraction of plants in the natural population, but vertebrate herbivory was critically important in isolated experimental plants. Pollinators, mainly two species of native bees in the genus *Andrena*, proved to be necessary for *F. purdyi* seed production, and the plant appeared to be both pollinator-limited and partially self-incompatible. The highly aggregated spatial distribution of *F. purdyi* may be influenced in part by these biotic interactions. Transplants located 25–35 m beyond the limits of the population—a reasonable distance for natural long-distance dispersal by *F. purdyi* seeds—suffered elevated rates of leaf and flower loss to vertebrates and, simultaneously, received substantially reduced visitation by potential pollinators.

The positive density dependence seen in bee visitation is a commonly observed pattern predicted by the economics of pollinator foraging behavior (Hixon et al., 1983; Schmitt, 1983; Byers, 1995). We suggest that *F. purdyi* seeds landing a few tens of meters outside a dense population might experience both lower survival probability (due to higher herbivory) and lower probability of reproduction (due to lower visitation by potential pollinators). These dual biotic interactions might play an important role in maintaining the spatially aggregated distribution of *F. purdyi*. New populations of *F. purdyi* are probably initiated by chance long-distance dispersal to favorable new sites, but as the population grows, aggregation may be reinforced by spatially dependent interactions with both herbivores and pollinators. We found that *F. purdyi* distributions are associated negatively with shade and positively with pH (unpublished data), but many apparently suitable sites are unoccupied across the species's range. In other words, we found little evidence that physical factors dictate the local distribution of this species in serpentine chaparral. Instead, demographic processes (dispersal, germination, etc.) and biotic interactions (herbivory and pollination) seem to be largely responsible for the aggregated patterns found in serpentine populations of *F. purdyi*.

BEE DISTRIBUTIONS IN SERPENTINE LANDSCAPES

During 1999, we trapped bees in standard bowl traps (LeBuhn et al., 2007) at 33 sites in 4 different habitat types (serpentine meadow, serpentine chaparral, nonserpentine meadow near [<250 m] serpentine outcrops, and nonserpentine meadows far [>1 km] from serpentine outcrops) in the McLaughlin Natural Reserve, the primary location of our serpentine plant studies. Serpentine meadows in the North Coast Range are vegetated by a mixture of mostly native grasses and forbs (Harrison et al., 2000b) on deep clays or shallow rocky slopes, formed by weathering of serpentine and related parent materials. Serpentine meadows typically are embedded in a landscape of serpentine chaparral, a shrubby habitat dominated by *Arctostaphylos viscida* (white-leaved manzanita), *Ceanothus jepsonii* (Jepson's ceanothus), and *Q. durata* (leather oak). Serpentine chaparral generally occurs on thin, rocky soils. Nonserpentine grasslands or meadows occur on loams formed from sandstones, mudstones, and shales. These grasslands are interspersed within blue oak (*Quercus douglasii*) savanna and woodland and often are dominated by non-native, annual grasses (e.g., *Avena fatua*, *Avena barbata*, *Bromus hordeaceus*, *Lolium multiflorum*, *Taeniatherum caput-medusae*) mixed with native and non-native forbs (Harrison et al., 2000b). Nonserpentine meadows within 250 m of serpentine outcrops generally are surrounded by nonserpentine chaparral, dominated by shrubs such as *Adenostoma fasciculatum* (chamise), *Quercus berberidifolia* (scrub oak), *Arctostaphylos manzanita* (Parry manzanita), *Ceanothus sorediatus* (Jim bush), *Pickeringia montana* (chaparral pea), and *Heteromeles arbutifolia* (toyon).

Bees were trapped by placing an array of 10 traps 0.25 m apart at each site. Ten sites were evaluated for each habitat type except nonserpentine meadow far (>1 km) from serpentine outcrops, which was represented by only three sites. Traps consisted of yellow plastic bowls filled with water and a drop of liquid detergent. Traps were deployed at each site before 0800, and all bees were collected from traps and placed in vials after 1600 on the same day. Trapping was conducted twice, once on May 8, 1999, and again on May 16, 1999. We also recorded the presence/absence and abundance (on a log scale) of all flowering plant species along a 50 m × 2 m transect at each trap location during each of the trap periods.

Overall we collected 63 species of bees in 23 genera and 5 families. Comparisons among sites (Figure 13.3) showed that bee assemblages from serpentine chaparral were generally different from bee assemblages in meadow (open) habitats, but few differences were observed between serpentine and nonserpentine meadows. In other words, the bee species assemblages on serpentine outcrops were not much different than those on nonserpentine meadows in the same area. Likewise, neither bee species richness nor the average numbers of bee species collected during our constant-effort samples differed significantly between serpentine and nonserpentine meadows, although species richness was lower in serpentine chaparral. Coupled with studies of serpentine plant distributions in the same region (e.g., Wolf, 2001; Harrison et al., 2006), these results suggest that habitat specialization in serpentine outcrops applies to plants but not necessarily to the bees that pollinate them. The bee diversity of Mediterranean landscapes in California is relatively high (Moldenke, 1979), but our findings suggest that serpentine specialists contribute little if any to this regional diversity.

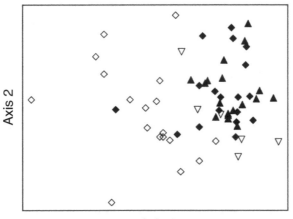

HABITAT
◆ serpentine meadow
◇ serpentine chaparral
▲ nonserpentine meadow (near)
▽ nonserpentine meadow (far)

FIGURE 13.3. First two axes of a nonmetric multidimensional scaling ordination of bee assemblages from different habitats in the McLaughlin Natural Reserve in Napa and Lake Counties, California. Each symbol represents a collection of bees from an array of 10 traps 0.25 m apart on May 8 or May 16, 1999.

SERPENTINE AS A MODEL LANDSCAPE

Do serpentine landscapes provide special opportunities for studying terrestrial ecosystems? In the case of plant–pollinator mutualisms, the answer is yes, but not because the interactions are themselves unique. Serpentine habitats are particularly interesting and useful for study because the harsh edaphic settings are discrete and repeated at different scales, including small, isolated patches and extensive areas of hundreds or thousand square kilometers (Wolf, 2001). Physical conditions consisting of inhospitable soil chemistry and frequent water deficits create a patchy environment inhabited by a distinct set of geographically restricted plant species. Strong selection in this harsh environment promotes local adaptation (e.g., Sambatti and Rice, 2006) and a fascinating arena for studying ecological and evolutionary dynamics.

Perhaps the most important general message from studies of plant–animal mutualisms on serpentine, including ours, is that pollination networks must be viewed from a community perspective and in a landscape context. Clearly, bee populations that pollinate serpentine flowers are not limited to the patches in which these flowers occur. For example, nesting sites for many (if not most) bee visitors occur outside serpentine flower patches. We have shown that several species of serpentine flowering plants are pollen-limited and sensitive to the configuration of conspecifics at spatial scales of 100 m or less. Bee pollinators, on the other hand, are known to cover much larger areas (Osborne et al., 1999; Gathman and Tscharntke, 2002; Greenleaf et al., 2007). To fully understand the dynamics of plant–pollinator mutualisms in serpentine, the larger scale configuration of pollinator (bee) nesting habitats and the availability of alternative nectar or pollen resources need to be evaluated. Unfortunately, few pollination studies, including investigations of habitat fragmentation, pay adequate attention to the spatial configuration of pollinator nesting habitats and other large-scale life history considerations (Kearns and Inouye, 1997; Cane, 2001; Spira, 2001).

A multiscale perspective of plant–pollinator mutualisms (i.e., a perspective that accounts for both local and regional influences) suggests a role for many species in maintaining and modifying plant–pollinator networks (Olesen and Jordano, 2002; Lopezaraiza-Mikel et al., 2007). Drought-prone serpentine meadows lack flower resources for much of the year, so flowering plants in the matrix separating serpentine patches might be needed to sustain pollinator populations (Schmitt, 1983). Westerbergh and Saura (1994) have shown that the physical structure of adjacent vegetation, like the biological matrix, can affect pollinator movements and gene flow between isolated populations. The idea that regional factors may affect local plant–pollinator interactions becomes especially relevant when one considers both temporal and spatial scales.

Recent quantitative approaches to coevolutionary networks (Bascompte et al., 2003; Fortuna et al., 2008; and others) have established an innovative future direction for understanding plant-pollinator systems. Bascompte et al. (2006), for example, demonstrated that most species' dependencies in mutualism networks are weak. Critical ecosystem services therefore might depend on the collective contributions of many species and conditions, suggesting that we have only scratched the surface in exploring the ecology of plant–pollinator interactions. Serpentine landscapes, where physical conditions and species distributions can be relatively easily documented, would seem to provide excellent opportunities for analyses of these ecological networks.

The multiscale ecology of serpentine plant–pollinator interactions is also relevant to the notion of habitat fragmentation, one of today's most important conservation challenges (Lindenmayer and Fischer, 2006). Serpentine habitats clearly are insular, yet our field studies have demonstrated that plant species distributions are patchy at even smaller scales within these serpentine habitats. In some cases (e.g., *Helianthius exilis*), patchiness of local plant populations is due to microhabitat specialization, whereas in other cases (*C. collina* and *F. purdyi*), historical or ecological factors seem to be involved. Because serpentine environments are edaphically distinct and are relatively sparsely vegetated, the spatial processes and interactions underlying multiscale geographic patterns can be observed quite clearly. Such processes and interactions might be easily overlooked or oversimplified in less tractable systems.

Studies over the past several decades have yielded important insights into the effects of habitat fragmentation on species, communities, and ecosystems in human-dominated landscapes (e.g., Powell and Powell, 1987; Klein, 1989; Stouffer and Bierregaard, 1995; Davies and Margules, 1998; Laurance et al., 1998; Carvalho and Vasconcelos, 1999; Tabarelli et al., 2004; Damschen et al., 2008). Conclusions from many of these studies are similar to those emerging from studies of serpentine landscapes. Specifically, investigations on habitat remnants and serpentine outcrops both have shown that important population and community processes occur at spatial scales that do not necessarily overlap among interacting species (Wolf and Harrison, 2001; Tscharntke and Brandl, 2004; Harrison et al., 2006). Likewise, the matrix separating habitat patches may play a critical role in ecological interactions, even when the species of interest are restricted to discrete, easily recognized sites (Ricketts, 2001; Steffan-Dewenter et al., 2002).

Plant–pollinator interactions in serpentine landscapes offer opportunities for exploring many other ecological questions. Gardner and Mcnair (2000), for example, showed that differences in pollinators can lead to asymmetric pollen transfer between a serpentine endemic species (*Mimulus nudatus*) and its presumed ancestor (*M. guttatus*). Coupled with differences in drought tolerance, the imbalance in pollen transfer contributes to long-term coexistence of these two plant species. Gardner and Mcnair's (2000) study addresses one of ecology's most important historical questions: how do species coexist in nature? Their results suggest that plant–pollinator interactions might play a significant role in the coexistence of closely related species pairs. Other important ecological questions involving plant–pollinator interactions include the following. How do alien species affect native plant–pollinator interactions (Lopezaraiza-Mikel et al., 2007)? What are the likely effects of global climate change on plant–pollinator interactions (Memmott et al., 2007; Hegland et al., 2009)? How important is local adaptation in the spatial dynamics of plant–pollinator interactions (Sambatti and Rice, 2006; Gomez et al., 2009)? For each of these topics, serpentine landscapes provide excellent settings for future studies because of their severe edaphic constraints, spatial patchiness, high degree of plant endemism, and (in many regions) relatively undisturbed conditions.

In conclusion, our understanding of plant–pollinator mutualisms in serpentine landscapes is only superficial, yet preliminary research suggests that serpentine might provide a model system for understanding the multiscale dynamics and other important aspects of species interactions. Although serpentine plants experience strong selection for survival in a harsh environment (Kruckeberg, 1984; Brady et al., 2005; and others), at least some serpentine species, including two of the three described here, are obligate outcrossers and require animal vectors for sexual reproduction. Hence, pollination is an important feature of plant ecology in serpentine landscapes. The discrete nature of serpentine substrates and the typically sparse and distinctive vegetation simplifies the spatial analysis of plant–pollinator dynamics. The dynamics themselves, however, are hardly simple. Instead, serpentine landscapes help us appreciate the complex, multiscale character of plant–pollinator interactions. With future study, species interactions on serpentine have great potential for improving our understanding of spatial ecology, species diversification, and biological conservation.

Acknowledgments

Work described in this chapter was conducted at the University of California, Davis, with the support and guidance of Susan Harrison. We are truly grateful for her inspiration and vision in the study of serpentine ecology. Jennifer Williamson Burt was a major contributor to the field work on *Fritillaria purdyi*. Other significant field contributors included Nicole Jurjavcic, Paul Brodmann, and Joe Callizo, whose insights on serpentine plant ecology were important for all of us. Robert Howe, Sharon Strauss, Martha Groom, and Carol Goodwillie provided valuable comments and advice during various stages of the serpentines field studies. We are also grateful to the editors and two anonymous reviewers for constructive comments on earlier drafts of this chapter. Voucher bee specimens collected during our investigation have been deposited in the R. M. Bohart Museum of Entomology at the University of California, Davis.

LITERATURE CITED

Aguilar, R., Ashworth, L., Galetto, L., and Aizen, M. A. (2006) Plant reproductive suscepti- bility to habitat fragmentation: Review and synthesis through a meta-analysis. *Ecology Letters*, 9, 968–80.

Aizen, M. A., Ashworth, L., and Galetto, L. (2002) Reproductive success in fragmented habitats: Do compatibility systems and pollination specialization matter? *Journal of Veg- etation Science*, 13, 885–92.

Aizen, M. A., and Feinsinger, P. (1994) Forest fragmentation, pollination, and plant repro- duction in a Chaco dry forest, Argentina. *Ecology*, 75, 330–51.

Bascompte, J., Jordano, P., Melian, C. J., and Olesen, J. M. (2003) The nested assembly of plant-animal mutualistic networks. *Proceedings of the National Academy of Sciences, USA*, 100, 9383–87.

Bascompte, J., Jordano, P., and Olesen, J. M. (2006) Asymmetric coevolutionary networks facilitate biodiversity maintenance. *Science*, 312, 431–33.

Beattie, A. J. (1976) Plant dispersion, pollination and gene flow in *Viola*. *Oecologia*, 25, 291–300.

Bowers, M. A. (1985) Bumble bee colonization, extinction, and reproduction in subalpine meadows in northeastern Utah. *Ecology*, 66, 914–27.

Boyd, R. S. (2009) High-nickel insects and nickel hyperaccumulator plants: A review. *Insect Science*, 16, 19–31.

Boyd R. S., and Martens, S. N. (1998) The significance of metal hyperaccumulation for bi- otic interactions. *Chemoecology*, 8, 1–7.

Boyd, R. S., Wall, M. A., and Jaffré, T. (2006) Nickel levels in arthropods associated with Ni hyperaccumulator plants from an ultramafic site in New Caledonia. *Insect Science*, 13, 271–77.

Brady, K. U., Kruckeberg, A. R., and Bradshaw, H. D. Jr. (2005) Evolutionary ecology of plant adaptation to serpentine soils. *Annual Review of Ecology, Evolution, and Systemat- ics*, 36, 243–66.

Bronstein, J. L. (1995) The plant/pollinator landscape. In *Mosaic Landscapes and Ecological Processes* (eds. L. Fahrig, L. Hansson, and G. Merriam), pp. 256–88. Chapman and Hall, New York.

Brooks, R. R. (1987) *Serpentine and Its Vegetation*. Dioscorides Press, Portland, OR.

Byers, D. L. (1995) Pollen quantity and quality as explanations for low seed set in small populations exemplified by *Eupatorium* (Asteraceae). *American Journal of Botany*, 82, 1000–1006.

Cane, J. H. (2001) Habitat fragmentation and native bees: A premature verdict? *Conserva- tion Ecology*, 5, 2. Available online at http://www.consecol.org/vol5/iss1/art3.

Cane, J. H., and Sipes, S. (2006) Floral specialization by bees: Analytical methods and a re- vised lexicon for oligolecty. In *Plant-Pollinator Interactions: From Specialization to Generalization* (eds. N.M. Waser and J. Ollerton), pp. 99–122. University of Chicago Press, Chicago.

Carvalho, K. S., and Vasconcelos, H. L. (1999) Forest fragmentation in central Amazonia and its effects on litter-dwelling ants. *Biological Conservation*, 91, 151–57.

Charnov, E. L. (1976) Optimal foraging, the marginal value theorem. *Theoretical Population Biology*, 9, 129–36.

Connell, J. H. (1971) On the role of natural enemies in preventing competitive exclusion in some marine animals and in rain forest trees. In *Dynamics of Populations* (eds. P. J. den Boer and G. R. Gradwell), pp. 298–313. Centre for Agricultural Publishing and Documentation, Wageningen, The Netherlands.

Damschen, E. I., Brudvig, L. A., Haddad, N. M., Levey, D. J., Orrock, J. L., and Tewksbury, J. J. (2008) The movement ecology and dynamics of plant communities in fragmented landscapes. *Proceedings of the National Academy of Sciences, USA*, 105, 19078–83.

Davies, K. F., and Margules, C. R. (1998) Effects of habitat fragmentation on carabid beetles: Experimental evidence. *Journal of Animal Ecology*, 67, 460–71.

Dawson, J. W. (1981) The species rich, highly endemic serpentine flora of New Caledonia. *Tuatara* (Wellington), 25, 1–6.

Fortuna, M. A., Garcia, C., Guimaraes, P. R. Jr., and Bascompte, J. (2008) Spatial mating networks in insect-pollinated plants. *Ecology Letters*, 11, 490–98.

Gailing, O., Macnair, M. R., and Bachmann, K. (2004) QTL mapping for a trade-off between leaf and bud production in a recombinant inbred population of *Microseris douglasii* and *M. bigelovii* (Asteraceae, Lactuceae): A potential preadaptation for the colonization of serpentine soils. *Plant Biology*, 6, 440–46.

Galen, C. (1999) Why do flowers vary? The functional ecology of variation in flower size and form within natural plant populations. *Bioscience*, 49, 631–40.

Gardner, M., and Macnair, M. (2000) Factors affecting the co-existence of the serpentine endemic *Mimulus nudatus* Curran and its presumed progenitor, *Mimulus guttatus* Fischer ex DC. *Biological Journal of the Linnean Society*, 69, 443–59.

Gathmann, A., and Tscharntke, T. (2002) Foraging ranges of solitary bees. *Journal of Animal Ecology*, 71, 757–64.

Gomez, J. M., Abdelaziz, M., Camacho, J. P. M., Munoz-Pajares, A. J., and Perfectti, F. (2009) Local adaptation and maladaptation to pollinators in a generalist geographic mosaic. *Ecology Letters*, 12, 672–82.

Goulson, D. (2000) Why do pollinators visit proportionally fewer flowers in large patches? *Oikos*, 91, 485–92.

Goverde, M., Schweizer, K., Baur, B., and Erhardt, A. (2002) Small-scale habitat fragmentation effects on pollinator behaviour: Experimental evidence from the bumblebee *Bombus veteranus* on calcareous grasslands. *Biological Conservation*, 104, 293–99.

Greenleaf, S. S., Williams, N. M., Winfree, R., and Kremen, C. (2007) Bee foraging ranges and their relationship to body size. *Oecologia*, 153, 589–96.

Groom, M. J. (1998) Allee effects limit population viability of an annual plant. *American Naturalist*, 151, 487–96.

Harrison, S., and Cornell, H. (2008) Toward a better understanding of the regional causes of local community richness. *Ecology Letters*, 11, 969–79.

Harrison, S., Maron, J., and Huxel, G. (2000a) Regional turnover and fluctuation in populations of five plants confined to serpentine seeps. *Conservation Biology*, 14, 769–79.

Harrison, S., Safford, H. D., Grace, J. B., Viers, J. H., and Davies, K. F. (2006) Regional and local species richness in an insular environment: Serpentine plants in California. *Ecological Monographs*, 76, 41–56.

Harrison, S., Viers, J. L., and Quinn, J. F. (2000b) Climatic and spatial patterns of diversity in the serpentine plants of California. *Diversity and Distributions* 6, 153–61.

Hegland, S. J., Nielsen, A., Lazaro, A., Bjerknes, A., and Totland, O. (2009) How does climate warming affect plant-pollinator interactions? *Ecology Letters*, 12, 184–95.

Heinrich, B. (1976) Foraging specializations of individual bumblebees. *Ecological Monographs*, 46, 105–28.

Heinrich, B. (1979) Resource heterogeneity and patterns of foraging in bumblebees. *Oecologia* 140, 235–45.

Hickman, J. C. (1993) *The Jepson Manual: Higher Plants of California*. University of California Press, Berkeley.

Hixon, M. A., Carpenter, F. L., and Paton, D. C. (1983) Territory area, flower density, and time budgeting in hummingbirds: An experimental and theoretical analysis. *American Naturalist*, 122, 366–91.

Janzen, D. H. (1970) Herbivores and the number of tree species in tropical forests. *American Naturalist*, 104, 501–28.

Johnson, S. D., and Steiner, K. E. (2000) Generalization versus specialization in plant pollination systems. *Trends in Ecology and Evolution*, 15, 140–43.

Kearns, C. A., and Inouye, D. W. (1997) Pollinators, flowering plants, and conservation biology. *BioScience*, 47, 297–307.

Kearns, C. A., Inouye, D. W., and Waser, N. M. (1998) Endangered mutualisms: The conservation of plant–pollinator interactions. *Annual Review of Ecology and Systematics*, 29, 83–112.

Klein, B. C. (1989) Effects of forest fragmentation on dung and carrion beetle communities in central Amazonia. *Ecology*, 70, 1715–25.

Kolb, A. (2008) Habitat fragmentation reduces plant fitness by disturbing pollination and modifying response to herbivory. *Biological Conservation*, 141, 2540–49.

Kremen, C., Williams, N. M., Aizen, M. A., Gemmill-Harren, B., LeBuhn, G., Minckley, R., Packer, L., Potts, S. G., Roulston, T., Steffen-Dewenter, I., Vazquez, D. P., Winfree, R., Adams, L. Crone, E. E., Greenlead, S. S., Keitt, T. H., Klein, A. M., Regetz, J., and Ricketts, T. H. (2007) Pollination and other ecosystem services produced by mobile organisms: A conceptual framework for the effects of land-use change. *Ecology Letters*, 10, 299–314.

Kruckeberg, A. R. (1954) The ecology of serpentine soils. III. Plant species in relation to serpentine soils. *Ecology*, 35, 267–74.

Kruckeberg, A. R. (1984) *California Serpentines: Flora, Vegetation, Geology, Soils and Management Problems*. University of California Press, Berkeley.

Kruckeberg, A. R. (2004) The influences of lithology on plant life. In *Geology and Plant Life: The Effects of Landforms and Rock Type on Plants*. University of Washington Press, Seattle.

Laurance, W. F., Ferreira, L. V., Rankin-de Merona, J. M., and Laurance, S. G. (1998) Rain forest fragmentation and the dynamics of Amazonian tree communities. *Ecology*, 79, 2032–40.

LeBuhn, G., Droege, S., and Carboni, M. (2007) Monitoring methods for solitary bee species using bee bowls in North America. In *State of the World's Pollinators Report*, coordinated by the United Nations Food and Agriculture Organisation. Available online at http://www.fao.org/ag/AGP/agps/C-CAB/Castudies/pdf/1-007.pdf. Accessed October 2009.

Lindenmayer, D. B., and Fischer, J. (2006) Habitat Fragmentation and Landscape Change: An Ecological and Conservation Synthesis. Island Press, Washington.

Lloyd, D. G. (1989) The reproductive ecology of plants and eusocial animals. In *Towards a More Exact Ecology, 75th Jubilee Symposium, British Ecological Society* (eds. P. J. Grubb and J. B. Whittaker), pp. 185–287. Blackwell Scientific, Oxford.

Lomolino, M. V. (2000) Ecology's most general, yet protean pattern: The species-area relationship. *Journal of Biogeography*, 27, 17–26.

Lopezaraiza-Mikel, M. E., Hayes, M. E., Whalley, M. R., and Memmott, J. (2007) The impact of an alien plant on native plant pollinator network: An experimental approach. *Ecology Letters*, 10, 539–50.

MacArthur, R. H., and Pianka, E. R. (1966) On optimal use of a patchy environment. *American Naturalist*, 100, 603–9.

Marlin, J., and LaBerge, W. (2001) The native bee fauna of Carlinville, Illinois, revisited after 75 years: A case for persistence. *Conservation Ecology*, 5, 9. Available online at http://www.ecologyandsociety.org/vol5/iss1.

Memmott, J., Craze, P. G., Waser, N. M., and Price, M. V. (2007) Global warming and the disruption of plant-pollinator interactions. *Ecology Letters*, 10, 710–17.

Minckley, R. L., and Roulston, T. H. (2006) Incidental mutualisms and pollen specialization among bees. In *Plant-Pollinator Interactions: From Specialization to Generalization* (eds. N. M. Waser and J. Ollerton), pp. 69–98. University of Chicago Press, Chicago.

Moldenke, A. R. (1979) Host-plant coevolution and the diversity of bees in relation to the flora of North America. *Phytologia*, 43, 357–419.

Olesen, J. M., and Jordano, P. (2002) Geographic patterns in plant pollinator mutualistic networks. *Ecology*, 83, 2416–24.

Osborne, J. L., Clark, S. J., Morris, R. J., Williams, I. H., Riley, J. R., Smith, A. D., Reynolds, D. R., and Edwards, A. S. (1999) A landscape-scale study of bumble bee foraging range and constancy, using harmonic radar. *Journal of Applied Ecology*, 36, 519–33.

Parker, G. A., and Maynard Smith, J. (1990) Optimality theory in evolutionary biology. *Nature*, 348, 27–33.

Pierce, G. J., and Ollason, J. G. (1987) Eight reasons why optimal foraging theory is a complete waste of time. *Oikos*, 49, 111–17.

Pleasants, J. M. (1989) Optimal foraging by nectarivores: A test of the marginal-value theorem. *American Naturalist*, 134, 51–71.

Pleasants, J. M., and Zimmerman, M. (1983) The distribution of standing crop of nectar: What does it really tell us? *Oecologia*, 57, 412–14.

Powell, A. H., and Powell, G. V. N. (1987) Population dynamics of male euglossine bees in Amazonian forest fragments. *Biotropica*, 19, 176–79.

Proctor, J., and Woodell, S. R. J. (1975) The ecology of serpentine soils. *Advances in Ecological Research*, 9, 255–365.

Pyke, G. H. (1978) Optimal foraging in hummingbirds: Testing the marginal value theorem. *American Zoologist*, 18, 739–52.

Pyke, G. H. (1984) Optimal foraging theory: A critical review. *Annual Review of Ecology and Systematics*, 15, 523–75.

Reeves, R. D., Baker, A. J. M., Borhidi, A., and Berazain, R. (1999) Nickel hyperaccumulation in the serpentine flora of Cuba. *Annals of Botany*, 83, 29–38.

Richards, C. M. S., Church, S., and McCauley, D. E. (1999) The influence of population size and isolation on gene flow by pollen in *Silene alba*. *Evolution*, 53, 63–73.

Ricketts T. H. (2001) The matrix matters: Effective isolation in fragmented landscapes. *American Naturalist*, 158, 87–99.

Safford, H. D., Viers, J. H., and Harrison, S. P. (2005) Serpentine endemism in the California flora: A database of serpentine affinity. *Madroño*, 52, 222–57.

Sambatti, J. B. M., and Rice, K. J. (2006) Local adaptation, patterns of selection, and gene flow in the Californian serpentine sunflower (*Helianthus exilis*). *Evolution*, 60, 696–710.

Schmitt, J. (1983) Flowering plant density and pollinator visitation in Senecio (Compositae). *Oecologia*, 60, 97–102.

Sih, A., and Baltus, M. (1987) Patch size, pollinator behavior, and pollinator limitation in catnip. *Ecology*, 68, 1679–90.

Skinner, M. W., and Pavlik, B. M. (1994) California Native Plant Society's inventory of rare and endangered plants of California. *CNPS Special Publication No. 1*, Sacramento, CA.

Spira, T. P. (2001) Plant-pollinator interactions: A threatened mutualism with implications for the ecology and management of rare plants. *Natural Areas Journal*, 21, 78–88.

Stearns, S. C., and Schmid-Hempel, P. (1987) Evolutionary insights should not be wasted. *Oikos*, 49, 118–25.

Stebbins, G. L. (1970) Adaptive radiation of reproductive characteristics in angiosperms, I: Pollination mechanisms. *Annual Review of Ecology and Systematics*, 1, 307–26.

Steffan-Dewenter, I., Munzenberg, U., Burger, C., Thies, C., and Tscharntke, T. (2002) Scale dependent effects of landscape structure on three pollinator guilds. *Ecology*, 83, 1421–32.

Stouffer, P. C., and Bierregaard, R. O. Jr. (1995) Effects of forest fragmentation on understory hummingbirds in Amazonian Brazil. *Conservation Biology*, 9, 1085–94.

Strauss, S. Y., and Whitall, J. B. (2006) Non-pollinator agents of selection on floral traits. In *Ecology and Evolution of Flowers* (eds. L. D. Harder and S. C. H. Barrett), pp. 120–38. Oxford University Press, New York.

Tabarelli, M., Da Silva, J. M. C., and Gascon, C. (2004) Forest fragmentation, synergisms and the impoverishment of neotropical forests. *Biodiversity and Conservation*, 13, 1419–25.

Thorp, R. W. (2007) Biology of specialist bees and conservation of showy vernal pool flowers. A review. In *Vernal Pool Landscapes* (eds. R. A. Schlising and D. G. Alexander), pp. 51–57. Studies from the Herbarium, 14, California State University, Chico.

Thorp, R. W., and Leong, J. M. (1998) Specialist bee pollinators of showy vernal pool flowers. In *Conservation, and Management of Vernal Pool Ecosystems* (eds. C. W. Witham, E. T. Bauder, D. Belk, W. R. Ferren Jr., and R. Orduff), pp. 169–79. Proceedings from a 1996 conference. California Native Plant Society, Sacramento, CA.

Tscharntke, T., and Brandl, R. (2004) Plant-insect interactions in fragmented landscapes. *Annual Review of Entomology*, 49, 405–30.

Wall, M. A., and Boyd, R. S. (2006) *Melanotrichus boydi* (Heteroptera: Miridae) is a specialist on the nickel hyperaccumulator *Streptanthus polygaloides* (Brassicaceae). *Southwestern Naturalist* 51, 481–89.

Waser, N. M., Chittka, L., Price, M. V., Williams, N., and Ollerton, J. (1996) Generalization in pollination systems, and why it matters. *Ecology*, 77, 279–96.

Westerbergh, A., and Saura, A. (1994) Gene flow and pollinator behaviour in *Silene dioica* populations. *Oikos*, 71, 215–24.

Williamson, S. D., and Balkwill, K. (2006) Factors determining levels of threat to serpentine endemics. *South African Journal of Botany*, 72, 619–26.

Williamson, J., and Harrison, S. (2002) Biotic and abiotic limits to the spread of exotic revegetation species. *Ecological Applications*, 12, 40–51.

Wolf, A. T. (2001) Conservation of endemic plants in serpentine landscapes. *Biological Conservation*, 100, 35–44.

Wolf, A. T., Brodmann, P. A., and Harrison, S. P. (1999) Distribution of the rare serpentine sunflower (*Helianthus exilis* Gray, Asteraceae): The roles of habitat availability, dispersal limitation and species interactions. *Oikos*, 84, 69–76.

Wolf, A. T., and Harrison, S. P. (2001) Natural habitat patchiness affects reproductive success of serpentine morning glory (*Calystegia collina*, Convolvulaceae) in northern California. *Conservation Biology*, 15, 111–21.

Wolf, A. T., Harrison, S. P., and Hamrick, J. L. (2000) The influence of habitat patchiness on genetic diversity and spatial structure of a serpentine endemic plant. *Conservation Biology*, 14, 454–63.

Wright, J. W., and Stanton, M. L. (2007) *Collinsia sparsiflora* in serpentine and nonserpentine habitats: Using F2 hybrids to detect the potential role of selection in ecotypic differentiation. *New Phytologist*, 173, 354–66.

Wright, J. W., Stanton, M. L., and Scherson, R. (2005) Local adaptation to serpentine and non-serpentine soils in *Collinsia sparsiflora*. *Evolutionary Ecology Research*, 8, 1–21.

Young, A., Boyle, T., and Brown, T. (1996) The population genetic consequences of habitat fragmentation for plants. *Trends in Ecology and Evolution*, 11, 413–18.

14

Spatial Ecology

The Effects of Habitat Patch Size, Shape, and Isolation on Ecological Processes

Susan Harrison, *University of California, Davis*

Serpentine outcrops, with their distinctive floras strongly contrasting with those of surrounding landscapes, bear as strong a resemblance to islands as any habitat on land. As Kruckeberg (1991) first suggested, it seems intuitive that the diversity of life on serpentine outcrops may be shaped by the same spatial attributes that are so important in islands—areas, distances from one another and from a mainland, and the qualities of the "matrix" through which organisms must move to reach them. This chapter reviews all the available studies that address effects of the size and isolation of serpentine outcrops on ecology of their flora and fauna and compares the findings with predictions from the subdiscipline of spatial ecology.

Spatial ecology addresses the effects of large-scale habitat structure, such as the area and isolation of habitat patches, on population and community processes. Ever since its origins in the influential model of island diversity by MacArthur and Wilson (1967), spatial ecology has enjoyed a well-developed mathematical basis (e.g., Hanski, 1999; Holyoak et al., 2005) and strong linkages to conservation biology (e.g., Hansson et al., 1995; McCullough, 1996; Collinge, 2009). Although spatial ecological theory has become highly diverse, a consistent central principle is that the dispersal of organisms across landscapes and among distantly separated habitat patches can influence population survival, outcomes of species interactions, and community diversity. Isolation and small patch size are generally seen in a negative light; extinction is more frequent and diversity is lower on small islands far from mainlands (MacArthur and Wilson, 1967), and species may be unable to persist in landscapes that fall below thresholds of patch area and isolation (Hanski, 1999). When competitive or predatory interactions are strong, however, intermediate levels of patch isolation may promote population persistence and community diversity (Holyoak et al., 2005).

Empirical tests of the influence of habitat structure on ecological processes are challenging because processes at large spatial scales usually take place over long time scales, and both factors tend to limit the possibilities for either experimental manipulation or direct observation. Some of the most successful tests of spatial theory have taken place in laboratory microcosms, outdoor mesocosms such as grasslands mown into patches of different sizes, and, in a few cases, experimentally fragmented forests (see reviews in Holyoak et al., 2005; Collinge, 2009). Important insights have also come from observational analyses of unmanipulated natural systems (e.g., Thomas and Hanski, 1997). However, the ability of spatial ecological theory to interpret and predict processes in large-scale natural systems is still open to question (Eriksson, 1996; Harrison and Taylor, 1997; Freckleton and Watkinson, 2002; Baguette, 2004).

This chapter asks how has spatial ecological theory helped us understand serpentine biotas, and what have we learned from studies of serpentine about spatial ecological theory? By necessity, most of the studies reviewed herein concern the Californian serpentine flora, although a few animal studies are covered, too, and the majority come from my and my collaborators' work. In addition to studies focusing on serpentine outcrop area and isolation, this review covers several studies of spatially isolated "seep" habitats within serpentine outcrops, and the one study I am aware of that measured the dispersal of organisms among serpentine outcrops and the effects of the surrounding matrix habitat on such dispersal. This chapter does not consider some issues that might be considered to fall within spatial ecology, including the spatial patterning of communities within continuous serpentine habitats (Wu and Levin, 1994), the spread of exotic species (Chapter 9 in this volume), seed limitation within serpentine habitat (Chapter 10), spatial relationships of native and exotic species richness (Chapter 11), and evolutionary processes across serpentine–nonserpentine boundaries (Chapters 3–7).

POPULATION PERSISTENCE

Metapopulation theory has become a major paradigm for understanding the dynamics of populations in naturally discontinuous or human-fragmented landscapes (Hanski, 1999; Holyoak et al., 2005; Collinge, 2009). Species are viewed as "populations of populations" existing on networks of habitat patches. Each patch can either be occupied or unoccupied by the species, and moves between these two states according to continuous rates of colonization (c) and extinction (e). A species never persists for long at the local (patch) scale, where constant turnover occurs; it can only persist at the regional (metapopulation) scale as long as c exceeds e, which would be the case, for example, if patches are not too far from one another or are connected by habitat corridors. Moreover, even if c does exceed e, a very small metapopulation (too few patches) can still go extinct by chance. The overall view is that individual populations are transient, but species can survive regionally in sufficiently large and well-connected universes of patches. The potential applications of this model to conservation biology, wildlife management, and restoration ecology have become the subjects of an extensive body of literature (e.g., McCullough, 1996; Hanski, 1999; Collinge, 2009).

One of the first tests of metapopulation theory in a natural system concerned the Bay checkerspot butterfly (*Euphydryas editha bayensis*), found on serpentine soils in the San Francisco Bay area, California. Classic work by Paul Ehrlich and colleagues (Ehrlich 1961, 1965; Singer and Ehrlich, 1979) showed that adjacent populations of the Bay checkerspot fluctuated and went extinct independently, an observation that contributed to the early development of metapopulation ideas. By the late 1980s, many Bay checkerspot populations had gone extinct because of urbanization and its side effects, except in the relatively rural southern end of its range, where a population of roughly 10^6 butterflies inhabited a large serpentine outcrop called Coyote Ridge, and a handful of smaller populations were also known.

In metapopulation studies, it is necessary to measure the habitat suitability, degree of isolation, and occupied or unoccupied status of all habitat patches within a landscape. Thanks to the detailed geologic maps for the region, I was able to locate about 60 serpentine outcrops in a 50×30 km area and search them for the Bay checkerspot and its host plants (annual plantain, *Plantago erecta*; owl's clover, *Castilleja densiflora*) and nectar plants (goldfields, *Lasthenia californica*). In addition to the Coyote Ridge population, I observed 9 other populations of 10^1–10^2 butterflies and 17 serpentine outcrops on which host and nectar plants were present but butterflies were absent. I built statistical models that accurately predicted butterfly presence or absence from two independent factors: habitat suitability (a combination of outcrop area, topography, and host and nectar plant abundance) and isolation (distance of each outcrop from the large Coyote Ridge population). Only patches within 4.5 km of Coyote Ridge had butterfly populations, even though many suitable outcrops existed at greater distances (Harrison et al., 1988).

This observational analysis, combined with simulation modeling, showed that the large Coyote Ridge population functioned as a stable "mainland" to an archipelago of smaller, transient "island" populations (Harrison et al., 1988). Dispersal experiments and other evidence indicated that the checkerspot was an ineffective navigator and colonist, and irregular topography in the matrix between outcrops further inhibited dispersal (Harrison, 1989). The butterfly did not depend for its persistence on the extinction–colonization balance portrayed in the classic metapopulation model, but only on the continued existence of the mainland, Coyote Ridge. In turn, several factors contributed to the size and stability of the Coyote Ridge population: a very large serpentine outcrop, a grazing regime that benefited the host and nectar plants, and finely dissected topography that buffered the butterfly population through cooler and warmer years (Weiss, 1988).

Not completely deterred by this experience, I subsequently tested metapopulation theory by resurveying around 140 populations of five rare plants in a serpentine landscape that had been surveyed 17 years earlier for a mining company's environmental impact study (Harrison et al., 2001). These plants were summer-flowering habitat specialists found in widely scattered small wetlands, or seeps, in an otherwise summer-dry serpentine environment. Numbers of populations remained roughly constant between the two time periods (1980–81 versus 1997–99), although around 40% population turnover had occurred. As expected under metapopulation theory, less isolated seeps were more likely to have become colonized, and populations in more isolated seeps were more likely to have gone extinct. However, molecular analysis later showed that in one of the species (*serpentine sunflower, Helianthus exilis*), levels of genetic variation were too high in the seemingly recolonized populations to be consistent with a recent colonization event and were more consistent with reappearance from a dormant seedbank (Sambatti et al., unpublished data). The role of seedbanks in preventing local extinction is a widespread finding in plant metapopulation studies (Eriksson, 1996; Freckleton and Watkinson, 2002).

In hindsight, it seems almost predictable that these particular butterflies and plants did not fit the classic metapopulation model well. Their patchy serpentine habitat is fixed in ecological time, rather than being a transient resource such as gopher mounds in a meadow or treefall gaps in a forest. The greatest natural threat these species face is multiple years of drought, which affects large regions synchronously. There is therefore little chance for these organisms to attain higher fitness through dispersing among habitat patches; both the sedentary behavior of the butterfly and the long-term seed dormancy of the plants make good evolutionary sense. For these and other reasons, some authors have concluded that even though many organisms occupy patchy habitats, there are few examples of metapopulations in which regional persistence depends on a delicate balance between local extinction and colonization (Eriksson, 1996; Harrison and Taylor, 1997; Freckleton and Watkinson, 2002; Baguette, 2004).

SPECIES INTERACTIONS

Theory suggests that patchy environments may help stabilize antagonistic interactions, such as those between predators and prey, diseases and hosts, or competing species (Holyoak et al., 2005). Isolated patches can serve as temporary refuges for the host, prey, or inferior competitor, where they survive until the disease, predator, or superior competitor catches up. This body of spatial theory has been tested in a number of experimental systems but seldom in any large-scale natural habitat such as serpentine (although good candidates might be the serpentine-endemic wild flaxes, *Hesperolinon* spp., and their rust fungus *Melampsora lini*; Springer, 2007).

With regard to mutualistic interactions, such as pollination or mycorrhizal associations, the main prediction theory makes about patchy habitats is that colonization and persistence in isolated habitats become even more challenging. For example, many studies have examined whether plants suffer reduced reproductive success as a result of lower pollinator visitation in fragmented habitats (reviewed in Harrison and Bruna, 1999). Two specific examples are briefly summarized here, although Chapter 13 in this volume describes them in more detail.

Serpentine morning glory (*Calystegia collina* ssp. *collina*) is a widespread serpentine endemic in northern California, growing in dense clonal patches. Its insect pollinators include both specialists and generalists. Plants on small (<5 ha) outcrops produced substantially less seed than plants on large (>300 ha) outcrops, even though they received no fewer visits by pollinators and did not have lower heterozygosity or genotypic diversity indicative of inbreeding depression. Pollen transplant experiments showed that plants were self-incompatible, and on small outcrops, their isolation from conspecifics caused them to receive too little nonself pollen (Wolf et al., 1999, 2000; Wolf and Harrison, 2001). Serpentine sunflowers (*H. exilis*) did not suffer reduced pollinator visitation as a result of spatial isolation either among serpentine outcrops or among seeps within outcrops. However, rates of seed predation by a tephritid fly were lower for *H. exilis* in more isolated seeps. The absence of *H. exilis* from the smallest serpentine outcrops (<1 ha) could be explained by the absence of suitable habitat at those sites (Wolf et al., 1999). These studies support the idea that living in small and isolated patches affects species interactions and plant reproductive success, but only for some species, and not necessarily via the mechanisms that ecologists have traditionally thought about.

COMMUNITY DIVERSITY

Diversity of communities in patchy habitats might be expected to reflect the processes of local extinction and recolonization. Small and isolated habitat patches are expected to have lower local diversity (fewer species per patch) than larger or less isolated ones, as a result of higher extinction and lower colonization rates (MacArthur and Wilson, 1967). But an entire ensemble of patches might have a greater regional diversity (aggregate number of species) than an undivided habitat of equal area because of the potential for competing species to coexist in a subdivided environment (Holyoak et al., 2005). Higher regional diversity in spite of lower local diversity would mean that patchy habitats have higher beta diversity, or differentiation among sites in species composition, than undivided ones.

As a historical note, the importance of considering spatial scale in studies of community diversity was first recognized by Robert Whittaker (1960) in a comparison of serpentine and nonserpentine communities along microclimatic gradients (not among discrete habitat patches) in the Siskiyou Mountains. Whittaker proposed the terms *alpha diversity* for the number of species at a local site, *beta diversity* for the differentiation in species composition among sites, and *gamma diversity* for the number of species in an entire region. In his original definition, *(average) alpha* × *beta* = *gamma*. Modifications of this system have been widely used by ecologists ever since.

To test the influence of spatial subdivision on patterns of diversity, I compared local (alpha), regional (gamma), and among-site (beta) plant species diversity on 24 small serpentine outcrops (<3 ha) and 24 sampling sites with the same spacing within four large (>500 ha) serpentine outcrops. As expected, for serpentine-restricted (endemic) species, average local diversity was lower, but among-site diversity was higher in the patchy sites than in the continuous ones. Regional diversity was identical in the two sets of sites, and there were no particular species restricted to either the patchy sites or the continuous ones. It appeared that the patchiness of small serpentine outcrops did not alter overall endemic diversity but redistributed it from the local scale to the among-site scale. For the species not restricted to serpentine, diversity was higher at both the local and the regional scale in the patchy sites than in the continuous ones, consistent with edge effects. This was particularly true for exotic species, which were much more numerous on the small outcrops than the large ones (Harrison, 1997, 1999). Later work suggested that the edges of small serpentine outcrops, surrounded by exotic-rich nonserpentine habitats, may be ideal settings for the evolution of serpentine tolerance in exotics, such as the grasses *Avena fatua* and *Bromus hordeaceus* (Harrison et al., 2001; Chapter 9 in this volume).

For the plant communities of serpentine seeps, Freestone and Inouye (2006) used randomization tests to show that among-site differentiation (beta diversity) was significantly higher than expected due to chance alone, both among separate seeps within the same serpentine outcrop and among seeps on separate outcrops. However, the community differentiation among seeps within the same outcrop could be explained by environmental variation such as soil chemistry differences, whereas the differentiation of communities among seeps on separate outcrops could only be explained statistically by geographic distance. They concluded that at the scale of seeps on separate serpentine outcrops (although not at smaller scales), the spatial isolation of seeps limits plant dispersal and thereby affects the composition of plant communities.

Based on these studies, there seemed a clear potential for spatial structure to play a role in shaping the overall distribution of plant diversity on serpentine outcrops across California. To test this, my colleagues and I measured regional plant diversity in 78 serpentine-containing regions (81–5306 km²), and local diversity at 109 sites (1000 m² plots) nested within those regions (Harrison et al., 2006). We expected that the diversity of serpentine endemic plants would be lower at local sites on smaller and more isolated serpentine outcrops, yet potentially higher in regions with more spatially complex serpentine (e.g., regions with greater numbers of separate outcrops for a given total area of serpentine). These expectations were not met. In a multivariate model considering >70 regional and local variables, the only significant spatial influence was that endemic diversity at the regional scale was higher in regions with a greater total area of serpentine. Endemic diversity at the regional scale was also highest in regions with abundant rainfall and in which the serpentine has been exposed longest in geologic time. Endemic diversity at the local scale was highest at rocky sites within endemic-rich regions (Harrison et al., 2006). By far the strongest predictor of endemic diversity, total species diversity, and community composition on serpentine was mean annual precipitation (or its correlate, the remotely sensed index of plant productivity known as NDVI or normalized difference vegetation index; Grace et al., 2007).

It might appear contradictory that the spatial structure of serpentine habitats affected plant diversity within individual regions but not across the entire realm of Californian serpentine. Once again, a possible resolution seems obvious in hindsight. In models of single species, Fahrig (2002) showed that the spatial structure of habitats may only be important under a narrow range of conditions: for example, when the species is neither too poor nor too good a disperser relative to distances among patches, and when the habitat is neither too scarce nor too common (e.g., 15–20% of the landscape). Conversely, she found that when dispersal is high or habitat is more common than 20% of the landscape, species persist regardless of spatial structure, and when dispersal is low or habitat is scarce, species become extinct regardless of spatial structure. Her result implies that it may be possible to find significant spatial effects by looking for them in settings where they are likeliest to be important (e.g., the small versus large outcrop comparison already described), yet not to find them in a more heterogeneous set of locations that are intended to be representative of the study system as a whole (e.g., the statewide study described above).

SERPENTINE ISLANDS IN EVOLUTIONARY TIME?

The island-like nature of serpentine might be important because of its evolutionary as well as its ecological consequences (Kruckeberg, 1991; Ackerly, 2003). Plant evolution on serpentine is a vast topic that is mostly beyond the scope of this chapter (see Chapters 3–9 in this volume). However, it is relevant to ask whether the occurrence of serpentine as many separate outcrops has been an important factor in patterns of adaptation and/or speciation. Analogously to the well-known evolutionary radiations on oceanic islands, the colonization of serpentine might be associated with increased diversification, and closely related and ecologically similar species might be found on separate outcrops, suggesting a role for spatial structure in stimulating allopatric speciation. Little evidence for such patterns on serpentine has been found so far, however (see Chapter 3 in this volume).

Another long-term consequence of an island-like distribution might be a higher rate of extinction under a changing climate. Warming and drying trends in recent millennia have been associated with latitudinal range shifts of hundreds of kilometers (Raven and Axelrod, 1978) and elevational range shifts of hundreds of meters (e.g., Briles et al., 2005). If climatic changes do not greatly affect the degree to which species are restricted to serpentine—which is admittedly questionable—it is hard to imagine how serpentine endemics can have survived while confined to small outcrops with little elevational or latitudinal room to migrate. One possibility is that extinction has sorted the serpentine flora and eliminated those species that inhabited marginal climates or lacked climate-resistant traits. In California, we found that triply rare species—restricted to serpentine, locally sparse, and having small geographic ranges—tend to be found in wetter climates and regions with larger areas of serpentine, suggesting a possible enhanced role for past extinctions in shaping the present-day distribution of the serpentine flora (Harrison and Inouye, 2002; Harrison et al., 2008). The consequences of modern climate change for serpentine endemics are considered in Chapter 17.

CONCLUSIONS

What Have We Learned from Studies of Serpentine about Spatial Ecological Theory?

The serpentine environment includes many small and isolated outcrops, and work in California suggest that these may be valuable settings for studying issues related to small population size, reduced colonization and gene flow, altered rates of interaction among species, and lower community diversity. The studies reviewed here have confirmed some straightforward predictions (e.g., lower endemic plant diversity on small outcrops) and contradicted others in interesting ways (e.g., lower pollen quality, rather than lower pollinator visitation or inbreeding depression, reduces plant reproductive success on small outcrops). Given the current emphasis on the consequences of biodiversity for ecosystem function, serpentine might provide a good setting to test for effects of habitat area and isolation on functional attributes of plant communities.

How Has Spatial Ecological Theory Helped Us Understand
Serpentine Biotas?

The answer to this question is less clear, in part because a considerable amount of the serpentine in California consists of outcrops that are not terribly small. Of the total of 5761 km², approximately 44%, 82%, and 99.5% is found in outcrops of greater than 100, 10, and 1 km², respectively (using Jennings, 1977). It is difficult to find regions with numerous small outcrops within a few kilometers of one another but no large ones, the circumstances that would give rise to the maximum significance for processes that depend on dispersal among small outcrops. Like the mainlands in island biogeography theory, the larger exposures of serpentine within any given region may be the storehouses of the majority of serpentine endemic plant diversity in that region. Serpentine endemic diversity is highest in regions with more serpentine and wetter climates, and the high diversity in California as a whole probably largely reflects the occurrence of serpentine across a wide range of climates and source floras.

Another issue is that metapopulation theory may not apply well in circumstances where large-scale climatic events such as droughts are the major causes of local extinctions of populations. Such circumstances should select for adaptations to "escape in time" rather than disperse in space. Strategies for escape in time include seed banks, below-ground storage organs, selfing, and pollinator redundancy. These have the side effect of reducing the sensitivity of population persistence to habitat area and isolation, as others have also concluded (Eriksson, 1996; Freckleton and Watkinson, 2002). Serpentine endemics that occur only on one or a few small outcrops (e.g., *Clarkia franciscana* in the San Francisco Presidio, *Calochortus tiburonensis* and *Streptanthus niger* on the Tiburon Peninsula) illustrate the ability of some plants to persist for a very long time in spite of small population size and total isolation.

Mark Williamson (1989) said that island biogeography theory was "true but trivial," meaning that he believed only transient species undergo population turnover on islands in ecological time, while resident species diversity is higher on larger islands because of habitat heterogeneity rather than because of extinction and colonization dynamics. It seems possible that spatial ecological theory as applied to serpentine systems may also be true but trivial, meaning that some of its predictions may be successfully tested, and some interesting insights may be gained about limitations of and potential improvements to theory, but paradoxically, the theory has not explained much about major patterns in the ecology of serpentine systems.

A good model system in the conventional sense is one that varies only in the aspect of interest and has particular attributes that make the aspect of interest likely to be important. In the case of spatial theory, this might mean a patchy habitat of uniform quality, varying only in patch area and isolation, occupied by organisms with moderate dispersal abilities and no dormancy, and subject to random and independent local disturbances. This kind of model system enables us to ask whether a particular theory or idea *can* work, but not whether it actually *does* help explain and predict phenomena in complex natural systems. Serpentine has been a good system for better recognizing the difference between those two goals.

LITERATURE CITED

Ackerly, D. D. (2003) Community assembly, niche conservatism, and adaptive evolution in changing environments. *International Journal of Plant Sciences*, 164, 165–84.

Baguette, M. (2004) The classical metapopulation theory and the real, natural world: A critical appraisal. *Basic and Applied Ecology*, 5, 213–24.

Briles, C. E., Whitlock, C., and Bartlein, P. J. (2005) Postglacial vegetation, fire, and climate history of the Siskiyou Mountains, Oregon, USA. *Quaternary Research*, 64, 44–56.

Collinge, S. K. (2009) *Ecology of Fragmented Landscapes*. Johns Hopkins University Press, Baltimore, MD.

Ehrlich, P. R. (1961) Intrinsic barriers to dispersal in the checkerspot butterfly. *Science*, 134, 108–9.

Ehrlich, P. R. (1965) The population biology of the butterfly *Euphydryas editha*. II. The structure of the Jasper Ridge colony. *Evolution*, 19, 327–36.

Eriksson, O. (1996) Regional dynamics of plants: A review of evidence for remnant, source-sink, and metapopulations. *Oikos*, 77, 248–58.

Fahrig, L. (2002) Effect of habitat fragmentation on the extinction threshold: A synthesis. *Ecological Applications*, 12, 346–53.

Freckleton, R. P., and Watkinson, A. R. (2002) Large-scale spatial dynamics of plants: Metapopulations, regional ensembles and patchy populations. *Journal of Ecology*, 90, 419–34.

Freestone, A. T., and Inouye, B. D. (2006) Dispersal limitation and environmental heterogeneity shape scale-dependent diversity patterns in plant communities. *Ecology*, 87, 2425–32.

Grace, J. B., Safford, H. D., and Harrison, S. (2007) Large-scale causes of variation in the serpentine vegetation of California. *Plant and Soil*, 293, 121–32.

Hanski, I. (1999) *Metapopulation Ecology*. Oxford University Press, New York.

Hansson, L., Fahrig, L., and Merriam, G. (1995) *Mosaic Landscapes and Ecological Processes*. Studies in Landscape Ecology series. New York, Chapman and Hall.

Harrison, S. (1989) Long-distance dispersal and colonization in the bay checkerspot butterfly, *Euphydryas editha bayensis*. *Ecology*, 70, 1236–43.

Harrison, S. (1997) How natural habitat patchiness affects the distribution of diversity in Californian serpentine chaparral. *Ecology*, 78, 1898–906.

Harrison, S. (1999) Local and regional diversity in a patchy landscape: Native, alien and endemic herbs on serpentine soils. *Ecology*, 80, 70–80.

Harrison, S., and Bruna, E. M. (1999) Habitat fragmentation and large-scale conservation: What do we know for sure? *Ecography*, 22, 1–8.

Harrison, S., and Inouye, B. D. (2002) High beta diversity in the flora of Californian serpentine "islands." *Biodiversity and Conservation*, 11, 1869–76.

Harrison, S., Murphy, D. D., and Ehrlich, P. R. (1988) Distribution of the bay checkerspot butterfly, *Euphydryas editha bayensis*: Evidence for a metapopulation model. *American Naturalist*, 132, 360–82.

Harrison, S., Rice, K., and Maron, J. (2001) Habitat patchiness promotes invasions by alien grasses on serpentine soil. *Biological Conservation*, 100, 45–53.

Harrison, S., Safford, H. D., Grace, J. B., Viers, J. H., and Davies, K. F. (2006) Regional and local species richness in an insular environment: Serpentine plants in California. *Ecological Monographs*, 76, 41–56.

Harrison, S., and Taylor, A. D. (1997) Empirical evidence for metapopulation dynamics. In *Metapopulation Dynamics: Ecology, Genetics and Evolution* (eds. I. Hanski and M. E. Gilpin), pp. 27–42. Academic Press, New York.

Harrison, S., Viers, J. H., Thorne, J. H., and Grace, J. B. (2008) Favorable environments and the persistence of naturally rare species. *Conservation Letters*, 1, 65–74.

Holyoak, M. A., Leibold, M. A., and Holt, R. D. (2005) *Metacommunities: Spatial Dynamics and Ecological Communities*. University of Chicago Press, Chicago.

Jennings, C. W. (1977) *Geologic Map of California*. U.S. Geologic Survey, Menlo Park, CA.

Kruckeberg A. R. (1991) An essay: Geoedaphics and island biogeography for vascular plants. *Aliso*, 13, 225–38.

MacArthur, R., and Wilson, E. O. (1967) *The Theory of Island Biogeography*. Princeton University Press, Princeton, NJ.

McCullough, D. (1996) *Metapopulations and Wildlife Conservation Management*. Island Press, Covelo, CA.

Raven, P. J., and Axelrod, D. (1978) *Origin and Relationships of the California Flora*. University of California Publications in Botany 72.

Singer, M. C., and Ehrlich, P. R. (1979) Population dynamics of the checkerspot butterfly *Euphydryas editha*. *Fortschritte der Zoologie*, 25, 53–60.

Springer, Y. P. (2007) Clinal resistance structure and pathogen local adaptation in a serpentine flax–flax rust interaction. Evolution, 61, 1812–22.

Thomas, C. D., and Hanski, I. (1997) Butterfly metapopulations. In: *Metapopulation Dynamics: Ecology, Genetics and Evolution* (eds. I. Hanski and M. E. Gilpin), pp. 359–86. Academic Press, New York.

Weiss, S. B., Murphy, D. D., and White, R. R. (1988) Sun, slope, and butterflies: Topographic determinants of habitat quality for *Euphydryas editha*. *Ecology*, 69, 1486–96.

Whittaker, R. H. (1960) Vegetation of the Siskiyou Mountains, Oregon and California. *Ecological Monographs*, 30, 279–338.

Williamson, M. (1989) The MacArthur and Wilson theory today: True but trivial. *Journal of Biogeography*, 16, 3–4.

Wolf, A. T., Brodmann, P. A., and Harrison, S. (1999) Distribution of the rare serpentine sunflower (*Helianthus exilis* Gray, Asteraceae): The roles of habitat availability, dispersal limitation and species interactions. *Oikos*, 84, 69–76.

Wolf, A. T., and Harrison, S. (2001) Natural habitat patchiness affects reproductive success of serpentine morning glory (*Calystegia collina*, Convolvulaceae) in northern California. *Conservation Biology*, 15, 111–21.

Wolf, A. T., Harrison, S., and Hamrick, J. L. (2000) The influence of habitat patchiness on genetic diversity and spatial structure of a serpentine endemic plant. *Conservation Biology*, 14, 454–63.

Wu, J., and Levin, S. A. (1994) A spatial patch dynamic modeling approach to pattern and process in an annual grassland. *Ecological Monographs*, 64, 447–64.

15

Systematic Conservation Planning

Protecting Rarity, Representation, and Connectivity in Regional Landscapes

James H. Thorne, Patrick R. Huber, and Susan Harrison
University of California, Davis

Ecologists have long recognized the need for a strategy of land-based conservation aimed at protecting the full spectrum of biological diversity. The first effort at land protection for conservation in the United States, the 1872 creation of the National Park system, was motivated mainly by appreciation of beauty and outdoor recreation; later, lands were set aside for timber supply (the Forest Reserve Act of 1891) and huntable wildlife (the National Wildlife Refuge system, 1903). By the mid-twentieth century, it was clear that many species and natural communities were unprotected and at risk. The Nature Conservancy formed in 1946 as an offshoot of the Ecological Society of America to preserve "the last of the least and the best of the rest," focusing much initial effort on creating small preserves for rare and noncharismatic species. The advent of geographic information system (GIS) technology in the 1980s brought sharpened focus to the pronounced "rock and ice" bias of the existing network of conserved lands and the consequent gaps in protection of biodiversity (Scott et al., 1993; Kiester et al., 1996). Finally, the many controversies following the 1973 passage of the Endangered Species Act, such as the battles over snail darters, furbish louseworts, and spotted owls, further stimulated conservation biologists to develop methods for the systematic conservation of native species and communities, as a better alternative than waiting until species are near extinction to take action (Pressey et al., 1993; Anderson et al., 1999; Groves et al., 2002).

Systematic conservation planning today takes advantage of the availability of extensive geographic databases, such as land cover maps and species occurrence records, combined with a variety of so-called reserve selection algorithms reliant on fast computing, to analyze the biodiversity of regional landscapes and identify the most important areas within them for protection (reviewed in Groves et al., 2002; Williams et al., 2005; Moilanen et al., 2009). Reserve selection algorithms can also accommodate other important geographic information, such as land prices and the degree of development threat. Although regional land conservation in practice is, of course, nearly always driven to a high degree by social and political considerations (Prendergast et al., 1999), the systematic approach to reserve selection can provide a metric of success and a guide to future land acquisition priorities.

Biodiversity itself is not a single or simple thing, however, and although technical tools can help elucidate competing goals, only human decision making can resolve them (e.g., Meynard et al., 2009; Huber et al., 2010). One goal concerns *rarity*—the conservation of the most unique species and communities within a region, especially those found nowhere else on Earth. The search for biodiversity hot spots is an example of a rarity-based approach (Reid, 1998; Myers et al., 2000; Stein et al., 2000). Another goal is *representation*—the protection of excellent examples of all of the natural communities within a planning region, including the most common and widespread ones (Anderson et al., 1999; Groves et al., 2002). A third goal is *connectivity*—the preservation of large, undisturbed areas, well connected to one another by habitat corridors, widely considered necessary to maintain large and mobile wildlife (Noss and Harris, 1986; Soulé and Terborgh, 1999; Beier and Noss, 1998). Complicating the search for a resolution to these trade-offs is one of the most vexing problems in conservation biology: no good way exists to estimate either the minimum area necessary to support a viable species or community or the exact amount that a habitat corridor adds to the viability of a species or community. Answers to these key questions still lie in the realm of well-educated guesswork.

Serpentine floras around the world are emblematic of the rarity component of biodiversity. In California it has been estimated that serpentine comprises 1–2% of the state's land area yet contributes >10% of the plant species found nowhere else on Earth (Kruckeberg, 1984; Safford et al., 2005). Serpentine endemic plants have strikingly small geographic ranges, and thus the serpentine flora has extremely high beta diversity or geographic turnover in species composition (Harrison, 1997, 1999; Harrison and Inouye, 2002; Freestone and Inouye, 2006). These patterns suggest that relative to other floras and faunas, greater numbers of more widely dispersed reserves may be needed to protect serpentine floras, a goal that might be especially incompatible with connectivity for large animals. Indeed, entire preserves have been created or managed to protect serpentine endemic plant species with very narrow distributions, such as the Presidio clarkia (*Clarkia franciscana*), Tiburon jewelflower (*Streptanthus niger*), Ring Mountain fairy lantern (*Calochortus tiburonensis*), and Yreka phlox (*Phlox hirsuta*). Considering that there are several hundred rare plants with moderate to strong affinities to serpentine in California (Safford et al., 2005), one preserve for every species is not likely to be a viable strategy. It seems important to explore the compatibility of serpentine plant conservation with other conservation goals, such as representation and connectivity.

This chapter uses serpentine plants to explore the trade-offs among rarity, representation, and connectivity in systematic land-based conservation. We analyze a serpentine-rich landscape in the region of Napa County, California, from three perspectives: (1) the existing network of protected areas, relatively little of which is the product of biodiversity considerations; (2) the program Marxan (Ball et al., 2009), a widely used reserve design algorithm that identifies spatial solutions to user-determined objectives, one of which is typically the representation of all plant community types; and (3) least-cost corridor analysis (Theobald, 2006; Huber et al., 2010), an algorithm that identifies large blocks of undisturbed habitat and the linkages between them that are most likely to provide connectivity for wildlife movement. By comparing how well the plant communities and known occurrences of rare plant species on serpentine and non-serpentine soils are protected under each of these approaches, we can assess the degree to which serpentine floras have unique conservation requirements.

All systematic conservation planning depends on geographically explicit data, and the quality of the results is likely to depend on the quality of the input. Our rare species data come from the state Natural Diversity Database (California Department of Fish and Game, 2010), a compilation of publicly registered observations of rare and endangered species. These data are highly incomplete, because botanists have mainly had access to public lands and those private lands that were environmentally surveyed prior to development projects. However, we were fortunate to have an excellent regional map of plant communities. We began our analysis by comparing the coverage of serpentine plant communities on this map to several less detailed land cover maps for the region that are more typical of what is generally available.

STUDY REGION

Our study region is Napa County and adjacent parts of Lake, Yolo, Solano, and Colusa Counties in the Coast Ranges of northern California, totaling 292,400 ha (Figure 15.1). The Napa Valley in the center of Napa County is mainly agricultural with several small to medium-sized towns, but the remainder of the region is lightly populated. The less developed eastern part was named the Blue Ridge Berryessa Natural Area by a regional consortium of conservation stakeholders (see

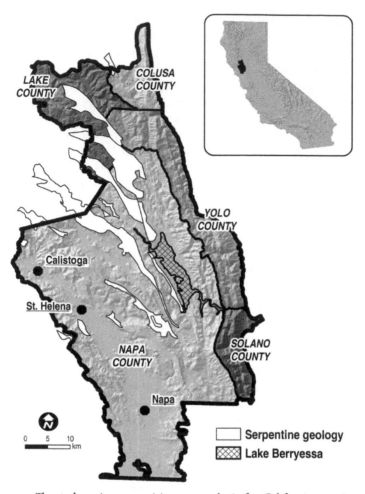

FIGURE 15.1. The study region, comprising 292,400 ha in five California counties.

www.brbna.org). The bedrock geology includes extensive areas of serpentine, volcanic rock, and sandstone (Lambert and Kashiwagi, 1978; Norris and Webb, 1990).
The climate is Mediterranean with a moderating maritime influence in its western
portion; moving eastward, precipitation declines, temperatures increase, and vegetation changes markedly.

Major nonserpentine vegetation types in mesic areas include evergreen forests
with live oak (*Quercus wislizenii* and *Q. agrifolia*) and black oak (*Q. kelloggii*); in
drier areas this gives way to blue oak (*Q. douglasii*) woodland, chaparral dominated by chamise (*Adenostoma fasciculatum*) and other evergreen shrubs, and

grasslands dominated by European annual grasses. Serpentine supports extensive areas of chaparral with mixtures of leather oak (*Q. durata*), whiteleaf manzanita (*Arctostaphylos manzanita*), and other shrubs, interspersed by stands of MacNab's cypress (*Cupressus macnabiana*) and by montane or riparian stands of Sargent's cypress (*C. sargentii*), and small pockets of grassland dominated by native forbs and perennial grasses.

The region has long been noted for its botanical diversity, rarity, and ende-mism (Stebbins and Major, 1965; Stein, 2002; Thorne et al., 2009), much of which is associated with its climatic, topographic, and edaphic diversity (Ornduff et al., 2003). The state Natural Heritage Database lists 171 observations of 23 rare plants associated with serpentine in the region, and 183 observations of 36 other rare plants. Important habitats for serpentine rare plants are steep talus slopes (bar-rens), small riparian wetlands (seeps), and clay-rich meadows. Rare plants are also found on nonserpentine soils, often in vernal pools, tidal wetlands, native grasslands, or volcanic outcrops (e.g., three rare species of California lilac, *Ceanothus*).

MAPS AND THEIR COVERAGE OF
SERPENTINE HABITATS

Interest in regional conservation led this area to be used as a test case for the de-velopment of high-quality regional vegetation mapping which we refer to as MCV (Thorne et al., 2004). This mapping is based on the system of plant community classification described in the Manual of California Vegetation (Sawyer and Keeler-Wolf, 1995) and was implemented using digital orthophoto interpretation and ex-tensive ground truthing. We used the MCV map in all reserve design modeling. We compared the extent and detail of serpentine plant community coverage on MCV to three more widely available but less detailed maps: the U.S. GAP pro-gram's second vegetation map of California (GAP II; Lennartz et al., 2008), the U.S. Geological Survey's Landfire Existing Vegetation Type dataset (Landfire; Rol-lins, 2009), and the fuels-oriented land cover map developed by the California Department of Forestry and Fire Protection (FRAP; 2002). Earlier vegetation mapping of the region was done by Wieslander (1935), Kuchler (1988), Schwind and Gordon (2001), and Davis et al., (1998).

We cross-tabulated the four land cover maps, compared them to identify the number of native vegetation classes and their extent, and measured the congru-ence of lands classified as serpentine. We also compared the serpentine land cover classes to mapped serpentine geology (Jennings, 1977). All map analysis and mod-eling operations were conducted using ArcMap 9.3 (ESRI, 2009). The study area was rendered to hexagons of 10 ha size to permit the reserve design models to run on standardized units, creating an effective analysis area of 284,650 ha.

We found the MCV map identified 66 land cover classes in the region, of which 62 were natural terrestrial classes, including 9 serpentine types covering 29,128 ha or 10.2% of the region. The FRAP map identified only 29 land cover classes with no serpentine types, and the Landfire map identified 35 land cover classes with 1 serpentine type measuring 3.9 ha. The GAP II map identified 36 land cover classes with 2 serpentine types covering 24,215 ha (8.5% of the region); this map distinguished serpentine chaparral and woodlands but missed serpentine grasslands, barrens, and riparian areas, and classified the Sargent's and MacNab's cypress alliances as part of "Mediterranean California Mesic Serpentine Woodland and Chaparral." Only 3.5% of the study region is classified as serpentine vegetation by both MCV and GAP II (Figure 15.2). Although the MCV serpentine vegetation matches the geology fairly well, GAP II shows a pattern typical of remote sensing imagery classification, with many small polygons distributed across the region including in areas where no serpentine is found.

Using MCV, we classified plant communities as rare (<200 ha), intermediate (<5697 ha or 2% of the region), or common (>2% of the region), so we could compare the protection of serpentine and nonserpentine plant communities with comparable degrees of rarity or commonness.

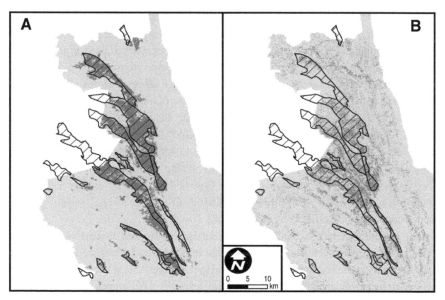

FIGURE 15.2. Serpentine vegetation as mapped by (A) the Manual of California Vegetation (MCV) classification and (B) the National GAP Analysis program; extent of serpentine from the state geologic map is shown in the crosshatch overlay.

HISTORY AND OPPORTUNISM: SERPENTINE IN
EXISTING PROTECTED AREAS

About 27% of the region, or 77,000 ha, is in some form of land protection; the majority of this, 49,400 ha (66%), is in the domain of the U.S. Bureau of Land Management (BLM) (www.blm.gov). The BLM inherited the responsibility of managing all lands not privatized either during the Spanish and Mexican land grant era (ending in 1848) or the homesteading era (ending in 1976), and thus its holdings are largely rugged, inaccessible, and/or unproductive. Extensive areas of serpentine are found in the BLM's Cache Creek and Cedar Roughs wilderness areas, managed for wildlife and low-impact recreation, and its Walker Ridge and Knoxville public lands, managed for multiple use including motorized recreation and energy development. Another 11,400 ha (14% of all protected land) is managed by the state Fish and Game Department, mostly in the Knoxville Wildlife Area (purchased from a rancher in 2002). Other protected areas include the U.S. Bureau of Reclamation land surrounding the Berryessa Reservoir (created in 1957); the land and easement holdings of the Land Trust of Napa County (www.napalandtrust. org), including a serpentine preserve known as Missimer Snell Valley Wildflower Preserve (1999); and three units of the University of California Natural Reserve System (www.nrs.ucop.edu): Stebbins Cold Canyon (1979), Quail Ridge (1991), and McLaughlin (1992), the last of which contains extensive areas of serpentine.

We found that existing protected areas (GreenInfo Network, 2009) encompassed 18,356 ha or 63% of serpentine vegetation as recorded by MCV (Table 15.1; Figure 15.3A). This included 69% of rare (restricted range), 63% of intermediate, and 54% of common (broadly distributed) plant communities. The corresponding protection for nonserpentine vegetation is 15%, 32%, and 27% of the rare, intermediate, and common communities, respectively. Existing areas protect an average 32% of serpentine and 30% of nonserpentine rare plant localities (Table 15.2; Figure 15.4A).

EQUAL REPRESENTATION FOR VEGETATION:
THE MARXAN APPROACH

Reserve selection programs such as Marxan use maps that depict a region as a set of polygons (usually equal area), where each polygon contains variable quantities of diverse features, such as species and vegetation types. The user specifies a set of conservation targets, in terms of which features are important and how much of each feature is desired. The program then analyzes a large number of potential combinations of polygons to find a set of polygons that comes closest to representing the total desired array of conservation targets with a low "cost" (usually determined by total area selected). The results of multiple runs of the model are combined to find

TABLE 15.1 Levels of Protection for Each Serpentine Vegetation Type as Classified by the Manual of California Vegetation

Rarity	Vegetation Type	Area (ha)	Current %	Marxan %	Connect %	Mean %
Rare	Brewer willow alliance	174	55.3	99.6	32.1	62.3
	Serpentine barren	61	82.4	99.8	25.0	69.1
Intermediate	Serpentine (mean)		68.9	99.7	28.6	65.7
	Nonserpentine (mean)		15.3	99.0	13.7	42.7
	California bay, leather oak, *Rhamnus* spp., mesic serpentine superalliance	4187	52.0	43.6	35.7	43.8
	Leather oak, California bay, *Rhamnus* spp., mesic serpentine alliance	3123	56.8	45.6	33.6	45.3
	McNab cypress alliance	1435	74.5	42.6	18.8	45.3
	Sargent cypress alliance	826	95.7	56.1	94.4	82.1
	Serpentine grasslands superalliance	1078	32.0	43.9	38.6	38.2
	Whiteleaf manzanita, leather oak, Chamise–Ceanothus spp., xeric serpentine superalliance	4067	46.8	43.3	35.1	41.7
	Whiteleaf manzanita alliance	392	80.6	77.5	9.3	55.8
Common	Serpentine (mean)		62.6	50.4	37.9	50.3
	Nonserpentine (mean)		32.3	56.3	19.7	36.1
	Leather oak, whiteleaf manzanita, Chamise xeric serpentine superalliance	13,786	54.1	25.3	46.4	41.9
	Nonserpentine (mean)		26.9	29.1	31.0	29.0

NOTE: "Area" refers to the total amount of each type within the study area. "Current," "Marxan," and "Connect" refer to the currently protected areas and the two reserve designs. "Non-serpentine" refers to all vegetation types within the study area that are not found on serpentine. Rarity classes are <200 ha, 200–5697 ha, and >5697 ha.

SOURCE: Thorne et al., 2004.

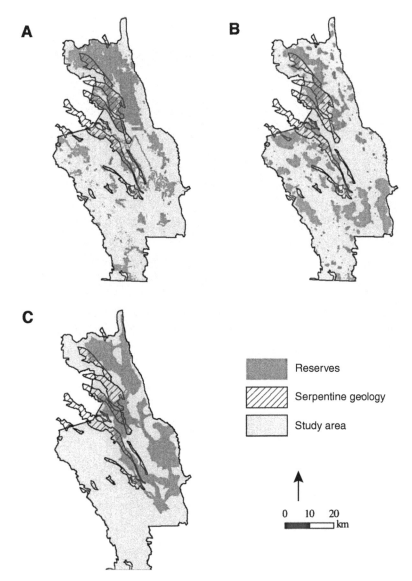

FIGURE 15.3. Geologic map for serpentine overlain with (A) existing protected lands, (B) the Marxan reserve design, and (C) the core-corridor design.

TABLE 15.2 Rare Serpentine Plant Species Occurrences in the State Natural Heritage Database and the Percentage of These Occurrences in Current Protection and Each of the Reserve Designs

Species	Occ.	Current %	Marxan %	Connect %	Mean %
Astragalus claranus	5	40.0	20.0	0.0	20.0
Astragalus rattanii var. jepsonianus	5	40.0	40.0	20.0	33.3
Balsamorhiza macrolepis var. macrolepis	1	0.0	0.0	0.0	0.0
Brodiaea californica var. leptandra	12	33.3	58.3	0.0	30.5
Calystegia collina ssp. oxyphylla	3	33.3	66.7	0.0	33.3
Castilleja affinis ssp. neglecta	1	0.0	0.0	0.0	0.0
Castilleja rubicundula ssp. rubicundula	1	0.0	0.0	0.0	0.0
Cryptantha clevelandii var. dissita	1	0.0	0.0	0.0	0.0
Erigeron greenei	6	33.3	33.3	16.7	27.8
Eriogonum nervulosum	3	66.7	66.7	0.0	44.5
Fritillaria pluriflora	21	19.0	57.1	23.8	33.3
Harmonia hallii	7	28.6	57.1	0.0	28.6
Hesperolinon bicarpellatum	8	12.5	75.0	25.0	37.5
Hesperolinon breweri	6	33.3	50.0	33.3	38.9
Hesperolinon drymarioides	5	40.0	40.0	0.0	26.7
Hesperolinon sp. nov. "serpentinum"	30	23.3	46.7	16.7	28.9
Layia septentrionalis	15	20.0	73.3	13.3	35.5
Leptosiphon jepsonii	4	25.0	25.0	0.0	16.7
Navarretia rosulata	3	33.3	66.7	33.3	44.4
Streptanthus brachiatus ssp. brachiatus	1	100.0	0.0	0.0	33.3
Streptanthus brachiatus ssp. hoffmanii	1	100.0	100.0	0.0	66.7
Streptanthus breweri var. hesperidis	13	30.8	61.5	7.7	33.3
Streptanthus morrisonii	17	29.4	58.8	17.6	35.3
Serpentine (mean)		32.3	43.3	9.0	28.2
Nonserpentine (mean)		30.1	42.2	7.5	26.6

NOTE: Mean protection of nonserpentine rare plant occurrences within the study area is also noted.

SOURCE: California Department of Fish and Game, 2010.

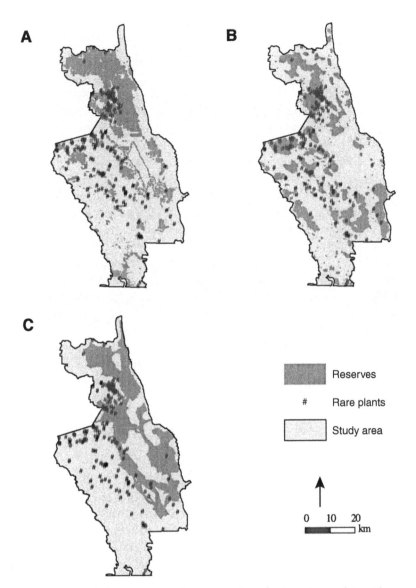

FIGURE 15.4. Recorded occurrences of rare serpentine plant species overlain with (A) the existing protected lands, (B) the Marxan reserve design, and (C) the core-corridor design.

either the best identified reserve design or the summed results of multiple reserve designs. Often, however, the design produced by this model is a set of excessively small and widely scattered reserves. To achieve some degree of spatial coherence, a boundary modifier is added to the model to give greater value to polygons next to already selected polygons, thus creating a set of fewer and larger reserves, although at some cost to representation (Ball and Possingham, 2000; Briers, 2002; Fischer and Church, 2005; Williams et al., 2005; Ball et al., 2009; Moilanen, 2009).

We used Marxan to select approximately the same amount of area as already protected in the study region but with the goal of systematically representing the region's natural communities as portrayed by MCV. We used 10-ha hexagons, a boundary modifier of 0.001 (subjectively selected as the most appropriate after a series of trial runs), and 100 model runs. Our specific targets were 100% of rare communities, 30% of intermediate communities, and 15% of common communities. We summed the number of times each hexagon was selected out of 100 runs, and the collection of highest scoring hexagons that equaled the area of the current reserve network was selected as the final Marxan reserve network.

The Marxan reserve design covered 16,815 ha (58%) of the MCV serpentine vegetation, including 99.7% of the rarest serpentine communities, 43–77% of the intermediate communities, and 25% of the common community (Table 15.1; Figure 15.3B). Not surprisingly given the intent of the model, these percentages were quite similar for rare (99%), intermediate (56%), and common (29%) nonserpentine communities. This design also captured an average of 43% of the known occurrences of serpentine rare plants and 42% of the nonserpentine rare plants (Table 15.2; Figure 15.4B).

CORES AND CONNECTIVITY: LEAST-COST CORRIDOR ANALYSIS

The core-corridor or ecological network approach focuses on the needs of wide-ranging animals (Noss and Harris, 1986; Beier and Noss, 1998). Large core reserves are identified by landscape attributes such as lack of roads or other human impacts, and dispersal corridors connecting these cores are identified using assumptions about the ecological cost of animal dispersal through different land cover types (Thorne et al., 2006). Confirming the functioning of habitat corridors is an area of considerable interest (Hilty and Merenlender, 2004; Bennett et al., 2006), which has been made even more critical by the possibility that species will need to shift their ranges in the face of the effects of changing climate (Araujo et al., 2004; Hannah, 2010). However, proposed wildlife corridors have sometimes been controversial because they traverse expensive, heavily disturbed lowland real estate lying between undeveloped upland core habitats, and thus they carry a large opportunity cost in terms of meeting other conservation goals (Mann and Plummer, 1995).

We modeled an ecological network using mountain lion (*Puma concolor*) as a focal species because of its large area requirements (Beier and Noss, 1998; Morrison and Boyce, 2009). Areas >2000 ha in extent and >1.5 km from roads were selected as cores. The MCV map was converted to a cost surface by applying statewide mountain lion–specific habitat ratings to the vegetation types (California Department of Fish and Game, 2008). Polygons were converted to a raster with 30 m cell size, and the inverse of this habitat suitability rating was then used as the movement cost for each raster cell. Major roads were converted from line features to raster cells with the highest cost value associated with them. Corridors were identified between the centroids of adjacent cores using the ArcGIS Least Cost Corridor tool (ESRI, 2009), with the total network approximately equaling the area of the existing protected lands.

The core-corridor reserve design protects 10,748 ha (37%) of the MCV serpentine lands, including 29% of rare, 37% of intermediate, and 46% of common communities (Table 15.1; Figure 15.3C). In contrast, 14% of rare, 20% of intermediate, and 31% of common nonserpentine communities are covered. This design protects an average 12% of serpentine and 8% of nonserpentine rare plant localities (Table 15.2; Figure 15.4C).

REGIONAL CONSERVATION AND SERPENTINE: CONCLUSIONS

Serpentine-containing landscapes provide opportunities to test how well conservation strategies can meet multiple objectives. Our analyses suggest that there are indeed strong differences between the geographic strategies that would best protect unusual landscape elements such as serpentine vegetation and rare plants, all natural community types as in Marxan, and connectivity for large wildlife as in the core-corridor approach. However, systematic conservation analyses can help identify specific actions that best contribute to combinations of such objectives. To illustrate this, we overlaid existing conserved areas with the two reserve designs to identify lands that are currently not protected, contain a high diversity of vegetation types as identified by Marxan, and contribute importantly to regional connectivity according to the core-corridor model (Table 15.3, Figure 15.5). These conservation opportunity areas total 9859 ha, including 1068 ha of serpentine plant communities. This general type of approach can be used to identify lands of high priority when opportunities for land or easement acquisition arise (see Kremen et al., 2008, for a recent example from Madagascar).

Our most striking finding with respect to serpentine vegetation is how well it is already protected in this region. Serpentine plant communities are consistently two to four times better conserved than nonserpentine communities, even those within the same rarity class; also, they are nearly always better protected than they

TABLE 15.3 Conservation Opportunities

Vegetation Type	Area (ha)
Brewer willow alliance	22
California bay, leather oak, *Rhamnus* spp. mesic serpentine NFD super alliance	152
Leather oak, California bay, *Rhamnus* spp. mesic serpentine NFD alliance	196
Leather oak, whiteleaf manzanita, Chamise xeric serpentine NFD super alliance	429
McNab cypress alliance	3
Sargent cypress alliance	<1
Serpentine barren	4
Serpentine grasslands NFD super alliance	78
Whiteleaf manzanita, leather oak, Chamise–Ceanothus spp. xeric serpentine NFD super alliance	184
Total	1068

Species	Occ.
Castilleja rubicundula ssp. *rubicundula*	1
Fritillaria pluriflora	1
Hesperolinon bicarpellatum	2
Hesperolinon sp. nov. "*serpentinum*"	3
Monardella villosa ssp. *globosa* (nonserpentine)	1

would be under either systematic reserve design (Table 15.1). The only exceptions are the two most restricted serpentine communities in the MCV map, barrens and Brewer willow alliance, which are 82% and 55% protected, while Marxan prescribed >99% protection; and serpentine grasslands, which are only 32% protected, whereas Marxan called for 44% and the core-corridor design called for 39% protection. Particularly outstanding for high levels of existing protection were Sargent's cypress woodlands (96%), MacNab's cypress woodlands (75%), and whiteleaf manzanita chaparral (81%). Nearly all protected serpentine is on BLM lands, and this public ownership is a common phenomenon reflecting the relative lack of economic value of serpentine. The cypress woodland and chaparral communities form dense scrubby vegetation on rocky upland soils with no agricultural potential; thus, like other upland vegetation, they tend to fall into the public domain much more often than do more fertile and accessible lowland habitats (Pressey, 1994; Underwood et al., 2008). Similarly, across the western United States, a considerable proportion of serpentine is under the jurisdiction of the BLM and the U.S. Forest Service. Serpentine grasslands are a revealing exception, because they are used for livestock grazing and occasional agriculture (usually with limited success), and hence were underrepresented on the public lands in our study region.

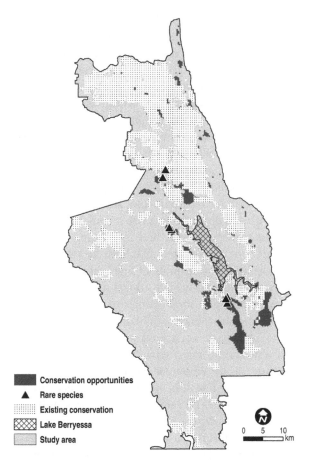

FIGURE 15.5. Potential conservation opportunities within the study region. Dark gray areas indicate overlap between the Marxan and core-corridor reserve designs, and triangles represent occurrences of rare serpentine species within the overlap area. The speckled pattern denotes existing conservation areas.

Serpentine supports 40% of the rare plant species in the study region although it forms only 10% of its area. Known occurrences of rare serpentine plants are substantially worse protected (mean 28%) than serpentine vegetation (63%) in the existing landscape, as well as in the two conservation designs, though this under-protection of rare species relative to vegetation was not the case for nonserpentine plants (Tables 15.1, 15.2). The relative protection of known rare plant occurrences in the existing landscape differed little between serpentine and nonserpentine rare plants (28% versus 27%), even though serpentine vegetation types were considerably better protected than nonserpentine ones. These findings suggest that rare plant species on serpentine do indeed have special conservation needs that are harder to meet in a reserve system than those of rare plants in other habitats, perhaps related to being more widely distributed across the landscape. Small preserves and other forms of special protection may continue to be needed more often for serpentine rare plants than for other species.

Our analysis of serpentine conservation would have been impossible without the high-quality MCV map, which was the only map to fully recognize some of the most important serpentine plant communities in the region (grasslands, barrens, conifer stands, and riparian habitats). Rare species data were of lesser quality, as is often the case due to survey incompleteness on private lands (Thorne et al., 2009). A rapidly developing technique to address this problem is species distribution modeling (SDM), in which the environmental attributes of sites where a species is known to occur are used to project where else it could occur, with a wide array of applications including reserve design and forecasting climate change impacts (Guisan and Thuiller, 2005). Two factors that limit the efficacy of species distribution modeling are the limited availability of fine-scale geographic data on important habitat quality variables (many SDMs thus use only broad-scale climate variables) and the limited ability of some species to disperse to all of their suitable habitats. Rare serpentine plants may be difficult to model for both reasons: that is, they are sensitive to subtle aspects of habitat quality, and their patchy habitat makes them likely to be absent from many environmentally suitable sites within their ranges (Harrison, 1997, 1999; Freestone and Inouye, 2006). Nonetheless, species distribution modeling using fine-scale habitat data has been successfully used to improve the efficacy of searches for rare serpentine plants (Williams et al., 2009).

Our analysis points to a degree of optimism about the future of serpentine plant communities both in California and elsewhere in the world (see Kruckeberg, 2004; Whiting et al., 2004; Rajakaruna et al., 2009, for other reviews). Because of their generally low economic value, they may be in less danger of outright land conversion than many other natural communities, although important exceptions occur in tropical regions where laterite soils are mined for nickel (see Chapter 18 in this volume). Globally, serpentine exposures are likely to present many outstanding opportunities for the protection of relatively less disturbed natural communities. Based on our analysis, we suggest the planned conservation of serpentine-containing landscapes can be undertaken using geographic analyses of biodiversity data such as land cover maps and rare species occurrences. Unfortunately, the requisite data are still far from being available in most cases.

We also note the pressing need for better stewardship of existing protected lands. For example, in our study region, serpentine grasslands are being invaded by barbed goatgrass (*Aegilops triuncialis*), with drastic negative effects on native diversity (Batten et al., 2006). On multiple-use public lands, serpentine riparian areas and adjacent habitats are often badly damaged by off-highway vehicle recreation.

Excessively frequent fire caused by recreationists is another potential threat on these lands, especially for slow-maturing cypress forests which may be >100 years old (Safford and Harrison, 2004). Industrial-scale wind energy development is a new threat to extensive areas of serpentine vegetation rich in rare species. We suggest that another excellent use for systematic conservation planning would be to help inform management policies and priorities on the public's existing protected lands, so that high-quality examples of natural communities are treated as being of high intrinsic value.

LITERATURE CITED

Anderson, M., Comer, P., Grossman, D., Groves, C., Poiani, K., Reid, M., Schneider, R., Vickery, B., and Weakley, A. (1999) *Guidelines for Representing Ecological Communities in Ecoregional Conservation Plans*. Nature Conservancy, Arlington, VA.

Araujo, M. B., Cabezas, M., Thuiller, W., Hannah, L., and Williams, P. (2004) Would climate change drive species out of reserves? An assessment of existing reserve-selection methods. *Global Change Biology*, 10, 1618–26.

Ball, I. R., and Possingham, H .P. (2000) *MARXAN (V1.8.2): Marine Reserve Design Using Spatially Explicit Annealing, a Manual*. Available online at http://www.uq.edu.au/marxan/docs/marxan_manual_1_8_2.pdf.

Ball, I. R., Possingham, H. P., and Watts, M. (2009) Marxan and relatives: Software for spatial conservation prioritization. In *Spatial Conservation Prioritization: Quantitative Methods and Computational Tools* (eds. A. Moilanen, K. A. Wilson, and H. P. Possingham), pp. 185–95. Oxford University Press, Oxford.

Batten, K. M., Scow, K. M., Davies, K. F., and Harrison, S. (2006) Two invasive plants alter soil microbial community composition in serpentine grasslands. *Biological Invasions*, 8, 217–30.

Beier, P., and Noss, R. F. (1998) Do habitat corridors provide connectivity? *Conservation Biology*, 12, 1241–52.

Bennett, A. F., Crooks, K. R., and Sanjayan, M. (2006) The future of connectivity conservation. In *Connectivity Conservation* (eds. K. R. Crooks and M. Sanjayan), pp. 676–94. Cambridge University Press, Cambridge.

Briers, R. A. (2002) Incorporating connectivity into reserve selection procedures. *Biological Conservation*, 103, 77–83.

California Department of Fish and Game. (2008) *CWHR Version 8.2 Personal Computer Program*. California Interagency Wildlife Task Group, Sacramento.

California Department of Fish and Game. (2010) *California Natural Diversity Data Base*. California Resources Agency, Sacramento.

California Department of Forestry and Fire Protection. (2002) *Multi-Source Land Cover Data*. California Resources Agency, Sacramento.

Davis, F. W., Stoms, D. M., Hollander, A. D., Thomas, K. A., Stine, P. A., Odion, D., Borchert, M. I., Thorne, J. H., Grey, M. V., Walker, R. E., Warner, K., and Graae, J. (1998) *The California Gap Analysis Project: Final Report*. University of California, Santa Barbara.

ESRI. (2009) *ArcGIS 9.3*. Environmental Systems Research Institute, Redlands, CA.

Fischer, D. T., and Church, R. L. (2005) The SITES reserve selection system: A critical review. *Environmental Modeling and Assessment*, 10, 215–28.

Freestone, A. T., and Inouye, B. D. (2006) Dispersal limitation and environmental heterogeneity shape scale-dependent diversity patterns in plant communities. *Ecology*, 87, 2425–32.

GreenInfo Network. (2009) *California Protected Areas Database (CPAD), Version 1.2, Database Manual.* GreenInfo Network, San Francisco.

Groves, C. R., Jensen, D. B., Valutis, L. L., Redford, K. H., Shaffer, M. L., Scott, J. M., Baumgartner, J. V., Higgins, J. V., Beck, M. W., and Anderson, M. G. (2002) Planning for biodiversity conservation: Putting conservation science into practice. *BioScience*, 51, 499–512.

Guisan, A., and Thuiller, W. (2005) Predicting species distributions: Offering more than simple habitat models. *Ecology Letters*, 8, 993–1009.

Hannah, L. (2010) A global conservation system for climate-change adaptation. *Conservation Biology*, 24, 70–77.

Harrison, S. (1997) How natural habitat patchiness affects the distribution of diversity in Californian serpentine chaparral. *Ecology*, 78, 1898–906.

Harrison, S. (1999) Local and regional diversity in a patchy landscape: Native, alien and endemic herbs on serpentine soils. *Ecology*, 80, 70–80.

Harrison, S., and Inouye, B. D. (2002) High beta diversity in the flora of Californian serpentine "islands." *Biodiversity and Conservation*, 11, 1869–76.

Hilty, J. A., and Merenlender, A. M. (2004) Use of riparian corridors and vineyards by mammalian predators in northern California. *Conservation Biology*, 18, 126–35.

Huber, P. R., Greco, S. E., and Thorne, J. H. (2010) Spatial scale effects on conservation network design: Tradeoffs and omissions in regional versus local-scale planning. *Landscape Ecology*.

Jennings, C. W. (1977) *Geologic Map of California.* U.S. Geological Survey, Menlo Park, CA.

Kiester, A. R., Scott, J. M., Csuti, B., Noss, R. F., Butterfield, B., Sahr, K., and White, D. (1996) Conservation prioritization using GAP data. *Conservation Biology*, 10, 1332–42.

Kremen, C., Cameron, A., Moilanen, A., Phillips, S. J., Thomas, C. D., Beentje, H., Dransfield, J., Fisher, B. L., Glaw, F., Good, T. C., Harper, G. J., Hijmans, R. J., Lees, D. C., Louis, E., Nussbaum, R. A., Raxworthy, C. J., Razafimpahanana, A., Schatz, G. E., Vences, M., Vieites, D. R., Wright, P. C., and Zjhra, M. L. (2008) Aligning conservation priorities across taxa in Madagascar with high-resolution planning tools. *Science*, 320, 222–25.

Kruckeberg, A. R. (1984) *California Serpentines: Flora, Vegetation, Geology, Soils and Management Problems.* University of California Press, Berkeley.

Kruckeberg, A. R. (2004) The status of conservation of ultramafic sites in the USA and Canada. In *Ultramafic Rocks: Their Soils, Vegetation, and Fauna* (eds. R. S. Boyd, A. J. M., Baker and J. Proctor), pp. 311–14. Proceedings of the 4th International Conference on Serpentine Ecology. Science Reviews 2000, St. Albans, Herts.

Kuchler, A. W. (1988) The map of the natural vegetation of California. In *Terrestrial Vegetation of California* (eds. M. G. Barbour and J. Major), pp. 909–38. California Native Plant Society, Sacramento.

Lambert, G., and Kashiwagi, J. (1978) *Soil Survey of Napa County, California*. USDA, Soil Conservation Service, in cooperation with University of California Agricultural Experiment Station, Davis.

Lennartz, S., Bax, T., Aycrigg, J., Davidson, A., Reid, M., and Congalton, R. (2008) *Final Report on Land Cover Mapping Methods for California Map Zones 3, 4, 5, 6, 12, and 13*. National Biological Information Infrastructure, Idaho.

Mann, C. C., and Plummer, M. L. (1995) Are wildlife corridors the right path? *Science*, 270, 1428–30.

Meynard, C. N., Howell, C. A., and Quinn, J. F. (2009) Comparing alternative systematic conservation strategies against a politically driven conservation plan. *Biodiversity Conservation*, 18, 3061–83.

Moilanen, A., Wilson, K. A., and Possingham, H. P., eds. (2009) *Spatial Conservation Prioritization*. Oxford University Press, Oxford.

Morrison, S. A., and Boyce, W. M. (2009) Conserving connectivity: Some lessons from mountain lions in southern California. *Conservation Biology*, 23, 275–85.

Myers, N., Mittermeier, R. A., Mittermeier, C. G., da Fonseca, G. A. B., and Kent, J. (2000) Biodiversity hotspots for conservation priorities. *Nature*, 403, 853–58.

Norris, R. M., and Webb, R. W. (1990) *Geology of California*. Wiley, New York.

Noss, R., and Harris, L. (1986) Nodes, networks and MUMs: Preserving diversity at all scales. *Environmental Management*, 10, 299–309.

Ornduff, R., Faber, P., and Keeler-Wolf, T. (2003) *California Plant Life*. University of California Press, Berkeley.

Prendergast, J., Quinn, R. M., and Lawton, J. H. (1999) The gaps between theory and practice in selecting nature reserves. *Conservation Biology*, 13, 484–92.

Pressey, R. L. (1994) Ad hoc reservations: Forward or backward steps in developing representative reserve systems? *Conservation Biology*, 8, 662–68.

Pressey, R. L., Humphries, C. J., Margules, C. R., Vane-Wright, R. I., and Williams, P. H. (1993) Beyond opportunism: Key principles for systematic reserve selection. *Trends in Ecology and Evolution*, 8, 124–28.

Rajakaruna, N., Harris, T. B., and Alexander, E. B. (2009) Serpentine geoecology of eastern North America: A review. *Rhodora*, 111, 21–108.

Reid, W. V. (1998) Biodiversity hotspots. *TREE*, 13, 275–80.

Rollins, M. G. (2009) LANDFIRE: A nationally consistent vegetation, wildland fire, and fuel assessment. *International Journal of Wildland Fire*, 18, 235–49.

Safford, H. D., and Harrison, S. (2004) Fire effects on plant diversity in serpentine versus sandstone chaparral. *Ecology*, 85, 539–48.

Safford, H. D., Viers, J. H., and Harrison, S. (2005) Serpentine endemism in the California flora: A database of serpentine affinity. *Madroño*, 52, 222–57.

Sawyer, J. O., and Keeler-Wolf, T. (1995) *A Manual of California Vegetation*. California Native Plant Society, Sacramento.

Schwind, B., and Gordon, H. (2001) *Calveg Geobook: A Comprehensive Information Package Describing California's Wildland Vegetation, Version 2*. USDA Forest Service, Pacific Southwest Region, Remote Sensing Lab, Sacramento. CD ROM.

Scott, J. M., Davis, F., Csuti, B., Noss, R., Butterfield, B., Caicco, S., Groves, C., Edwards, T. C. Jr., Ulliman, J., Anderson, H., D'Erchia, F., and Wright, R. G. (1993) Gap analysis: A geographic approach to protection of biological diversity. *Wildlife Monographs*, 123, 1–41.

Soulé, M. E., and Terborgh, J., eds. (1999) *Continental Conservation: Scientific Foundations of Regional Reserve Networks*. Island Press, Washington.

Stebbins, G. L., and Major, J. (1965) Endemism and speciation in the California flora. *Ecological Monographs*, 35, 1–35.

Stein, B. A., Kutner, L. S., and Adams, J. S. (2000) *Precious Heritage: The Status of Biodiversity in the United States*. Oxford University Press, Oxford, UK.

Stein, B. A., Kutner, L. S., and Adams, J. S. (2000) *Precious Heritage: The Status of Biodiversity in the United States*. Oxford University Press, Oxford.

Theobald, D. M. (2006) Exploring the functional connectivity of landscapes using landscape networks. In *Connectivity Conservation* (eds. K. R. Crooks and M. Sanjayan), pp. 416–43. Cambridge University Press, Cambridge.

Thorne, J. H., Cameron, D., and Quinn, J. F. (2006) A conservation design for the Central Coast of California and the evaluation of mountain lion as an umbrella species. *Natural Areas Journal*, 26, 137–48.

Thorne, J. H., Kennedy, J. A., Quinn, J. F., McCoy, M., Keeler-Wolf, T., and Menke, J. (2004) A vegetation map of Napa County using the Manual of California Vegetation classification and its comparison to other digital vegetation maps. *Madroño*, 51, 343–63.

Thorne, J. H., Viers, J. H., Price, J., and Stoms, D. M. (2009) Spatial patterns of endemic plants in California. *Natural Areas Journal*, 29, 344–66.

Underwood, E. C., Klausmeyer, K. R., Cox, R. L., Busby, S. L., Morrison, S. C., and Shaw, R. (2008) Expanding the global network of protected areas to save the imperiled Mediterranean biome. *Conservation Biology*, 23, 43–52.

Whiting, S. N., Reeves, R. D., Richards, D., Johnson, M. S., Cooke, J. A., Malaisse, F., Paton, A., Smith, J. A. C., Angle, J. S., Chaney, R. L., Ginocchio, R., Jaffre, T., Johns, R., McIntyre, T., Purvis, O. W., Salt, D. E., Schat, H., Zhao, F. J., and Baker, A. J. M. (2004) Research priorities for conservation of metallophyte biodiversity and their potential for restoration and site remediation. *Restoration Ecology*, 12, 106–16.

Wieslander, A. E. (1935) A vegetation type map for California. *Madroño*, 3, 140–44.

Williams, J. C., ReVelle, C. S., and Levin, S. A. (2005) Spatial attributes and reserve design models: A review. *Environmenral Modelling and Assessment*, 10, 163–81.

Williams, J. N., Seo, C., Thorne, J., Nelson, J. K., Erwin, S., O'Brien, J. M., and Schwartz, M. W. (2009) Using species distribution models to predict new occurrences for rare plants. *Diversity and Distributions*, 15, 565–76.

16

Biodiversity, Ecosystem Functioning, and Global Change

David Hooper, *Western Washington University*

Environmental changes, including elevated CO_2, climate change, enhanced nutrient deposition, change in land use type and intensity, and species invasions, are contributing to worldwide loss of biodiversity (Vitousek, 1994). Ecosystem responses may either buffer or exacerbate these changes through various feedback loops, with important implications for the services ecosystems provide to humanity (Daily, 1997; Millennium Ecosystem Assessment, 2005; Suding et al., 2008). Whole-ecosystem studies provide a complement to modeling approaches and smaller scale studies to test whether our understanding of the underlying mechanisms scale up, and if not, which other processes need more consideration (Field et al., 1996; Vitousek, 2004). This chapter evaluates use of serpentine grasslands as model systems for understanding whole-ecosystem responses to two different global changes: loss of biodiversity and elevated atmospheric CO_2.

California serpentine grasslands have several useful attributes as a model system for whole-ecosystem questions. First, the small stature of the vegetation allows multiple experimental treatments in plots of tractable size (Field et al., 1996). Moderate-sized experimental plots ($1-2$ m^2) can contain hundreds to thousands of individuals of each species, so each plot has potentially viable populations over multiyear time scales (Hooper, 1998). Second, there is a diverse native flora of mostly annual species with a variety of functional characteristics. Annual life cycles allow for population turnover and shifts in species composition, which can greatly influence whole-ecosystem responses. Within the annuals, relatively well-delineated functional groups based on resource acquisition strategies have the potential to partition resources both in time (early versus late season) and space (differences in rooting depth) (Gulmon et al., 1983; Mooney et al., 1986). There are additional functional differences among perennials based on morphology and life history, for example, grass versus forb, phenology, and rooting structure (geophytes versus taproot), though these have been investigated less extensively (but see Hobbs and Mooney, 1991, 1995; Hobbs et al., 2007). Another benefit of serpentine grasslands is their proximity to ecosystems on contrasting soil types, providing useful comparisons of ecosystem responses to global changes (e.g., elevated CO_2) in systems with different degrees of resource limitation (Field et al., 1996). The relative importance of the basic mechanisms that control processes at the ecosystem scale will certainly differ across ecosystem types, so experiments in multiple systems are necessary (Field et al., 1996; Hooper, 1998).

PLANT FUNCTIONAL TRAITS AND ECOSYSTEM PROPERTIES

Ecosystem ecologists have long recognized that the functional traits of organisms are key drivers of ecosystem properties (e.g., Jenny, 1980; Jones and Lawton, 1995). Understanding how ecosystem properties are likely to respond to the suite of global changes now in effect will require understanding both how communities are likely to change due to altered environments and how the altered environment and resulting biotic community are likely to affect the ecosystem properties in question. Understanding potential changes in ecosystem properties thus depends on understanding the traits that determine a species's response to environmental parameters (response traits), the traits that influence ecosystem properties (effect traits), and how they are related (Walker et al., 1999; Hooper et al., 2002; Lavorel and Garnier, 2002). Understanding the potential responses for different processes in response to different drivers of global change is key question at the intersection of community and ecosystem ecology.

Serpentine grasslands have helped refine our understanding of differences between response and effect traits and of how diversity of functional traits can influence ecosystem properties. One reason serpentine grasslands are useful for studies of ecosystem functioning is that some plant functional groups are relatively clearly defined. Phenology plays a particularly strong role and correlates with a number of other functional traits. In the Mediterranean-like climate of California, annual plants in serpentine grasslands show a continuum of peak flowering from February to as late as October or November (Mooney et al., 1986). Perennials follow a similar temporal sequence of flowering and above-ground senescence, though they are typically neither as diverse nor abundant as the annuals (Gonzalez, 2007;

Hobbs et al., 2007). In annuals, rooting depth and litter quality also roughly correlate with phenological differences, with early-season species having much shallower roots and relatively decomposable litter compared with late-season species (Gulmon et al., 1983; Hooper and Vitousek, 1998). Perennial species also vary in storage strategies, including fibrous roots (perennial grasses), bulbs (geophytes), and tap roots (e.g., *Eschscholzia californica*). This variation suggests substantial potential for resource partitioning that can affect species coexistence (Gonzalez, 2007; Levine and HilleRisLambers, 2009), invasions (Dukes, 2001a) (see also Chapter 11 in this volume), productivity (Dukes, 2001b; Hooper and Dukes, 2004), nutrient retention (Hooper and Vitousek, 1998), and ecosystem responses to global change (Field et al., 1997; Cleland et al., 2006).

Findings from Serpentine Studies

Functional classifications commonly used in biodiversity and ecosystem experiments are plant morphotypes, such as grass, forb, and legume. Though relatively clear-cut, these may not be any more predictive of ecosystem properties than random groupings (Wright et al., 2006), except where they correlate with strong differences in functional traits, such as litter quality (e.g., Hobbie, 1992). Legumes have particularly strong effects on productivity due to their effects on nitrogen availability (Tilman et al., 2001; Spehn et al., 2002, 2005), as in serpentine (Hooper and Dukes, 2004). Phenological groupings have been tested in some other systems but are variable in their degree of complementarity relative to serpentine grasslands. On one hand, phenological complementarity holds some of the strongest gains in production observed in intercropping studies (Steiner, 1982; Vandermeer, 1990), plays a large role in plant–animal interactions (e.g., pollination, Vazquez et al., 2009), and can influence animal effects on ecosystem properties (Byrnes and Stachowicz, 2009). Counterexamples also exist, however: spring ephemerals were thought to have a strong effect on nutrient retention in eastern deciduous forests before trees leaf out, but microbial immobilization appeared to have a stronger effect than plant uptake (Zak et al., 1990; Rothstein, 2000). Similarly, warm-season (C4) and cool-season (C3) grasses in prairie appear less complementary than the warm-season grasses and legumes (Tilman et al., 2001; Fornara and Tilman, 2009). Further investigation of the role of diverse phenologies in complementary resource uptake should help understand conditions under which phenology is a particularly important functional grouping.

Classifications used to assess effects of functional diversity on ecosystem properties may differ from those used to assess species responses to resource gradients. One common axis for functional classification has been the so-called leaf economic spectrum (Wright et al., 2004). This spectrum recognizes a widespread trade-off wherein plants growing in low-resource environments have traits leading to higher resource conservation (low tissue turnover, high C:N ratios, slow growth rates) compared to those living in more resource-rich environments, with subsequent feedbacks on primary production, decomposition, and nutrient cycling (Chapin, 1980; Díaz et al., 1999; Grime, 2001). The leaf economics spectrum reflects common responses to environmental gradients that should result in relatively similar traits across species within a site. However, diversity of these particular traits may not be predictive of opportunities for complementary resource use within a site (Petchey et al., 2004; Ackerly and Cornwell, 2007). Opportunities for complementarity should arise from traits that are overdispersed (more different than expected) among species within a site (Weiher and Keddy, 1998). For example, infertile serpentine soils strongly select for species on the resource conservation end of the leaf economic spectrum, but phenological diversification among those species is a more important component of complementarity (Gulmon et al., 1983; Hooper and Dukes, 2004).

The distinction between responses to environmental gradients and effects on ecosystem properties (i.e., response versus effect traits) has important implications for multivariate metrics of functional diversity. These metrics have the advantage of avoiding arbitrary cutoffs for grouping across continuous axes of variation and of incorporating a wide variety of functional traits (e.g., Petchey, 2002; Botta-Dukát, 2005; Cornwell et al., 2006; Petchey and Gaston, 2006). However, as illustrated in the serpentine example, distinguishing among traits more or less likely to influence a particular ecosystem process needs to guide the weighting of traits included in multivariate metrics. Until this is done, coarse functional groups based on known response and effect traits may prove as useful as complex multivariate metrics in predicting ecosystem properties (Petchey et al., 2004).

BIODIVERSITY AND ECOSYSTEM FUNCTIONING

As concern about the functional consequences of loss of biodiversity rose to prominence in the early 1990s, the first key issue was determining the shape of the response of ecosystem properties to changing levels of species richness and how it differed among ecosystems (Schulze and Mooney, 1993). Soon, however, the debate turned to underlying mechanisms (Givnish, 1994; Huston, 1997; Wardle, 1999; Huston and McBride, 2002). If species are complementary in use of limiting resources, then having more species or more functionally different species should increase both resource retention and ecosystem productivity (Tilman et al., 1997b; Hooper, 1998). However, if one or a few species are extremely competitive, then more diverse communities are likely to contain these effective species, especially in experiments. This could lead to similar patterns of increasing productivity and resource use via the so-called sampling or selection effect (Huston, 1997; Tilman et al., 1997b; Loreau and Hector, 2001). Another possibility is that nitrogen fixers are the prime drivers of increased productivity in biodiversity ecosystem functioning experiments and that other opportunities for complementarity are limited. Substantial effort has since focused on trying to distinguish among these mechanisms (e.g., Chapin et al., 2000; Loreau et al., 2001, 2002; Kinzig et al., 2002; Hooper et al., 2005; Cardinale et al., 2006a, 2007). This section highlights key insights about biodiversity and ecosystem functioning that have emerged from experiments in serpentine grasslands.

Productivity

Recent meta-analyses have found consistent increases in average productivity in response to increased species richness across multiple studies (Cardinale et al., 2007). Whereas plant diversity tends to be a significant positive predictor of productivity, the amount of variation it explains is commonly quite low (typically <10%), with high variability even within levels of richness (Naeem et al., 1995; Tilman, 1997; Hector et al., 1999). Because few studies have replicated species composition within levels of richness, the ability to understand the mechanisms underlying such variability has been limited. In experimental studies in serpentine grasslands, we found that community composition mattered as much as (or more than) plant diversity in predicting ecosystem productivity. With replicated treatments of functional groups, we found up to fourfold differences in above-ground net primary production and nitrogen retention among different functional group compositions, even within levels of functional group richness (Hooper and Vitousek, 1997, 1998; Hooper, 1998). Community composition also explained more variation than diversity.

Early results suggested that productivity saturated at relatively low levels of plant richness (Tilman et al., 1996, 1997a; Hector et al., 1999). Skeptics pointed out that nutrient additions by N-fixing species might be responsible for effects seen in mixtures (Huston, 1997; Wardle, 1999), an effect already well known from intercropping studies (Vandermeer, 1989). Analyses in Minnesota grasslands indicated that interactions between N-fixers and C4 grasses were the primary mechanism leading to enhanced productivity at high diversity (Tilman et al., 2001; Dybzinski et al., 2008). We analyzed the productivity of functional group mixtures in serpentine grassland, using relative yield totals (Harper, 1977), differences from monoculture expectations (Loreau, 1998), and additive partitioning (Loreau and Hector, 2001) to ask which combinations of species within mixtures were complementary (Hooper, 1998; Dukes, 2001b; Hooper and Dukes, 2004). We found that complementarity was not restricted to mixtures with N-fixers; early- and late-season annuals were also strongly complementary, and early-season annuals and perennial bunchgrasses were complementary in at least some years. However, late-season annuals and perennial bunchgrasses tended to have negative competitive effects on one another (Hooper and Dukes, 2004). Other studies have similarly found complementarity among non–N-fixing species (van Ruijven and Berendse,

2003), though most include N-fixers in their treatments, which could bias meta-analyses toward finding consistent effects of diversity on productivity (Cardinale et al., 2006a, 2007). Studies are needed that explicitly separate effects of N-fixers (Spehn et al., 2002, 2005; Fornara and Tilman, 2009) from other mechanisms of complementarity and rigorously evaluate the contribution of particular species (or groups of species) to complementarity in different ecosystems.

Considering the degree of complementarity we observed, and in contrast to some other long-term experiments (Tilman et al., 2001; Cardinale et al., 2007), it was surprising that functional diversity only led to significantly greater total productivity in serpentine grasslands in one of three years assessed (Hooper and Dukes, 2004). The reason appears to be negative selection effects. Early considerations of mechanisms postulated that a positive effect of diversity on productivity could arise simply from the greater probability of including highly productive species in high-diversity treatments (Huston, 1997; Tilman et al., 1997b). However, this assumes that the species that are highly productive in monoculture also competitively dominate in mixtures—a positive selection effect (Loreau and Hector, 2001). This may not be true on nutrient-limited soils where competitive success depends more on effective nutrient acquisition and retention than on rapid growth rates (Chapin, 1980; Wedin and Tilman, 1993; Grime, 2001)—traits that are typically negatively correlated (Wright et al., 2004). In serpentine grasslands, early-season annuals strongly deplete soil nitrogen during the early growing season (Hooper and Vitousek, 1998) but are also the smallest-statured species, investing much of their carbon gain in reproduction rather than growth (Armstrong, 1991). In contrast, perennial bunchgrasses, which often had the highest productivity in monocultures, did relatively poorly in mixtures (Hooper and Dukes, 2004). On a proportional basis, mixtures of early-season annuals and other functional groups were complementary because they produced more than expected based on monoculture productivity. However, in terms of absolute productivity, the fact that the unproductive early annuals were competitively dominant in mixtures partially offset the effects of complementarity and led such mixtures to have intermediate rather than high productivity (Loreau and Hector, 2001; Hooper and Dukes, 2004).

Nutrient Retention

Plant diversity could influence nutrient retention by a variety of mechanisms, the most common hypothesis being complementary resource use, just as with productivity (Tilman, 1999). However, several other pathways could have equal or greater effects. Plant litter quantity and quality influence the capacity for microbial immobilization, which can rival or exceed plant uptake in effects on total N retention (Vitousek and Matson, 1985; Zak et al., 1990; Hobbie, 1992; Currie and Nadelhoffer, 1999; Rothstein, 2000). For nitrogen, the major loss pathways are denitrification and leaching, both of which depend on availability of NO_3^-. High net rates of nitrification can therefore provide suitable substrate for losses (e.g., Scherer-Lorenzen et al., 2003). Leaching losses also depend on water flux through the soil to groundwater. To the extent that plants might influence water flux (e.g., through altered leaf area index and transpiration rates; Field et al., 1997), they could also influence loss rates by water uptake as well as nutrient uptake. The question is how these mechanisms may be influenced by plant diversity.

Most studies relating plant diversity to nutrient uptake and retention have examined the pool sizes of inorganic nitrogen (ammonium and nitrate). Some of these experiments have found evidence for complementarity (e.g., Tilman et al., 1996; Cardinale et al., 2006a), whereas others have found stronger effects of particular species or functional groups (Hooper and Vitousek, 1997, 1998; Hiremath and Ewel, 2001). However, these spot measurements of available soil pools are unlikely to give a complete picture of actual losses, because the timing of plant growth versus rainfall is critical. In climates where most soil water flow occurs outside of the growing season, direct plant uptake may play a relatively small role in ecosystem nitrogen retention compared to microbial immobilization. Relatively few biodiversity experiments have directly measured actual losses of nitrogen (but see Scherer-Lorenzen et al., 2003).

In Californian serpentine grasslands, soil drainage occurs only during the wet season. Biomass of the dominant annual plants is often small early in the growing season, with concomitantly small effects on water and nutrient fluxes (Field et al., 1997). In studies of nitrogen losses using lysimeters, we found that complementarity among plant functional groups had relatively minor effects on nitrogen retention. Bare plots had the highest nitrogen leaching losses, and loss varied among different functional groups (N-fixers > late-season annuals ≥ early-season annuals ≥ perennial bunchgrasses). Except for the step from bare to vegetated plots, however, average losses did not decrease with increasing plant functional diversity (Hooper and Vitousek, 1998). Losses were highest in January in all treatments due to low plant cover early in the growing season in the Mediterranean-like climate. By February, however, losses were very low in all treatments except bare plots and N-fixers alone, illustrating both the increasing plant influence with increasing biomass and of N-fixers on N supply, uptake, or both (Hooper and Vitousek, 1998). Studies in other systems have confirmed the strong effects of N-fixers on N availability (Palmborg et al., 2005; Dybzinski et al., 2008) and leaching losses (Scherer-Lorenzen et al., 2003). Increasing diversity decreased losses relative to N-fixer-only plots because of lower biomass of N-fixers, greater uptake by other species, or both (Scherer-Lorenzen et al., 2003; Dybzinski et al., 2008).

Microbial Immobilization

Although microbial immobilization is an important component of ecosystem nutrient retention, potential mechanisms by which plant diversity might influence immobilization are unclear. On one hand, plant litter quality is well known to affect immobilization, with litter of low nitrogen content, high C:N, or high lignin:N leading to greater microbial immobilization (Chapin et al., 2002). On the other hand, effects of plant litter *diversity* on decomposition and immobilization have been mixed. Studies assessing mixtures of litter have typically found idiosyncratic results, in which net effects of mixtures on decomposition and immobilization rates have been positive, negative, or neutral, depending on the litter quality of the component species (Wardle et al., 1997; Hector et al., 2000; Srivastava et al., 2009). The outcome of plant competition with microbes for nutrients could also influence diversity effects on immobilization. If plants compete effectively with microbes for inorganic nitrogen, greater plant diversity could lead to lower immobilization if more diverse plant communities have greater total nitrogen uptake (Hooper, 1996). On the other hand, if plants get the leftover nutrients after microbes have gotten what they need (e.g., Hart et al., 1993; Harte and Kinzig, 1993), microbial immobilization may be unaffected by plant diversity (e.g., Fornara et al., 2009). Finally, if plants and microbes interact positively in nutrient retention via plant root exudation and microbial turnover, then greater plant diversity could also lead to greater immobilization.

Studies in serpentine grassland with ^{15}N tracer suggested no net effect of plant diversity on nitrogen retention except in going from bare to vegetated plots (Figure 16.1). Microbial immobilization had as great an effect on nitrogen retention, as did plant uptake, in this experiment (Hooper and Vitousek, 1997, 1998). Lower retention in bare plots compared to vegetated plots (vegetation effect, $p < 0.001$ in both years) likely resulted from lack of plant uptake but also lack of plant inputs of organic matter for fueling microbial population growth and immobilization of inorganic nitrogen (Hooper, 1996) (Figure 16.2). In contrast, all vegetated treatments had similar overall patterns of tracer recovery and turnover (Figure 16.2). Recovery of tracer in soil organic matter (SOM) was about twice that in live plants, and SOM accumulated more ^{15}N in the second year of the experiment. Such net transfer of retained nitrogen to soil pools is common in other tracer studies as well (Currie et al., 1999; Barrett and Burke, 2002).

Because sampling in our experiment occurred three months after tracer addition, high recovery of tracer in soil could have resulted from three potential mechanisms, alone or in combination: (1) direct microbial uptake of tracer from the inorganic N pool, then transfer to the SOM pool with microbial turnover; (2) plant uptake followed by tissue turnover that became incorporated into SOM either directly or following microbial processing; or (3) direct abiotic fixation on minerals and organic matter. We did not measure abiotic immobilization, and although it is considered more important for ammonium than nitrate (the form of our added tracer), we cannot rule it out (Barrett and Burke, 2002; Davidson et al., 2003). However, we found evidence for both mechanisms 1 and 2. Immediate microbial

FIGURE 16.1. Effects of plant functional diversity on ecosystem nitrogen retention. Tracer amounts of ^{15}N were added to serpentine grassland communities of different functional composition and diversity, and the fate of that nitrogen was assessed after 3 months and 15 months in separate cores. Cores (30 cm deep × 30 cm diameter) were inserted into exper-imental plots composed of different combinations of early-annual (E), late-annual (L), and perennial bunchgrass (P) functional groups. Tracer amounts of ^{15}N were added as $K^{15}NO_3$ (66% enriched) to the soil surface near the start of the growing season in January 1993. Cores were harvested after 3 months (April 1993) and 15 months (April 1994) to estimate N retention both within and across growing seasons. Core extraction apparatus provided courtesy of Louise Jackson (University of California, Davis). Photo by Lydia Chu.

immobilization (mechanism 1) was supported by rates of gross microbial immo-bilization (measured separately; Hooper and Vitousek, 1998) that corresponded with SOM tracer recovery and the fact that tracer was added in January, when plant biomass was relatively low (Hooper and Vitousek, 1998). Substantial ^{15}N in litter supports the pathway including plant uptake and transfer to SOM (mechanism 2)

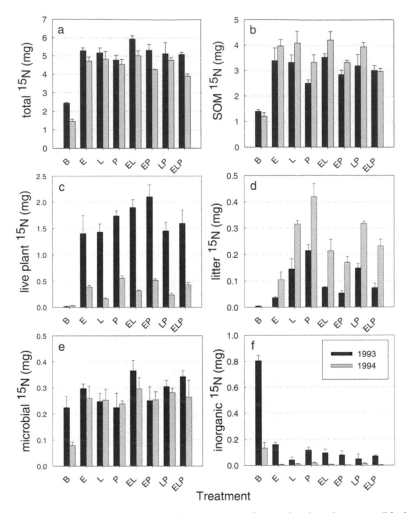

FIGURE 16.2. Distribution of recovered ^{15}N tracer in plant and soil pools in 1993 (black bars) and 1994 (gray bars) to 30 cm depth. Total ^{15}N (a) is the sum of (b) soil organic matter (SOM), (c) live plant (roots + shoots), (d) litter, (e) microbial biomass (chloroform labile), and (f) inorganic (ammonium + nitrate) ^{15}N pools. Bars are means ± 1 SE, $n = 3$. Treatments are a three-way factorial combination of functional groups: E = early-season annuals, L = late-season annuals, and P = perennial bunchgrasses, and combinations thereof; B = bare plots. Data were analyzed using a three-way factorial analysis of variance (ANOVA) with presence/absence of each functional group (E, L, or P) as treatment variables using the general linear model procedure of SYSTAT (SYSTAT, 1992). We evaluated treatment main effects and interactions as well as a suite of a priori comparisons, including vegetation effects (bare plots versus mean of all vegetated treatments); differences among single functional group treatments (E versus L versus P); and averaging effects (differences between mixtures and the mean of the component single functional groups,

in the first year. Also, tracer in SOM pools increased from the first to the second year of the experiment, at least partially from losses from the plant pool (Figure 16.2) during that time period. Shorter term (24–48-hour) experiments would be necessary to more fully resolve these mechanisms. However, in both mechanisms 1 and 2, microbial immobilization likely plays a pivotal role via both competition and synergism with plants.

Among vegetated treatments, differences in plant composition and diversity had little effect on total tracer retention. This was very surprising, given the large potential for complementary interactions (Gulmon et al., 1983; Hooper and Dukes, 2004). Of the differences that did arise, plant composition appeared to have greater effects than diversity (Figure 16.2). In the first year, differences among treatments in tracer recovered in soils (perennial bunchgrasses [P] < early annuals [E], $p = 0.046$; Figure 16.2) or plants (late annual $\times$ perennial bunchgrass interaction, $p = 0.041$; Hooper, 1996) roughly balanced out, leading to small overall differences in total recovery. By the second year, significant differences in live plant recovery developed among the single functional group treatments (single functional group tests: $P \geq E$, $p = 0.007$, $E > L$, $p < 0.003$), but again these differences were counterbalanced by litter and soil recovery, leading to no significant differences in total recovery (Hooper, 1996). The only difference that developed among vegetated treatments was a trend toward lower SOM and total recovery in treatments containing perennial bunchgrasses (alone and in mixtures) compared to those with annuals only (perennial bunchgrass partial main effect, $p = 0.037$). This was the opposite of what we expected. Because perennial bunchgrasses were the only functional group with live plant biomass at the start of the fall rains in this experiment, we had hypothesized that they would help retain nitrogen from leaching while the annuals were just germinating (analogous to the hypothesized "vernal dam" by spring ephemerals in temperate forest systems; Zak et al., 1990). However, similar to tests of the vernal dam hypothesis (Zak et al., 1990; Rothstein, 2000), microbial immobilization appeared to play a larger role than plant uptake, potentially driven by litter inputs from the previous growing season (e.g., Scherer-Lorenzen et al., 2003).

---

for example, EP versus average[E, P]). For vegetation variables (live plant ^{15}N and litter ^{15}N), where zero values for the bare plots are meaningless, we analyzed the data using a partial means model ANOVA without the bare plots (SYSTAT, 1992; Kirby, 1993). Where previous experiments indicated that soil variables might differ between treatments with and without perennial bunchgrasses, the appropriate contrast for a partial P main effect was: $3^*P + 3^*EP + 3^*LP + 3^*ELP - 4^*E - 4^*L - 4^*EL = 0$. Effects were considered statistically significant if p-values fell below family-wide confidence levels using Kimball's inequality (also called the Dunn-Šidak correction; Neter et al., 1990). Nonsignificant trends were those with p-values between 0.05 and the corrected p-value. Statistical details are in Hooper (1996) and Hooper and Vitousek (1998).

Community Stability

Effects of plant diversity on stability of ecosystem properties could be as important as effects on process rates in any given year (Tilman, 1999). Compensatory growth among competing species may lead to lower temporal variability in production in more diverse communities, because as conditions change and some species decline, others rebound (Cottingham et al., 2001). Uncorrelated species responses to environmental fluctuation could also lead to higher stability via the portfolio effect, where, in an analogy to stock portfolios, random fluctuations in many individual species decreases overall fluctuations at the community level (Yachi and Loreau, 1999). Assessment of such effects and their mechanisms requires long-term data that control for several potentially confounding variables, including underlying environmental gradients, differences in species composition, and correlation of compositional similarity with species richness in experimental manipulations (Hooper et al., 2005).

Patterns across many different types of experiments are consistent with expectations for diversity increasing stability, but teasing apart mechanisms is more challenging (Hooper et al., 2005). In serpentine grasslands at Jasper Ridge Biological Preserve, long-term data indicate lower variation in community productivity than in abundances of individual species, as would be expected if asynchronous species-level fluctuations help stabilize overall community production (McNaughton, 1977; Hobbs et al., 2007). Though no formal statistical tests were done, at least some species (e.g., *Microseris douglasii*) appeared to peak when others were crashing, potentially in response to long-term drought. Whether such a response is sufficient to account for buffering of ecosystem properties is unclear, however, because at least one study suggests most annual species respond similarly to interannual variation in temperature and precipitation in California grasslands (Elmendorf and Harrison, 2009). Other long-term grassland experiments have noted a decrease in the temporal coefficient of variation in production in plots with higher species richness (Tilman, 1996) and some negative correlations among dominant and subdominant species in response to climate variation (Bai et al., 2004). Further analysis of these patterns could help shed more light on underlying mechanisms of stability. Given the large annual fluctuations in climate, the serpentine grassland seems like a very promising system for further tests of the relationship between diversity and stability.

Functional Consequences of Diversity:
Lessons from Studies on Serpentine

Serpentine grasslands have provided an excellent experimental system for understanding the effects of plant diversity on primary production and nutrient retention. In terms of production, they have helped elucidate functional complementarity and the importance of negative selection effects (dominance in mixtures by lower productivity species) (e.g., Hooper and Dukes, 2004). Meta-analyses suggest that although positive effects of complementarity on productivity are common, increased diversity only infrequently has led to productivity greater than the most productive monoculture (transgressive overyielding; Cardinale et al., 2006b, 2007). Indeed, the strength and direction of selection effects varied widely across experiments. On average, selection effects were positive, but 43% of selection effects were negative, suggesting that dominance by less productive species is a very common phenomenon across a variety of systems (Cardinale et al., 2007). Increasing productivity is therefore not a necessary consequence of increasing diversity, even where complementarity occurs. Whether this pattern is typical of infertile systems, such as serpentine, remains to be determined. One important note, however, is that most studies only assess above-ground net primary productivity (NPP). Theoretically, total NPP and gross primary production should increase with complementary resource use. Although some studies have also assessed root biomass (e.g., Hooper, 1998), only one has assessed total allocation to below-ground production, including root growth, turnover, and exudation (Adair et al., 2009). Such a study would be a challenge in rocky serpentine soils.

For nitrogen retention, differences in litter quality and quantity among plant functional groups likely outweighed any effects of complementary plant uptake. High rates of fine root turnover in annual-dominated plots, compared to the coarser roots of perennial bunchgrasses, may have provided abundant labile carbon to fuel high rates of microbial immobilization in the fall. This is supported by lower inorganic nitrogen pool sizes in early annual versus perennial bunchgrass plots after fall wetup but before leaching (Hooper and Vitousek, 1998). Plant uptake could not have been responsible because the annuals had not yet germinated. Such a mechanism is opposite to that postulated for Minnesota grasslands, where coarse roots of C3 and C4 grasses contributed more to accumulation of soil carbon, but fine roots of legumes and forbs contributed to enhanced nitrogen mineralization (Fornara et al., 2009). Species richness and productivity in that system appeared to influence mineralization more than microbial immobilization (Zak et al., 2003). Clearly, multiple mechanisms operating simultaneously complicate simple predictions of nutrient retention based only on plant diversity and complementarity. Understanding interactions among species composition, diversity, litter (root and shoot) quantity and quality, mineralization, and microbial immobilization will be critical for a full assessment of effects of changing plant composition and diversity on nitrogen retention.

RESPONSES TO ELEVATED CO_2

The potential for ecosystems to sequester anthropogenic CO_2 will influence the extent to which they will act as positive or negative feedbacks to climate warming.

Most studies at all scales (growth chamber, greenhouse, field chamber, and Free Air CO_2 Exchange) have found at least temporary increases in plant photosynthesis resulting from experimental enhancement of CO_2 concentrations (Körner, 2000). The key questions involve the timing and fate of that enhanced carbon uptake. First, some species show acclimation to enhanced CO_2 such that photosynthetic rates eventually decrease, whereas others continue to sequester additional CO_2 throughout the length of the experiment (Vitousek, 1994). Even with acclimation, however, there is the opportunity for enhanced production if water is limiting and acclimation results from stomatal closure. To the extent that lower plant water use in wetter periods translates to higher soil moisture in drier periods, plant growth can be extended (Field et al., 1997). Second, if additional fixed carbon is stored in plant biomass or soil carbon pools with long turnover times, higher net ecosystem production could act as a negative feedback to elevated CO_2 by sequestering more anthropogenically released carbon. On the other hand, if additional production is rapidly decomposed and released back into the atmosphere, little additional storage will result. Finally, if elevated temperatures increase decomposition of soil organic matter more than they and elevated CO_2 increase plant photosynthesis, ecosystems could act as a positive feedback to atmospheric and climate change (Denman et al., 2007).

The Jasper Ridge CO_2 experiment (1992–1997) investigated the potential for grasslands to sequester anthropogenic CO_2. Based at Stanford University's Jasper Ridge Biological Preserve in Palo Alto, California, this experiment used grasslands on adjacent serpentine and sandstone-derived soils to evaluate a variety of potential mechanisms influencing net ecosystem CO_2 flux (Figures 16.3 and 16.4). The close proximity of both grassland types at the same site allowed the experiment to keep climate factors consistent across these natural soil fertility treatments (Figure 16.3) (Field et al., 1996). In addition to greenhouse, growth chamber, and modeling components to the project, there were two primary outdoor components: (1) a field experiment using open-topped chambers in a randomized block design with two plant communities (serpentine and sandstone grassland) and three CO_2 treatments: no chamber/ambient CO_2, chamber/ambient CO_2, and chamber/elevated CO_2 (approximately doubling CO_2 concentrations from 1992 levels to 70 Pa, or ~690 ppm); and (2) microcosms in PVC tubes (20 cm diameter $\times$ 95 cm deep) in a three-way factorial design of soil type/grassland community (serpentine versus sandstone) crossed with CO_2 concentration (ambient versus elevated) crossed with nutrient availability (unfertilized versus +NPK) (Fredeen et al., 1998). These microcosms were referred to as MECCAs (microecosystems for climate change analysis) and contained plant communities composed of both mixtures and monocultures of representative species of from each soil type. This combination of experiments allowed evaluation of responses to elevated CO_2 in natural and seminatural conditions. Natural conditions allowed assessment of the integrated

FIGURE 16.3. Effects of elevated CO_2 on whole ecosystem carbon balance in the Jasper Ridge CO_2 experiment (Field et al., 1996). Control plots (no chamber), control chambers (no added CO_2), and elevated CO_2 chambers were placed over grassland communities on serpentine- (background) and sandstone- (foreground) derived soils. Rapid transition from one soil type to the other allowed control of other environmental factors while testing effects of community and soil types on whole ecosystem responses. Photo by D. Hooper.

ecosystem response to elevated CO_2, which could be buffered or enhanced by a variety of processes operating above the scale of individual plants assessed in greenhouses, including shifts in plant density, species composition, litter feedbacks, and below-ground respiration (both plant and soil) (Fredeen et al., 1995). At the same time, the microcosms permitted direct manipulations (e.g., species composition) and measurements (e.g., soil water flux) that would have been much more difficult in the natural grasslands (Field et al., 1996).

Effects on Plant Production

Overall, elevated CO_2 increased C flux into both field and MECCA experimental systems (Figure 16.4). *Avena barbata*, the dominant grass in sandstone grassland, had increased leaf level photosynthesis (~70%) at peak physiology in spring 1993 (Jackson et al., 1994; Fredeen et al., 1995). Over the longer term, it appeared that CO_2 uptake was down-regulated by plants under elevated CO_2 by a decrease in rubisco levels in leaves (Fredeen et al., 1995). Increases in whole-ecosystem net CO_2 exchange (NCE, plant photosynthesis minus total ecosystem respiration) in

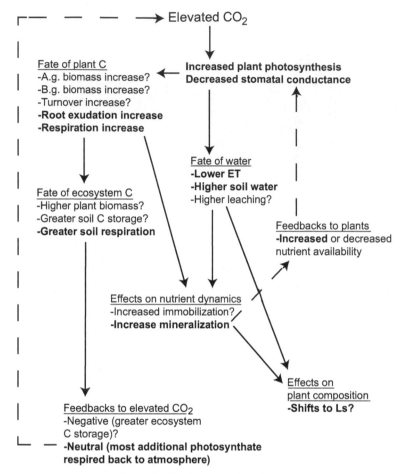

FIGURE 16.4. Summary of main feedbacks assessed by the Jasper Ridge CO_2 experiment. Overall, most additional CO_2 uptake was respired back to the atmosphere by fast-turnover pathways; relatively little remained stored in plant and soil pools. Bold type indicates the primary responses, with other potential responses also shown. The "Shift to Ls" under Effects on plant composition is uncertain. g_s = stomatal conductance; C = carbon; A.g. = above ground; B.g. = below ground; ET = evapotranspiration; L = late season annuals; ps = photosynthesis. Bold arrows show primary and secondary responses to elevated CO_2; dashed arrows indicate feedbacks to those responses.

the sandstone grassland were more modest (10–30%) than increases in photosynthesis. In serpentine grasslands, increases in NCE ranged from 25% to 117% depending on year and time of year. Rates of NCE measured by chambers in this experiment (2–5 mmol*m^{-2}*sec^{-1}) were similar to those measured by eddy covariance in the natural serpentine grasslands (Valentini et al., 1995).

Increases in leaf-level photosynthesis clearly did not translate into equivalent increases in carbon uptake at the ecosystem level. Several mechanisms could be responsible. Per amount of photosynthesis, elevated CO_2 treatments had lower plant biomass in the sandstone grassland, suggesting that a smaller percentage of photosynthetically fixed carbon was incorporated into plant growth under elevated CO_2 than in ambient conditions. Indeed, total above-ground biomass, though up slightly under elevated CO_2 in serpentine grasslands, did not increase significantly in 1992 or 1993. Similar responses were seen in MECCAs (Fredeen et al., 1998). NCE increased under elevated CO_2 and fertilization in the later part of the growing season when plant biomass was higher. Single species mixtures had higher NCE under elevated CO_2 only when fertilized. This resulted from a combination of higher plant biomass with fertilization and, presumably, greater initial rates of plant photosynthesis to achieve that higher biomass in the first place. In both single- and mixed-species communities, however, acclimation occurred, such that plants under elevated CO_2 had lower foliar rubisco activity and hence lower NCE than under ambient CO_2 when measured under equivalent conditions (e.g., ambient-grown and elevated-grown communities measured at both ambient and elevated CO_2). Overall, in both natural and microcosm communities, elevated CO_2 stimulated total CO_2 uptake, but not to the extent that would be expected in a short-term experiment, because both of plant acclimation and shifts in allocation that apparently resulted in a lower efficiency of conversion of plant photosynthesis to above-ground biomass. Although fertilization in the microcosms did stimulate NCE, the differences in response by serpentine and sandstone grassland were not as great as expected based on the natural differences in fertility between these soils.

Below-Ground Respiration and Total Carbon Storage

Below-ground respiration in both serpentine and sandstone grasslands increased in response to elevated CO_2, and this response had large effects on the potential for these systems to sequester atmospheric CO_2 in the long term. Yearly total below-ground respiration in sandstone grasslands was about a third higher than in serpentine (485 versus 346 g $C^*m^{-2*}yr^{-1}$) (Luo et al., 1996). Overall, below-ground CO_2 efflux correlated with plant biomass. However, plants on serpentine soils allocated a much larger percentage of their production below ground: the ratio of below-ground respiration to above-ground production was 3.8–7.7 in serpentine compared to ~3 in sandstone. This pattern is compatible with serpentine plants needing to expend more energy to cope with the more stressful serpentine soil environment. Below-ground respiration increased in response to elevated CO_2 to similar degrees proportionally on serpentine and sandstone in one year (1993, a wet year), but to a much greater degree in serpentine (139%) than in sandstone (~33%) the following relatively dry year (Luo et al., 1996). Soil respiration in elevated CO_2 treatments was higher than control treatments even in the fall (before much plant production), likely because of increased growth and litter inputs from the previous growing season. Pools of rapidly recycling carbon (plants, detritus, and soil microbes) increased 25% and 37% in the serpentine and sandstone, respectively. However, total soil carbon didn't change detectably (Hungate et al., 1996), so that proportional gains in total ecosystem carbon were much lower (8–9%) (Hungate et al., 1997b).

The increased below-ground respiration resulted from additional root respiration and exudation. This was determined by a combination of techniques in both microcosms and field chambers, including mass balance, minirhizotrons and root in-growth cores to determine root turnover, pulse labeling with 99% ^{13}C in microcosms, and the ^{13}C signature of fossil fuel–derived CO_2 used in the experiment (Hungate et al., 1997b). Increased below-ground allocation and turnover likely caused the discrepancy between responses of carbon uptake to elevated CO_2 at the leaf level and those at the whole ecosystem level, such that most of the increased photosynthesis seen under elevated CO_2 was released back into the atmosphere by allocation to these fast-turnover pools (Figure 16.4) (Hungate et al., 1997b).

Soil Water Balance

Stomatal closure under elevated CO_2 allows plants to save water while still getting the same amount of CO_2. The extent to which the resulting decreases in plant transpiration affect overall ecosystem water balance then depends on the relative importance of that flux pathway compared to other ecosystem water fluxes, including evaporation from the soil surface, storage in soil, leaching losses to groundwater, and overland flow. In the Jasper Ridge elevated CO_2 experiment, reduced stomatal conductance and plant transpiration led to savings in soil water during the middle of the growing season, which then translated into greater late season soil moisture (Field et al., 1996, 1997; Fredeen et al., 1997). Although both serpentine and sandstone grasslands had similar qualitative responses, the effect was much greater in sandstone than in serpentine. This was attributed to greater plant biomass and leaf area in sandstone, which meant that plant transpiration was a much greater percentage of the ecosystem water budget on sandstone than serpentine soils (Field et al., 1997). On serpentine, limited leaf area meant that evaporation dominated the ground-to-atmosphere water flux, so any decrease in transpiration due to plant stomatal closure had a much smaller effect on total water balance (Jackson et al., 1998). Similarly, the amount and timing of rainfall also influenced the soil moisture effect: years with less and more evenly distributed rainfall saw greater extensions of growing season water availability under elevated CO_2 because transpiration exerted a proportionally greater influence on total water flux. In contrast, large rainfall events overwhelmed plant uptake capacity and resulted in greater amounts of leaching and overland flow (Jackson et al., 1998).

The increased late-season water availability had several follow-on effects. First, nitrogen mineralization rates increased, which led to greater plant N uptake (Hungate et al., 1997a). Second, the increased nutrient availability, potentially combined with greater water availability, led to higher plant production in the late season (230–730% increases) (Field et al., 1996; Hungate et al., 1997a). In the sandstone grassland, the absolute production by late-season annuals was still a relatively small proportion of total plant production. In the serpentine grassland, however, the increases in late-season production represented major shifts in community composition (Field et al., 1996). Given the complementary nature of early- and late-season functional types, it is not clear that the increase in late-season annuals resulted in a concomitant decrease in early season annuals.

Ecosystem Storage of CO_2: Lessons from Studies on Serpentine

Serpentine grasslands have played a central role in at least three general lessons about ecosystem storage of CO_2. First, increasing flux to high-turnover pools of carbon (e.g., root respiration and root exudates; Field et al., 1996) can return extra fixed carbon from higher photosynthesis under elevated CO_2 to the atmosphere very quickly. This could lead to little net carbon storage in terrestrial ecosystems if the increased allocation to these pathways is a common response in other ecosystems as well. Net ecosystem production, not plant photosynthesis or net primary production, is the bottom line for ecosystem carbon sequestration of anthropogenic CO_2 (Chapin et al., 2002). Second, nutrient limitation can restrict the potential for ecosystems to sequester additional CO_2 from the atmosphere (Reich et al., 2006). Ecosystem carbon sequestration will also be limited to the extent that plants need to invest additional carbon in root respiration and exudation to acquire resources or that lack of those resources restricts the amount of new tissue plants can produce. Lack of accounting for such nutrient limitation led to overestimates in many global models of the ability for ecosystems to sequester anthropogenic CO_2 (Hungate et al., 2003; Luo et al., 2004). These effects, as well as the potential for nitrogen deposition to partially alleviate that limitation in some regions, are now widely recognized and complicate predictions of sequestration by terrestrial ecosystems (Denman et al., 2007). Third, decreased stomatal conductance in response to elevated CO_2 can influence soil water flux, with several results, including increased nutrient availability (Hungate et al., 1997a; Luo et al., 2006), increased plant production (Field et al., 1996), and potential shifts in species composition (Joel et al., 2001). Shifts in seasonal water availability are expected in any systems in which soil water savings through stomatal closure are not completely offset by increases in leaf area index. This could occur if plants are co-limited by other resources (e.g., nutrients) or are adapted to conservative resource use by low relative growth rates (Chapin, 1980; Field et al., 1997; Fredeen et al., 1997; Grime, 2001).

Comparing responses of serpentine and sandstone grasslands showed that plant canopy characteristics can have a strong influence on the strength of the foregoing responses. In serpentine, the percentage response of both net CO_2 exchange (Field et al., 1996) and below-ground respiration (Luo et al., 1996; Hungate et al., 1997b) were similar to or greater than those in sandstone grassland. However, serpentine grassland, with lower plant biomass and lower leaf area index (LAI), had smaller absolute effects on late-season soil moisture because transpiration was such a small component of the ecosystem water budget (Field et al., 1996, 1997). These findings help provide a basis for evaluating the potential for plant acclimation and resource use patterns to influence ecosystem carbon, water, and nutrient flux. Not surprisingly, low biomass systems such as serpentine are much less likely to show strong shifts in plant-mediated resource fluxes. These systems may therefore not serve as the best models for extrapolating mechanisms to the high biomass systems more likely to influence carbon fluxes at the global scale (Field et al., 1996). Subsequent studies of factorial effects of global change (nutrients, CO_2, and temperature) at Jasper Ridge Biological Preserve have focused primarily on the sandstone grassland as a model (Zavaleta et al., 2003; Dukes et al., 2005).

CONCLUSIONS: GENERALITY OF SERPENTINE STUDIES AND QUESTIONS FOR THE FUTURE

Studies of serpentine grassland have shed light on several issues of general importance to the study of ecosystem ecology.

Utility of functional classifications. It is unlikely that the particular functional groups of serpentine grasslands will be general across systems. However, the more important point is that studies assessing effects of functional diversity need to focus more strongly on traits such as phenology, canopy structure, and rooting depth that provide clear opportunities for resource partitioning. These may or may not follow familiar delineations such as grasses versus forbs or trade-offs in specific leaf area (SLA) and leaf nitrogen. Studies using quantitative characteristics may find different patterns when functional traits are weighted accordingly (e.g., Petchey et al., 2004).

Effects of negative selection and facilitation on diversity–productivity relationships. In systems strongly limited by resources, we might expect that species with resource-conserving strategies will dominate in long-term competition in mixtures, as in serpentine. In such systems, we would expect negative selection effects, leading to less gain in production as diversity increases than otherwise expected from complementarity alone. Strong facilitation, as with N-fixers, might be the only way to get transgressive overyielding (greater production than the most productive monocultures) in such systems (Hooper and Dukes, 2004). Future studies on biodiversity-ecosystem functioning in serpentine grassland could usefully address at least two further issues: integrating multiple trophic levels (e.g., Raffaelli et al., 2002; Schmitz, 2006) and assessing consequences of realistic patterns of species loss for ecosystem properties (e.g., Zavaleta and Hulvey, 2004).

Importance of microbial immobilization for nitrogen retention. The effects of plant diversity on nitrogen retention need more study. First, more studies need to investigate actual nitrogen losses, not just available pool sizes, for a clearer understanding of the effects of changes in plant composition and diversity on nitrogen retention (e.g., Scherer-Lorenzen et al., 2003). Second, although nitrogen sequestration in soil pools appears to be common in serpentine, as in other systems (including forests; Currie et al., 1999), the extent to which this is driven by direct immobilization versus uptake by and turnover of plant tissue remains unresolved.

Responses of ecosystems to elevated CO_2. Comparisons between serpentine systems and more fertile adjacent grasslands have been important in recognizing the relative role of plant-mediated and abiotic fluxes in controlling whole-system responses to elevated CO_2. However, the relative effects of these mechanisms in low-productivity systems, such as serpentine, differ from the relative effects in more productive grasslands and forests. Therefore, serpentine grasslands are less useful than more productive systems for projecting the global response of carbon storage to elevated CO_2.

With respect to all these issues, the mechanistic understanding allowed by serpentine grassland model systems has helped point the way for better understanding of the diversity of ways ecosystem-scale processes may respond to global environmental changes.

Acknowledgments

This manuscript benefited greatly from comments by Susan Harrison, Bruce Hungate, and two anonymous reviewers. Financial support for field and lab work by D. Hooper was provided by NSF grants DEB-9212995 and DEB-9974159. Site access and logistical support were provided by WMI and the staff at the Kirby Canyon Sanitary Landfill. Numerous undergraduate and graduate students assisted with field work, lab work, and data analysis.

LITERATURE CITED

Ackerly, D. D., and Cornwell, W. K. (2007) A trait-based approach to community assembly: Partitioning of species trait values into within- and among-community components. *Ecology Letters*, 10, 135–45.

Adair, E. C., Reich, P. B., Hobbie, S. E., and Knops, J. M. H. (2009) Interactive effects of time, CO_2, N, and diversity on total belowground carbon allocation and ecosystem carbon storage in a grassland community. *Ecosystems,* 12, 1037–52.

Armstrong, J. (1991) *Rainfall Variation, Life Form and Phenology in California Serpentine Grassland.* PhD diss., Stanford University, Stanford, CA.

Bai, Y. F., Han, X. G., Wu, J. G., Chen, Z. Z., and Li, L. H. (2004) Ecosystem stability and compensatory effects in the inner Mongolia grassland. *Nature,* 431, 181–84.

Barrett, J. E., and Burke, I. C. (2002) Nitrogen retention in semiarid ecosystems across a soil organic-matter gradient. *Ecological Applications,* 12, 878–90.

Botta-Dukát, Z. (2005) Rao's quadratic entropy as a measure of functional diversity based on multiple traits. *Journal of Vegetation Science,* 16, 533–40.

Byrnes, J., and Stachowicz, J. J. (2009) Short and long term consequences of increases in exotic species richness on water filtration by marine invertebrates. *Ecology Letters,* 12, 830–41.

Cardinale, B. J., Srivastava, D. S., Duffy, J. E., Wright, J. P., Downing, A. L., Sankaran, M., and Jouseau, C. (2006a) Effects of biodiversity on the functioning of trophic groups and ecosystems. *Nature,* 443, 989–92.

Cardinale, B. J., Weis, J. J., Forbes, A. E., Tilmon, K. J., and Ives, A. R. (2006b) Biodiversity as both a cause and consequence of resource availability: A study of reciprocal causality in a predator-prey system. *Journal of Animal Ecology,* 75, 497–505.

Cardinale, B. J., Wright, J. P., Cadotte, M. W., Carroll, I. T., Hector, A., Srivastava, D. S., Loreau, M., and Weis, J. J. (2007) Impacts of plant diversity on biomass production increase through time because of species complementarity. *Proceedings of the National Academy of Science, USA,* 104, 18125–28.

Chapin, F. S. III (1980) The mineral nutrition of wild plants. *Annual Review of Ecology and Systematics,* 11, 233–60.

Chapin, F. S. III, Matson, P. A., and Mooney, H. (2002) *Principles of Terrestrial Ecosystem Ecology.* Springer-Verlag, New York.

Chapin, F. S. III, Zavaleta, E. S., Eviner, V. T., Naylor, R. L., Vitousek, P. M., Reynolds, H. L., Hooper, D. U., Lavorel, S., Sala, O. E., Hobbie, S. E., Mack, M. C., and Diaz, S. (2000) Consequences of changing biodiversity. *Nature,* 405, 234–42.

Cleland, E. E., Chiariello, N. R., Loarie, S. R., Mooney, H. A., and Field, C. B. (2006) Diverse responses of phenology to global changes in a grassland ecosystem. *Proceedings of the National Academy of Sciences, USA,* 103, 13740–44.

Cornwell, W. K., Schwilk, D. W., and Ackerly, D. D. (2006) A trait-based test for habitat filtering: Convex hull volume. *Ecology,* 87, 1465–71.

Cottingham, K. C., Brown, B. L., and Lennon, J. T. (2001) Biodiversity may regulate the temporal variability of ecological systems. *Ecology Letters,* 4, 72–85.

Currie, W. S., and Nadelhoffer, K. J. (1999) Dynamic redistribution of isotopically labeled cohorts of nitrogen inputs in two temperate forests. *Ecosystems,* 2, 4–18.

Currie, W. S., Nadelhoffer, K. J., and Aber, J. D. (1999) Soil detrital processes controlling the movement of ^{15}N tracers to forest vegetation. *Ecological Applications,* 9, 87–102.

Daily, G. (1997) *Nature's Services. Societal Dependence on Natural Ecosystems.* Island Press, Washington, DC.

Davidson, E. A., Chorover, J., and Dail, D. B. (2003) A mechanism of abiotic immobilization of nitrate in forest ecosystems: The ferrous wheel hypothesis. *Global Change Biology*, 9, 228–36.

Denman, K. L., Brasseur, G., Chidthaisong, A. , Ciais, P., Cox, P. M., Dickinson, R. E., Hauglustaine, D., Heinze, C., Holland, E., Jacob, D., Lohmann, U., Ramachandran, S., da Silva Dias, P. L., Wofsy, S. C., and Zhang, X. (2007) Couplings between changes in the climate system and biogeochemistry. In *Climate Change 2007: The Physical Science Basis. Contribution of Working Group I to the Fourth Assessment Report of the Intergovernmental Panel on Climate Change* (eds. S. Solomon, D. Qin, M. Manning, Z. Chen, M. Marquis, K. B. Averyt, M. Tignor, H. L. Miller), pp. 499–588. Cambridge University Press, Cambridge.

Díaz, S., Cabido, M., and Casanoves, F. (1999) Functional implications of trait-environment linkages in plant communities. In *Ecological Assembly Rules: Perspectives, Advances, Retreats* (eds. E. Weiher and P. Keddy), pp. 338–62. Cambridge University Press, Cambridge.

Dukes, J. S. (2001a) Biodiversity and invasibility in grassland microcosms. *Oecologia*, 126, 563–68.

Dukes, J. S. (2001b) Productivity and complementarity in grassland microcosms of varying diversity. *Oikos*, 94, 468–80.

Dukes, J. S., Chiariello, N. R., Cleland, E. E., Moore, L. A., Shaw, M. R., Thayer, S., Tobeck, T., Mooney, H. A., and Field, C. B. (2005) Responses of grassland production to single and multiple global environmental changes. *PLOS Biology*, 3, 1829–37.

Dybzinski, R., Fargione, J. E., Zak, D. R., Fornara, D., and Tilman, D. (2008) Soil fertility increases with plant species diversity in a long-term biodiversity experiment. *Oecologia*, 158, 85–93.

Elmendorf, S. C. and Harrison, S. P. (2009) Temporal variability and nestedness in California grassland species composition. *Ecology*, 90, 1492–97.

Field, C. B., Chapin, F. S. III, Chiariello, N. R., Holland, E. A., and Mooney, H. A. (1996) The Jasper Ridge CO_2 experiment: Design and motivation. In *Carbon Dioxide and Terrestrial Ecosystems* (eds. G. W. Koch and H. A. Mooney), pp. 121–45. Academic Press, San Diego, CA.

Field, C. B., Lund, C. P., Chiariello, N. R., and Mortimer, B. E. (1997) CO_2 effects on the water budget of grassland microcosm communities. *Global Change Biology*, 3, 197–206.

Fornara, D. A., and Tilman, D. (2009) Ecological mechanisms associated with the positive diversity-productivity relationship in an N-limited grassland. *Ecology*, 90, 408–18.

Fornara, D. A., Tilman, D., and Hobbie, S. E. (2009) Linkages between plant functional composition, fine root processes and potential soil N mineralization rates. *Journal of Ecology*, 97, 48–56.

Fredeen, A. L., Koch, G. W., and Field, C. B. (1995) Effects of atmospheric CO_2 enrichment on ecosystem CO_2 exchange in a nutrient and water limited grassland. *Journal of Biogeography*, 22, 215–19.

Fredeen, A. L., Koch, G. W., and Field, C. B. (1998) Influence of fertilization and atmospheric CO_2 enrichment on ecosystem CO_2 and H_2O exchanges in single- and multiple-species grassland microcosms. *Environmental and Experimental Botany*, 40, 147–57.

Fredeen, A. L., Randerson, J. T., Holbrook, N. M., and Field, C. B. (1997) Elevated atmospheric CO_2 increases water availability in a water-limited grassland ecosystem. *Journal of the American Water Resources Association*, 33, 1033–39.

Givnish, T. J. (1994) Does diversity beget stability? *Nature*, 371, 113–14.

Gonzalez, L. M. (2007) *Plant Species and Functional Diversity across Gradients of Resource Availability and Grazing in a California Serpentine Grassland.* MS thesis, Western Washington University, Bellingham.

Grime, J. P. (2001) *Plant Strategies, Vegetation Processes and Ecosystem Properties.* Wiley, Chichester.

Gulmon, S. L., Chiariello, N. R., Mooney, H. A., and Chu, C. C. (1983) Phenology and resource use in three co-occurring grassland annuals. *Oecologia*, 58, 33–42.

Harper, J. L. (1977) *Population Biology of Plants.* Academic Press, London.

Hart, S. C., Firestone, M. K., Paul, E. A., and Smith, J. L. (1993) Flow and fate of soil nitrogen in an annual grassland and a young mixed-conifer forest. *Soil Biology and Biochemistry*, 25, 431–42.

Harte, J., and Kinzig, A. P. (1993) Mutualism and competition between plants and decomposers—implications for nutrient allocation in ecosystems. *American Naturalist*, 141, 829–46.

Hector, A., Beale, A., Minns, A., Otway, S., and Lawton, J. H. (2000) Consequences of loss of plant diversity for litter decomposition: Mechanisms of litter quality and microenvironment. *Oikos*, 90, 357–71.

Hector, A., Schmid, B., Beierkuhnlein, C., Caldeira, M. C., Diemer, M., Dimitrakopoulos, P. G., Finn, J. A., Freitas, H., Giller, P. S., Good, J., Harris, R., Högberg, P., Huss-Danell, K., Joshi, J., Jumpponen, A., Körner, C., Leadley, P. W., Loreau, M., Minns, A., Mulder, C. P. H., O'Donovan, G., Otway, S. J., Pereira, J. S., Prinz, A., Read, D. J., Scherer-Lorenzen, M., Schulze, E.-D., Siamantziouras, A.-S. D., Spehn, E. M., Terry, A. C., Troumbis, A. Y., Woodward, F. I., Yachi, S., and Lawton, J. H. (1999) Plant diversity and productivity experiments in European grasslands. *Science*, 286, 1123–27.

Hiremath, A. J., and Ewel, J. J. (2001) Ecosystem nutrient use efficiency, productivity, and nutrient accrual in model tropical communities. *Ecosystems*, 4, 669–82.

Hobbie, S. E. (1992) Effects of plant species on nutrient cycling. *Trends in Ecology and Evolution*, 7, 336–39.

Hobbs, R. J., and Mooney, H. A. (1991) Effects of rainfall variability and gopher disturbance on serpentine annual grassland dynamics. *Ecology*, 72, 59–68.

Hobbs, R. J., and Mooney, H. A. (1995) Spatial and temporal variability in California annual grassland: Results from a long-term study. *Journal of Vegetation Science*, 6, 43–56.

Hobbs, R. J., Yates, S., and Mooney, H. A. (2007) Long-term data reveal complex dynamics in grassland in relation to climate and disturbance. *Ecological Monographs*, 77, 545–68.

Hooper, D. U. (1996) *The Effects of Plant Functional Group Diversity on Nutrient Cycling in a California Serpentine Grassland.* PhD diss., Stanford University, Stanford, CA.

Hooper, D. U. (1998) The role of complementarity and competition in ecosystem responses to variation in plant diversity. *Ecology*, 79, 704–19.

Hooper, D. U., Chapin, F. S. III, Ewel, J. J., Hector, A., Inchausti, P., Lavorel, S., Lawton, J. H., Lodge, D., Loreau, M., Naeem, S., Schmid, B., Setälä, H., Symstad, A. J., Vandermeer, J., and Wardle, D. A. (2005) Effects of biodiversity on ecosystem functioning: A consensus of current knowledge. *Ecological Monographs,* 75, 3–35.

Hooper, D. U., and Dukes, J. S. (2004) Overyielding among plant functional groups in a long-term experiment. *Ecology Letters,* 7, 95–105.

Hooper, D. U., Solan, M., Symstad, A., Díaz, S., Gessner, M. O., Buchmann, N., Degrange, V., Grime, P., Hulot, F., Mermillod-Blondin, F., Roy, J., Spehn, E., and van Peer, L. (2002) Species diversity, functional diversity and ecosystem functioning. In *Biodiversity and Ecosystem Functioning: Synthesis and Perspectives* (eds. M. Loreau, S. Naeem and P. Inchausti), pp. 195–208. Oxford University Press, Oxford.

Hooper, D. U., and Vitousek, P. M. (1997) The effects of plant composition and diversity on ecosystem processes. *Science,* 277, 1302–5.

Hooper, D. U., and Vitousek, P. M. (1998) Effects of plant composition and diversity on nutrient cycling. *Ecological Monographs,* 68, 121–49.

Hungate, B. A., Chapin, F. S. III, Zhong, H., Holland, E. A., and Field, C. B. (1997a) Stimulation of grassland nitrogen cycling under carbon dioxide enrichment. *Oecologia,* 109, 149–53.

Hungate, B. A., Dukes, J. S., Shaw, M. R., Luo, Y., and Field, C. B. (2003) Nitrogen and climate change. *Science,* 302, 1512–13.

Hungate, B. A., Holland, E. A., Jackson, R. B., Chapin, F. S. III, Mooney, H. A., and Field, C. B. (1997b) The fate of carbon in grasslands under carbon dioxide enrichment. *Nature,* 388, 576–79.

Hungate, B. A., Jackson, R. B., Field, C. B., and Chapin, F. S. III (1996) Detecting changes in soil carbon in CO_2 enrichment experiments. *Plant and Soil,* 187, 135–45.

Huston, M. A. (1997) Hidden treatments in ecological experiments: Re-evaluating the ecosystem function of biodiversity. *Oecologia,* 110, 449–60.

Huston, M. A., and McBride, A. C. (2002) Evaluating the relative strengths of biotic versus abiotic controls on ecosystem processes. In *Biodiversity and Ecosystem Functioning: Synthesis and Perspectives* (eds. M. Loreau, S. Naeem and P. Inchausti), pp. 47–60. Oxford University Press, Oxford.

Jackson, R. B., Sala, O. E., Field, C. B., and Mooney, H. A. (1994) CO_2 alters water use, carbon gain, and yield for the dominant species in a natural grassland. *Oecologia,* 98, 257–62.

Jackson, R. B., Sala, O. E., Paruelo, J. M., and Mooney, H. A. (1998) Ecosystem water fluxes for two grasslands in elevated CO_2: A modeling analysis. *Oecologia,* 113, 537–46.

Jenny, H. (1980) *Soil Genesis with Ecological Perspectives.* Springer-Verlag, New York.

Joel, G., Chapin, F. S., Chiariello, N. R., Thayer, S. S., and Field, C. B. (2001) Species-specific responses of plant communities to altered carbon and nutrient availability. *Global Change Biology,* 7, 435–50.

Jones, C. G., and Lawton, J. H. (1995) *Linking Species and Ecosystems.* Chapman and Hall, New York.

Kinzig, A. P., Pacala, S. W., and Tilman, D. (2002) The functional consequences of biodiversity: Empirical progress and theoretical extensions. *Monographs in Population Biology.* Princeton University Press, Princeton, NJ.

Kirby, K. N. (1993) *Advanced Data Analysis with SYSTAT.* Van Nostrand Reinhold, New York.

Körner, C. (2000) Biosphere responses to CO_2 enrichment. *Ecological Applications,* 10, 1590–619.

Lavorel, S., and Garnier, E. (2002) Predicting the effects of environmental changes on plant community composition and ecosystem functioning: Revisiting the Holy Grail. *Functional Ecology,* 16, 545–56.

Levine, J. M., and HilleRisLambers, J. (2009) The importance of niches for the maintenance of species diversity. *Nature,* 461, 254–57.

Loreau, M. (1998) Separating sampling and other effects in biodiversity experiments. *Oikos,* 82, 600–602.

Loreau, M., and Hector, A. (2001) Partitioning selection and complementarity in biodiversity experiments. *Nature,* 412, 72–76.

Loreau, M., Naeem, S., and Inchausti, P., eds. (2002) *Biodiversity and Ecosystem Functioning: Synthesis and Perspectives.* Oxford University Press, Oxford.

Loreau, M., Naeem, S., Inchausti, P., Bengtsson, J., Grime, J. P., Hector, A., Hooper, D. U., Huston, M. A., Raffaelli, D., Schmid, B., Tilman, D., and Wardle, D. A. (2001) Biodiversity and ecosystem functioning: Current knowledge and future challenges. *Science,* 294, 804–8.

Luo, Y. Q., Field, C. B., and Jackson, R. B. (2006) Does nitrogen constrain carbon cycling, or does carbon input stimulate nitrogen cycling? *Ecology,* 87, 3–4.

Luo, Y. Q., Jackson, R. B., Field, C. B., and Mooney, H. A. (1996) Elevated CO_2 increases belowground respiration in California grasslands. *Oecologia,* 108, 130–37.

Luo, Y., Su, B., Currie, W. S., Dukes, J. S., Finzi, A., Hartwig, U., Hungate, B., McMurtrie, R. E., Oren, R., Parton, W. J., Pataki, D. E., Shaw, M. R., Zak, D. R., and Field, C. B. (2004) Progressive nitrogen limitation of ecosystem responses to rising atmospheric carbon dioxide. *Bioscience,* 54, 731–39.

McNaughton, S. J. (1977) Diversity and stability of ecological communities: A comment on the role of empiricism in ecology. *American Naturalist,* 111, 515–25.

Millennium Ecosystem Assessment. (2005) *Ecosystems and Human Well-Being: Biodiversity Synthesis.* World Resources Institute, Washington, DC.

Mooney, H. A., Hobbs, R. J., Gorham, J., and Williams, K. (1986) Biomass accumulation and resource utilization in co-occurring grassland annuals. *Oecologia,* 70, 555–58.

Naeem, S., Thompson, L. J., Lawler, S. P., Lawton, J. H., and Woodfin, R. M. (1995) Empirical evidence that declining species diversity may alter the performance of terrestrial ecosystems. *Philosophical Transactions of the Royal Society of London B,* 347, 249–62.

Neter, J., Wasserman, W., and Kutner, M. H. (1990) *Applied Linear Statistical Models: Regression, Analysis of Variance, and Experimental Designs.* Richard D. Irwin, Homewood, IL.

Palmborg, C., Scherer-Lorenzen, M., Jumpponen, A., Carlsson, G., Huss-Danell, K., and Hogberg, P. (2005) Inorganic soil nitrogen under grassland plant communities of different species composition and diversity. *Oikos,* 110, 271–82.

Petchey, O. L. (2002) Functional diversity (FD), species richness and community composition. *Ecology Letters,* 5, 402–11.

Petchey, O. L., and Gaston, K. J. (2006) Functional diversity: Back to basics and looking forward. *Ecology Letters,* 9, 741–58.

Petchey, O. L., Hector, A., and Gaston, K. J. (2004) How do different measures of functional diversity perform? *Ecology*, 85, 847–57.

Raffaelli, D., Putten, W. H. v. d., Persson, L., Wardle, D. A., Petchey, O. L., Koricheva, J., Heijden, M. G. A. v. d., Mikola, J., and Kennedy, T. (2002) Multi-trophic dynamics and ecosystem processes. In *Biodiversity and Ecosystem Functioning: Synthesis and Perspectives* (eds. M. Loreau, S. Naeem, and P. Inchausti), pp. 147–54. Oxford University Press, Oxford.

Reich, P. B., Hungate, B. A., and Luo, Y. Q. (2006) Carbon-nitrogen interactions in terrestrial ecosystems in response to rising atmospheric carbon dioxide. *Annual Review of Ecology Evolution and Systematics*, 37, 611–36.

Rothstein, D. E. (2000) Spring ephemeral herbs and nitrogen cycling in a northern hardwood forest: An experimental test of the vernal dam hypothesis. *Oecologia*, 124, 446–53.

Scherer-Lorenzen, M., Palmborg, C., Prinz, A., and Schulze, E. D. (2003) The role of plant diversity and composition for nitrate leaching in grasslands. *Ecology*, 84, 1539–52.

Schmitz, O. J. (2006) Predators have large effects on ecosystem properties by changing plant diversity, not plant biomass. *Ecology*, 87, 1432–37.

Schulze, E. D., and Mooney, H. A. (1993) *Biodiversity and Ecosystem Function*. Springer-Verlag, Berlin.

Spehn, E. M., Hector, A., Joshi, J., Scherer-Lorenzen, M., Schmid, B., Bazeley-White, E., Beierkuhnlein, C., Caldeira, M. C., Diemer, M., Dimitrakopoulos, P. G., Finn, J. A., Freitas, H., Giller, P. S., Good, J., Harris, R., Hogberg, P., Huss-Danell, K., Jumpponen, A., Koricheva, J., Leadley, P. W., Loreau, M., Minns, A., Mulder, C. P. H., O'Donovan, G., Otway, S. J., Palmborg, C., Pereira, J. S., Pfisterer, A. B., Prinz, A., Read, D. J., Schulze, E. D., Siamantziouras, A. S. D., Terry, A. C., Troumbis, A. Y., Woodward, F. I., Yachi, S., and Lawton, J. H. (2005) Ecosystem effects of biodiversity manipulations in European grasslands. *Ecological Monographs*, 75, 37–63.

Spehn, E. M., Scherer-Lorenzen, M., Schmid, B., Hector, A., Caldeira, M. C., Dimitrakopoulos, P. G., Finn, J. A., Jumpponen, A., O'Donovan, G., Pereira, J. S., Schulze, E.-D., Troumbis, A. Y., and Körner, C. (2002) The role of legumes as a component of biodiversity in a cross-European study of grassland biomass nitrogen. *Oikos*, 98, 205–18.

Srivastava, D. S., Cardinale, B. J., Downing, A. L., Duffy, J. E., Jouseau, C., Sankaran, M., and Wright, J. P. (2009) Diversity has stronger top-down than bottom-up effects on decomposition. *Ecology*, 90, 1073–83.

Steiner, K. (1982) *Intercropping in Tropical Smallholder Agriculture with Special Reference to West Africa*. German Agency for Technical Cooperation (GTZ), Eschborn, Germany.

Suding, K. N., Lavorel, S., Chapin, F. S., Cornelissen, J. H. C., Diaz, S., Garnier, E., Goldberg, D., Hooper, D. U., Jackson, S. T., and Navas, M. L. (2008) Scaling environmental change through the community-level: A trait-based response-and-effect framework for plants. *Global Change Biology*, 14, 1125–40.

SYSTAT. (1992) *SYSTAT for Windows: Statistics, Version 5 Edition*. Evanston, IL.

Tilman, D. (1996) Biodiversity: Population versus ecosystem stability. *Ecology*, 77, 350–63.

Tilman, D. (1997) Distinguishing between the effects of species diversity and species composition. *Oikos*, 80, 185.

Tilman, D. (1999) Ecological consequences of biodiversity: A search for general principles. *Ecology,* 80, 1455–74.

Tilman, D., Knops, J., Wedin, D., Reich, P., Ritchie, M., and Siemann, E. (1997a) The influence of functional diversity and composition on ecosystem processes. *Science,* 277, 1300–1302.

Tilman, D., Lehman, C. L., and Thomson, K. T. (1997b) Plant diversity and ecosystem productivity: Theoretical considerations. *Proceedings of the National Academy of Sciences, USA,* 94, 1857–61.

Tilman, D., Reich, P. B., Knops, J., Wedin, D., Mielke, T., and Lehman, C. (2001) Diversity and productivity in a long-term grassland experiment. *Science,* 294, 843–45.

Tilman, D., Wedin, D., and Knops, J. (1996) Productivity and sustainability influenced by biodiversity in grassland ecosystems. *Nature,* 379, 718–20.

Valentini, R., Gamon, J. A., and Field, C. B. (1995) Ecosystem gas-exchange in a California grassland—seasonal patterns and implications for scaling. *Ecology,* 76, 1940–52.

Vandermeer, J. H. (1989) *The Ecology of Intercropping.* Cambridge University Press, Cambridge.

Vandermeer, J. H. (1990) Intercropping. In *Agroecology* (eds. C. R. Carrol, J. H. Vandermeer, P. M. Rosset), pp. 481–516. McGraw-Hill, New York.

van Ruijven, J., and Berendse, F. (2003) Positive effects of plant species diversity on productivity in the absence of legumes. *Ecology Letters,* 6, 170–75.

Vazquez, D. P., Chacoff, N. P., and Cagnolo, L. (2009) Evaluating multiple determinants of the structure of plant-animal mutualistic networks. *Ecology,* 90, 2039–46.

Vitousek, P. M. (1994) Beyond global warming: Ecology and global change. *Ecology,* 75, 1861–76.

Vitousek, P. M. (2004) *Nutrient Cycling and Limitation: Hawai'i as a Model System.* Princeton University Press, Princeton, NJ.

Vitousek, P. M., and Matson, P. A. (1985) Disturbance, nitrogen availability, and nitrogen losses in an intensively managed loblolly pine plantation. *Ecology,* 66, 1360–76.

Walker, B., Kinzig, A., and Langridge, J. (1999) Plant attribute diversity, resilience, and ecosystem function: The nature and significance of dominant and minor species. *Ecosystems,* 2, 95–113.

Wardle, D. A. (1999) Is "sampling effect" a problem for experiments investigating biodiversity—ecosystem function relationships? *Oikos,* 87, 403–7.

Wardle, D. A., Bonner, K. I., and Nicholson, K. S. (1997) Biodiversity and plant litter: Experimental evidence which does not support the view that enhanced species richness improves ecosystem function. *Oikos,* 79, 247–58.

Wedin, D., and Tilman, D. (1993) Competition among grasses along a nitrogen gradient—initial conditions and mechanisms of competition. *Ecological Monographs,* 63, 199–229.

Weiher, E., and Keddy, P. (1998) Assembly rules, null models, and trait dispersion: New questions from old patterns. *Oikos,* 74, 159–64.

Wright, I. J., Reich, P. B., Westoby, M., Ackerly, D. D., Baruch, Z., Bongers, F., Cavender-Bares, J., Chapin, T., Cornelissen, J. H. C., Diemer, M., Flexas, J., Garnier, E., Groom, P. K., Gulias, J., Hikosaka, K., Lamont, B. B., Lee, T., Lee, W., Lusk, C., Midgley, J. J.,

Navas, M. L., Niinemets, U., Oleksyn, J., Osada, N., Poorter, H., Poot, P., Prior, L., Pyankov, V. I., Roumet, C., Thomas, S. C., Tjoelker, M. G., Veneklaas, E. J., and Villar, R. (2004) The worldwide leaf economics spectrum. *Nature,* 428, 821–27.

Wright, J. P., Naeem, S., Hector, A., Lehman, C., Reich, P. B., Schmid, B., and Tilman, D. (2006) Conventional functional classification schemes underestimate the relationship with ecosystem functioning. *Ecology Letters,* 9, 111–20.

Yachi, S., and Loreau, M. (1999) Biodiversity and ecosystem productivity in a fluctuating environment: The insurance hypothesis. *Proceeding of the National Academy of Sciences, USA,* 96, 1463–68.

Zak, D. R., Groffman, P. M., Pregitzer, K. S., Christensen, S., and Tiedje, J. M. (1990) The vernal dam: Plant-microbe competition for nitrogen in northern hardwood forests. *Ecology,* 71, 651–56.

Zak, D. R., Holmes, W. E., White, D. C., Peacock, A. D., and Tilman, D. (2003) Plant diversity, soil microbial communities, and ecosystem function: Are there any links? *Ecology,* 84, 2042–50.

Zavaleta, E. S., and Hulvey, K. B. (2004) Realistic species losses disproportionately reduce grassland resistance to biological invaders. *Science,* 306, 1175–77.

Zavaleta, E. S., Shaw, M. R., Chiariello, N. R., Thomas, B. D., Cleland, E. E., Field, C. B., and Mooney, H. A. (2003) Grassland responses to three years of elevated temperature, CO_2, precipitation, and N deposition. *Ecological Monographs,* 73, 585–604.

17

Climate Change and Plant Communities on Unusual Soils

Ellen I. Damschen, *University of Wisconsin, Madison*
Susan Harrison, Barbara M. Going, and Brian L. Anacker
University of California, Davis

Human-caused climate change is altering patterns of temperature, precipitation, and net primary productivity across the globe (Nemani et al., 2003; IPCC, 2007), and many ecological responses have already been detected (Parmesan and Yohe, 2003). One of the greatest challenges facing conservation biologists today is predicting the fates of species and communities under altered climates, which is difficult for at least three reasons. First, the direction and magnitude of climatic changes vary across regions (Nemani et al., 2003; IPCC, 2007). Second, species responses are highly variable (IPCC, 2007). Third, climate change can have complex indirect effects through altering species interactions, such as competition, and through synergisms with other human impacts (Kareiva et al., 1992). Given the severity and rapidity of global climate change, it is imperative that we develop a better predictive framework for how suites of species and habitats are likely to be affected.

Special-soil floras are one group of species and habitats that merit particular attention. Although the focus of this book is on serpentine, we consider serpentine to be one example of a broader phenomenon in which soils that are unusually harsh support distinctive floras that may be rich in specialized (endemic) plant species (Anderson et al., 1999; Kruckeberg, 2005). These are sometimes referred to as "azonal" or "edaphic climax" floras, as contrasted with the "zonal" or "climatic climax" floras found on more fertile soils within the same regions. Such floras contribute disproportionately to the world's total biodiversity. For example, the limestone grasslands of southern Europe, the dolomite glades of the Ozarks, the shale barrens of Appalachia, and the serpentine floras of the Mediterranean, Cuba, New Caledonia, and California (Anderson et al., 1999; Myers et al., 2000; Stein et al., 2000; Kruckeberg, 2005) all contain high numbers of endemic plants. In California, 612 of the state's 1742 rare plants are associated with serpentine, limestone, volcanic outcrops, vernal pools, or other special substrates (Skinner and Pavlik, 1994). Plants that are strongly restricted to serpentine in California form >10% of the species unique to the state even though serpentine is <2% of its surface area (Kruckeberg, 1984; Safford et al., 2005). At a global scale, 68% of the biodiversity hot spots identified by Myers et al. (2000) and Mittermeier et al. (2000, 2005) include floras associated with serpentine, limestone, sandstone, or other special soils (Figure 17.1, Table 17.1).

We propose that plant species and communities found on serpentine and other special soils share a number of ecological attributes that may cause them to respond differently to climate change than more typical species and communities in the same regions:

1. They are confined to relatively small and spatially isolated outcrops, where migrating to track the shifting climate will be exceptionally difficult.

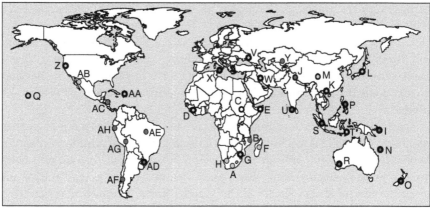

FIGURE 17.1. Biodiversity hot spots, special soils, and climate change impacts on productivity. Thirty-four biodiversity hot spots defined by Myers et al. (2000) and Mittermeier et al. (2000, 2005) are shown as well as whether serpentine or other special soils occur within these regions (width of circles). Shading indicates whether the region each hot spot is found in has shown an increase (dark gray), decrease (light gray), or no change (white) in net primary productivity from 1982 to 1999 (Nemani et al., 2003; IPCC, 2007).

TABLE 17.1 The Contribution of Serpentine and Other Special Soils to Biodiversity Hot Spots:
Countries That Have Serpentine Soils are in Bold

	Biodiversity Hot Spot	# End	% End	Serp	Other Soils	Countries
	Africa					
A	Cape Floristic Region	6210	69.0	No	No	South Africa
B	Coastal forests of Eastern Africa	1750	43.8	No	No	Kenya, Mozambique, Somalia, Tanzania
C	Eastern Afromontane	2356	31.0	No	Yes[a,f]	Burundi, Democratic Republic of Congo, Eritrea, Ethiopia, Malawi, Mozambique, Rwanda, Saudi Arabia, Tanzania, Uganda, Yemen, Zimbabwe
D	Guinean forests of West Africa	1800	20.0	Yes[b]	No	Benin, Cameroon, Côte d'Ivoire, Equatorial Guinea, **Ghana**, Guinea, Liberia, Nigeria, São Tomé and Príncipe, Sierra Leone, Togo
E	Horn of Africa	2750	55.0	Yes[b]	No	Djibouti, Eritrea, Ethiopia, Kenya, **Oman**, Saudi Arabia, Somalia, Sudan, Yemen
F	Madagascar and the Indian Ocean Islands	11600	89.2	No	No	Comoros, Madagascar, Mauritius, Reunion (France), Seychelles
G	Maputaland-Pondoland-Albany	1900	23.5	Yes[c,d,e]	No	Mozambique, **South Africa, Swaziland**
H	Succulent Karoo	2439	38.4	No	No	Namibia, South Africa
	Asia Pacific					
I	East Melanesian Islands	3000	37.5	Yes[b]	No	**Papua New Guinea, Solomon Islands**, Vanuatu
J	Himalaya	3160	31.6	No	Yes[a,f]	Bangladesh, Bhutan, China, India, Myanmar, Nepal, Pakistan
K	Indo-Burma	7000	51.9	Yes[b]	Yes[a,g]	Cambodia, China, Bangladesh, India, Laos, Malaysia, **Myanmar (Burma)**, Thailand, Vietnam
L	Japan	1950	34.8	Yes[b]	Yes[a,g]	**Japan**
M	Mountains of southwest China	3500	29.2	No	Yes[a,i]	China, Myanmar (Burma)
N	New Caledonia	2432	74.4	Yes[b,c]	No	**New Caledonia (France)**

(continued)

TABLE 17.1 *(continued)*

Biodiversity Hot Spot	# End	% End	Serp	Other Soils	Countries
O	New Zealand	1865	81.1	Yes[b,c]	No	**New Zealand**
P	Philippines	6091	65.8	Yes[b,c]	No	**Philippines**
Q	Polynesia-Micronesia	3074	57.7	No	Yes[a,j]	Fiji, Hawaii, Micronesia, Palau, Polynesia
R	Southwest Australia	2948	52.9	Yes[b,c]	No	**Australia**
S	Sundaland	15,000	60.0	Yes[b]	Yes[a,l,m,n]	**Borneo**, Brunei, Indonesia, Malaysia, Philippines, Singapore, Sumatra, Thailand
T	Wallacea	1500	15.0	Yes[b]	No	**Indonesia: Lesser Sundas, Moluccas (Spice Islands), Timor Leste, Sulawesi**
U	Western Ghats and Sri Lanka	3049	51.5	Yes[b]	No	India, **Sri Lanka**
	Europe and Central Asia					
V	Caucasus	1600	25.0	Yes[b]	Yes[o]	Azerbaijan, **Georgia, Iran, Russian Federation, Turkey**
W	Irano-Anatolian	2500	41.7	Yes[b]	Yes[p]	Azerbaijan, **Iran**, Iraq, **Georgia, Turkey**, Turkmenistan
X	Mediterranean Basin	11,700	52.0	Yes[b,c]	No	Algeria, Egypt, **France, Greece, Italy**, Libya, Morocco, **Portugal, Spain**, Syria, **Turkey**, Tunisia
Y	Mountains of central Asia	1500	27.3	No	No	Afghanistan, China, Kazakhstan, Kyrgyzstan, Tajikistan, Turkmenistan, Uzbekistan
	North and Central America					
Z	California Floristic Province	2124	60.9	Yes[b,c]	No	United States **(California and Oregon)**
AA	Caribbean Islands	6550	50.4	Yes[b,c]	No	Bahamas, **Cuba**, Dominican Republic, Haiti, Jamaica, **Puerto Rico (USA)**
AB	Madrean pine-oak woodlands	3975	75.0	No	No	Mexico, United States (Baja, but not same range as serpentine)
AC	Mesoamerica	2941	17.3	No	No	Belize, Costa Rica, El Salvador, Guatemala, Honduras, Nicaragua, Panama

South America

AD	Atlantic forest	8000	40.0	Yes[b,c]	No	Argentina, **Brazil**, Paraguay, Uruguay
AE	Cerrado	4400	44.0	No	No	Brazil, Bolivia, Paraguay
AF	Chilean winter rainfall-Valdivian forests	1957	50.3	No	No	Argentina, Chile
AG	Tropical Andes	15,000	50.0	No	No	Argentina, Bolivia, Brazil, Chile, Colombia, Ecuador, Panama, Venezuela
AH	Tumbes-Choco-Magdalena	2750	25.0	No	No	Colombia, Ecuador, Galápagos Islands, Panama, Peru

NOTE: Biodiversity hot spots (Mittermeier et al., 2000, 2005; Myers et al., 2000) are listed by continental region. Letters correspond to locations in Figure 17.1. This table includes the number and percentage of endemic plant species for each hot spot (Mittermeier et al., 2000, 2005; Myers et al., 2000), whether serpentine or other special soils are found within the hot spot, and what countries are a part of each hot spot.

SOURCES: [a]Myers (2000) and Mittermeier et al. (2000, 2005), [b]Brooks (1987), [c]Alexander et al. (2006), [d]Smith et al. (2001), [e]Witkowski et al. (2001), [f]high-elevation rocky outcrops, [g]karst limestone, [h]rocky cliffs, [i]alpine scree, [j]limestone, [m]granite, [n]sandstone, [o]colchic limestone, [p]salt steppes.

2. They occur on infertile soils, where nutrients may be more limiting to plant growth than temperature or water, and where stress-resistant plant traits may confer resistance to climate change.

3. Their distribution and composition are the products of a competitive balance between soil "generalists" and soil "specialists" that are likely to be altered under a changing climate.

After elaborating on each of these issues, we consider preliminary evidence from a suite of related studies that compare the responses of serpentine and non-serpentine plant communities to climate change.

SPATIAL ISOLATION, LIMITED DISPERSAL, AND CLIMATE CHANGE SENSITIVITY

Considerable concern has been expressed for how climatic warming will affect the fates of mountaintop-dwelling species, such as the pika (Moritz et al., 2008) and high alpine plants (Grabherr et al., 1994; Walther et al., 2005; Parolo and Rossi, 2008). These species are seen as being trapped on shrinking islands of habitat, unable to reach newly suitable climates by moving either upward in elevation or poleward in latitude. Although little concern has yet been given to plants endemic to serpentine or other special soils, they may face a closely similar situation. Individual serpentine outcrops are usually no more than a few square kilometers in area, so they contain only a limited range of elevations, thus restricting the possibilities for plants to survive through upward movement. Attaining a suitable climate by moving from one outcrop to another, perhaps farther north or south, would also be difficult for most soil-specialist plants because it would require crossing large expanses of unsuitable habitat.

Some indirect evidence supports the idea that soil-specialist plants are less able to survive climatic fluctuations than other species. In California, serpentine endemics inhabit environments with higher average annual and summer rainfall and less extreme seasonal temperatures compared with other species; these differences persist when only congeners are compared, and are even more pronounced when considering those serpentine endemics that have particularly small geographic ranges and low local abundances (Harrison et al., 2008). These patterns suggest that serpentine endemics, especially the rarest ones, may have been filtered out of relatively warm and dry regions by natural extinctions during climatic fluctuations in the past.

Considering this issue calls attention to an important general shortcoming of current models that predict the fate of biodiversity under climate change. These models begin by estimating for each species the set of climate parameters that best describes its current distribution (its "climate envelope"); then a map is generated of where that combination of climate parameters is expected to occur, say, 10, 50, or 100 years into the future, as projected by global climate models. It is assumed that even without dispersal, a species will survive as long as there is geographic overlap between its present and future climate envelopes (e.g., Thomas et al., 2004; Schwartz et al., 2006; Loarie et al., 2008). However, this approach does not consider the fact that species have other factors besides climate that define their niches. It may be considerably more difficult for species to remain in or migrate to locations that have both suitable climates and, for example, suitable soils than locations with suitable climates alone.

Dispersal ability is critical for predicting the effects of global climate change for any species (Engler et al., 2009), perhaps even more so for soil-specialist plants with their naturally fragmented distributions. The evolutionary ecology of dispersal strategies is shaped by environmental attributes such as habitat age, area, isolation, and stability (Southwood, 1977; Roff, 1990; Denno et al., 1996). Thus, it is likely that plants specialized on small outcrops of low-productivity soils have different arrays of dispersal syndromes (e.g., different frequencies of animal, wind, and passive dispersal) than their nonspecialist counterparts. However, we know of no studies that have compared dispersal rates or syndromes in soil specialists versus other species. Pärtel and Zobel (2007) found that species without special dispersal syndromes prevail at low productivity levels. Preliminary evidence from the Klamath-Siskiyou region of southern Oregon suggests that bird and mammal dispersal are much less common, and passive and wind dispersal are much more common, in serpentine woodlands compared with adjacent nonserpentine forests (Damschen and Harrison, unpublished data).

NUTRIENT LIMITATION AND CLIMATE CHANGE SENSITIVITY

Considering the importance of special-soil floras to biodiversity, and the problems posed by their naturally fragmented distributions, it may seem surprising that little alarm has been raised yet about their fates under climate change. In fact, the few mentions of special-soil floras in the climate change literature generally express optimism. For example, referring to the European Alps, where limestone is widespread, Theurillat and Guisan (2001: 90) argue that "climatic climax plant communities (i.e., zonal vegetation) and edaphic climax plant communities may behave differently. . . . For edaphic climaxes, plant communities could sustain a climatic change provided that their limiting factors were not modified." In other words, a given amount of change to temperature or moisture may have a lesser effect in environments in which plant growth is strongly limited by soil fertility, as compared with more fertile environments.

Experiments by Grime and colleagues (2000, 2008) in British limestone grasslands appear to support this optimistic view. Both temperature and precipitation were manipulated at two sites: Buxton, an "ancient" grassland that had been used only for traditional grazing, and Wytham, a grassland recovering from cultivation, where the soil still showed traces of fertilizer addition. After 5 years, changes in species composition and biomass in response to the climate treatments were considerably greater at fertile Wytham than infertile Buxton (Grime et al., 2000), and even after 13 years of warming and drought the Buxton grassland remained relatively little altered (Grime et al., 2008). The authors attributed this result not simply to nutrient limitation itself but to the dominance of nutrient-poor communities by particular types of species whose longevity, low intrinsic growth rates, and other life history traits make them tolerant of abiotic stresses. The authors concluded that unproductive ecosystems are relatively stable refuges for biodiversity under climate change, unless subjected to land use transformation.

In one of the few studies of the paleoecology of a special-soil flora, Briles (2008) compared 15,000-year pollen sequences from three lakes on ultramafic (serpentine and peridotite) soils and three lakes on normal (granitic) soils in the Klamath Mountains of California. The woody vegetation on serpentine soils, dominated by Jeffrey pine and huckleberry oak, remained more constant on centennial-to-millennial time scales during the climatic fluctuations of the Pleistocene and post-Pleistocene eras than did the dominant woody vegetation on granitic soils.

Plant species that occur on serpentine often exhibit slow growth rates, high root-to-shoot ratios, and small, thick, spiny, hairy, and/or waxy leaves (Brady et al., 2005). For example, Borhidi (1991) concluded that in terms of leaf morphology, plants growing on serpentine in any given climate are equivalent to plants growing off of serpentine in a climate with 50 cm less annual rainfall. Following the reasoning of Grime et al. (2000, 2008), such traits could be interpreted as adaptations to a nutrient-poor environment that have the fortuitous consequence of reducing the sensitivity of plant growth to climatic variability. However, a possible complication is that water availability is low in many special-soil environments because soils are shallow and rocky, and these traits therefore may be adaptations to drought stress. In environments where water and nutrients are jointly limiting to plant growth, further reductions in water availability—as when the climate becomes warmer but not wetter—may have consequences that are just as serious as on more fertile soils.

INDIRECT EFFECTS OF CLIMATE CHANGE IN SPECIAL-SOIL FLORAS

Warmer and wetter climates, which are predicted in many parts of the world, offer conditions that are more favorable to plant growth and productivity in general. If most expectations are for a dramatic loss of biodiversity, then the indirect effects of climate change bear a large share of the responsibility. For example, at high altitudes in the European Alps, overall plant species richness has increased as the climate has warmed, but some species have declined due to increased competition from invading lower elevation species (Grabherr et al., 1994; Theurillat and Guisan, 2001; Walther et al., 2005; Parolo and Rossi, 2008). The addition of late-season water to a Mediterranean-type grassland, to mimic a wetter future predicted climate, caused initial increases in the biomass of most species; however, a year or two later, native forbs declined because of enhanced competition from exotic grasses (Suttle et al., 2007). Under a changing climate, then, the composition of communities will be altered as species with the appropriate traits (e.g., wide climatic tolerances, high potential growth rates and competitive abilities, fast dispersal) gain a competitive advantage over species lacking such traits.

Changes in the competitive balance may have especially dramatic effects on the floras of low-fertility soils, because these soils are classically considered to be refuges from competition. Experiments by Kruckeberg (1954) and other evidence has suggested that in the absence of competition, species normally found only on serpentine or other special soils may grow equally well or better on more fertile soils (e.g., Tansley, 1917; Sharitz and McCormick, 1973; Harper, 1977; Ware, 1991; Baskin and Baskin, 1998). In other words, many special soil endemics are able to tolerate a wide range of soils but are confined to nutrient-poor soils by their inability to compete with faster growing species. Even the strictest serpentine or limestone glade "endemics" are not infrequently found on other soils (Nelson and Ladd, 1983; Kruckeberg, 1984; Safford et al., 2005), suggesting that the affinity of slow-growing, stress-tolerating species for a particular low-fertility soil is not an absolute requirement but the outcome of competition and other interacting ecological factors.

Experimental and observational evidence suggest that climate change may alter the outcome of competition between soil specialists and soil generalists. For example, in granite glades, experimental water addition shifted competitive dominance from the glade endemic *Minuartia uniflora* to the larger species that form the dominant community of surrounding nonglade soils (Sharitz and McCormick, 1973). The limestone endemic *Talinum calcaricum* was less affected by drought stress than its generalist competitors were (Ware, 1991). A series of especially rainy years led to a strong increase in *Bromus hordeaceus* in serpentine grasslands, but this exotic grass declined sharply during an ensuing drought (Hobbs and Mooney, 1991; Armstrong and Huenneke, 1992).

Geographic evidence also suggests that climate affects the competitive balance that causes species to be confined—or not—to special soils. In general, more favorable climates are associated with higher endemism (Brooks, 1987). So-called regional indicators, or species restricted to serpentine in only parts of their ranges, may be serpentine specialists in wet and productive regions but not in colder or more arid zones (Whittaker, 1960, Kruckeberg, 1984). Boreal and subarctic species may be restricted to serpentine only at their forested southern limits (Rune, 1953; Brooks, 1987), and montane species may be restricted to serpentine only at their lowest elevations (Whittaker, 1960; Brooks, 1987; Borhidi, 1991). Summarizing these patterns in a worldwide review, Brooks (1987) stated, "Competitive pressure restricts certain plants either to the edaphically harsh environment of serpentine, or to climatically harsh environments." Illustrating the possible historical origins of such patterns, Rune (1953) showed that *Arenaria norvegica* became confined to serpentine in southern Sweden after forests expanded in response to postglacial climatic warming, while it remained a soil generalist north of the latitudinal treeline.

Building on the foregoing evidence, we have suggested the following hypothesis (see Harrison et al., 2009). In regions where the future climate is warmer, wetter, and more productive, the majority of the direct benefits will be gained by soil-generalist species with the capacity for high intrinsic growth. The indirect effects of increased competition from these generalists will increasingly confine the slow-growing specialist species to the least productive soils, such as the shallowest and rockiest serpentine habitats, and these stress-tolerant species will tend to be eliminated from the less harsh habitats such as deeper serpentine grassland soils in which they formerly coexisted with generalists. The slow-growing soil specialists may also be eliminated from special-soil environments in which water is an important limiting factor but nutrients are not (e.g., sandstone glades; Nelson and Ladd, 1983). These threats posed to edaphic endemics by a more productive climate will be greatly exacerbated in cases where nutrient limitation is ameliorated on special soils, such as by atmospheric nitrogen deposition in urban areas (Weiss, 1999), and in cases where new exotic species are introduced that combine a high tolerance for nutrient-poor soils with the capacity for rapid resource acquisition and population growth (e.g., goatgrass, *Aegilops triuncialis*, Meimberg et al., 2009).

By the same reasoning, we also hypothesize that in regions where the future climate is drier and less productive, the slow-growing soil specialists may be more resistant to the direct effects of these changes than the faster growing generalists (Grime et al., 2000, 2008). The resulting reduced competition from generalists may allow the soil specialists to expand their distributions into a variety of less harsh soil environments (Harrison et al., 2009). The success of our predictions may depend critically, however, on whether plant growth on special soils is primarily limited by nutrient scarcity, "toxicity" (e.g., cation excess), water scarcity, or more than one of these factors; it may also depend on the functional traits of plants in the particular community in question.

Finally, another reason infertile systems may be resistant to climate change is because litter and organic matter produced by "stress-tolerant plants" is very recalcitrant, decomposes slowly, and promotes slow nutrient cycling (Wardle et al., 2004). It may take a long time for the whole soil system, including organic matter quality, microbial community composition, and so on, to change in a way to promote more rapid nutrient cycling and increased nutrient availability to plants.

INTERACTIONS BETWEEN CLIMATE CHANGE AND OTHER HUMAN IMPACTS

In all ecosystems, the effect of climate change on biodiversity will be modulated by its interactions with other human impacts. In special-soil ecosystems, nutrient enhancement is especially likely to amplify responses to climate change. Atmospheric nitrogen deposition from hydrocarbon combustion is a major threat to plant diversity worldwide, with some ecosystems receiving as much as 20 kg N ha^{-1} year^{-1} (Phoenix et al., 2006). Serpentine grasslands near San Francisco, California, have lost much of their native biodiversity as nitrogen deposition has promoted the invasion of the exotic grass *Lolium multiflorum* (Weiss, 1999). In addition to these direct effects, nutrient addition may increase the sensitivity of low-fertility ecosystems to climate change, as Grime et al. (2000) found in comparing fertilized and unfertilized limestone grasslands. Likewise, Klanderud and Totland (2005) found that experimental warming had no effect on plant species composition in nutrient-poor arctic tundra unless it was combined with fertilization.

Climate change effects can also interact with grazing practices, either natural ones or those controlled by humans. For instance, Post and Pedersen (2008) showed that experimental warming-induced changes in vegetation, associated with increased biomass, were counteracted by grazing. In general, high productivity may not lead to loss of species if there is intensive grazing. In serpentine grasslands, grazing by livestock has been used to counteract the negative effects of eutrophication and invasive species (Weiss, 1999). As a note, the results of climate change studies using field open-topped chambers may be seriously confounded by the fact that such chambers exclude mammalian grazers.

Habitat fragmentation is another human impact frequently noted for its potential to interact with climate change to reduce biodiversity. Fragmentation decreases the potential for species to migrate to adjust to climatic shifts, whereas climatic extremes may cause extinctions of local populations and thus increase the effective amount of fragmentation, finally leading to the regional collapse of species if a critical threshold is passed (Opdam and Wascher, 2004). It might be expected that special-soil ecosystems are particularly vulnerable to this downward spiral, because their naturally patchy distributions may place them closer to the critical fragmentation threshold. On the other hand, it may be that plants in naturally isolated habitats have especially effective adaptations for local persistence, such as long-lived seedbanks and the capacity for self-fertilization; such species may exist as disconnected regional ensembles with little dependency on dispersal among populations and hence relatively low sensitivity to further fragmentation (Freckleton and Watkinson, 2002).

Fire is another potentially important human-influenced process that may interact with climate change in many ecosystems. For example, climatic warming and drying have been shown to increase the size and destructiveness of forest fires, many of which are set by humans (Westerling et al., 2006; Malhi et al., 2008). Although fire is also a natural process that maintains biodiversity in many ecosystems, excessively frequent fire may cause irreversible degradation of vegetation, as is widely occurring in evergreen shrubland (chaparral) in southern California (Keeley and Fotheringham, 2001). The adverse effects of excessive fire frequency may be especially strong in low-nutrient ecosystems. Serpentine chaparral has been shown to support less intense and more heterogeneous fires and to recover much more slowly in terms of its prefire biomass and species composition than neighboring chaparral on more productive sandstone soils (Safford and Harrison, 2004). However, in other special-soil ecosystems, such as the Ozark glades, fire suppression is causing tree encroachment, leading to the loss of specialized herbaceous floras that depend on the combination of nutrient-poor soils and periodic fires (Nelson and Ladd, 1983; Anderson et al., 1999).

EMPIRICAL EVIDENCE FROM SERPENTINE–NONSERPENTINE COMPARISONS

Resampling Whittaker's Historical Plots in the Siskiyou Mountains, Oregon

In ongoing work, we are comparing the historical and modern composition of vegetation on a "special" versus a "normal" soil. In 1949–51, the eminent community ecologist Robert Whittaker sampled plant communities in several hundred 50 × 20 m plots on serpentine, gabbro, and diorite soils in the Klamath-Siskiyou region, Oregon (Whittaker, 1960). This region is the richest in North America for serpentine endemic plants, and it is considered a botanical diversity hot spot of global significance (Kruckeberg, 1984; Ricketts et al., 1999; Safford et al., 2005). Since Whittaker's time, the mean annual temperature in this region has risen almost 2°C, and mean precipitation has not increased (NOAA, 2009). In 2007, using Whittaker's records of road, elevation, slope, and aspect to identify sites as closely similar as possible to his, we resampled herb identity and cover at 55 of his sites in serpentine conifer woodland and 53 of his sites in mixed conifer-evergreen forest on diorite at comparable elevations (Damschen et al., in press).

On both serpentine and diorite soils, we found significant decreases in total herb cover since Whittaker's time (Figure 17.2), as well as changes in herb species

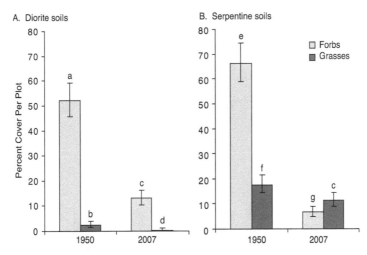

FIGURE 17.2. Changes in cover of forbs and grasses. Total percent forb and grass cover per plot decreased over time. This decline was greater on (A) diorite than (B) serpentine soils and was greater for forbs than grasses on both soils. Error bars represent 95% confidence limits. Small letters indicate significant differences among groups across both panels.

composition that appeared consistent with climatic warming (Figure 17.3). Sites in 2007 compared with 1949–51 had lower relative cover by herb species of north-temperate biogeographic origin, which tend to be found in cool and shady micro-habitats, and equal or higher relative cover by species of "southern" (semiarid or desert) biogeographic origin, which tend to be found in hot and dry microhabitats (Figure 17.4; Valiente-Banuet et al., 2006; Harrison and Grace, 2007; Ackerly, 2009). In addition, an ordination analysis showed that sites today have an herb composition more similar to that of warm south-facing slopes, and less similar to that of cool north-facing slopes, than they did in 1949–51. Neither fire, fire suppression, exotic species, grazing, nor logging appeared to explain these climate-consistent changes (Damschen et al., in press). Contrary to the expectations of our conceptual model and the conclusions of Grime et al. (2000, 2008), these changes toward "warmer" species composition were no less pronounced in the flora of nutrient-poor serpentine soils than in the flora of more fertile diorite soils.

Long-Term Observational Data from California Grasslands

We have recently begun to use a long-term data set of community composition on serpentine and nonserpentine soils to ask how special soil floras respond to climatic fluctuations. The data set consists of nine years of presence-absence data for 80 grasslands sites on serpentine and nonserpentine soils (38 and 42 sites, respec-

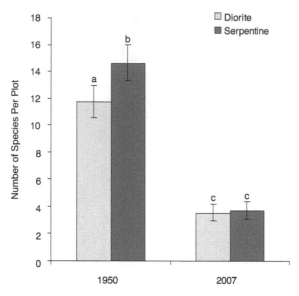

FIGURE 17.3. Forb species richness per plot decreased over time and this decline was smaller on diorite than it was on serpentine soils. Small letters indicate significant differences among groups. Error bars represent 95% confidence limits.

tively) located at McLaughlin Reserve in Lake County, California. Each site contains five 1 m² plots spaced 12.5 m apart along a 50 m transect and is sampled annually. Previous work with this site network found that local plant species richness was higher on serpentine, probably due to decreased competitive dominance by alien plants (Harrison, 1999), whereas beta diversity was higher for natives on nonserpentine, possibly reflecting higher total area of the habitat. In a second study, disturbances from fire and grazing increased plant diversity on both soil types but tended to increase native diversity more on serpentine and non-native diversity more on nonserpentine (Harrison et al., 2003).

An analysis of the stability of plant communities over time showed that temporal change, measured as the decay in the similarity species composition over time, is higher for nonserpentine than serpentine grassland sites (Figure 17.5). These results suggest that serpentine plant communities may be more resistant to environmental fluctuations than nonserpentine communities. The greater stability of serpentine community composition may be due to several interrelated diversity and spatial mechanisms that influence species immigration and extirpation. The patchiness and edaphic stress of serpentine communities may restrict the immigration of new species or prevent the success of colonists, respectively, adding to stability. In addition, edaphic stress and high species diversity limits

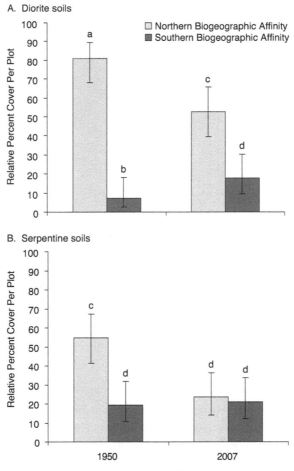

FIGURE 17.4. Changes in species with northern and southern biogeographic affinities. The relative percent cover out of the total herb cover per plot for species belonging to taxa of northern biogeographic affinities declined over time on (A) diorite and (B) serpentine soils, and the relative percent cover of species belonging to higher taxa with southern biogeographic affinities showed an increasing trend on diorite and did not change on serpentine. Small letters represent statistically significant differences among groups. Error bars represent 95% confidence limits. Small letters indicate significant differences among groups across both panels.

the presence and abundance of non-native species, which may have a more dynamic response to climatic variation. Other possible explanations include differences in species traits, such as a higher proportion of perennial herbs on serpentine, and differences in the level of habitat heterogeneity within serpentine versus

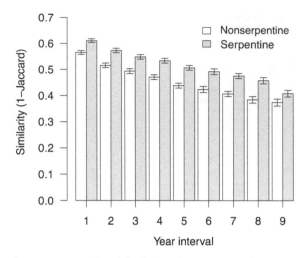

FIGURE 17.5. Average compositional similarity of serpentine and nonserpentine plant communities measured repeatedly in time. Higher values of similarity indicate lower change of community composition between survey at time 1 (t_1) and survey (t_1 + interval).

nonserpentine soil types. We are currently exploring the evidence for these potential mechanisms.

Experimental Climate Manipulations in Grasslands

In another ongoing study, we are experimentally comparing the responses of California serpentine and nonserpentine grasslands to altered spring precipitation. We are focusing on several soil-specialist annual herb species, which are planted into experimental plots in natural grasslands. We use a watering system to impose increases and rainout shelters to impose decreases of about 50% each in spring precipitation. We are also manipulating competition by removing (or not) the grassland vegetation immediately surrounding the focal experimental plants.

Our early findings suggest that the response to changes in spring precipitation may be species-specific (Going et al., unpublished data). For example, a reduction in spring precipitation nearly eliminated the effect of competition for one serpentine endemic, *Navarretia jepsonii*, but increased the effect of competition for another endemic, *Clarkia gracilis* ssp. *tracyii* (Figure 17.6). Of these two herbaceous annuals, *N. jepsonii* has leaf morphology more consistent with drought tolerance. It is also interesting that *N. jepsonii* is more productive when grown in serpentine soil than in normal soil, suggesting that it may have some physiological requirement for serpentine. Some other serpentine endemics have been shown to require very high concentrations of magnesium for optimal growth (Brady et al., 2005).

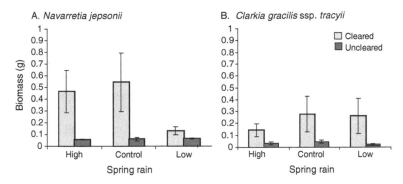

FIGURE 17.6. Biomass of two serpentine endemics, *Navarretia jepsonii* (A) and *Clarkia gracilis* ssp. *tracyii* (B), in response to manipulation of spring precipitation and competition in a serpentine grassland. Error bars represent ± 1 SE.

Rather than finding consistently high resistance to climate change in serpentine endemics, we may find there are "winners" and "losers" and that species responses depend on a particular trait or set of traits.

CONCLUSIONS AND RESEARCH PRIORITIES

Predicting the consequences of global climate change is a major challenge in any ecosystem, especially in the presence of such complexities as diverse topography, steep environmental gradients, and high levels of biodiversity and endemism (Fagre et al., 2003). In such systems, even "subtle environmental differences can have a major influence over the rate and direction of systems response and can lead to unanticipated consequences ('surprises') for those who manage biological resources" (Burkett et al., 2005: 385).

Special-soil environments provide excellent examples of these ecological complexities. We may expect that the rate and direction at which special-soil ecosystems respond to climate change will be greatly affected by the nutrient deficiency of their soils, possibly combined with water deficiency and/or cation excess, and also combined with the particular life history traits of plants that grow in these challenging environments. As this review illustrates, we do not yet have a conceptual framework well developed enough to predict successfully whether climate-related changes will be greater or lesser in these systems, let alone other details of how they may differ. Improved predictions are likely to require a more nuanced understanding of what limiting factors control plant growth on particular soils, how climate change will affect the levels of these limiting factors, and the ways plant responses depend on functional traits. Another complication we have not considered here is how climate change may directly affect the soils themselves; for example, Holzinger et al. (2008) found greater climate-related changes in alpine plant communities on limestone than siliceous soils, and they attributed this difference to the effects of climate change on the chemistry of the calcareous soils.

We believe that models, experiments, and observational studies have important roles to play in improving our understanding of climate change impacts on special soil ecosystems. Climate envelope models can be modified to ask how the inclusion of other factors, particularly the spatial distribution of suitable soils, alters predictions for species persistence or extinction. Climate manipulation experiments could be particularly revealing of how warming and other changes exert differential impacts on different species depending on their functional traits and on different soils depending on their levels of nutrients and other limiting factors. Observational studies such as our Siskiyou research will be helpful in documenting how already occurring changes have affected special-soil ecosystems but will continue to be severely limited by the availability of appropriate historical data sets. Setting up monitoring systems now in anticipation of future change is critically importance, and we hope that researchers will include both special-soil communities and nearby normal soil floras in their monitoring designs.

Evolutionary studies are also needed. Given their limited opportunities to migrate, local adaptation is one of the few options available for the survival of serpentine endemics and other plants on special soils. Potential avenues for such research include rapid evolution of altered climatic tolerances, seedbank adaptations, and selfing abilities; the influences of patch size and isolation on evolutionary potential; and comparisons with conspecifics or congeners on normal soils.

Assisted migration is a proposed but controversial climate-change intervention in which plants or animals may be moved to cooler environments in anticipation of their possible extinction in their home environments (McLachlan et al., 2007). If natural migration is near-impossible and local adaptation is judged unlikely, special-soil endemics may top priority candidates for such intervention. Research is needed on the feasibility and limitations of assisted migration, as well as its possible impacts on population genetic structure, local adaptation, and surrounding communities (McLachlan et al., 2007).

Until recently, the floras of serpentine, karstic limestone, shale barrens, and many other special-soil ecosystems were reasonably well protected by the relative uselessness of these soils for agriculture. Despite severe localized impacts from mining and urbanization, large areas of these habitats survived with considerably less human alteration than ecosystems on richer soils. Climate change may end the status of these ecosystems as stable refuges for biodiversity, however. Though we have noted some possible reasons for optimism (Grime et al., 2000, 2008; Briles, 2008), we believe that other evidence, such as our Siskiyou study, suggests it is unwise to consider special-soil ecosystems safe from climate change.

Ironically, in the case of serpentine, the most destructive human impact of all has recently been proposed as a geoengineering approach to halting climate change (Goff and Lackner, 1998; Howell, 2009; Krevor et al., 2009). Immense quantities of ultramafic rocks would be mined, ground, and mixed with HCl to form $MgCl_2$, which would be piped to fossil fuel power plants where it would react with CO_2 emissions to form thermodynamically stable magnesite ($MgCO_3$). The magnesite could be disposed in open-pit mines or possibly used to make building materials. This idea has attracted considerable interest from the media and environmental groups, and the U.S. Geologic Survey has gone as far as to map the ultramafic formations in the United States that could be consumed for this purpose (Howell, 2009; Krevor et al., 2009). We conclude with the sincere hope that climate change can be addressed successfully without using an approach that would be so devastating to the world's botanical diversity.

Acknowledgments

We thank David Ackerly, Anu Eskelinen, James Grace, and Scott Loarie for invaluable feedback and discussion. The National Science Foundation (DEB-0542451) provided funding for this work (www.nsf.gov).

LITERATURE CITED

Ackerly, D. D. (2009) Evolution, origin and age of lineages in the Californian and Mediterranean floras. *Journal of Biogeography*, 36, 1221–33.

Alexander, E. A., Coleman, R. G., Keeler-Wolf, T., and Harrison, S. (2006) *Serpentine Geoecology of Western North America*. Oxford University Press, Oxford, UK.

Anderson, R., Fralish, J., and Baskin, J. (1999) *Savannas, Barrens and Rock Outcrop Communities of North America*. Cambridge University Press, Cambridge.

Armstrong, J. K., and Huenneke, L. F. (1992) Spatial and temporal variation in species composition in California grasslands: The interaction of drought and substratum. In *Proceedings of the First International Conference on Serpentine Ecology*, pp. 213–33. Intercept, Andover.

Baskin, C. C., and Baskin, J. M. (1998) *Seeds: Ecology, Biogeography, Evolution of Dormancy and Germination*. Academic Press, New York.

Borhidi, A. (1991) *Phytogeography and Vegetation Ecology of Cuba*. Akademiai Kiado, Budapest.

Brady, K. U., Kruckeberg, A. R., and Bradshaw, H. D. (2005) Evolutionary ecology of plant adaptation to serpentine soils. *Annual Review of Ecology Evolution and Systematics*, 36, 243–66.

Briles, C. E. (2008) *Holocene Vegetation and Fire History from the Floristically Diverse Klamath Mountains, Northern California, USA*. PhD diss., University of Oregon.

Brooks, R. R. (1987) *Serpentine and Its Vegetation*. Dioscorides Press, Portland.

Burkett, V. R., Wilcox, D. A., Stottlemyer, R., Barrow, W., Fagre, D., Baron, J., Price, J., Nielsen, J. L., Allen, C. D., and Peterson, D. L. (2005) Nonlinear dynamics in ecosystem response to climatic change: Case studies and policy implications. *Ecological Complexity,* 2, 357–94.

Damschen, E. I., Harrison, S. P., and Grace, J. B. (In Press) Climate change effects on an endemic-rich edaphic flora. *Ecology,* 91.

Denno, R. F., Roderick, G. K., Peterson, M. A., Huberty, A. F., Dobel, H. G., Eubanks, M. D., Losey, J. E., and Langellotto, G. A. (1996) Habitat persistence underlies intraspecific variation in the dispersal strategies of planthoppers. *Ecological Monographs,* 66, 389–408.

Engler, R., Randin, C. F., Vittoz, P., Czáka, T., Beniston, M., Zimmermann, N. E., and Guisan, A. (2009) Predicting future distributions of mountain plants under climate change: Does dispersal capacity matter? *Ecography,* 32, 34–45.

Fagre, D. B., Peterson, D. L., and Hessl, A. E. (2003) Taking the pulse of mountains: Ecosystem responses to climatic variability. *Climatic Change,* 59, 263–82.

Freckleton, R. P., and Watkinson, A. R. (2002) The large scale spatial dynamics of plants: Metapopulations, regional ensembles and patchy populations. *Journal of Ecology,* 90, 419–34.

Goff, F., and Lackner, K. S. (1998) Carbon dioxide sequestering using ultramafic rocks. *Environmental Geosciences,* 5, 89–102.

Grabherr, G., Gottfried, M., and Pauli, H. (1994) Climate effects on mountain plants. *Nature,* 369, 448.

Grime, J. P., Brown, V. K., Thompson, K., Masters, G. J., Hillier, S. H., Clarke, I. P., Askew, A. P., Corker, D., and Kielty, J. P. (2000) The response of two contrasting limestone grasslands to simulated climate change. *Science,* 289, 762–65.

Grime, J. P., Fridley, J. D., Askew, A. P., Thompson, K., Hodgson, J. G., and Bennett, C. R. (2008) Long-term resistance to simulated climate change in an infertile grassland. *Proceedings of the National Academy of Sciences, USA,* 105, 10028–32.

Harper, J. L. (1977) *The Population Biology of Plants.* Academic Press, London.

Harrison, S. (1999) Native and alien species diversity at the local and regional scales in a grazed California grassland. *Oecologia,* 121, 99–106.

Harrison, S., Damschen, E., and Going, B. (2009) Climate gradients, climate change, and special edaphic floras. *Northeastern Naturalist,* 16, 121–30.

Harrison, S., and Grace, J. B. (2007) Biogeographic affinity helps explain productivity-richness relationships at regional and local scales. *American Naturalist,* 170, S5–S15.

Harrison, S., Inouye, B. D., and Safford, H. D. (2003) Ecological heterogeneity in the effects of grazing and fire on grassland diversity. *Conservation Biology,* 17, 837–45.

Harrison, S., Viers, J. H., Thorne, J. H., and Grace, J. B. (2008) Favorable environments and the persistence of naturally rare species. *Conservation Letters,* 1, 65–74.

Hobbs, R. J., and Mooney, H. A. (1991) Effects of rainfall variability and gopher disturbance on serpentine annual grassland dynamics. *Ecology,* 72, 59–68.

Holzinger, B., Hülber, K., Camenisch, M., and Grabherr, G. (2008) Changes in plant species richness over the last century in the eastern Swiss Alps: Elevational gradient, bedrock effects and migration rates. *Plant Ecology,* 195, 179–96.

Howell, K. (2009) Maps show rocks ideal for sequestering carbon. *New York Times.*

IPCC. (2007) *Climate Change 2007: Impacts, Adaptation and Vulnerability: Contribution of Working Group II to the Fourth Assessment Report of the Intergovernmental Panel on Climate Change.* Cambridge University Press, Cambridge.

Kareiva, P., Kingsolver, J., and Huey, R. (1992) *Biotic Interactions and Global Change.* Sinauer Press, Sunderland, MA.

Keeley, J. E., and Fotheringham, C. J. (2001) Historic fire regime in southern California shrublands. *Conservation Biology,* 15, 1536–48.

Klanderud, K., and Totland, O. (2005) Simulated climate change altered dominance hierarchies and diversity of an alpine biodiversity hotspot. *Ecology,* 86, 2047–54.

Krevor, S. C., Graves, C. R., Gosen, B. S. V., and McCafferty, A. E. (2009) Mapping the mineral resource base for mineral carbon-dioxide sequestration in the conterminous United States. U.S. Geologic Survey.

Kruckeberg, A. R. (1954) The ecology of serpentine soils: A symposium. III. Plant species in relation to serpentine soils. *Ecology,* 35, 267–74.

Kruckeberg, A. (1984) *California Serpentines: Flora, Vegetation, Geology, Soils, and Management Problems.* University of California Press, Berkeley.

Kruckeberg, A. (2005) *Geology and Plant Life.* University of Washington Press, Seattle.

Loarie, S. R., Carter, B. E., Hayhoe, K., McMahon, S., Moe, R., Knight, C. A., and Ackerly, D. D. (2008) Climate change and the future of California's endemic flora. *PLoS One,* 3, 1–10.

Malhi, Y., Roberts, J. T., Betts, R. A., Killeen, T. J., Li, W., and Nobre, C. A. (2008) Climate change, deforestation, and the fate of the Amazon. *Science,* 319, 169.

McLachlan, J. S., Hellmann, J. J., and Schwartz, M. W. (2007) A framework for debate of assisted migration in an era of climate change. *Conservation Biology,* 21, 297–302.

Meimberg, H., Rice, K. J., Milan, N. F., Njoku, C. C., and McKay, J. K. (2009) Multiple origins promote the ecological amplitude of allopolyploid *Aegilops* (Poaceae). *American Journal of Botany,* 96, 1262.

Mittermeier, R. A., Gil, P. R., Hoffman, M., Pilgrim, J., Brooks, T., Mittermeier, C. G., Lamoreux, J., and Fonseca, G. A. B. d. (2005) *Hotspots Revisited: Earth's Biologically Richest and Most Endangered Terrestrial Ecoregions.* University of Chicago Press, Chicago.

Mittermeier, R., Myers, N., and Mittermeier, C. (2000) Hotspots: Earth's Biologically Richest and Most Endangered Terrestrial Ecoregions. Conservation International.

Moritz, C., Patton, J. L., Conroy, C. J., Parra, J. L., White, G. C., and Beissinger, S. R. (2008) Impact of a century of climate change on small-mammal communities in Yosemite National Park, USA. *Science,* 322, 261.

Myers, N., Mittermeier, R. A., Mittermeier, C. G., da Fonseca, G. A. B., and Kent, J. (2000) Biodiversity hotspots for conservation priorities. *Nature,* 403, 853–58.

Nelson, P., and Ladd, D. (1983) Preliminary report on the identification, distribution, and classification of Missouri glades. In *Proceedings of the Seventh North American Prairie Conference* (ed. C. L. Kucera), pp. 59–76. Southwest Missouri State University, Springfield.

Nemani, R. R., Keeling, C. D., Hashimoto, H., Jolly, W. M., Piper, S. C., Tucker, C. J., Myneni, R. B., and Running, S. W. (2003) Climate-driven increases in global terrestrial net primary production from 1982–1999. *Science,* 300, 1560–63.

NOAA. (2009) National Environmental Satellite, Data, and Information Service U.S. Climate Data. National Oceanic and Atmospheric Administration.

Opdam, P., and Wascher, D. (2004) Climate change meets habitat fragmentation: Linking landscape and biogeographical scale levels in research and conservation. *Biological Conservation*, 117, 285–97.

Parmesan, C., and Yohe, G. (2003) A globally coherent fingerprint of climate change impacts across natural systems. *Nature*, 421, 37–42.

Parolo, G., and Rossi, G. (2008) Upward migration of vascular plants following a climate warming trend in the Alps. *Basic and Applied Ecology*, 9, 100–107.

Pärtel, M., and Zobel, M. (2007) Dispersal limitation may result in the unimodal productivity-diversity relationship: A new explanation for a general pattern. *Journal of Ecology*, 95, 90–94.

Phoenix, G. K., Hicks, W. K., Cinderby, S., Kuylenstierna, J. C. I., Stock, W. D., Dentener, F. J., Giller, K. E., Austin, A. T., Lefroy, R. D. B., and Gimeno, B. S. (2006) Atmospheric nitrogen deposition in world biodiversity hotspots: The need for a greater global perspective in assessing N deposition impacts. *Global Change Biology*, 12, 470–76.

Post, E., and Pedersen, C. (2008) Opposing plant community responses to warming with and without herbivores. *Proceedings of the National Academy of Sciences, USA*, 105, 12353–58.

Ricketts, T. H., Dinerstein, E., Olson, D., Loucks, C., Eichbaum, W., DellaSalla, D., Kavanagh, K., Hedao, P., Hurley, P., Carney, K., Abell, R., and Walters, S. (1999) *Terrestrial Ecoregions of North America: A Conservation Assessment*. Island Press, Washington, DC.

Roff, D. A. (1990) The evolution of flightlessness in insects. *Ecological Monographs*, 60, 389–421.

Rune, O. (1953) Plant life on serpentine and related rocks in the north of Sweden. *Acta Phytogeographica Suecica*.

Safford, H. D., and Harrison, S. (2004) Fire effects on plant diversity in serpentine vs. sandstone chaparral. *Ecology*, 85, 539–48.

Safford, H., Viers, J., and Harrison, S. (2005) Serpentine endemism in the Calfornia flora: A database of serpentine affinity. *Madrono*, 52, 222–57.

Schwartz, M. W., Iverson, L. R., Prasad, A. M., Matthews, S. N., and O'Connor, R. J. (2006) Predicting extinctions as a result of climate change. *Ecology*, 87, 1611–15.

Sharitz, R. R., and McCormick, J. F. (1973) Population dynamics of two competing annual species. *Ecology*, 54, 723–40.

Skinner, M., and Pavlik, B. (1994) *California Native Plant Society's Inventory of Rare and Endangered Plants in California*. California Native Plant Society, Sacramento.

Smith, S., Balkwill, K., and Williamson, S. (2001) Compositae on serpentine in the Barberton Greenstone Belt, South Africa. *South African Journal of Science*, 97, 518–20.

Southwood, T. R. E. (1977) Habitat, the templet for ecological strategies? *Journal of Animal Ecology*, 46, 337–65.

Stein, B., Kutner, L., and Adams, J. (2000) *Precious Heritage: The Status of Biodiversity in the United States*. Oxford University Press, Oxford.

Suttle, K. B., Thomsen, M. A., and Power, M. E. (2007) Species interactions reverse grassland responses to changing climate. *Science*, 315, 640.

Tansley, A. G. (1917) On competition between *Galium saxatile* L. (G. Hercynicum Weig.) and *Galium sylvestre* Poll. (*G. Asperum* Schreb.) on different types of soil. *Journal of Ecology*, 5, 173–79.

Theurillat, J. P., and Guisan, A. (2001) Potential impact of climate change on vegetation in the European Alps: A review. *Climatic Change*, 50, 77–109.

Thomas, C. D., Cameron, A., Green, R. E., Bakkenes, M., Beaumont, L. J., Collingham, Y. C., Erasmus, B. F. N., Siqueira, M. F. d., Grainger, A., Hannah, L., Hughes, L., Huntley, B., Jaarsveld, A. S. v., Midgley, G. F., Miles, L., Ortega-Huerta, M. A., Peterson, A. T., Phillips, O. L., and Williams, S. E. (2004) Extinction risk from climate change. *Nature*, 427, 145-148.

Valiente-Banuet, A., Rumebe, A. V., Verdú, M., and Callaway, R. M. (2006) Modern Quaternary plant lineages promote diversity through facilitation of ancient Tertiary lineages. *Proceedings of the National Academy of Sciences, USA*, 103, 16812–17.

Walther, G. R., Beiflner, S., Burga, C. A., and White, P. S. (2005) Trends in the upward shift of alpine plants. *Journal of Vegetation Science*, 16, 541–48.

Wardle, D. A., Walker, L. R., and Bardgett, R. D. (2004) Ecosystem properties and forest decline in contrasting long-term chronosequences. *Science*, 305, 509–13.

Ware, S. (1991) Influence of interspecific competition, light, and moisture levels on growth of rock outcrop *Talinum* (Portulacaceae). *Bulletin of the Torrey Botanical Club*, 118, 1–5.

Weiss, S. B. (1999) Cars, cows, and checkerspot butterflies: Nitrogen deposition and management of nutrient-poor grasslands for a threatened species. *Conservation Biology*, 13, 1476–86.

Westerling, A. L., Hidalgo, H. G., Cayan, D. R., and Swetnam, T. W. (2006) Warming and earlier spring increase western US forest wildfire activity. *Science*, 313, 940–43.

Whittaker, R. (1960) Vegetation of the Siskiyou Mountains, Oregon and California. *Ecological Monographs*, 30, 279–338.

Witkowski, E. T., Dahlmann, L. A., and Boycott, R. C. (2001) Conservation biology of *Kniphofia umbrina*, a critically endangered Swaziland serpentine endemic. *South African Journal of Science* 97, 609–16.

18

Restoration and Revegetation
of Harsh Soils

Ryan E. O'Dell, *Bureau of Land Management*
Victor P. Claassen, *University of California, Davis*

Human activities such as fire regime alteration, introduction of non-native plant and animal species, livestock grazing, agriculture, logging, pollution, motorized recreation, construction, and mining can dramatically alter ecosystem structure and function. Adverse impacts to ecosystems from human activities may include alteration of plant, animal, and microbial community cover/abundance, structure, composition, and diversity; soil removal and/or erosion; and landscape instability (mass wasting/landslide).

Restoration, reclamation, rehabilitation, and revegetation of degraded ecosystems are undertaken for the benefit of human health, environment, or both. Restoration is the process of returning an ecosystem to a condition very similar in ecological structure and function to what existed prior to disturbance (Society for Ecological Restoration, 2004). This typically involves manipulation of vegetation cover, structure, composition, and/or diversity and may also involve some minor manipulation of disturbance regime, soil conditions, and/or hydrology to restore original conditions. Restoration is typically undertaken on mildly or moderately degraded ecosystems where some to most ecological function is still intact. Reclamation, rehabilitation, or revegetation, in comparison, are typically undertaken on severely degraded ecosystems where most ecological function has been lost, such as construction sites and mined lands. These processes often involve intensive soil structure and function rebuilding, followed by replanting. Reclamation, rehabilitation, and revegetation have the common primary goal of landscape stability and useful function, rather than a high level of ecological function, therefore, the term *revegetation* is used to encompass all three processes in this chapter.

In this chapter, we focus on methods used in the restoration and revegetation of stressful edaphic environments. We begin with methods used in restoration or revegetation of nonserpentine soils and then describe how the same approaches are applied to the restoration or revegetation of serpentine soils. Discussion highlights soil and vegetation manipulation methods used to restore partially degraded serpentine ecosystems and then focuses on three steps, including physical site stabilization, substrate amendment, and plant and microbial materials selection, that are critical to successful revegetation of severely degraded serpentine ecosystems. We conclude by evaluating how well serpentine ecosystems serve as model environments for restoration and revegetation studies.

RESTORATION

Restoration methods for nonserpentine soils vary by ecosystem type, but most require some minor manipulation of vegetation, soil, and/or hydrology to regenerate original ecosystem conditions. Grasslands are one of the most imperiled ecosystems globally due largely to fire suppression, livestock grazing, invasive species, and conversion to agriculture (Gibson, 2009; Stromberg et al., 2007). As a result, grasslands are also one of the most commonly restored ecosystems. Methods used to restore grasslands have included tilling or ripping to relieve soil compaction; invasive species control or native plant species management with herbicide, livestock grazing, mechanical vegetation removal, and control burning; and replanting with native plant species (Gibson, 2009; Stromberg et al., 2007).

To date, most serpentine restoration we know of has involved grasslands on the Shetland Islands, Scotland, United Kingdom; California, Pennsylvania, Maryland, and North Carolina in the United States; Vysočina, Czech Republic; and Prato, Italy. Primary impacts to serpentine grassland communities in the United Kingdom have been from grazing and pollution (nitrogen and phosphorus) (Carter et al., 1988; Slingsby et al., 2001). In the United States, serpentine grasslands have been adversely impacted by fire regime alteration, grazing, pollution (nitrogen), and non-native species introduction (Weiss, 1999, 2002; Safford and Harrison, 2001; Gelbard and Harrison, 2003; Harrison et al., 2003; Weiss and Wright, 2005, 2006; Weiss et al., 2007). Serpentine grasslands in the Czech Republic and Italy have been impacted by fire regime alteration, grazing, and non-native species introduction, resulting in alteration of plant community composition and diversity (Veselý, 2002; HABIO, 2010).

Chemical fertilization (nitrogen, phosphorus, and potassium) of serpentine grasslands has a profound effect on plant cover, structure, composition, and diversity (Ferreira and Wormell, 1971; Turitzin, 1982; Carter et al., 1988; Huenneke et al., 1990; Slingsby et al., 2001; Going et al., 2009). Studies have shown that fertilization promotes invasion by non-native as well as native nonserpentine species, whereas abundances of native serpentine species decline due to increased competition. Fertilization to increase grassland productivity and manure from livestock grazing has increased eutrophication of rocky serpentine barrens and grasslands at the Keen of Hamar, Shetland Islands, Scotland (Slingsby and Carter, 1986, 1987; Carter et al., 1988; Slingsby et al., 2001). The result has been a dramatic increase in vegetative cover, mostly from increased abundance of species more common on nonserpentine soils surrounding the site and a corresponding decrease in the abundance of species typically restricted to serpentine soils, such as the classified vulnerable serpentine endemic Edmonston's chickweed (*Cerastium nigrescens*). Restoration at the Keen of Hamar has involved removal of nutrient-enriched soil layers and elimination of cattle grazing (Scottish Natural Heritage, 2008; Slingsby et al., 2010). Although historic soil conditions of the serpentine barrens have been restored, few Edmonston's chickweed or other rare plants have colonized the habitat to date (Slingsby et al., 2010), illustrating the difficulty of attaining full restoration.

Prior to widespread European settlement around 1850, California's grasslands were dominated by native, perennial, bunchgrass species (Huenneke and Mooney, 1989; Stromberg et al., 2007). Lightning and Native American–ignited fires were once a common occurrence in California's dry Mediterranean climate (Carle, 2008). Due to a combination of fire suppression, livestock grazing, and non-native species introduction from the Mediterranean region, California's once native perennial grasslands have been almost entirely converted to invasive annual grasslands (Huenneke and Mooney, 1989; Stromberg et al., 2009). Serpentine soils support some of the last remaining stands of California perennial grassland due to their inhospitable soil conditions, which inhibit invasion of non-native annual grass species (Turitzin, 1982; Huenneke et al., 1990; Harrison, 1999a, 1999b; Harrison et al., 2001; Jurjavcic et al., 2002; Williamson and Harrison, 2002; Seabloom et al., 2003; Gram et al., 2004). Dry atmospheric nitrogen deposition has had similar effects on serpentine grasslands in California as eutrophication did at the Keen of Hamar.

The San Francisco Bay area in California is densely populated and atmospheric nitrogen (N) deposition as nitrogen oxide (NOx) and ammonia (NH_3^+), mostly from automobile exhaust, is estimated at 10–20 kg N ha^{-1} year^{-1} (Weiss, 1999). Atmospheric N deposition has been positively correlated with increased abundance of non-native grasses and decreased abundance of native forbs and grasses in Bay area serpentine grasslands (Weiss, 1999; East Bay Regional Park District, 2009). These changes have negatively affected rare species, including the federally listed threatened Bay checkerspot butterfly (*Euphydryas editha bayensis*; Lepidoptera) and federally listed endangered Presidio clarkia (*Clarkia franciscana*; Onagraceae) (Weiss, 1999; East Bay Regional Park District, 2009).

Restoration methods used in these serpentine grasslands have included seasonal mowing, seasonal and limited livestock grazing, prescribed burning, and the removal of both native and non-native woody species (Weiss, 2002; Peters, 2003; East Bay Regional Park District, 2009). All three treatments have been effective at decreasing the abundance of non-native annual plant species and increasing the abundance of native serpentine grassland plant species, including those that host Bay checkerspot butterfly (Safford and Harrison, 2001; Weiss, 2002; Gelbard and Harrison, 2003; Harrison et al., 2003; Weiss and Wright, 2005, 2006; Weiss et al., 2007). Similar treatments have been used with some success to restore native perennial nonserpentine grasslands in California (Stromberg and Kephart, 1996; Bugg et al., 1997: Stromberg et al., 2007).

Invasion of both native and non-native woody species (mostly trees) into serpentine grasslands in Pennsylvania, Maryland, and North Carolina in the eastern United States has resulted in major reductions of grassland area (Tyndall, 1992; Tyndall and Hull, 1999). The primary reason identified for woody species invasion is fire suppression (Tyndall, 1992; Arabas, 2000; Latham, 2008). Historically, fires ignited by Native Americans maintained open grasslands on low-nutrient soils, but widespread European settlement in the region around 1750 decreased fire frequency. Livestock grazing partially replaced the effects of fire by reducing woody species establishment in the grasslands; however, elimination of grazing around 1900 allowed woody species to rapidly invade the grasslands. It is estimated that more than 80% of the Eastern serpentine grasslands have succeeded to woodland and forest since European settlement (Tyndall, 1992). Serpentine grasslands of the eastern United States harbor many rare plant species such as serpentine aster (*Symphyotrichum depauperatum*; Asteraceae) and the serpentine endemic Rhiannon's aster (*S. rhiannon*). Restoration techniques have included cutting native and non-native woody species, prescribed burning, seasonal livestock grazing, and removal of soil organic matter and plant litter (Tyndall, 1994, 2005; Marx, 2007; Latham, 2008). These treatments have been effective at reducing woody species cover and have greatly increased serpentine grassland area, native species diversity, and rare species abundance (Tyndall, 1994, 2005; Marx, 2007; Latham, 2008). The same treatments have been effectively used to reduce both native and non-native invading woody species cover and promote serpentine grassland diversity in the Czech Republic and Italy (Veselý, 2002; HABIO, 2010).

Natural vegetation patterns on serpentine soils demonstrate that harsh soils can naturally resist non-native plant species invasion. Any changes in soil conditions, such as nutrient pollution, that alleviate edaphic stress, however, increase invasion susceptibility. Once the edaphic stress is relieved and plant community changes have been allowed to progress, they can be very difficult to reverse. In these conditions, full restoration may be difficult or impossible to achieve.

REVEGETATION

Revegetation of nonserpentine soils is most often employed on drastically disturbed landscapes, including construction areas and mined lands. Construction and mining severely degrades ecosystems by removing both vegetation and soil down to bedrock, resulting in shallow, nutrient-poor substrate that is susceptible to drought (Williamson et al., 1982; Tordoff et al., 2000). Metal ore mine tailings have additional substrate stress factors, including low pH and high bioavailable metal concentrations (Williamson et al., 1982; Tordoff et al., 2000).

Revegetation of drastically disturbed substrates can ameliorate adverse environmental and human health impacts such as those described, but such extensive regeneration presents significant challenges. Successful revegetation of drastically disturbed substrates is dependent on three critical steps, including (1) physical site stabilization, (2) substrate amendment, and (3) plant and microbial materials selection. These processes are universal for approaching any revegetation project regardless of location or environmental conditions (Quilty, 1975; Williamson et al., 1982; Van Kererix and Kay, 1986; Bradshaw, 1997; Tordoff et al., 2000).

Mine pits and tailings piles are usually unstable and susceptible to erosion and mass wasting. Structural (geotechnically engineered) stabilization and control of surface erosion is the first step in revegetation (Williamson et al., 1982; Tordoff et al., 2000). Following stabilization, amendment of mine tailings is important to improve adverse substrate conditions for plant establishment and growth. Many studies have concluded that chemical (inorganic) and organic amendments (biosolids, compost, peat), especially when applied in combination, are the most effective treatments for improving the fertility of mine tailings (Goodman et al., 1973; Johnson et al., 1977; Smith and Bradshaw, 1979; Kramer et al., 2000a, 2000b; Tordoff et al., 2000; Yang et al., 2003; Clemente et al., 2006; O'Dell et al., 2007). Greatest rooting depth and plant growth is achieved when the amendments are incorporated into the tailings, as opposed to only surface application (Tordoff et al., 2000; von Willert and Stehouwer, 2003; O'Dell et al., 2007; Curtis and Claassen, 2009).

The importance of using locally adapted plant species and ecotypes tolerant of the adverse site conditions is widely recognized (Antonovics et al., 1971; Goodman et al., 1973; Smith and Bradshaw, 1979; Kruckeberg and Wu, 1992; Tordoff et al., 2000; Grant et al., 2002; Shu et al., 2005). Plant species and ecotypes growing on metal-contaminated soils and mine tailings typically display greater tolerance to adverse substrate conditions of the tailings than those collected from undisturbed soils (Wu et al., 1975; Smith and Bradshaw, 1979; Kruckeberg and Wu, 1992; Tordoff et al., 2000). Mycorrhizal fungi have also been shown to have distinct metal tolerant ecotypes (Gaur and Adholeya, 2004). Some plant species and ecotypes have specific metal tolerance mechanisms, such as metal exclusion and sequestration (Wu et al., 1975; Smith and Bradshaw, 1979; Punz and Sieghardt, 1993; Brooks, 1998). Hyperaccumulators, which are exemplified by their ability accumulate shoot tissue metal concentrations $>$1000 $\mu g\ g^{-1}$, are often used in the phytoremediation of industrial- and smelter-contaminated nonserpentine soils (Chaney et al., 1997; Brooks, 1998).

Serpentine substrates bear many similarities to mine tailings, including extreme adverse physical and chemical substrate conditions and physiologically adapted species and ecotypes that grow on them. As a source of many economically important minerals and metal ores, serpentine substrates also experience high levels of disturbance due to mining. As a result, serpentine substrates present model environments for studying how physical stabilization, soil amendment, and plant materials selection interact during restoration and revegetation.

SERPENTINE MINING IMPACTS

Due to its unique geologic origin and chemical composition, serpentine contains many commercially valuable minerals and metal ores. Semiprecious jadeite and rare gems mined from serpentine (such as garnet and diamond) are highly valued by mineral collectors. Industrial minerals commonly mined from serpentine include magnesite (magnesium carbonate), talc, and asbestos. High-quality verde antique (green marble) is quarried for building construction. Lower quality serpentine rock is occasionally quarried and crushed for road base aggregate. Serpentine also contains many world-class metal ore deposits that are mined for nickel laterite, millerite/pentlandite (nickel sulfide), chalcopyrite/bornite (copper sulfide), chromite (iron-magnesium-chromium oxide), platinum group metals, and cinnabar (mercury sulfide). Globally, the largest mining disturbances on serpentine are associated with asbestos, nickel laterite, and chromite mining, as described here.

Asbestos, a fibrous mineral prized for its heat, flame, and electrical resistance properties and high durability, was mined as early as 5,000 years ago in Finland and Greece (Ross and Nolan, 2003). In 2006, 76% of the 2.3 million metric ton global total production of asbestos was mined from serpentine. Inhalation of airborne asbestos fibers is known to have adverse impacts on human health, including asbestosis and mesothelioma (Ross and Nolan, 2003). Global asbestos production increased 5–10% annually up until about 1975, when the adverse effects of asbestos exposure on human health became widely recognized (Virta, 2006). Global production and its use in products declined dramatically thereafter and are expected to continue declining (Virta, 2006). Asbestos production and its use in products are currently banned by more than 40 countries. The global total estimated area of disturbance associated with asbestos mining alone is 11,700 ha (Table 18.1).

TABLE 18.1 Asbestos Production and Estimated Total Area of Drastically Disturbed Serpentine Substrate of Major Serpentine-Hosted Asbestos Mines in 2006

Mining Center	State/Province	Country	Location Coordinates		Current Status	Production (metric tons)[a]	% of Total Global Production[a]	Total Disturbance Area (ha)[b]
			Latitude	Longitude				
Calaveras Asbestos	California	U.S.	37° 56′ 16″	−120° 32′ 48″	Inactive	0	0	160
KCAC	California	U.S.	36° 20′ 21″	−120° 36′ 54″	Inactive	0	0	120
Atlas	California	U.S.	36° 19′ 20″	−120° 35′ 15″	Inactive	0	0	180
Belvidere	Vermont	U.S.	44° 45′ 47″	−72° 31′ 44″	Inactive	0	0	220
Cassiar	British Columbia	Canada	59° 19′ 42″	−129° 49′ 04″	Inactive	0	0	200
Clinton Creek	Yukon Territory	Canada	64° 27′ 03″	−140° 43′ 24″	Inactive	0	0	180
Jeffrey	Quebec	Canada	45° 46′ 18″	−71° 57′ 25″	Active ↗	185,000	8	740 ↑
Thetford	Quebec	Canada	46° 01′ 48″	−71° 22′ 16″	Active ↖	↑	↑	2150 ↑
Cana Brava	Goiás	Brazil	−13° 31′ 55″	−48° 14′ 29″	Active	227,304	11	460
Woodsreef	New South Wales	Australia	−30° 24′ 17″	150° 44′ 06″	Inactive	0	0	290
Shabani	Midlands	Zimbabwe	−20° 19′ 55″	30° 03′ 57″	Active ↗	96,956	4	160 ↑
Gaths	Masvingo	Zimbabwe	−20° 01′ 06″	30° 30′ 49″	Active ↖	↑	↑	190 ↑
Havelock	Hhohho	Swaziland	−25° 57′ 26″	31° 07′ 54″	Inactive	0	0	40
Balangero	Torino	Italy	45° 17′ 40″	7° 30′ 27″	Inactive	0	0	70
Canari	Corsica	France	42° 49′ 15″	9° 19′ 48″	Inactive	0	0	50
Korlace	Central Serbia	Serbia	43° 22′ 08″	20° 40′ 46″	Inactive	0	0	50

(continued)

TABLE 18.1 (continued)

| Mining Center | State/Province | Country | Location Coordinates | | Current Status | Production (metric tons)[a] | % of Total Global Production[a] | Total Disturbance Area (ha)[b] |
			Latitude	Longitude				
Zidani	Kozani	Greece	40° 05 16"	21° 51' 07"	Inactive	0	0	140
Amiantos	Cyprus	Greece	34° 55' 19"	32° 54' 51"	Inactive	0	0	650
Dzhetygara (Zetiqara)	Qostanay	Kazakhstan	52° 08' 25"	61° 15' 22"	Active	314,700	13	1300
Bazhenovo	Sverdlovsk	Russia	57° 01' 08"	61° 29' 48"	Active ↗			2950
Ak-Dovurak	Tuva	Russia	51° 11' 55"	90° 34' 07"	Active ↑↑	925,000 ↑	40 ↑	650
Kiembay	Orenburg	Russia	51° 00' 19"	59° 55' 59"	Active ↖			750
Global Total						2,300,000		11,700

[a] Data from Brown et al (2009).

[b] Estimated from aerial and satellite photography. Google Earth (2008).

Nickel laterite is a highly weathered tropical soil derived from serpentine. It consists of both Ni-rich limonite (soil) and saprolite (weathered serpentine bedrock). Nickel laterite is a relatively low-grade Ni ore (typically 0.5–2.0% Ni by mass) mined by stripping large areas of soil down to bedrock (Golightly, 1981). Nickel laterite mining provides 30–40% of the world's Ni and represents about 70% of the world's Ni reserves (Brown et al., 2009). Global Ni production has increased eightfold, or 5–15%, annually from 1950 to 2003 (Jessup and Mudd, 2008). The proportion of Ni sourced from nickel laterite increased from less than 10% to 33% during the same period (Kuck, 2006). The global total estimated area of disturbance associated with Ni laterite mining to present is at least 19,070 ha (Table 18.2).

Chromite is a mineral that occurs as stratiform (layer) or podiform (pod) deposits within serpentine bedrock. Chromite mining provides virtually all of the world's chromium (Cr) and represents nearly all of the world's Cr reserves. Global Cr production has increased sixfold from 1950 to 2003 with world production increasing 5–10% annually (Kelly and Matos, 2008). The global total estimated area of disturbance associated with chromite mining to present is at least 4,795 ha (Table 18.3).

Mining and construction activities typically remove serpentine topsoil down to bedrock. Removal of topsoil results in loss of the important biological features associated with it, including rooting depth, water holding capacity, cation exchange capacity, organic matter, plant essential nutrients, plant seed, mycorrhizal propagules, and associated microbial communities. These losses result in drastically disturbed, barren substrate that is erosive and fails to naturally revegetate, even after several decades. The barren substrate results in elevated sediment and heavy metal transport to watersheds and are a source of windborne heavy metal and asbestos pollution.

Step 1: Physical Site Stabilization

Physical stabilization is the critical first step in revegetation. This step consists of structural stabilization (geotechnical engineering) and control of surface erosion. During the mining process, tailings (tips, spoils) are often piled without planning for long-term stability or revegetation. Other unstable landscape features that may be created during mining include mine pit sidewalls and poorly engineered access roads. These features are susceptible to structural collapse and mass wasting, requiring structural stabilization through engineering techniques such as slope reprofiling (slope reduction <2:1 [horizontal:vertical]), contour terracing, benched lift construction, and/or landform grading (Williamson et al., 1982; Van Kekerix and Kay, 1986; Schor and Gray, 1995; Abramson et al., 2001). Slope reprofiling and benched, compacted lifts were successfully employed to stabilize tailings piles prior to substrate amendment and planting at the Amiantos asbestos mine,

TABLE 18.2 Nickel Production and Estimated Total Area of Drastically Disturbed Serpentine Substrate of Major Nickel Laterite Mining Centers in 2006

Mining Center	State/Province	Country	Location Coordinates		Current Status		Production (metric tons)[a]		% of Total Global Production[a]		Total Disturbance Area (ha)[b]
			Latitude	Longitude							
Nickel Mountain	Oregon	U.S.	42° 57' 57"	-123° 26' 17"	Inactive		0		0		180
Moa Bay	Holguín	Cuba	20° 37' 27"	-74° 56' 06"	Active	↗	74,000	↑	5	↑	4700
El Purio	Holguín	Cuba	20° 36' 14"	-75° 32' 00"	Active	↖				↑	1100
Falcondo	Monseñor Novel	Dominican Republic	18° 54' 10"	-70° 19' 19"	Active		46,500		3		1440
Cerro Matoso	Córdoba	Columbia	7° 56' 27"	75° 28' 50"	Active	↗	51,000[c]	↑	3[c]		50
Loma de Níquel	Aragua	Venezuela	10° 02' 22"	-67° 27' 10"	Active		18,200		1		50
Niquelândia	Goiás	Brazil	-14° 21' 20"	-48° 23' 50"	Active	↗				↑	360
Barro Alto	Goiás	Brazil	-15° 06' 01"	-49° 00' 32"	Active	↑	82,492	↑	5	↑	50
Codemin	Goiás	Brazil	-14° 09' 28"	-48° 21' 12"	Active	↖				↑	100
Sulawesi	Southeast Sulawesi	Indonesia	-4° 13' 22"	121° 36' 03"	Active	↗	150,000	↑	10	↑	710
Buli Serani	North Maluku	Indonesia	0° 50' 41"	128° 16' 22"	Active	↖				↑	1170
Cawse	Western Australia	Australia	-30° 22' 44"	121° 09' 21"	Inactive	↗	31,524[c]	↑	2[c]	↑	250
Murrin Murrin	Western Australia	Australia	-28° 47' 08"	121° 52' 33"	Active	↖				↑	580
Ravensthorpe	Western Australia	Australia	-33° 38' 37"	120° 24' 36"	Active		0		0		70

Site	Region	Country	Latitude	Longitude	Status						
Taganito	Caraga	Philippines	9° 31' 51"	125° 49' 02"	Active	↗				↑	170
Taganaan	Caraga	Philippines	9° 46' 02"	125° 43' 15"	Active	↗				↑	150
South Dinagat	Caraga	Philippines	9° 50' 09"	125° 37' 59"	Active	↑	64,705	↑	4	↑	280
Cagdianao	Caraga	Philippines	10° 09' 34"	125° 40' 06"	Active	↗				↑	30
Rio Tuba	Mimaropa	Philippines	8° 33' 55"	117° 25' 18"	Active	↖				↑	760
Manicani	Eastern Visayas	Philippines	10° 59' 32"	125° 38' 21"	Inactive		0		0	↑	390
Tiébaghi	North	New Caledonia	-20° 28' 17"	164° 13' 11"	Active	↗				↑	220
Kaala	North	New Caledonia	-20° 35' 51"	164° 22' 31"	Active	↗				↑	220
Kopéto	North	New Caledonia	-20° 44' 34"	164° 28' 58"	Active	↗				↑	590
Kouaoua	North	New Caledonia	-21° 24' 13"	165° 44' 28"	Active	↗				↑	2520
Thio	South	New Caledonia	-21° 36' 49"	166° 10' 51"	Active	↗				↑	1500
Ningua	South	New Caledonia	-21° 43' 32"	166° 08' 22"	Active	↑	102,988	↑	7	↑	420
Kongouhaou	South	New Caledonia	-21° 45' 55"	166° 10' 23"	Active	↖				↑	330
SMMO	South	New Caledonia	-21° 56' 11"	166° 12' 35"	Active	↖				↑	360
Mont Dore	South	New Caledonia	-22° 15' 29"	166° 36' 25"	Active	↖				↑	80
Plaine des Lacs	South	New Caledonia	-22° 18' 31"	166° 49' 03"	Active	↖				↑	190
Goro	South	New Caledonia	-22° 17' 12"	167° 00' 11"	Active	↖				↑	50
Global Total							1,567,000				19,070

NOTE: Cawse was operable in 2006, but is currently inactive. Ravensthorpe did not become active until 2007 (no productivity in 2006).

a Data from Brown et al. (2009) (except for Cerro Matoso, Columbia and Cawse and Murrin Murrin, Australia).

b Estimated from aerial and satellite photography. Google Earth (2008).

c Data from Kuck (2006).

TABLE 18.3 Chromium Production and Estimated Total Area of Drastically Disturbed Serpentine Substrate of Major Chromium Mining Centers in 2006

Mining Center	State/Province	Country	Location Coordinates		Current Status	Production (metric tons)[a]	% of Total Global Production[a]	Total Disturbance Area (ha)[b]
			Latitude	Longitude				
Grey Eagle	California	U.S.	39° 44' 17"	-122° 35' 30"	Inactive	0	0	10
Moa Bay	Holguín	Cuba	20° 37' 27"	-74° 56' 06"	Active	5,047	<0.1	[c]
Kemi	Lappi	Finland	65° 47' 02"	24° 42' 47"	Active	549,000	3	405
Bulqizë	Dibër	Albania	41° 29' 22"	20° 13' 23"	Active ↗			15
Qafe Bualli	Dibër	Albania	41° 29' 07"	20° 12' 03"	Active ↑↑ ↑	201,120	1 ↑	20 ↑
Batra	Dibër	Albania	41° 28' 49"	20° 13' 28"	Active ↖			15 ↑
Guleman	Maden and Alacakaya	Turkey	38° 30' 20"	38° 44' 41"	Active	1,059,901	5	80
Saranovskaya Rudnaya	Perm	Russia	58° 32' 33"	58° 45' 20"	Active	966,095	5	40
Luobusa	Tibet	China	29° 13' 56"	92° 11' 31"	Active	220,000	1	100
Donskoi (Khromtau)	Aqtobe	Kazakhstan	50° 14' 56"	58° 27' 50"	Active	3,336,078	17	1525
Furumad-Gaft	North Khorasan	Iran	36° 36' 48"	56° 58' 33"	Active ↗			10
Esfandegeh–Abdasht	Kerman	Iran	28° 23' 13"	56° 46' 52"	Active ↑↑ ↑	244,603	1 ↑	5
Faryab	Hormozgan and Kerman	Iran	27° 23' 13"	57° 24' 35"	Active ↖			20 ↑
Muslim Bagh	Balochistan	Pakistan	30° 43' 35"	67° 52' 57"	Active ↗			15
Khanozai	Balochistan	Pakistan	30° 37' 04"	67° 22' 31"	Active ↑↑ ↑	64,572	0.3 ↑	5
Nisai	Balochistan	Pakistan	30° 46' 28"	67° 57' 35"	Active ↖			5
Ingessana Hills	An Nil al Azraq (Blue Nile)	Sudan	11° 26' 12"	33° 57' 50"	Active	24,200	0.1	10

Oman Cr Company	Al Buraimi	Oman	24° 13' 58"	56° 04' 20"	Active		67,000		0.3		90
Ankazotaolana	Mahajanga	Madagascar	-17° 42' 13"	47° 34' 31"	Active		116,290		0.6		50
Shurugwi	Midlands	Zimbabwe	-19° 38' 58"	30° 01' 08"	Active	↗		↑		↑	50
Mutorashanga	Mashonaland West	Zimbabwe	-17° 06' 25"	30° 42' 47"	Active	↑	700,001		4	↑	110
Lalapanzi	Midlands	Zimbabwe	-19° 19' 54"	30° 09' 22"	Active	↖				↑	50
Grasvally	Limpopo	South Africa	-24° 19' 40"	28° 58' 23"	Active	↘				↑	20
Dilokong	Limpopo	South Africa	-24° 33' 05"	30° 08' 43"	Active	↘				↑	35
Tubatse (Steelpoort)	Mpumalanga	South Africa	-24° 44' 34"	30° 12' 52"	Active	↘				↑	105
Dwarsrivier	Mpumalanga	South Africa	-24° 55' 52"	30° 07' 21"	Active	↘				↑	90
Lyndenburg	Mpumalanga	South Africa	-25° 03' 53"	30° 28' 15"	Active	↘				↑	85
Thorncliffe-Helena	Mpumalanga	South Africa	-24° 56' 07"	30° 29' 24"	Active	↘				↑	60
Kennedy's Vale-Lion	Mpumalanga	South Africa	-24° 49' 02"	30° 06' 47"	Active	↘				↑	50
Lavino	Mpumalanga	South Africa	-24° 47' 11"	30° 10' 14"	Active	↑	7,418,326	↑	37	↑	40
Tweefontein	Mpumalanga	South Africa	-24° 53' 38"	30° 07' 14"	Active	↖				↑	20
Doombosch	Mpumalanga	South Africa	-24° 41' 52"	30° 12' 11"	Active	↖				↑	15
Machadodorp	Mpumalanga	South Africa	-25° 42' 46"	30° 13' 47"	Active	↖				↑	60
Kroondal	North West	South Africa	-25° 43' 10"	27° 19' 29"	Active	↖				↑	70
Wonderkop	North West	South Africa	-25° 42' 35"	27° 24' 07"	Active	↖				↑	80
Ruighoek	North West	South Africa	-25° 13' 49"	26° 54' 40"	Active	↖				↑	35
Mllsell	North West	South Africa	-25° 42' 46"	27° 17' 32"	Active	↖				↑	30
Waterkloof	North West	South Africa	-25° 41' 12"	27° 16' 07"	Active	↖				↑	25

(continued)

TABLE 18.3 (continued)

Mining Center	State/Province	Country	Location Coordinates		Current Status		Production (metric tons)[a]	% of Total Global Production[a]	Total Disturbance Area (ha)[b]
			Latitude	Longitude					
Sukinda	Orissa	India	21° 02' 07"	85° 46' 27"	Active	↗	4,095,513	21	↑ 905
Nuasahi	Orissa	India	21° 16' 31"	86° 19' 35"	Active	↖			↑ 55
Coto	Zambales	Philippines	15° 33' 39"	120° 05' 22"	Active		46,728	0.2	↑ 185
Coobina	Western Australia	Australia	-23° 29' 42"	120° 16' 29"	Active		107,103	0.5	45
Cascabulhos	Bahia	Brazil	-10° 33' 44"	-40° 26' 29"	Active	↗			↑ 15
Pedrinhas	Bahia	Brazil	-10° 32' 45"	-40° 25' 08"	Active	↗			↑ 35
Limoeiro	Bahia	Brazil	-10° 32' 48"	-40° 21' 20"	Unknown	↗	562,739	3	↑ 5
Mato Limpo	Bahia	Brazil	-10° 31' 45"	-40° 19' 35"	Unknown	↖			↑ 25
Mina De Ipueira	Bahia	Brazil	-10° 18' 49"	-39° 45' 39"	Active	↖			↑ 35
Mina De Medrado	Bahia	Brazil	-10° 20' 37"	-39° 45' 56"	Active	↖			↑ 40
Global Total							19,800,000		4,795

[a] Data from Brown et al. (2009).

[b] Estimated from aerial and satellite photography. Google Earth (2008).

[c] Complete area overlap with Ni laterite mining. See Table 18.2.

Cyprus, Greece (Kyrou and Petrides, 2004); Atlas asbestos mine, California (CH2MHill, 2006); Nickel Mountain laterite mine, Oregon (Arnoldt Engineering and Redwing Ecological Services, 1999); and the Tiébaghi, Kaala, Kopéto, Kouaoua, and Thio Ni laterite mines, Nord and Sud, New Caledonia (Société le Nickel, 2009).

Substrate porosity influences infiltration rate, and substrate depth determines infiltration capacity (Chong and Cowsert, 1997; Mukhlisin et al., 2008). Both infiltration rate and capacity strongly influence the resistance of substrate to erosion. Compacted substrates (low porosity and infiltration rate) are susceptible to saturation during intense rainfall events, resulting in overland flow and erosion. Compacted substrates may require deep ripping to increase substrate infiltration (Williamson et al., 1982; Luce, 1997).

Surface erosion control is highly beneficial to revegetation, particularly when seedlings are becoming established. Properly placed erosion control materials, including mulch (straw, wood chips), straw bales, straw rolls (wattles), and natural fiber geotextiles (straw, jute, coconut fiber), are effective in reducing surface disturbances, such as rain drop impact, overland flow, and sediment erosion or deposition, which may uproot or bury seedlings.

Step 2: Substrate Amendment

Like nonserpentine mine tailings, drastically disturbed serpentine substrates are severely nutrient deficient. Compared to serpentine topsoil, drastically disturbed serpentine substrates are much more deficient in N, phosphorus (P), potassium (K), and calcium (Ca). Most studies investigating the reasons for the poor productivity of serpentine soils have found primary nutrient limitations due to some combination of N, P, or K deficiency, along with secondary limitations due to a combination of Ca deficiency and magnesium (Mg) toxicity (Spence and Millar, 1963; Halstead, 1968; Ferreira and Wormell, 1971; Jones et al., 1977; Meyer, 1980; Turitzin, 1982; Smith and Kay, 1986; Carter et al., 1988; Huenneke et al., 1990; Nagy and Proctor, 1997a; Chiarucci et al., 1999; O'Dell et al., 2006). Serpentine substrates have also been found to be molybdenum (Mo) deficient, which may limit N-fixing nodules of plants that harbor *Frankia* or *Rhizobia* (White, 1967; Jones et al., 1977; Walker, 2001). Heavy metal toxicity can restrict plant growth on serpentine soils, but the general consensus in the literature is that high concentrations of heavy metals are not as severely limiting to plant growth as macronutrient deficiency (Johnston and Proctor, 1981; Robertson, 1992; Nagy and Proctor, 1997b; Boyd et al., 2000). Drastically disturbed serpentine substrates typically have extremely low levels of organic matter, cation exchange capacity, water holding capacity, and microbiological activity. Such extreme conditions prevent even native serpentine-tolerant plants species from recruiting onto and establishing on the disturbed substrate. The intent of substrate amendment is to restore some or all of the soil conditions originally provided by the serpentine topsoil, so that plants may establish and be productive.

Serpentine topsoil (mixed O, A, and B soil horizons) application is the most effective treatment to promote revegetation (Williamson et al., 1982; Van Kekerix and Kay, 1986; Bradshaw, 1997; Tordoff et al., 2000). Application of nonserpentine topsoil on drastically disturbed serpentine landscapes is not advocated, however, because it has the potential to introduce invasive species and completely alter the landscape's unique ecology from serpentine- to nonserpentine-based characteristics. Well-planned mining and construction projects remove and stockpile topsoil for reapplication after subsurface disturbance activities are ceased. Serpentine topsoil application returns most soil productivity and functional biological activity to the landscape. Recommended topsoil application depth is at least 30 cm to provide adequate rooting depth and nutrient cycling capacity (Williamson et al., 1982; Tordoff et al., 2000). Topsoiling with serpentine soil has been successfully employed to revegetate asbestos mine tailings at the Msauli mine, Mpumalanga, South Africa (Briers et al., 1988; Been, 1990; van Rensburg and Pistoruis, 1998, Morgenthal et al., 2004) and the Amiantos mine, Cyprus, Greece (Kyrou and Petrides, 2004). Topsoil application is not always possible, especially in the case of Ni laterite mining, where the soil itself is mined and removed as ore. Lacking topsoil, the only other alternative is to use commercially available soil amendments to alleviate soil nutrient deficiency and improve substrate physical characteristics. In recent decades, great advances in revegetation of drastically disturbed serpentine sites have been made with effective substrate amendment types, combinations, and application methods.

Commercially available substrate amendments are available as mineral, chemical (fertilizer), and organic amendments. Mineral amendments primarily consist of products produced from kaolinite clays, calcined volcanic ash, or calcined diatomaceous earth. Their primary benefits are increased cation exchange capacity and water holding capacity. Commercially available mineral amendments have been shown to vary greatly in both water holding capacity and plant available water (Curtis and Claassen, 2008). Mineral amendments are most effective when the substrate holds less than 10% plant-available moisture. Little research has been conducted regarding the use of mineral amendments to promote revegetation of disturbed serpentine substrates.

Most chemical amendments contain the primary macronutrients N, P, and K, and may contain lesser quantities of sulfur (S), Ca, Mg, and micronutrients. Chemical amendments may be formulated or polymer coated to provide slow nutrient release, which is often advantageous for increasing the efficiency of plant nutrient uptake. Surface application of inorganic NPK fertilizer, with or without Ca supplied as gypsum ($CaSO_4$), has typically failed to promote revegetation on drastically disturbed serpentine substrates, or it has only promoted poor growth that does not persist beyond the first year (Moore and Zimmermann, 1977; Smith and Kay, 1986; O'Dell and Claassen, 2006a, 2006b). The inability of chemical amendments to promote revegetation beyond the first year has been attributed to their soluble nature and short persistence in the substrate as the nutrients are rapidly taken up by plants or leached from the root zone. Polyurethane- or resin-coated fertilizers have a lower nutrient release rate than uncoated fertilizers, but typically do not last longer than 18 months (J.R. Simplot Company, 2010; Scotts Company, 2010). Application of soluble or even sparingly soluble Ca, such as gypsum, has often failed to produce a positive plant response (Meyer, 1980; Turitzin, 1982; Chiarucci et al., 1999; Reid and Naeth, 2005b). This has been attributed in some cases to the displacement of Mg by Ca from the substrate cation exchange complex into soil solution, thereby increasing Mg toxicity (Meyer, 1980). Chemical amendments also fail to correct poor soil physical characteristics, including low cation exchange capacity, low infiltration rates, and low water holding capacity.

Compared to chemical amendments, solid organic amendments promote the most vigorous and sustainable vegetative cover on disturbed serpentine substrates (Moore and Zimmermann, 1977, 1979; Liston and Balkwill, 1997; Reid and Naeth, 2005a, 2005b; CH2MHill, 2006; O'Dell and Claassen, 2006a, 2006b, 2006c). Organic amendments have the ability to act as a long-term nutrient pool, slowly releasing nutrients over an extended period of time as the organic matter decomposes. Most of the organic matter cation exchange complex is dominated by Ca (Ca:Mg molar ratio > 1), which may increase low substrate Ca:Mg ratios without displacing elevated levels of Mg from the serpentine substrate cation exchange complex (Halstead, 1968; O'Dell and Claassen, 2006a, 2006b). Organic matter has the ability to complex potentially phytotoxic heavy metals in serpentine, such as Ni, greatly reducing their bioavailability (Halstead, 1968). Organic amendments also increase cation exchange capacity, water holding capacity, and aeration (Curtis and Claassen, 2005, 2009).

Not all organic amendments provide equal performance in field conditions. Organic amendments may consist of animal waste and plant waste alone or in combination. Varying raw organic material sources and composting and finishing processes used greatly affect the properties of organic amendments (Chen and Inbar, 1993; Epstein, 1997; Adani et al., 1999). Potential organic amendments should be analyzed for their nutrient and physical properties before one is chosen and used for revegetation. Most bulk livestock waste composts use manure and feedstock waste as the raw organic material resulting in a low C:N ratio (10:1). The high urea content of manure compost generates high concentrations of ammonium (NH_4^+) and nitrate (NO_3^-), which are rapidly leached through the soil and removed in runoff. Biosolids (sewage sludge) are a source of highly mixed organics with the same issues as livestock waste compost, along with elevated levels of heavy metals and toxic compounds, if produced from industrial waste sources.

Plant waste composts tend to have higher C:N ratios. Peat moss is dried, partially decomposed plant material (primarily *Sphagnum* moss) commercially mined from bogs. It has a relatively high C:N ratio (60:1) and low decomposition rate as a result of anoxic decomposition in bogs. Raw materials for plant waste compost may include wood and paper waste, agricultural waste, or yard waste (garden clippings). Like peat, wood and paper waste has a very high C:N ratio (500:1), making it a poor source of N for microorganisms and plants. Moore and Zimmermann (1977) determined that sawdust was substantially less effective than manure at promoting revegetation on asbestos tailings, possibly due to the considerable differences in C:N ratio between the two amendment types. Likewise, van Rensberg and Morgenthal (2004) found that although wood chips promoted short-term revegetation of platinum mine tailings, their high C:N ratio impeded long-term revegetation vigor. Agricultural waste compost has a lower C:N ratio, but the lack of woody materials results in rapid decomposition. Yard waste compost is the most ideal organic amendment for revegetation due to a mixture of herbaceous and woody materials that produce a compost with a well-balanced C:N ratio (10:1 to 20:1) (Claassen and Carey, 2004).

Organic amendment application rates must be carefully chosen with consideration of organic amendment type and disturbed serpentine substrate properties. Overamendment of serpentine substrates can have undesirable results, particularly non-native plant species invasion (Turitzin, 1982; Smith and Kay, 1986; Koide et al., 1988; Huenneke et al., 1990; Weiss, 1999; Munzinger et al., 2004; O'Dell and Claassen, 2006b). The adverse chemical and physical conditions of serpentine soils have largely prevented them from being invaded by the same non-native species that have invaded adjacent nonserpentine areas (Turitzin, 1982; Huenneke et al., 1990; Harrison, 1999a, 1999b; Harrison et al., 2001; Jurjavcic et al., 2002; Williamson and Harrison, 2002; Seabloom et al., 2003; Gram et al., 2004). Any amendment of serpentine substrate alleviates the adverse substrate characteristics, making it easier for non-native species to invade. Small field trials should be used to assess the potential for adverse effects posed by candidate amendment types and application levels before the entire site is treated.

Most compost consists of both organic matter and inert or mineral material (up to 50% by mass). The amendment rate should be increased to compensate for inert and mineral material. Target organic matter amendment rate should be between 2.5% and 15% by volume of the total substrate volume (Moore and Zimmermann, 1977, 1979; Williamson et al., 1982; Reid and Naeth, 2005b; O'Dell and Claassen, 2006a, 2006b, 2006c). Application of chemical amendment as polymer-coated, slow-release NPK fertilizer in addition to compost provides an initial nutrient boost while vegetation is becoming established (Moore and Zimmermann, 1977, 1979; Reid and Naeth, 2005a, 2005b; CH2MHill, 2006; O'Dell and Claassen, 2006a, 2006b). Previous serpentine revegetation projects have used fertilizer application rates between 50 and 500 kg ha^{-1} each of N, P, and K (Moore and Zimmermann, 1977, 1979; Meyer, 1980; Reid and Naeth, 2005a, 2005b; CH2MHill, 2006). Fertilizer application rate is dependent on compost fertility and the risk of undesirable plant species invasion. As suggested before, small field trials may be required to determine optimal fertilizer application rates.

Amendment of the substrate surface does little to improve subsurface substrate chemical and physical properties (von Willert and Stehouwer, 2003). Incorporation of organic amendment into disturbed serpentine substrate significantly increases the rooting depth of revegetation species (Moore and Zimmermann, 1977; Reid and Naeth, 2005b; CH2MHill, 2006; O'Dell and Claassen, 2006c). This enhances the ability of vegetation to endure drought and provides greater soil stability. Incorporation also reduces loss of the amendment to erosion because it is distributed throughout the substrate rather than being exposed on the substrate surface. Although difficult, every effort should be made to incorporate the amendment into the substrate. This can be accomplished by broadcast application of the amendment to the substrate surface followed by tilling or ripping to incorporate it or mixing amendment and substrate in piles and then applying to the substrate surface. Recommended substrate rooting depth (amendment incorporation depth) for revegetation is at least 30 cm (Williamson et al., 1982; Tordoff et al., 2000). A wide variety of heavy equipment suitable for amendment application and tillage on slopes is summarized by Jennings et al. (2003).

Step 3: Plant and Microbial Materials Selection

Serpentine substrates support a high diversity of plant species and harbor a high proportion of serpentine endemics (Kruckeberg, 1984; Brooks, 1987; Baker et al., 1992). Like nonserpentine mine tailings, native plant species that naturally grow on serpentine substrates typically contain specialized physiological mechanisms that allow them to tolerate the adverse chemical conditions imposed by the substrate (Proctor and Woodell, 1975; Kruckeberg, 1984; Brooks, 1987; Baker et al., 1992; Brady et al., 2005; Kazakou et al., 2008). Common general stress tolerance features of serpentine plant species include high root:shoot biomass ratios and inherently slow growth rates (Kruckeberg, 1984; Westerbergh, 1994; Freitas and Mooney, 1996). Many species exhibit physiological regulation of Ca and Mg uptake and translocation to maintain adequate tissue Ca:Mg molar ratios (Walker et al., 1955; Lyon et al., 1971; Shewry and Peterson, 1975; Wallace et al., 1982; Lee and Reeves, 1989; O'Dell et al., 2006; O'Dell and Claassen, 2006a; Asemaneh et al., 2007). Root-level tolerance mechanisms include selective Ca uptake and translocation to the shoot, Mg exclusion, and Mg sequestration. In addition to general physiological heavy metal tolerance mechanisms such as root-level exclusion and sequestration (Baker, 1981, 1987; Shaw, 1990; Westerbergh, 1994; O'Dell et al., 2006; O'Dell and Claassen, 2006a), some serpentine plants bear the unique physiological uptake mechanism of hyperaccumulation, in which plants can actively uptake heavy metals, such as Ni, and translocate them to the shoot where they may accumulate to concentrations >1000 μg g⁻¹ (Brooks, 1998). Hyperaccumulators have been successfully used to revegetate serpentine substrates, but their most common and effective application is in heavy metal phytoremediation of industrially polluted nonserpentine soils (Chaney et al., 1997; Brooks, 1998).

Native plant species growing on undisturbed serpentine soils near revegetation sites are more likely to bear the local climate and edaphic tolerance mechanisms necessary for survival on serpentine at that location. Therefore, plant materials for revegetation should only be collected from local serpentine plant communities (Smith and Kay, 1986; Cherrier, 1990; Jaffré and Pelletier, 1992; Jaffré et al., 1994a, 1994b, 1994c, 1997; Luçon et al., 1997; Baker, 1999; Hoover et al., 1999; Sarrailh and Ayrault, 2001; Pelletier, 2003; Kyrou and Petrides, 2004; Morgenthal et al., 2004; O'Dell and Claassen, 2006a, 2006b, 2006c). It should be recognized that some species that appear to grow both on and off serpentine may have specialized serpentine tolerant ecotypes (Kruckeberg, 1950, 1951; Linston and Balkwill, 1997; O'Dell and Claassen, 2006a), while other species with the same on- and off-serpentine distribution may display complete indifference to serpentine substrates (Kruckeberg, 1950; Griffin, 1965). If no local serpentine-tolerant plant materials can be obtained and commercially available native plant materials must be used, tolerance of the plant materials to serpentine should be tested before they are used for revegetation (Wilkins, 1978; O'Dell and Claassen, 2006a).

Most plant species harbor mycorrhizal associations that have been shown to be highly beneficial to plant establishment and productivity, particularly in nutrient-deficient substrates such as serpentine (Chiariello et al., 1982; Hopkins, 1987). Some plant genera harbor N-fixing bacteria in addition to mycorrhizas. Drastically disturbed substrates may lack adequate levels of mycorrhizal spores and N-fixing bacteria to colonize plants (DeGrood et al., 2005). Therefore, it may be necessary to apply inoculum to the substrate. Like its unique flora and endemic species, microbial communities also appear to be unique in serpentine substrates (Maas and Stuntz, 1969; Gauthier et al., 1992; Amir et al., 1997; Gonçalves et al., 1997; Amir and Pineau, 1998a, 1998b; DeGrood et al., 2005; Moser et al., 2005, 2009; Chaintreuili et al., 2007; Schechter and Bruns, 2008). To provide the revegetation community with its native serpentine-tolerant microbial community, small quantities of local serpentine topsoil should be applied to the disturbed serpentine substrate and incorporated with amendments, or, in the case of container stock, a small quantity of the soil should be added to its planting hole. Seeding and planting should be conducted during seasons when most plant species for that region typically germinate (wet season). In regions subject to seasonal drought, care should be taken to seed and plant early in the wet season so that plants have sufficient time to become established before soils dry at the surface. This will re-duce the need for supplemental irrigation.

SUMMARY

Increasing global demand for industrial commodities has driven ever greater lev-els of mineral extraction at the same time that nations seek more ways to balance environmental protection with economic growth. Revegetation partially mitigates the adverse impacts of mining. Although revegetation can often be a daunting, dif-ficult, and costly task, successful revegetation of drastically disturbed serpentine and nonserpentine substrates can be achieved with careful planning and consider-ation of the three most important steps in revegetation, including physical stabili-zation, substrate amendment, and plant and microbial materials selection.

Acknowledgments

We thank Ben Mundie (Oregon Department of Geology and Mineral Industries, Albany) and Tanguy Jaffré (Institut de Recherche pour le Développement, Nou-méa, New Caledonia) for providing information on revegetation of the Nickel Mountain laterite mine and New Caledonia laterite mines.

LITERATURE CITED

Abramson, L., Sharma, S., Lee, T. S., and Boyce, G. M. (2001) *Slope Stability and Stabiliza-tion Methods*. Wiley, New York.

Adani, F., Genevini, P. L., Gasperi, F., and Tambone, F. (1999) Composting and humifica-tion. *Compost Science and Utilization*, 7, 24–33.

Amir, H., and Pineau, R. (1998a) Influence of plants and cropping on microbiological char-acteristics of some New Caledonian ultramafic soils. *Australian Journal of Soil Research*, 36, 457–71.

Amir, H., and Pineau, R. (1998b) Effects of metals on the germination and growth of fungal isolates from New Caledonian ultramafic soils. *Soil Biology and Biochemistry*, 30, 2043–54.

Amir, H., Pineau, R., and Violette, Z. (1997) Premiers résultats sur les endomycorhizes des plantes de maquis miniers de Nouvelle-Calédonie. In *Écologie des milieu sur roches ul-tramafiques et sur sols métallifères* (eds. T. Jaffré, R. D. Reeves, and T. Becquer), pp. 79–85. Documents Scientifiques et Techniques, Nouméa, Nouvelle-Calédonie Volume spécial III2.

Antonovics, J., Bradshaw, A. D., and Turner, R. G. (1971) Heavy metal tolerance in plants. *Advances in Environmental Science and Technology*, 7, 1–85.

Arabas, K. B. (2000) Spatial and temporal relationships among fire frequency, vegetation, and soil depth in an eastern North American serpentine barren. *Journal of the Torrey Botanical Society*, 1, 51–65.

Arnoldt Engineering and Redwing Ecological Services. (1999) *Reclamation Plan—Lower Ore Body*. Glenbrook Nickel Company. Riddle, OR. Submitted to Oregon Department of Geology and Mineral Industries, Albany.

Asemaneh, T., Ghaderian, S. M., and Baker, A. J. M. (2007) Responses to Mg/Ca balance in an Iranian serpentine endemic plant, *Cleome heratensis* (Capparaceae) and a related non-serpentine species, *C. foliolosa. Plant and Soil,* 293, 49–59.

Baker, A. J. M. (1981) Accumulators and excluders—strategies in the response of plants to heavy metals. *Journal of Plant Nutrition,* 3, 643–54.

Baker, A. J. M. (1987) Metal tolerance. *New Phytologist,* 106, 93–111.

Baker, A. J. M. (1999) Revegetation of asbestos mine wastes. In *ECO-TEC Architecture of the In-between* (ed. A. Marras), pp. 119–25. Princeton Architectural Press, New York.

Baker, A. J. M., Proctor, J., and Reeves, R. D. (1992) *The Vegetation of Ultramafic (Serpentine) Soils. Proceedings of the First International Conference on Serpentine Ecology.* Intercept, Andover.

Been, A. (1990) Greening South Africa: I. Asbestos tailings. *Restoration and Management Notes,* 8, 14–22.

Boyd, R. S., Wall, M. A., and Watkins, J. E. (2000) Correspondence between Ni tolerance and hyperaccumulation in *Streptanthus* (Brassicaceae). *Madroño,* 47, 97–105.

Bradshaw, A. D. (1997) Restoration after mining for metals—an ecological view. In *Écologie des milieu sur roches ultramafiques et sur sols métallifères* (eds. T. Jaffré, R.D. Reeves, and T. Becquer), pp. 239–48. Documents Scientifiques et Techniques, Nouméa, Nouvelle-Calédonie Volume spécial III2.

Brady, K. U., Kruckeberg, A. R., and Bradshaw, H. D. (2005) Evolutionary ecology of plant adaption to serpentine soils. *Annual Review of Ecology, Evolution, and Systematics,* 36, 243–66.

Briers, J. H., van Wyk, S., and Michael, M. D. (1988) Root development and growth of selected plant species under various fertilization conditions and with various soil structures in the asbestos tailings from the Msauli mine in east Transvaal, South Africa. *South African Journal of Science,* 84, 325–29.

Brooks, R. R. (1987) *Serpentine and Its Vegetation. A Multidisciplinary Approach.* Dioscorides Press, Portland, OR.

Brooks, R. R. (1998) *Plants that Hyperaccumulate Heavy Metals.* CAB International, Wallingford, Oxon, U.K.

Brown, T. J., Hetherington, L. E., Hannis, S. D., Bide, T., Benham, A. J., Idoine, N. E., and Lusty, P. A. J. (2009) *World Mineral Production 2003–07.* British Geological Survey, Keyworth, Nottingham.

Bugg, R. L., Brown, C. S., and Anderson, J. H. (1997) Restoring native perennial grasses to rural roadsides in the Sacramento Valley of California: Establishment and evaluation. *Restoration Ecology,* 5, 214–28.

Carle, D. (2008) *Introduction to Fire in California.* University of California Press, Berkeley.

Carter, S. P., Proctor, J., and Slingsby, D. R. (1988) The effects of fertilization on part of the Keen of Hamar serpentine, Shetland. *Transactions of the Botanical Society of Edinburgh,* 45, 97–105.

CH2MHill. (2006) *Five-Year Review Report for Atlas Asbestos Mine Superfund Site and Coalinga Asbestos Mine (Johns-Manville mill) Superfund Sites*. Fresno County, CA. CH2MHill, Oakland, CA. Available online at http://yosemite.epa.gov/r9/sfund/r9sfdocw.nsf/3dc283e6c5d6056f88257426007417a2/5f64d7d1be7f8c388825727b006e6ff1!OpenDocument. Accessed January 16, 2010.

Chaintreuli, C., Rigault, F., Moulin, L., Jaffré, T., Fardoux, J., Giraud, E., Dreyfus, B., and Bailly, X. (2007) Nickel resistance determinants in *Bradyrhizobium* strains from nodules of the endemic New Caledonia legume *Serianthes calycina*. *Applied and Environmental Microbiology*, 73, 8018–22.

Chaney, R. L., Malik, M., Li, Y. M., Brown, S. L., Angle, J. S., and Baker, A. J. M. (1997) Phytoremediation of soil metals. *Current Opinions in Biotechnology*, 8, 279–84.

Chen, Y., and Inbar, Y. (1993) Chemical and spectroscopical analyses of organic matter transformations during composting in relation to compost maturity. In *Science and Engineering of Composting: Design, Environmental, Microbiological, and Utilization Aspects* (eds. H. A. Hoitnik and H. Keener). Renaissance Publications, Worthington, OH.

Cherrier, J. F. (1990) Reconstitution of vegetative cover in mine areas in New Caledonia. *Bois et Foréts des Tropiques*, 225, 5–23.

Chiariello, N., Hickman, J. C., and Mooney, H. A. (1982) Endomycorrhizal role for interspecific transfer of phosphorus in a community of annual plants. *Science*, 217, 941–43.

Chiarucci, A., Maccherini, S., Bonini, I., and Dominicis, V. (1999) Effects of nutrient addition on community productivity and structure of serpentine vegetation. *Plant Biology*, 1, 121–26.

Chong, S. K., and Cowsert, P. T. (1997) Infiltration in reclaimed mined land ameliorated with deep tillage treatments. *Soil and Tillage Research*, 44, 255–64.

Claassen, V. P., and Carey, J. L. (2004) Regeneration of nitrogen fertility in disturbed soils using composts. *Compost Science and Utilization*, 12, 145–52.

Clemente, R., Almela, C., and Bernal, M. P. (2006) A remediation strategy on active phytoremediation followed by natural attenuation in a soil contaminated by pyrite waste. *Environmental Pollution*, 143, 397–406.

Curtis, M. J., and Claassen, V. P. (2005) Compost incorporation increases plant available water in a drastically disturbed serpentine soil. *Soil Science*, 170, 939–53.

Curtis, M. J., and Claassen, V. P. (2008) An alternative method for measuring plant available water in inorganic amendments. *Crop Science*, 48, 2447–52.

Curtis, M. J., and Claassen V. P. (2009) Regenerating topsoil functionality in four drastically disturbed soil types by compost incorporation. *Restoration Ecology*, 17, 24–32.

DeGrood, S. H., Claassen, V. P., and Scow, K. M. (2005) Microbial community composition on native and drastically disturbed serpentine soils. *Soil Biology and Biochemistry*, 37, 1427–35.

East Bay Regional Park District. (2009) *Serpentine Prairie Restoration Plan*. Redwood Regional Park. East Bay Regional Park District, Oakland, CA. Available online at http://www.ebparks.org/files/PLAN_SerpPrairieRestoration_web_final_05-01-09.pdf. Accessed January 16, 2010.

Epstein, E. (1997) *Science of Composting*. Technomic Publishing, Lancaster, PA.

Ferreira, R. E. C., and Wormell, P. (1971) Fertiliser response of vegetation on ultrabasic terraces on Rhum. *Transactions of the Botanical Society of Edinburgh*, 41, 149–54.

Freitas, H., and Mooney, H. (1996) Effects of water stress and soil texture on the performance of two *Bromus hordeaceus* ecotypes from sandstone and serpentine soils. *Acta Oecologia*, 17, 307–17.

Gaur, A., and Adholeya, A. (2004) Prospects of arbuscular mycorrhizal fungi in phytoremediation of heavy metal contaminated soils. *Current Science*, 86, 528–34.

Gauthier, D., Jaffré, T., and Rigault, F. (1992) *Importance and Diversity of Frankia-Causuarinaceae Symbiosis in New Caledonia*. 8th International conference on *Frankia* and actinorhizal plants. University of Lyon, France.

Gelbard, J. L., and Harrison, S. (2003) Roadless habitats as refuges for native grasslands: Interactions with soil, aspect, and grazing. *Ecological Applications*, 13, 404–15.

Gibson, D. J. (2009) *Grasses and Grassland Ecology*. Oxford University Press, New York.

Going, B. M., Hillerislambers, J., and Levine, J. M. (2009) Abiotic and biotic resistance to grass invasion in serpentine annual plant communities. *Oecologia*, 159, 839–47.

Golightly, J. P. (1981) Nickeliferous laterite deposits. *Economic Geology*, 75, 710–35.

Gonçalves, S. C., Gonçalves, M. T., Freitas, H., and Martins-Loução, M. A. (1997) Mycorrhizae in a Portuguese serpentine community. In *Écologie des milieu sur roches ultramafiques et sur sols métallifères* (eds. T. Jaffré, R. D. Reeves, and T. Becquer), pp. 79–85. Documents Scientifiques et Techniques, Nouméa, Nouvelle-Calédonie Volume spécial III2.

Goodman, G. T., Pitcairn, C. E. R., and Gemmell, R. P. (1973) Ecological factors affecting growth on sites contaminated with heavy metals. In *Ecology and Reclamation of Devastated Land*, volume 2. (eds. R. J. Hutnik and G. Davis). Gordon and Breach, New York.

Google Earth. (2008) *Google Earth Version 4.0.2742*. Google Corporation, Mountain View, CA.

Gram, W. K., Borer, E. T., Cottingham, K. L., Seabloom, E. W., Boucher, V. L., Goldwasser, L., Micheli, F., Kendall, B. E., and Burton, R. S. (2004) Distribution of plants in a California serpentine grassland: Are rocky hummocks spatial refuges for native species? *Plant Ecology*, 172, 159–71.

Grant, C. D., Campbell, C. J., and Charnock, N. R. (2002) Selection of species suitable for derelict mine site rehabilitation in New South Wales, Australia. *Water, Air, and Soil Pollution*, 139, 215–35.

Griffin, J. R. (1965) Digger pine seedling response to serpentinite and non-serpentinite soil. *Ecology*, 46, 801–7.

HABIO. (2010) Project life—Nature HABIO. Biodiversity protection in Calvana–Monteferrato area. Available online at http://habio.provincia.prato.it/habioen.htm. Accessed January 16, 2010.

Halstead, R. L. (1968) Effect of different amendments on yield and composition of oats grown on a soil derived from serpentine material. *Canadian Journal of Soil Science*, 48, 301–5.

Harrison, S. (1999a) Native and alien species diversity at the local and regional scales in a grazed California grassland. *Oecologia*, 121, 99–106.

Harrison, S. (1999b) Local and regional diversity in a patchy landscape: Native, alien, and endemic herbs on serpentine. *Ecology*, 80, 70–80.

Harrison, S., Inouye, B. D., and Safford, H. D. (2003) Ecological heterogeneity in the effects of grazing and fire on grassland diversity. *Conservation Biology*, 17, 837–45.

Harrison, S., Rice, K., and Maron, J. (2001) Habitat patchiness promotes invasion by alien grasses on serpentine soil. *Biological Conservation*, 100, 45–53.

Hoover, L. D., McRae, J. D., McGee, E. A., and Cook, C. (1999) Horse Mountain Botanical Area serpentine revegetation study. *Natural Areas Journal*, 19, 361–67.

Hopkins, N. A. (1987) Mycorrhizae in a California serpentine grassland community. *Canadian Journal of Botany*, 65, 484–87.

Huenneke, L. F., Hamburg, S. P., Koide, R., Mooney, H. A., and Vitousek, P. M. (1990) Effects of soil resources on plant invasion and community structure in California serpentine grassland. *Ecology*, 71, 478–91.

Huenneke, L. F., and Mooney, H. A. (1989) *Grassland Structure and Function: California Annual Grassland*. Kluwer Academic Publishers, Dordrecht.

Jaffré, T., Gauthier, D., Rigault, F., and McCoy, S. (1994a) La réhabilitation des sites miniers; les Casuarinacées endemiques. *Bois et Forêts des Tropiques*, 242, 31–43.

Jaffré, T., McCoy, S, Rigault, F., and Dagostini, G. (1997) Quelle méthod de végétalisation pour la réhabilitation des anciens sites miniers de Nouvelle-Calédonie. In *Écologie des milieu sur roches ultramafiques et sur sols métallifères* (eds. T. Jaffré, R. D. Reeves, and T. Becquer), pp. 285–88. Documents Scientifiques et Techniques, Nouméa, Nouvelle-Calédonie Volume spécial III2.

Jaffré, T., Morat, P., and Veillon, J.-M. (1994b). La réhabilitation des sites miniers; la flore. *Bois et Forêts des Tropiques*, 242, 7–30.

Jaffré, T., and Pelletier, B. (1992) *Plantes de Nouvelle-Calédonie permettant de revégétaliser des sites miniers*. Société le Nickel, SLN, Nouméa, New Caledonia.

Jaffré, T., Rigault, F., and Sarrailh, J.-M. (1994c) La réhabilitation des sites miniers; la végétalisation des anciens sites miniers. *Bois et Forêts des Tropiques*, 242, 45–57.

Jennings, S. R., Goering, J. D., Blicker, P. S., and Taverna, J. J. (2003) *Evaluation of organic matter compost addition and incorporation on steep cut slopes, Phase I: Literature review and potential applicable equipment evaluation*. Montana State University, Bozeman. Available online at http://www.mdt.mt.gov/research/docs/research_proj/organic_matter/final_report.pdf. Accessed Janaury 16, 2010.

Jessup, A., and Mudd, G. M. (2008) Environmental sustainability metrics for nickel sulphide versus nickel laterite. In *3rd International Conference on Sustainability Engineering and Science: Blueprints for Sustainable Infrastructure*. Aukland, New Zealand. Available online at http://civil.eng.monash.edu.au/about/staff/muddpersonal/2008-NZ-SustEng-Sci-Jessup-Mudd-SustMetrics-v-Nickel-Sulf-v-Lat.pdf. Accessed January 16, 2010.

Johnson, M. S., McNeilly, T., and Putwain, P. D. (1977) Revegetation of metalliferous mine spoil contaminated by lead and zinc. *Environmental Pollution*, 12, 261–77.

Johnston, W. R., and Proctor, J. (1981) Growth of serpentine and nonserpentine races of *Festuca rubra* in solutions simulating the chemical conditions in a toxic serpentine soil. *Journal of Ecology*, 69, 855–70.

Jones, M. B., Williams, W. A., and Ruckman, J. E. (1977) Fertilization of *Trifolium subterraneum* L. growing on serpentine soils. *Soil Science Society of America Journal*, 41, 87–89.

J.R. Simplot Company. (2010) *APEX Nursery fertilizer*. Available online at http://www.simplot.com/turf/apex/index.cfm. Accessed January 16, 2010.

Jurjavcic, N. L., Harrison, S., and Wolf, A. T. (2002) Abiotic stress, competition, and the distribution of the native annual grass *Vulpia microstachys* in a mosaic environment. *Oecologia*, 130, 555–62.

Kazakou, E., Dimitrakopoulos, P. G., Baker, A. J. M, Reeves, R. D., and Troumbis, A. Y. (2008) Hypotheses, mechanisms and trade-offs of tolerance and adaptation to serpentine soils: From species to ecosystem level. *Biological Reviews*, 83, 495–508.

Kelly, T. D., and Matos, G. R. (2008) *Historical Statistics for Mineral and Material Commodities in the United States*. U.S. Geological Survey, Reston, VA. Available online at http://minerals.usgs.gov/ds/2005/140/chromium.pdf. Accessed January 16, 2010.

Koide, R. T., Huenneke, L. F., Hamburg, S. P., and Mooney, H. A. (1988) Effects of applications of fungicide, phosphorus and nitrogen on the structure and productivity of an annual serpentine plant community. *Functional Ecology*, 2, 335–44.

Kramer, P. A., Zabowski, D., Scherer, G., and Everett, R. L. (2000a) Native plant restoration of copper mine tailings: I. Substrate effect on growth and nutritional status in a controlled environment. *Journal of Environmental Quality*, 29, 1762–69.

Kramer, P. A., Zabowski, D., Scherer, G., and Everett, R. L. (2000b) Native plant restoration of copper mine tailings: II. Field survival, growth, and nutrient uptake. *Journal of Environmental Quality*, 29, 1770–77.

Kruckeberg, A. R. (1950) *An Experimental Inquiry into the Nature of Endemism on Serpentine soils*. PhD diss., University of California, Berkeley.

Kruckeberg, A. R. (1951) Intraspecific variability in the response of certain native plant species to serpentine soil. *American Journal of Botany*, 38, 408–16.

Kruckeberg, A. R. (1984) *California Serpentines: Flora, Vegetation, Geology, Soils, and Management Problems*. University of California Press, Berkeley.

Kruckeberg, A. L., and Wu, L. (1992) Copper tolerance and copper accumulation of herbaceous plants colonizing inactive California copper mines. *Ecotoxicology and Environmental Safety*, 23, 307–19.

Kuck, P. H. (2006) *Nickel. 2006 Minerals Yearbook*. U.S. Department of the Interior, U.S. Geological Survey, Washington, DC.

Kyrou, K., and Petrides, G. (2004) *The Rehabilitation of the Asbestos Mine, Cyprus*. Water Development Department, Republic of Cyprus, Greece. Available online at http://www.moa.gov.cy/moa/wdd/Wdd.nsf/All/DE3FF3A33B41610BC2256FE000281065/$file/Amiantos_Published.pdf?OpenElement. Accessed January 16, 2010.

Latham, R. (2008) *Pink Hill Serpentine Barrens Restoration and Management Plan*. Continental Conservation, Rose Valley, PA. Available online at http://www.tylerarboretum.org/arboretum/natural-areas/documents/Latham-PinkHillReport-2008.pdf. Accessed January 16, 2010.

Lee, W. G., and Reeves, R. D. (1989) Growth and chemical composition of *Celmisia spedenii*, an ultramafic endemic, and *Celmisia markii* on ultramafic soil and garden loam. *New Zealand Journal of Botany*, 27, 595–98.

Liston, R. J., and Balkwill, K. (1997) Suitability of serpentine plants for the revegetation of chrysotile asbestos tailings. In *Écologie des milieu sur roches ultramafiques et sur sols métallifères* (eds. T. Jaffré, R. D. Reeves, and T. Becquer), pp. 275–83. Documents Scientifiques et Techniques, Nouméa, Nouvelle-Calédonie Volume spécial III2.

Luce, C. H. (1997) Effectiveness of road ripping in restoring infiltration capacity of forest roads. *Restoration Ecology,* 5, 265–70.

Luçon, S., Marion, F., Niel, J. F., and Pelletier, B. (1997) Réhabilitation des sites miniers sur roches ultramafiques en Nouvelle-Calédonie. In *Écologie des milieu sur roches ultramafiques et sur sols métallifères* (eds. T. Jaffré, R. D. Reeves, and T. Becquer), pp. 297–303. Documents Scientifiques et Techniques, Nouméa, Nouvelle-Calédonie Volume spécial III2.

Lyon, G. L., Peterson, P. J., Brooks, R. R., and Butler, G. W. (1971) Calcium, magnesium, and trace elements in a New Zealand serpentine flora. *Journal of Ecology,* 59, 421–29.

Maas, J. L., and Stuntz, D. E. (1969) Mycoecology on serpentine soil. *Mycologia,* 61, 1106–16.

Marx, E. (2007) *Vegetation Dynamics of the Buck Creek Serpentine Barrens, Clay County, North Carolina.* BS thesis, University of North Carolina, Chapel Hill.

Meyer, D. R. (1980) Nutritional problems associated with the establishment of vegetation on tailings from an asbestos mine. *Environmental Pollution (Series A),* 23, 287–98.

Moore, T. R., and Zimmermann, R. C. (1977) Establishment of vegetation on serpentine asbestos mine wastes, southeastern Quebec, Canada. *Journal of Applied Ecology,* 14, 589–99.

Moore, T. R., and Zimmermann, R. C. (1979) Follow-up studies of vegetation establishment on asbestos tailings, southeastern Quebec, Canada. *Reclamation Review,* 2, 143–46.

Morgenthal, T., Maboeta, M., and van Rensburg, L. (2004) Revegetation of heavy metal contaminated mine dumps using locally serpentine-adapted grassland species. *South African Journal of Botany,* 70, 784–89.

Moser, A. M., Frank, J. L., D'Allura, J. L., and Southworth, D. (2009) Ectomycorrhizal communities of *Quercus garryana* are similar on serpentine and nonserpentine soils. *Plant and Soil,* 315, 185–94.

Moser, A. M., Petersen, C. S., D'Allura, J. A., and Southworth, D. (2005) Comparison of ectomycorrhizas of *Quercus garryana* (Fagaceae) on serpentine and non-serpentine soils in southwestern Oregon. *American Journal of Botany,* 62, 224–30.

Mukhlisin, M., Taha, M. R., and Kosugi, K. (2008) Numerical analysis of effective soil porosity and soil thickness effects on slope stability at a hillslope of weathered granitic soil formation. *Geosciences Journal,* 12, 401–10.

Munzinger, J., Dagostini, G., and Jaffré, T. (2004) *Suivi de plantations d'espèces introduites sur le massif du Kopéto, dans le cadre de project de re-végétalisation.* In Rapport d'Expertise IRD/SIRAS Pacifique SA. Nouméa, New Caledonia.

Nagy, L., and Proctor, J. (1997a) Plant growth and reproduction on a toxic alpine ultramafic soil: Adaptation to nutrient limitation. *New Phytologist,* 137, 267–74.

Nagy, L., and Proctor, J. (1997b) Soil Mg and Ni as causal factors of plant occurrence and distribution at the Meikle Kilrannoch ultramafic site in Scotland. *New Phytologist,* 135, 561–66.

O'Dell, R. E., and Claassen, V. P. (2006a) Serpentine and nonserpentine *Achillea millefolium* accessions differ in serpentine substrate tolerance and response to organic and inorganic amendments. *Plant and Soil,* 279, 253–69.

O'Dell, R. E., and Claassen, V. P. (2006b) Vertical distribution of organic amendment influences the rooting depth of revegetation species on barren, subgrade serpentine substrate. *Plant and Soil,* 285, 19–29.

O'Dell, R. E., and Claassen, V. P. (2006c) Relative performance of native and exotic grass species in response to amendment of drastically disturbed serpentine substrates. *Journal of Applied Ecology,* 43, 898–908.

O'Dell, R. E., James, J. J., and Richards, J. H. (2006) Congeneric serpentine and nonserpentine shrubs differ more in leaf Ca:Mg than in tolerance of low N, low P, or heavy metals. *Plant and Soil,* 280, 49–64.

O'Dell, R., Silk, W., Green, P., and Claassen V. (2007) Compost amendment of Cu-Zn minespoil reduced toxic bioavailable heavy metal concentrations and promotes establishment and biomass production of *Bromus carinatus* (Hook and Arn.). *Environmental Pollution,* 148, 115–24.

Pelletier, B. (2003) *Les methods d'exploitation et de revegetalisation mises en place depuis les annees 70 sur les mines de nickel de Nouvelle Caledonie.* Bulletin de l'Union Française des Géologues.

Peters, R. (2003) Inspired change at San Francisco's Presidio. *INTERSCI,* 21, 2–4.

Proctor, J., and Woodell, S. J. R. (1975) The ecology of serpentine soils. *Advances in Ecological Research,* 9, 256–365.

Punz, W. F., and Sieghardt, H. (1993) The response of roots of herbaceous plant species to heavy metals. *Environmental and Experimental Botany,* 33, 85–98.

Quilty, J. A. (1975) Guidelines for rehabilitation of tailings dumps and open cuts. *Journal of the Soil Conservation Service of New South Wales,* 31, 95–107.

Reid, N. B., and Naeth, M. A. (2005a) Establishment of a vegetation cover on tundra kimberlite mine tailings: 1. A greenhouse study. *Restoration Ecology,* 13, 594–601.

Reid, N. B., and Naeth, M. A. (2005b) Establishment of a vegetation cover on tundra kimberlite mine tailings: 2. A field study. *Restoration Ecology,* 13, 602–8.

Robertson, A. I. (1992) The relation of nickel toxicity to certain physiological aspects of serpentine ecology: Some facts and a new hypothesis. In *The Vegetation of Ultramafic (Serpentine) Soils* (eds. A. J. M Baker, J. Proctor, and R. D. Reeves). Intercept, Andover.

Ross, M., and Nolan, R. P. (2003) History of asbestos discovery and use and asbestos related disease in context with the occurrence of asbestos within ophiolite complexes. *Geological Society of America Special Paper,* 373, 447–70.

Safford, H., and Harrison, S. (2001) Grazing and substrate interact to affect native vs. exotic diversity in roadside grasslands. *Ecological Applications,* 11, 1112–22.

Sarrailh, J.-M., and Ayrault, N. (2001) Rehabilitation of nickel mining sites in New Caledonia. *Unasylva 207.* Available online at http://www.fao.org/docrep/004/y2795e/y2795e05.htm#e. Accessed January 16, 2010.

Schechter, S. P., and Bruns, T. D. (2008) Serpentine and non-serpentine ecotypes of *Collinsia sparsifolia* associate with distinct arbuscular mycorrhizal fungal assemblages. *Molecular Ecology,* 17, 3198–210.

Schor, H. J., and Gray, D. H. (1995) Landform grading and slope evolution. *Journal of Geotechnical Engineering,* 121, 729–34.

Scottish Natural Heritage. (2008) *The Story of the Keen of Hamar*. Scottish Natural Heritage, Lerwick, Shetland. Available online at http://www.nnr-scotland.org.uk/downloads/publications/The_Story_of_Keen_of_Hamar_National_Nature_Reserve.pdf. Accessed January 16, 2010.

Scotts Company. (2010) *Scotts Controlled Release Fertilizers*. Available online at http://scottspro.com/specialty-agriculture/nursery-greenhouse-crop-solutions/osmocote. Accessed January 16, 2010.

Seabloom, E. W., Borer, E. T., Boucher, V. L., Burton, R. S., Cottingham, K. L., Goldwasser, K. L., Gram, W. K., Kendall, B. E., and Micheli, F. (2003) Competition, seed limitation, disturbance, and reestablishment of California native annual forbs. *Ecological Applications*, 13, 572–92.

Shaw, A. J. (1990) *Heavy Metal Tolerance in Plants: Evolutionary Aspects*. CRC Press, Boca Raton, FL.

Shewry, P. R., and Peterson, P. J. (1975) Calcium and magnesium in plants and soil from a serpentine area on Unst, Shetland. *Journal of Applied Ecology*, 12, 381–91.

Shu, W. S., Ye, Z. H., Zhang, Z. Q., Lan, C. Y., and Wong, M. H. (2005) Natural colonization of plants on five lead/zinc mine tailings in southern China. *Restoration Ecology*, 13, 49–60.

Slingsby, D. R., and Carter, S. P. (1986) *The Ecological Effects of Eutrophication on the Keen of Hamar SSSI, Shetland*. Report to the Nature Conservancy Council (now the Scottish Natural Heritage), Inverness, Scotland.

Slingsby, D. R., and Carter, S. P. (1987) *Experimental Study into the Restoration of Serpentine Debris Habitat Damaged by Autrophication*. Report to the Nature Conservancy Council (now the Scottish Natural Heritage), Inverness, Scotland.

Slingsby, D. R., Hopkins, J., Carter, S., and Dalrymple, S. (2010) *Change and Stability. Monitoring the Keen of Hamar: 1978–2006*. Report to the Scottish Natural Heritage, Inverness, Scotland. Available online at: http://www.snh.org.uk/nnr-scotland/downloads/publications/Keen_of_Hamar_NNR_Change_and_Stability.pdf. Accessed January 16, 2010.

Slingsby, D. R., Proctor, J., and Carter, S. P. (2001) Stability and change in ultramafic fellfield vegetation at the Keen of Hamar, Shetland, Scotland. *Plant Ecology*, 152, 157–65.

Smith, R. A. H., and Bradshaw, A. D. (1979) The use of metal tolerant plant populations for the reclamation of metalliferous wastes. *Journal of Applied Ecology*, 16, 595–612.

Smith, R. F., and Kay, B. L. (1986) Revegetation of serpentine soils: Difficult but not impossible. *California Agriculture*, January–February, 18–19.

Société le Nickel. (2009) *Revégétalisation*. Available online at http://www.sln.nc/content/view/71/43/lang,french. Accessed January 16, 2010.

Society for Ecological Restoration. (2004) *The SER International Primer on Ecological Restoration*. Available online at http://www.ser.org/content/ecological_restoration_primer.asp. Accessed January 16, 2009.

Spence, D. H. N., and Millar, E. A. (1963) An experimental study of the infertility of a Shetland serpentine soil. *Journal of Ecology*, 51, 333–43.

Stromberg, M. R., Corbin, J. D., and D'Antonio C. M. (2007) *California Grasslands: Ecology and Management*. University of California Press, Berkeley.

Stromberg, M. R., and Kephart, P. (1996) Restoring native grasses in California old fields. *Restoration Management Notes*, 14, 102–11.

Tordoff, G. M., Baker, A. J. M., and Willis, A. J. (2000) Current approaches to the revegetation and reclamation of metalliferous mine wastes. *Chemosphere*, 41, 219–28.

Turitzin, S. N. (1982) Nutrient limitations to plant growth in a California serpentine grassland. *American Midland Naturalist*, 107, 95–99.

Tyndall, R. W. (1992) Historical considerations of conifer expansion in Maryland serpentine "barrens." *Castanea*, 57, 123–31.

Tyndall, R. W. (1994) Conifer clearing and prescribed burning effects to herbaceous layer vegetation on a Maryland serpentine "barren." *Castanea*, 59, 255–73.

Tyndall, R. W. (2005) Twelve years of herbaceous vegetation change in oak savanna habitat on a Maryland serpentine barren after Virginia pine removal. *Castanea*, 70, 287–97.

Tyndall, R. W., and Hull, J. C. (1999) Vegetation, flora, and plant physiological ecology of serpentine barrens of eastern North America. In *Savannas, Barrens, and Rock Outcrop Plant Communities of North America* (eds. R. C. Anderson, J. S. Fralish, and J. M. Baskin). Cambridge University Press, Cambridge.

Van Kererix, L., and Kay, B. L. (1986) *Revegetation of Disturbed Land in California: An Element of Mined-Land Reclamation*. California Department of Conservation, Sacramento.

van Rensburg, L., and Morgenthal, T. (2004) The effect of woodchip waste on vegetation establishment during platinum tailings rehabilitation. *South African Journal of Science*, 100, 294–300.

van Rensburg, L., and Pistoruis, L. (1998) An investigation into the problems associated with revegetating chrysotile tailings. *South African Journal of Plant and Soil*, 15, 130–40.

Veselý, P. (2002). *Mohelenská hadcová step—historie vzniku reservace a jejího výzkumu.* AMAPRINT Kerndl s.r.o, Třebíč, Czech Republic.

Virta, R. L. (2006) *Worldwide Asbestos Supply and Consumption Trends from 1900 through 2003*. U.S. Department of the Interior, U.S. Geological Survey, Reston, VA. Available online at http://pubs.usgs.gov/circ/2006/1298/c1298.pdf. Accessed January 16, 2010.

von Willert, F. J., and Stehouwer, R. C. (2003) Compost and calcium surface treatment effects on subsoil chemistry in acidic minespoil columns. *Journal of Environmental Quality*, 32, 781–88.

Walker, R. B. (2001) Low molybdenum status of serpentine soils of western North America. *South African Journal of Science*, 97, 565–68.

Walker, R. B., Walker, H. M., and Ashworth, P. R. (1955) Calcium-magnesium nutrition with special reference to serpentine soils. *Plant Physiology*, 30, 214–21.

Wallace, A., Jones, M. B., and Alexander, G. V. (1982) Mineral composition of native woody plants growing on a serpentine soil in California. *Soil Science*, 134, 42–44.

Weiss, S. B. (1999) Cars, cows and checkerspot butterflies: Nitrogen deposition and management of nutrient-poor grasslands for a threatened species. *Conservation Biology*, 13, 1476–86.

Weiss, S. B. (2002) *Final Report on NFWF Grant for Habitat Restoration at Edgewood Natural Preserve, San Mateo County, CA*. Creekside Center for Earth Observation, Menlo Park, CA. Available online at http://www.creeksidescience.com/files/weiss_2002_edgewood1

.pdf and http://www.creeksidescience.com/files/weiss_2002_edgewood2.pdf. Accessed January 16, 2010.

Weiss, S. B., and Wright, D. H. (2005) *Serpentine Vegetation Management Project. 2005 Report.* Creekside Center for Earth Observation, Menlo Park, CA. Available online at http://www.creeksidescience.com/files/weiss_et_al_2005_usfws_serpveg.pdf. Accessed January 16, 2010.

Weiss, S. B., and Wright, D. H. (2006) *Serpentine Vegetation Management Project Interim Report. 2006 Report.* Creekside Center for Earth Observation, Menlo Park, CA. Available online at http://www.creeksidescience.com/files/weiss_et_al_2006_usfws_serpveg.pdf. Accessed January 16, 2010.

Weiss, S. B., Wright, D. H., and Niederer, C. (2007) *Serpentine Vegetation Management Project. 2007 Final Report.* Creekside Center for Earth Observation, Menlo Park, CA. Available online at http://www.creeksidescience.com/files/weiss_et_al_2007_usfws_serpveg.pdf. Accessed January 16, 2010.

Westerbergh, A. (1994) Serpentine and nonserpentine *Silene dioica* plants do not differ in nickel tolerance. *Plant and Soil,* 167, 297–303.

White, C. D. (1967) Absence of nodule formation on *Ceanothus cuneatus* in serpentine soils. *Nature,* 215, 875.

Wilkins, D. A. (1978) The measurement of tolerance to edaphic factors by means of root growth. *New Phytologist,* 80, 623–33.

Williamson, J., and Harrison, S. (2002) Biotic and abiotic limits to the spread of exotic revegetation species. *Ecological Applications,* 12, 40–51.

Williamson, N. A., Johnson, M. S., and Bradshaw, A. D. (1982) *Mine Wastes Reclamation.* Mining Journal Books, London.

Wu, L., Thurman, D. A., and Bradshaw, A. D. (1975) The uptake of copper and its effect upon respiratory processes of roots of copper-tolerant and non-tolerant clones of *Agrostis stolonifera. New Phytologist,* 75, 225–29.

Yang, B., Shu, W. S., Ye, Z. H., Lan, C. Y., and Wong, M. H. (2003) Growth and metal accumulation in vetiver and two *Sesbania* species on lead/zinc mine tailings. *Chemosphere,* 52, 1593–600.

PART THREE

Synthesis

19

What Have We Learned from Serpentine in Evolution, Ecology, and Other Sciences?

Susan Harrison, *University of California, Davis*
Nishanta Rajakaruna, *College of the Atlantic*

We conclude this book by briefly reviewing some of the most provocative conclusions from the foregoing chapters. We hope these examples, though not exhaustive, illustrate the wealth of general scientific understanding that has come from studying serpentine ecosystems. If this book provides inspiration for future collaborations between disparate workers—earth and life scientists, evolutionary biologists and gene-free ecologists, naturalists and theoreticians, basic scientists and those who aim to save the world—then serpentine will have served as a good model system for the excitement and synergy that comes from crossing sharp boundaries.

HOW DID GEOLOGISTS DEDUCE THE PROCESS OF SEAFLOOR SPREADING?

Prior to the plate tectonics revolution, geologists believed that ocean basins were old and unchanging, but we now know the ocean crust is the youngest and most dynamic part of the Earth's crust. New ocean crust forms at seafloor spreading centers, and most of it later disappears via subduction, or the disappearance of one crustal plate beneath another. The nature of seafloor spreading was one of the key unanswered questions as evidence mounted for plate tectonics in the 1950s and 1960s. As Chapter 1 describes in historical detail, the decisive step in resolving this question came from the reinterpretation of ophiolites, terrestrial assemblages of rock that include large amounts of ultramafic rock (serpentinite, peridotite), as well as lesser amounts of gabbro, diorite, basalt, and chert. Beginning in the mid-1960s, these long-observed assemblages came to be understood as segments of oceanic crust and underlying mantle stranded on continents during subduction. Examination of ophiolites led to the development of models for oceanic crust formation that have since been tested and refined through deep-ocean studies and other means. Ophiolites provide a window in time, because oceanic crust not stranded on land seldom lasts more than 200 million years.

Although ophiolites are the major form in which serpentine (ultramafic rock) occurs on land, serpentine also occurs in other interesting settings described in Chapter 1: mélanges scraped off during subduction, exposed mantle from beneath continental crusts, and the somewhat mysterious stratiform complexes that may represent an early era in Earth surface history. We also know that ultramafic rocks and minerals compose the Earth's entire mantle and nearly all of its ocean crust. The fact that we think of their chemistry as strange is the result of our dwelling on the abnormally light, silica-rich rocks of the comparatively thin continental crust.

WHAT LIFE IS FOUND IN THE "DEEP BIOSPHERE" OF EARTH AND PERHAPS OTHER PLANETS?

Extremophile biology is the study of the microbes (Bacteria and Archaea) inhabiting hot springs, deep sea vents, mine effluents, and other environments of extreme temperature, pressure, pH, and/or chemical composition. They include the deep biosphere, subsurface zones such as rock and sediment interiors, where chemosynthesis supports a substantial fraction of Earth's biomass. Studies of the deep biosphere's geochemical, metabolic, genetic, and evolutionary processes are important sources of evidence concerning the origin of life on Earth and its potential for existence on other planets.

The serpentinizing subsurface is the "large-volume reaction zone in the planetary interior" where mantle meets water (Chapter 2). Temperatures are amenable to microbial life, and serpentinization produces reducing solutions enriched in Ca^{2+} and OH^- and possibly in dissolved hydrocarbons from abiotic sources. Microbes evidently make use of hydrogen oxidation coupled to methanogenesis and/or the reduction of sulfate, nitrate, and iron or other metals, as deduced through the identification of functional DNA sequences from these fluids. Challenges to life include high pH, scarce carbon, and fluctuations in chemistry and temperature. The microbiology of the serpentinizing subsurface has been explored at the recently discovered Lost City deep sea hydrothermal fields, the Mariana Trench, and several terrestrial serpentinite settings (Chapter 2).

Other planets, including Mars, are known to support the ingredients for serpentinization, so the study of the serpentinizing subsurface is closely linked to the generation of predictions for extraterrestrial life. Studies at Lost City have also led to speculation about the origin of life on Earth, as well as to evidence for lateral gene transfer as an evolutionary mechanism of possible importance.

DO ISLAND-LIKE TERRESTRIAL HABITATS GIVE RISE TO EVOLUTIONARY RADIATIONS?

Some of the most spectacular examples of rapid evolution and morphological change come from lineages that colonized oceanic islands and diversified to fill empty niche space. Serpentine outcrops are so island-like in appearance that it is tempting to look for similar phenomena. However, these outcrops are only truly insular from the perspective of the endemic species, which are usually a minority of the flora. When a new endemic lineage arrives on serpentine, rather than encountering unused resources, it may face substantial competition from the nonendemic (tolerator) flora. Thus, it is perhaps not surprising that transitions to serpentine endemism are not associated with increased rates of evolutionary diversification (Chapter 3).

Although it has been suggested that geographic isolation among separate serpentine outcrops could contribute to speciation, there are few examples, with the exception of the *Streptanthus glandulosus* complex (Chapters 3–5). This may be due in part to the shortage of biosystematic studies on the comparatively small number of genera that contain numerous closely related serpentine endemics (in California, candidates would include *Streptanthus*, *Hesperolinon*, *Allium*, and *Calochortus*; Chapter 3). It is also possible that the lack of examples is a real phenomenon reflecting the highly limited abilities of endemics to disperse among outcrops.

HOW EASILY CAN NATURAL SELECTION LEAD TO SPECIATION?

Early evolutionary biologists, most notably Darwin, assumed that the origin of new species was a direct outcome of natural selection, arising readily from adaptation to novel environments. Later theorists realized that even when divergent selection is strong, however, modest levels of gene flow can undermine the evolution of reproductive isolation and thus prevent such ecological speciation (Chapter 4). Geographic isolation has therefore come to be widely regarded as a key ingredient in most cases of speciation. The question remains: to what extent and under what conditions can divergent selection lead to speciation in the presence of gene flow? Strong selective gradients at edaphic boundaries, as well as the abundance of serpentine-endemic species that may have evolved from serpentine-intolerant ancestors, have made serpentine a classic model system for this central unanswered question (Chapter 4).

Recent studies on the genera *Leptosiphon* and *Layia* appear to illustrate, respectively, the early and the final (speciation) stages of divergence across serpentine–nonserpentine boundaries. Other serpentine studies have illustrated some of the prerequisite mechanisms: strong fitness trade-offs between soil genotypes, which reduce the effective gene flow between soils, and divergent selection on flowering times, which contributes directly to reproductive isolation (Chapter 4). Although serpentine studies have not fully resolved the major questions about speciation in the face of gene flow, they provide some of the most compelling potential examples and research opportunities.

A quicker route from adaptation to reproductive isolation is the classic idea of catastrophic selection, in which adaptation to a new environment such as serpentine entails a major genomic reorganization that confers immediate intersterility between an ancestor (in this case, a serpentine-intolerant species) and a descendant (in this case, a serpentine endemic). Though this has been proposed in the past for serpentine *Clarkia* species, there is little current evidence to support it (Chapter 4).

Ecological speciation is not the only mode of origin for serpentine endemics, because they arise as often from serpentine-tolerant as serpentine-intolerant taxa (Chapter 3). The evolution of endemics from serpentine-tolerant (bodenvag) ancestors—in other words, speciation associated with the loss of ability to grow off of serpentine—has not been widely studied but is presumably at least as strongly related to change in climate and the competitive environment as to any genetic change in the lineage. In the *S. glandulosus* complex, several endemics appear to have arisen through the breakup of a widespread bodenvag ancestor during a period of environmental change (Chapters 3–5). In Sweden, *Arenaria norvegica* became confined to serpentine in southern regions after forests expanded during postglacial climatic warming, but it remained a soil generalist north of the latitudinal treeline (Chapter 17).

DOES EVOLUTION REPEAT ITSELF?

Because they are found throughout the world and are globally consistent in their chemistry, serpentine environments provide a much repeated natural experiment in plant adaptation. Global comparisons demonstrate that tolerance to serpentine has evolved many times in unrelated plant families and genera (Chapter 3). Pathways to serpentine tolerance vary among families and species, involving different mixtures of low intrinsic growth rate, high allocation to roots, early flowering time, and selective uptake, exclusion, or internal translocation of Ca, Mg, and metals (Chapter 5).

Local adaptation and ecotypic differentiation are nearly ubiquitous in species that grow both on and off serpentine (Chapters 5, 7), and serpentine tolerance has sometimes evolved repeatedly within such species, as shown by closer relatedness between serpentine and nonserpentine ecotypes within the same region than between serpentine ecotypes in different regions (Chapter 5). The tools of genomics and proteomics, applied to serpentine and nonserpentine ecotypes, are just beginning to reveal some of the specific genes and metabolic mechanisms underlying serpentine tolerance (Chapters 6, 7).

Preadaptation to environmental challenges such as serpentine is also common, as evidenced by high prevalence of tolerance within a taxonomic group; tolerance to metal-rich environments is unusually common in Brassicaceae and Euphorbiaceae, for example. Mechanisms underlying preadaptation to serpentine are just beginning to be understood. It may be easier for grasses than other vascular plants to adapt to serpentine because the structure of their cell walls requires less calcium (Chapter 5).

WHY IS THE WORLD GREEN?

This classic ecological question refers to the observation that most plant biomass is not consumed by herbivores in most ecosystems, and one equally classic answer has been that plants are under strong pressure to evolve chemical and physical defenses making them tough, spiny, toxic, and/or indigestible. Plant antiherbivore defenses inspired some of the earliest developments in coevolutionary theory and continue to be its premier subjects. Some theory predicts that higher levels of defense are favored in harsh and unproductive ecosystems, because the relative fitness costs of herbivory are greater where resources are scarce. The resulting diminished food quality for herbivores could reinforce a lesser flow of energy and a simplified food web structure in unproductive environments. Soil-specific selection for antiherbivore defenses could also contribute to the fitness trade-offs that promote the evolution of edaphic specialization, as could other soil-specific selective pressures related to mycorrhizae, pollinators, and pathogens. Plant–animal and plant–microbe studies on serpentine have been surprisingly few, but existing studies have tended to support these predictions, as well as demonstrate that plants may be more visible and vulnerable in open environments, and that plants may coopt soil elements such as Ni as defenses (Chapter 8).

HOW DO SPECIES RANGES EVOLVE?

Invasion biologists have long studied the dynamics of species range boundaries, including the potential role of rapid evolution, and climate change has brought renewed interest to these questions (Chapter 9). One long-standing theory predicts that species ranges are limited by the fact that gene flow emanating outward from core populations in benign environments inhibits the potential for local adaptation in marginal, stressful environments. This model has been relatively little tested, but biological invasions in serpentine mosaic environments provide nearly ideal settings. In invasive *Aegilops triuncialis* (barbed goatgrass), dense core populations are found in serpentine grasslands, and sparser marginal areas occur where grassland soils give way to shallow rocky soils. Although conditions in the core and the margin created strong differences in selection, overall performance was found to be equal in the margin and the core; also, because plant density inhibited dispersal, migration was not higher from core to edge than the reverse. Studies of *Aegilops* also demonstrated some novel mechanisms promoting invasion success, including adaptive transgenerational plasticity, in which goatgrass growing on serpentine produces offspring with shortened flowering time and lower photosynthetic rates that are beneficial to fitness on serpentine. Finally, research on serpentine and nonserpentine populations of *Erodium cicutarium* examined how dispersal evolves in a mosaic environment; on serpentine, there is greater localized variation in fitness, which selects for lower dispersal (Chapter 8).

IS COMPETITION WEAKER IN UNPRODUCTIVE ENVIRONMENTS?

Ecologists disagree over whether and how the strength of plant interactions changes along gradients related to productivity. One viewpoint holds that in harsh and unproductive environments, competition decreases and may even give way to facilitation; another holds that competition must be at least as important in unproductive environments as in richer ones, although it may shift from primarily above-ground (light) to below-ground (nutrients) as productivity decreases. According to the second viewpoint, if unproductive environments demonstrate either higher diversity or lower resistance to the addition of new species than richer ones, it must be because spatial heterogeneity in resources is greater, rather than because competition is weaker (Chapters 10, 11).

Serpentine studies support the notion that competition is weaker in less productive environments, both within serpentine along gradients of soil depth and chemistry and between serpentine and nonserpentine environments (Chapters 10–12). Experimental neighbor removal had weaker effects on the success of target species at the harsher end of a natural productivity gradient (Chapter 10). Within harsh serpentine sites, numbers of exotic species were positively correlated with numbers of native species even at a very local scale, in contrast to the competition-based negative correlation found on richer sites (Chapter 11). Also, overall species richness was lower at harsh sites, contradicting the idea that higher spatial heterogeneity at the harsh sites could explain the coexistence of natives and exotics. Finally, a review of the available evidence suggests that disturbance plays a significantly weaker role in maintaining diversity in serpentine than nonserpentine environments, as would be expected if competition played a lesser role in limiting diversity in these environments (Chapter 12).

Exactly why plant competition appears to be weaker on serpentine remains unresolved. It seems unlikely that nutrient scarcity alone could reduce the intensity of competition. It may be that water and nutrients, which tend to be the limiting resources in unproductive environments, are less easily monopolized by one or a few species than light, which tends to be the limiting resource in more fertile environments. Plant functional traits (see below) and seasonality may also play roles in reducing the potential for competitive dominance on unproductive soils.

IS DISTURBANCE LESS IMPORTANT IN UNPRODUCTIVE ENVIRONMENTS?

Ecologists tend to assign disturbance—defined as the removal of above-ground biomass—a central role in structuring natural communities. At a local scale, disturbance temporarily shifts dominance to fast-dispersing and fast-growing (weedy) species; it may reduce local diversity, especially if it is excessively frequent or intense. At a regional scale, disturbance is likely to increase diversity, especially if it is heterogeneous in its frequency, intensity, and location. Because the primary ecological effect of disturbance is to reduce competition, its effects on communities should vary along environmental gradients that influence competitive intensity. Multiple theories therefore predict that disturbance should play a lesser role in maintaining diversity in unproductive environments. This prediction is generally borne out by studies showing lesser effects of fire, grazing, or soil disturbance on the diversity of plant communities on serpentine compared with nonserpentine soils (Chapter 12).

Lower disturbance dependency in unproductive ecosystems has both ecological and evolutionary facets. Ecologically, an individual species may respond less to a given fire on serpentine, where the vegetation is already more sparse and open, than on nonserpentine, where the same fire results in a larger pulse of available resources. Evolutionarily, species are less likely to show specialized adaptations for postfire regeneration in serpentine communities, where space and light are less limiting, than in nonserpentine communities. In addition, the lower biomass of unproductive communities may result in lower natural frequencies and intensities of fire and grazing, and thus to a lesser degree of adaptation to these disturbances (Chapter 12).

Recovery from natural or human disturbances may be exceptionally slow in environments such as serpentine (Chapters 12, 18). This suggests that management of nutrient-poor habitats may need to take account of a lower natural disturbance regime and a greater sensitivity to excessive disturbance, such as the increased frequency of fire caused by climate change (Chapter 17). Although it is possible to restore or revegetate environments such as serpentine once they are heavily altered, this requires striking a balance between amending sites enough to allow native species to establish and not amending them so much that exotic species take over (Chapter 18).

Low-nutrient systems may be dramatically altered by anthropogenic nutrient deposition, especially in conjunction with invasions by exotic species with functional traits conducive to fast growth (Chapters 12, 16–18). Mowing, grazing, and burning may be necessary to maintain and restore native vegetation in these circumstances (Chapter 18).

HOW DO PLANT FUNCTIONAL TRAITS MEDIATE THE OUTCOMES OF ECOLOGICAL PROCESSES?

Plants in unproductive environments show a consistent syndrome of traits, including slow intrinsic rate of growth, high relative allocation to below-ground structures, and slow rates of leaf turnover. Underlying all studies that compare ecological processes along productivity gradients is a fundamental question: are differences in response, such as the differences in competitive strength or disturbance outcomes between serpentine and nonserpentine soils, the result of the properties of the environments themselves, or of the traits of species inhabiting those environments? This question is sometimes overlooked, as when reviews of competition strength along productivity gradients fail to distinguish between studies that vary soil fertility but strictly control the identities of competitors versus studies that use competitor removal within naturally variable communities (Chapter 10).

Growing evidence supports the idea that plant functional traits play a key role in determining the ecological properties of soil fertility gradients (Chapters 10, 12, 16–18). For example, the best predictor of the strength of competition across a natural fertility gradient was a multivariate measure of plant community composition (Chapter 10). Target species benefited the most from competitor removal at sites dominated by tall annual grasses and forbs characteristic of fertile soils, and least at sites dominated by short-statured serpentine endemics; soil fertility itself was a weaker predictor of competitive strength. Likewise, the slower recovery of biomass following fire in serpentine than nonserpentine chaparral is partly the result of the lower frequency of postfire resprouting by shrubs, and the lesser postfire increase in diversity is related to the lack of obligately fire-stimulated germination in herbs, both of which reflect an evolutionary history of lower fire frequency. Conversely, the greater enhancement of native species diversity by grazing in serpentine (compared to nonserpentine) grasslands is caused by a greater prevalence of short-statured native annual forbs, the functional group that benefits most by the reduction in biomass of tall exotic annual grasses caused by grazing.

Plant traits are a critical issue in studies of the functional effects of biodiversity and the consequences of its loss (Chapter 16). Serpentine grassland studies have been valuable in demonstrating that community biomass is enhanced more by the diversity of functional types, for example, N-fixers, early- and late-season annuals, and perennials, than by the diversity of species per se (Chapter 16). However, an unexpected and potentially general finding from serpentine studies is negative selection, in which the species that competitively dominate mixtures are ones that have relatively low biomass when grown in monocultures. This effect, which in hindsight might be expected in a low-nutrient environment, tends to limit the degree to which functional diversity enhances productivity.

Global change effects may also depend on plant functional traits. One experimental study concluded that low-nutrient environments are relatively invulnerable to changes in temperature and precipitation because plant growth responses are so strongly limited by the slow-growing, stress-tolerating trait syndrome; some serpentine studies support this conclusion and others contradict it, and evidence remains scarce (Chapter 17). The low biomass and slow-growing plant traits found in low-nutrient environments also reduce the degree to which these environments sequester carbon and exert other biotic feedbacks on global atmospheric change (Chapter 16).

DO DIFFERENT LIMITING FACTORS HAVE SIMILAR ECOLOGICAL CONSEQUENCES?

In generalizing from serpentine to other low-productivity environments, it is important to consider differences in limiting factors. Scarcities of resources for which plants compete, such as nutrients and water, are unlikely to have the same ecological effects as salinity, unfavorable temperatures, cation (e.g., magnesium) excess, or other limiting conditions for which plants do not compete. The serpentine syndrome involves multiple limiting factors, including nutrient and water scarcity and cation excess, and the same may be true of some other special edaphic environments, such as limestone, dolomite, alkali sinks, and acid heaths. In other unproductive environments, such as granitic, sandstone, and shale barrens, the main limiting factor is poor water retention in shallow, rocky soils, and in others, such as deserts and alpine zones, plant growth is limited by climate (e.g., the length of the snow-free season or the rainy season). It would be interesting to investigate the extent to which the common features of all these environments—slow growth and its associated plant traits—do or do not lead to ecological similarities, such as the reduced roles for competition, disturbance, and herbivory observed in studies of serpentine.

HOW ARE ECOLOGICAL PROCESSES AFFECTED BY
LANDSCAPE HETEROGENEITY?

Landscape ecology investigates how ecological processes are affected by the spatial complexity of landscapes. Examples include edge effects, source-sink dynamics, and matrix effects, in all of which the flow of organisms through landscapes affects population abundances and community structure. Serpentine-containing landscapes show striking heterogeneity along gradients from shallow and rocky to deeper and finer soils, between adjacent communities such as serpentine chaparral and serpentine grassland, and along the boundaries between serpentine and other soils (Chapters 10–12). This heterogeneity has been shown to contribute to the coexistence of potentially competing native and exotic species, and hence to landscape-scale diversity (Chapter 11). The influx of species from surrounding nonserpentine habitats appears to elevate diversity at the edges of serpentine outcrops (Chapter 14). There have been surprisingly few studies yet that have actually measured or manipulated the flow of organisms in serpentine mosaic landscapes and attempted to determine the ecological consequences of such movement (but see Chapter 8).

HOW ARE INTERACTIONS AND DIVERSITY AFFECTED
BY THE SPATIAL ISOLATION OF HABITATS?

Island biogeography theory and, more recently, metapopulation and metacommunity theory examine the ecological consequences of spatially isolated habitats (Chapter 14). Small and isolated habitats are generally expected to support transient populations and low diversity, but species may survive in a well-connected regional network of small habitat patches. "Fugitive" species may coexist with their predators, competitors, or diseases by being able to disperse faster through such a regional patch network. Serpentine outcrops, with their discrete, island-like distributions and their specialized floras often rich in rare species, are an attractive setting for testing these ideas.

One question that has been of great interest to conservation biologists is whether plants on isolated habitat patches suffer reduced reproductive success as a result of pollinator limitation (Chapter 13). Several studies in serpentine environments have found evidence that plant fitness is sensitive to spatial isolation, but the patterns are less straightforward than expected. Pollinators may be too mobile or too unspecialized to produce the predicted effects of isolation on visitation rates. Plants in resource-poor environments such as serpentine may also show low degrees of pollinator specialization, high levels of self-compatibility or clonal reproduction, and highly persistent seedbanks or below-ground structures, all of which tend to buffer populations against pollinator scarcity.

Another question is how overall species diversity is affected by habitat configuration (Chapter 14). In a comparison of small serpentine outcrops to sites within very large outcrops, small outcrops had lower local (alpha) diversity and higher among-site (beta) diversity of serpentine specialist plants, much as expected. Small outcrops also had higher local diversities of generalist species, suggestive of edge effects. However, when plant diversity was studied across a much larger array of serpentine outcrops chosen more representatively, few effects of spatial habitat structure were detected. It appears that serpentine might provide opportunities for testing some aspects of spatial ecological theory, but that the theory may have more limited value for understanding plant diversity on serpentine—an interesting comment on the model system concept.

WHAT ARE APPROPRIATE CONSERVATION STRATEGIES FOR SPATIALLY COMPLEX LANDSCAPES?

The design of reserve networks has become a cornerstone of modern conservation biology. Goals include the protection of rare species and communities, the representation of all species and communities, and the provision of corridors for movement across landscapes (Chapter 15), which is especially important in light of climate change (Chapter 17). Serpentine landscapes, with their widely dispersed rare species and communities, illustrate the value of new analytical tools for reconciling these potentially conflicting conservation goals (Chapter 15). Because they have narrow geographic ranges and are confined to small outcrops, serpentine-endemic plants may be exceptionally vulnerable to extinction under climate change, and serpentine may prove to be a good model system for experimenting with more active approaches, such as managed relocation (Chapter 17).

One recently proposed solution to climate change involves mining massive amounts of serpentine for carbon sequestration (Chapters 1, 17). This will certainly create a need for the large-scale revegetation of devastated landscapes on harsh soils (Chapter 18). We conclude in the sincere hope that a solution gentler to some of the Earth's most botanically fascinating ecosystems can be found.

SPECIES INDEX

The letter *f* following a page number denotes a figure; the letter *t* following a page number denotes a table.

SUBJECT INDEX

The letter *f* following a page number denotes a figure; letter *t* following a page number denotes a table.

Indexer: Leonard Rosenbaum
Composition: Bytheway Publishing Services
Text: 10/12.5 Minion Pro
Display: Minion Pro

Milton Keynes UK
Ingram Content Group UK Ltd.
UKHW020236121223
434189UK00006B/183